Computer Science, Technology and Application

Computer Science, Technology and Application

Proceedings of the 2016 International Conference on Computer Science, Technology and Application (CSTA2016)

Changsha, China, 18–20 March, 2016

Editor

Xing Zhang

Wuhan University, China

World Scientific

NEW JERSEY • LONDON • SINGAPORE • BEIJING • SHANGHAI • HONG KONG • TAIPEI • CHENNAI • TOKYO

Published by

World Scientific Publishing Co. Pte. Ltd.
5 Toh Tuck Link, Singapore 596224
USA office: 27 Warren Street, Suite 401-402, Hackensack, NJ 07601
UK office: 57 Shelton Street, Covent Garden, London WC2H 9HE

British Library Cataloguing-in-Publication Data
A catalogue record for this book is available from the British Library.

COMPUTER SCIENCE, TECHNOLOGY AND APPLICATION
Proceedings of the 2016 Joint International Conference on Computer Science, Technology and Application (CSTA2016)

ISBN 978-981-3200-43-2

Printed in Singapore

Preface

The 2016 International Conference on Computer Science, Technology and Application (CSTA2016) have been successfully held in Changsha, China during March 18-20, 2016. The main objective of this conference is to provide a platform for scientists, researchers, and engineers from different part of the world to get together under one roof to discuss, present and share their research findings and results in the development in computing technology and its application.

CSTA2016 received over more than 150 submissions, but only 67 articles are included in this proceedings after a rigorous process of double peer review. The proceedings is organized into 5 specialize areas, which include image and signal processing, computer network and information security, algorithm and simulation, data mining and cloud computing, computer system and applications.

Finally, we would like to take this opportunity to express our sincere gratitude to all members of the conference committee and reviewers for their effort and comments on articles under their review. Without them, we would not able to pull out this proceedings so quickly after our conference.

Guest Editor
Dr. Xing Zhang (Wuhan University)
On behalf of the CSTA2016 Organizing Committee
Changsha, China

CSTA2016 Committees

General Chair

Bing-Lan Shui
Ph.D., Hubei Engineering University, China

Co-Chair

Xiao-Shan Ma
Ph.D., Yangzhou University, China

Technical Program Committee

Huai-Qiu Zhu
Professor, Department of Biomedical Engineering,
College of Engineering, Peking University,China

Jie Zhu
Professor, School of Information, Beijing Wuzi University, China

Bao-lin Sun
Professor, Computer Science and Technology,
Hubei University of Economics, Hubei, China

Hong-Zhi Fu
Professor, College of Physics and Electronic Information,
Luoyang Normal College, Luoyang, P. R. China

Kai-Qi Zou,
Professor, Dalian University, China

Jin-Xing Shen,
Ph.D. School of Computer Science, Guangdong Polytechnic Normal University, China

Deng Wu
Professor, Dalian Jiaotong University, China

Xue-Qian Fang
Professor, Shijiazhuang Tiedao University, China

Bo Sun
Professor, The International Outsourcing Research Institute of Guangdong University of Foreign Languages, China

Yan Cheng
Professor, Jiangxi Normal University, China Ph.D. Hanri Cao, Donghua University, China

Qian-Liu Li
Ph.D., Wenzhou University, China

Heng-Qing Cang
Ph.D., Hohai University, China

Xiu Xu
Ph.D., Shanghai Dian Ji University, China

Lv-Ben Wu
Ph.D., North China Institute of Aerospace Engineering, China

Zi-Chen Ning
Ph.D., North China University of Technology, China

Qiao Yuan
Ph.D., Shanxidatong University, China

Zheng-Hai Zhang
Ph.D., Changchun University of Science and Technology, China

Jia-Ding Feng
Ph.D., Shandong Agricultural University, China

Dong-Xuan Cui
Ph.D., Changchun University of Science and Technology, China

Ning Wan
Ph.D., Northeast Agricultural University, China

Fanchou Wu
Ph.D., Shanghai University of Electric Power, China

Qin Zheng
Ph.D., North University of China, China

Zhi-Hui Sun
Ph.D., Polytechnic School of Shenyang Ligong University, China

Kun-Hao Zhou
Ph.D., Taiyuan Institute of Technology, China

Xue-Qing Wang
Ph.D., North China Institute of Aerospace Engineering, China

Lin-Shan Fang
Ph.D., Beijing Institute of Technology, China

Mu He
Ph.D., Northeast Agricultural University, China

Chu-Wei Ren
Ph.D., Hubei Engineering University, China

De Yan
Ph.D., North University of China, China

Man-Di Zhao
Ph.D., Inner Mongolia Normal University, China

Xia-Rong Yin
Ph.D., Wuhan Business University, China

Rong-Xing Bao
Ph.D., Shanghai University of Electric Power, China

Di-Zhu Le
Ph.D., Yantai University, China

Yun-Yao Liu
Ph.D., Liaocheng University, China

Muhammad Naufal Bin Mansor
Ph.D., Universiti Malaysia Perlis, Malaysia

Mohd Helmy Abd Wahab
Ph.D., Universiti Tun Hussein Onn Malaysia, Malaysia

Imran Memon,
PhD. Zhejiang University, Computer Engineering, China

KEYNOTE SPEECH

Name: Prof. Huaiqiu Zhu
Department of Biomedical Engineering, College of Engineering, Peking University, China

Prof. Zhu's research interests include the computational methods and tools in bioinformatics based on the genome and genomic sequences, and protein sequences and structures. Recently Prof. Zhu and his lab are focusing on the problems of metagenomics issues driven by human health and environments, and data mining in medicine informatics. In the post-genome era, especially with the next generation sequencing (NGS) technologies, it is the great mission for bioinformatics to carry out the data intensive computation and analysis to meet the needs of high-throughput genomic data. Since 2004 Zhu's lab has addressed to the bioinformatics methods of genome analysis and annotation, and developed a series of methods of microbial gene prediction, genome comparison, and protein molecular dynamics. Since 2009, Prof. Zhu developed the tools such as sequence assembly, gene prediction and microbial community comparison for metagenomic sequence produced by NGS. Most of these methods and tools were demonstrated with leading performances in the field, and they have been applied in several genome projects and were incited or highlighted by Cell and Nature etc. Currently, the lab has close collaborations with Gastroenterology lab and Critical Care Medicine lab of Peking University Third Hospital in Beijing, and BME lab of GaTech, and Bioengineering lab of UIUC in USA.

KEYNOTE SPEECH

Name: Dr. Amol R. Madane
Video Security experts, TCS, Pune, India

Dr. Amol R. Madane has completed Doctorate of Philosophy (Ph.D.) under the guidance of Dr. Uttam D. Kolekar in Image/ Video Security area in May 2015. He holds Masters of Engineering degree in Electronics Engineering from Mumbai University, India. He is an associate of Tata Consultancy Services Ltd. from last 9+ years. His main area of research is Digital Image Processing, Digital Video Processing and Digital Signal Processing. He has published various research papers in international conferences and journals. He is “editor in chief” of various international journals. Recently, he has published book titled “Design and Performance Evaluation of Image Encoder in View of 4G Applications” by Lulu book publication, Raleigh, United States. He is member of professional institutes such as IEEE, IETE, ISTE, IEI, ISTD, CSI, ACM etc.

KEYNOTE SPEECH

Name: Prof. Yan Cheng
Professor, Jiangxi Normal University, China

Yan Cheng holds a Bachelor's degree, Master's degree and Ph.D. degrees from Tongji University in Shanghai and won Guanghua Ph.D. scholarship, national doctoral academic best paper award. From July 2013 to March 2014, she went to the California University as a visiting scholar. Currently she is a professor in Jiangxi Normal University. Currently she is the leader of national NSFC (Natural Science Foundation of China) Plan Project and more than Provincial projects. She also has undertaken provincial key projects, one Key Projects of Science and Technology in Jiangxi Province, One Jiangxi science and technology support project (the Areas of Social Development), one Natural Science Fund Projects in Jiangxi Province, one Provincial Science and Technology Support Project (Industry) and many Social Sciences Planning Projects. She published academic monograph (Beijing, Science Press, 2014) on educational data mining, computer professional teaching materials.

Contents

Chapter 1

Image and Signal Processing

Image denoising based on a new thresholding function

Jun-Ke Kou*
Department of Applied Mathematics, Beijing University of Technology,
Beijing, 100124, China
E-mail: xingfu36524@163.com

Yan-Ling Fu
Department of Information Engineering, Henan Finance and Taxation College,
Zhengzhou, Henan, 451464, China
E-mail: fylhappy@126.com

In this paper, a new continuous thresholding function is proposed. This new thresholding function does not only retain the advantages of conventional standard thresholding functions, but also overcome their shortcomings. The larger (than λ) coefficients are often taken as effective signal while the smaller (than λ) are noise. The proposed method of removing noise from the disturbed signal is to keep the larger coefficients and suppress the smaller coefficients. The simulation results also show that the proposed function can achieve better denoising performance than traditional methods.

Keywords: Image Denoising; Thresholding Function; Wavelet Transform.

1. Introduction

Images are usually disturbed by noise. What's more, the larger noise may be cover the original images. So, image denoising is a important step in the image processing application.

With the development of wavelet analysis, wavelet analysis has better abilities to describe the local characteristics of image in time-frequency domain. Owing to this property of wavelet, many simple and effective wavelet thresholding methods are taken by [1-2]. The most useful hard and soft thresholding functions are analysed by [3]. A wavelet-based denoising method with principal component analysis is given by [4]. The reference [5] introduced a 3D extension of the wavelet nanlysis based bilateral filtering for rician noise removal. A nonlinear suboptimal detector was suggested by [6], this detector

* Corresponding author.

consists of a nonlinear wavelet denoising filter, followed by a replica correlator. Silva, Dutra and Rodrigo[7] proposed an adaptive threshold denoising method based on edge strength. [8] discussed How much influence of the choice of mother wavelet functions in image denoising. Nasri and Pour [9] introduced the adaptive thresholding function into wavelet domain by using the thresholding neural network. Recently, A pre-processing hard thresholding algorithm is proposed by [10]. From those results, we can see that the thresholding function is a critical factor in image denoising. In this paper, we construct a new thresholding function. This function overcome the disadvantage of conventional thresholding functions in theory. What's more, the simulation results also indicate the new thresholding function has better denosing effect.

2. Wavelet-Based Noise Reduction

In this section, we will simply introduce the basis of signal denoising problem. Taking the original signal is x. Then the disturbed signal y is

$$y = x + n, \tag{1}$$

where n is the noise. In practice, we always assume the signal is a finite data. This means

$$y_i = x_i + n_i, \quad i = 1, 2, \ldots N-1. \tag{2}$$

The aim of the noise reduction is obtain an estimated signal $\hat{x}$ from the disturbed signal y. In other words, the propose of this problem is to remove the noise in y and make the estimate $\hat{x}$ as close to x as possible. The objective of noise reduction is to estimate the real signal x from y to minimize the MSE risk.

Noise reduction by wavelet can be described as following. First, one obtains the empirical wavelet coefficients by discrete wavelet transform (DWT). Second, detailed coefficients are revised based on a thresholding function and an appropriate threshold value. Finally, using inverse wavelet transform for the modified coefficients, the reconstructed image is achieved.

In noise reduction, the thresholding function is very important. The most commonly-used thresholding functions are the soft-thresholding function and the hard-thresholding function, which are shown in Fig.1. The standard soft-thresholding function is defined as

$$\eta_s(y,\lambda)=\begin{cases} y+\lambda & y<-\lambda \\ 0 & |y|\le\lambda \\ y-\lambda & y>\lambda \end{cases} \tag{3}$$

where $\lambda>0$ is the threshold. The hard-thresholding function is defined as

$$\eta_h(y,\lambda)=\begin{cases} y & |y|>\lambda \\ 0 & |y|\le\lambda \end{cases} \tag{4}$$

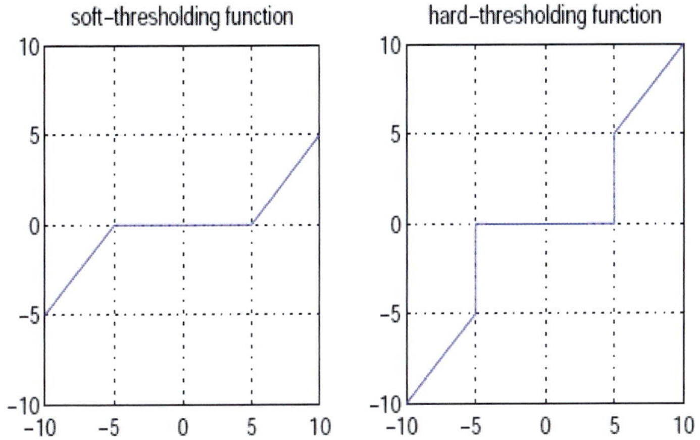

Fig. 1. soft and hard thresholding function ($\lambda=5$).

It is well known that the energy of the signal is usually focused on a few coefficients while the energy of noise is spread among all coefficients in the wavelet domain. So, the larger (than λ) coefficients are often taken as effective signal while the smaller (than λ) are noise. If we want to remove the noise from the disturbed signal, we should kept the larger and suppress the smaller.

Although the smaller coefficients are frequent noise, they may be some important high-frequency components. Therefore, the conventional soft and hard thresholding function can possibly remove some important components when the smaller (than λ) coefficients are modified to zero. On the other hand, because of the discontinuity, hard thresholding function may result in ringing artifact, pseudo-gibbs and image visual distortion.

3. A New Thresholding Function

In order to overcome the standard thresholding functions' weakness and improve the denoising performance, a new high-order differentiable thresholding function is constructed as follow:

$$\eta(y,\lambda) = y\, e^{-(\frac{\lambda}{y^2+1})^2} \tag{5}$$

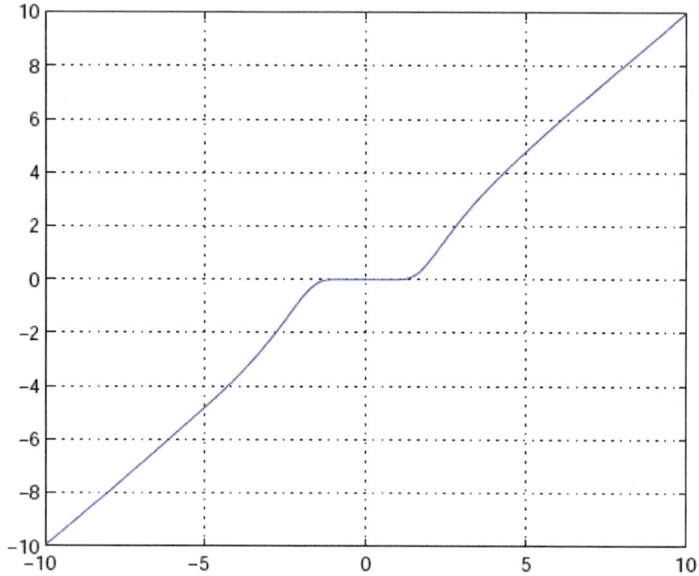

Fig. 2 New thresholding function ($\lambda = 5$).

It is easy to see from (5) and Fig.2 that this thresholding function is always continuous. It is similar with the standard hard-thresholding function when $|y| \to +\infty$. On the other hand, it shrinks the wavelet coefficient to a smaller value but not to zero with $|y| \le \lambda$. This mean it will keep some important high-frequency components (those may be the detail information of image). Thus, this new thresholding function is not only keep the advantage of the conventional standard soft (hard) thresholding function, but also avoid the bad influence of the conventional thresholding function.

4. Experimental Results

To analyze the denoising effect of the proposed approach, standard images (facets, wbarb, woman) have been used. This images are initially disturbed by Gaussian random noise with different standard deviation of 10, 15, 20. To investigate the effectiveness of the algorithm, commonly-used performance Peak Signal to Noise Ratio (PSNR) has been used. In the proposed approach, the threshold value $\lambda = \sigma\sqrt{2\log(MN)}$, M, N is the size of the image, σ is the noise standard deviation. The performance of image denoising by hard thresholding function, soft thresholding function and our proposed method are shown in Table 1. From Table 1, it is easy to see that PSNR of the image is decreasing when the standard deviation increasing. This means the image is More seriously corrupted by noise when the standard deviation increasing. On the other hand, we can see that our proposed approach has a higher PSNR than other image denoising methods. In other words, the proposed approach is very close to the optimal solutions. The noise can be removed effectively.

Table 1. Comparison of the different methods.

Image	σ	Hard PSNR	Soft PSNR	New PSNR
	10	27.6172	28.8905	30.1029
facets	15	25.8996	24.1970	29.2716
	20	24.9002	25.8977	29.1844
	10	30.1098	30.6038	33.7912
wbarb	15	28.2000	27.9145	30.7974
	20	24.2312	26.3308	29.8349
	10	32.2204	36.2308	38.0391
woman	15	30.3379	33.0391	35.3047
	20	29.0005	29.3089	30.1833

5. Summary

In this paper, a new thresholding function has been proposed for image denoising. As a result, the numerical results show that the proposed method has better denoising effect.

References

1. D. L. Donoho, L. M. Johnstone. Adaptating to unknown smoothness via wavelet shrinkage. *Journal of the American Statistical Association.* 90(432): 1200-1224, 1995.

2. D. L. Donoho. Denosing by soft thresholding. *IEEE Transactions on information theorey.* 41(3): 613-627, 1995.
3. D. L. Donoho, L. M. Johnstone. Ideal spatial adaptation by wavelet shrinkage. *Biometrika.* 81(3): 425-455, 1998.
4. R. G. Yang, M. G. Ren. Wavelet denoising using principal component analysis. *Expert systems with applications.* 38(1): 1073-1076, 2011.
5. Y. Wang, X. Q. Che, S. L. Ma. Multi-channel registered data denoising using wavelet transform. *Eurasip journal on advances in signal processing.* 1-14, 2012.
6. Z. Madadi, G. V. Anand, A. B. Premkumar. Signal detection in generalized Gaussian noise by nonlinear wavelet denoising. *IEEE transactions on circuits and systems iregular papers.* 60(11): 2973-2986, 2013.
7. D. Silva, R. Dutra, M. Rodrigo. Adaptive edge-preserving image denoising using wavelet transforms. *Pattern analysis and applications.* 16(4): 567-580, 2013.
8. F. Admo, A. Gregorio, A. Filippo. A comparative study on mother wavelet selection in ultrasound image denoising. *Measurement.* 46(8): 2447-2456, 2013.
9. M. Nasri, H. N. Pour. Image denoising in the wavelet domain using a new adaptive thresholding function. *Neurcomputing.* 72: 1012-1025, 2009.
10. S. Mina, B. Soosan, E. Pegah et al. Similarity validation based nonlocal means image denoising. *IEEE signal processing letters.* 22(12): 2185-2188, 2015.

Age estimation based on image feature fusion method

Lei Wang*
School of Software Engineering, Beijing University of Technology,
Beijing, China
E-mail: laithwang@qq.com
www.bjut.edu.cn

Qing Zhu
School of Software Engineering, Beijing University of Technology,
Beijing, China
E-mail: ccgszq@bjut.edu.cn

Factual recognition in association with age estimation has been studied widely. In this experiment, age estimation is performed by extracting and analyzing the image features such as contour and wrinkle of the human faces using Local Binary Pattern (LBP) algorithm and Canny operator for image detection. In this study, we deployed uniform LBP algorithm to extract the feature information of face wrinkles, and used the Canny operator to enhance the face contour and wrinkles information before integrating the two features of face images together to estimate the age of the subject. In this paper, we used the FG-NET Face Age libraries which are widely used for age experiments on groups of subjects using every ten years as the classification. The experiments showed that uniform-LBP algorithm and Canny algorithm for feature extraction and fusion have achieved results of 72% to 83% of correct classification based on recognition of contour and the wrinkle information on the human face, as the means to estimate the age of the candidates.

Keywords: Age Estimation; Face Image; Feature Fusion; Machine Learning.

1. Introduction

Human face contains a wealth of information, people's gender, age, emotion, race can be obtained through the face information. Since 1990s, with the development of human-computer interaction technology, the image processing, pattern recognition, computer vision and other fields have been widely used. In 1993, Kwon and Lobo [1] begin the first experiment to study the age of the face .Age estimation has been widely applied in identification, criminal investigation,

* Corresponding author.

shopping guide etc. There are some difficulties in the age estimation. Illumination, shooting angle, facial expression, wearing clothing are likely to have a certain impact on the prediction. At the same time, the different individual factors are also increase the uncertainty for age estimation.

In this paper, we use the uniform LBP algorithm to reduce image noise and extract the image feature. Then, we use Canny algorithm for feature extraction. The edge features widely contains the contour of the face and wrinkles information which have a good characteristics of the age. What's more, we reduce the feature dimensions through the Principal Component Analysis (PCA) algorithm and estimate the age using support vector machine (SVM) method. The rest of this paper is organized as follows: Section 2 describes the feature extraction and fusion. Section 3 presents experimental design and result, the conclusion is showed in Section 4.

2. Feature Extraction and Fusion

2.1. *Image preprocessing*

We use FG-NET database for the experiment. In this database, many picture are not formal. We change the original images to the grayscale images and then shear the pictures. In this way, we eliminate the useless information.

2.2. *Uniform LBP feature*

In 2002, Timo Ojala and Matti PietikaÈinen proposed the Local Binary Pattern to extract the texture features from images. Its main idea is: comparing the center pixel value with the other P pixels values. If the adjacent pixel value is greater than the central pixel value, this position marked 1, otherwise marked 0. For each pixel in the picture, its eigenvalues can be described as in Eq. (1):

$$LBP_{p,r}(x_c) = \sum_{p=0}^{p-1} u(x_p - x_c) * 2^p \quad (1)$$

Where x_c is the center pixel's gray value, x_p is the adjacent pixel's gray value. The algorithm is shown in Figure 1.

40	8	26
8	**20**	32
24	14	15

1	0	1
0		1
1	0	0

Fig. 1. A sample of LBP algorithm.

The uniform mode [2] requires binary sequence jumps from 0 to 1 or from 1 to 0 no more than twice. The uniform LBP can effectively describe the most of the texture features in the image, and reduce the number of features synchronously. For the 8 neighborhoods (P=8), the histogram of the image decrease from the original 256 dimensions to 59 dimensions. The general LBP algorithm and the uniform LBP algorithm are compared in Figure 2.

Fig. 2. The general LBP algorithm and the uniform LBP algorithm.

2.3. *Canny feature*

Canny edge detection method is proposed by F. Canny John[3] in 1986. A double threshold method is adopted to reduce the number of false edges in Canny algorithm. According to the high threshold, we can get an edge image which may contains a few false edges, the image edge may not be closed. The problem is solved by another low threshold. The inappropriate thresholds may infect the image features, as is shown in Figure 3. The inappropriate thresholds may not express facial features or over expression the facial details, they are all not conducive to feature extraction. Experiment by statistical method, we try to control the detect feature points between 10%-20%.

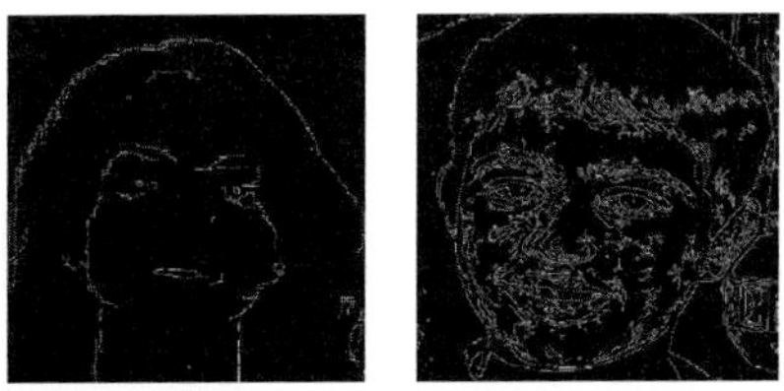

Fig. 3 The invalid feature extraction by inappropriate thresholds.

2.4. *Feature fusion*

Two kinds of strategies are adopted in the experiments. In experiment 1, we set uniform-LBP feature as L, Canny algorithm feature as C, the fusion feature as M. For every pixel (x, y) on the image, as in Eq. (2).

$$\mathrm{M(x,y)} = \begin{cases} \mathrm{C(x,y)}, & C(x,y) > 0 \\ \mathrm{L(x,y)}, & C(x,y) = 0 \end{cases} \quad (2)$$

In experiment 2, firstly, we put the uniform LBP features and Canny features into the classifier respectively and get the average correct rate as $P_1\%$ and $P_2\%$. Then, fused the two features with weighted fusion average method, as in Eq. (3).

$$\mathrm{M(x,y)} = \frac{P_1}{P_1+P_2}\mathrm{L(x,y)} + \frac{P_2}{P_1+P_2}\mathrm{C(x,y)} \tag{3}$$

3. Experiments

3.1. *Experimental setup*

The experiments using the FG-NET database for training and testing. In this paper, we use the face images from 0 to 59 age as the basic data and divide them into 6 categories, every ten years as a classification .The training set and the testing set accounted for 4:1. The main experimental steps in this article are as follows:

(1)Image preprocessing. Change the original images to the grayscale images and then shear the images.

(2)Image feature extraction. Set uniform-LBP feature as L, where p=8, r=1. Set Canny feature as C, the threshold t_1=10, t_2=3t_1 and the increments value of t_1 is 5.

(3)Feature Fusion. Experiments were carried out by the two kinds of fusion strategies. The fusion steps is shown in Figure 4.

(4)Using PCA for further dimensions reduction. Keep the sum of the eigenvalues more than 90%.

(5)Learning and forecasting by SVM classifier. Using RBF as the kernel function and "one to one" classification strategy.

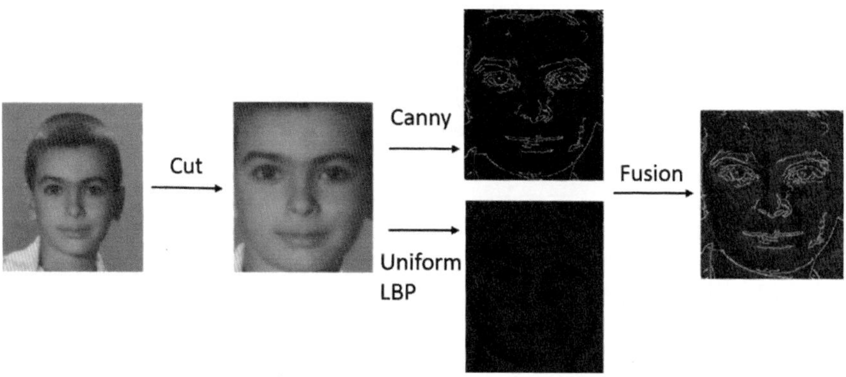

Fig. 4. A sample of the image fusion.

3.2. *Result and analysis*

Through the cross-test experiments, we got the results in Table 1.

Table 1. The statistics of accuracy rate.

Groups(age)	Strategy 1	Strategy 2
1 (00-09)	78.62%	80.92%
2 (10-19)	72.54%	76.25%
3 (20-29)	73.28%	77.64%
4 (30-39)	68.43%	74.28%
5 (40-49)	75.64%	78.61%
6 (50-59)	79.45%	83.66%
Average	74.66%	78.56%

Through the statistical table, the effect of the strategy 2 is better than strategy 1. Strategy 1 failed to reflect the differences in the performance of the two characteristics and the experimental results. Strategy two considered the effects of the two kinds of characteristics on the correct rate. The feature of the images can be made more meaningful by the weighted fusion method, which can better represent the age of the face.

As can be seen in figure 5, the impact of uniform LBP on the results is more important than the Canny feature. But the Canny feature still has a vital role in the experiments.

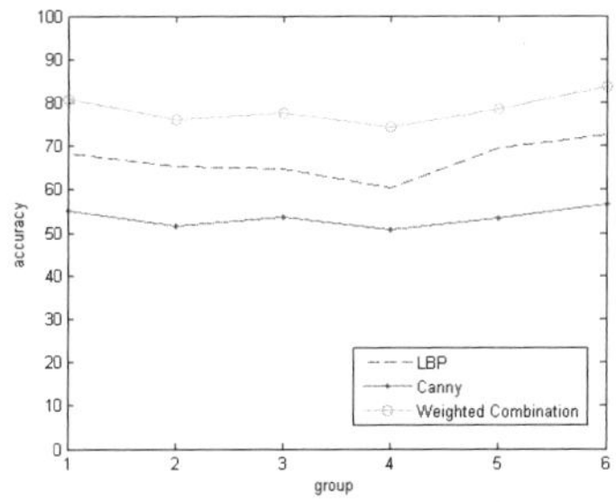

Fig. 5. The effects of the two characteristics on the results.

The average correct rate of this paper is 78.56%, as shown in Table 2. Compared with the similar studies, we still have a lot of room for improvement.

Table 2. compare with other methods.

Method	Accuracy
AAM+PCA+MDA[4]	72.52%
10-FR+RBF-RSVM [5]	78.15%-84.74%
2DPCA+Gabor [6]	83.84%-87.46%
This paper	78.56%

4. Conclusion

In this experiment, we use uniform LBP algorithm and Canny algorithm for feature extraction and fusion and get better results. It shows that the contour information and the wrinkle information of the human face are important to age estimation. The appropriate edge extraction and strengthening can reflect the age characteristics of the face. In the experiment, we also found some problems to be solved, such as the selected thresholds of Canny algorithm and the penalty factor of SVM classifier. At the same time, it is necessary to establish a standardized human face age database, which will play a better role in predicting the age of human faces. Next, we still need a lot of experiments and try to find the best feature extraction method and classification standard to improve the accuracy.

Acknowledgments

This research is partially sponsored by the Beijing Natural Science Foundation under contract No. 4152008.

References

[1] Kwon Y H, Lobo N D. Locating Facial Features For Age Classification [M]. Intelligent Robots and Computer Vision Xii: Algorithms and Techniques. Bellingham: Spie - Int Soc Optical Engineering, 1993: 2055, 62-72.

[2] MALLAT S. A theory for multiresolution signal decom-position: the wavelet representation [J].IEEE Trans on PAM, I1989, 11(7): 674-693.

[3] John Canny. A computational approach to edge detection. Transactions on pattern analysis and machine intelligence.Vol.PAM1-8.No.6.November 1986.

[4] Txia J D, Huang Ch L. Age estimation using AAM and local facial features[C]/ / Proceedings of Fifth International Conference on Intelligent Information Hiding and Multimedia Signal Processing. Kyoto: IEEE Press, 2009: 885-888

[5] Yong Ma Tao Xiong Yanming Zou. Person-specific Age Estimation under Ranking Framework[C]// ACM ICMR '11 Proceedings of the 1st ACM International Conference on Multimedia Retrieval, ACM New York, NY, USA. Article No. 38.

[6] Lu Li. A study of gender classification and age estimation based on face image[D].Shanghai: Shanghai Jiao Tong University,2010

The intelligent video surveillance and recognition system based on summarization video

Wan-Jia Qu* and Zhong-Lin Xu
Aviation University of Air Force,
Changchun, Jilin 130022, China
**E-mail: 1259737654@qq.com*

Yu-Wei Yuan
Naval Aeronautical and Astronautical University,
Yantai, Shandong 264001, China
E-mail: yyw57156@hotmail.com

As the fast development of computer science and the wide application of multimedia technology, a mass of digital videos have become an indispensable part of our lives. In this paper, we first introduce the research background, development progress and current researches of video abstraction and videos intelligent recognition. Then, we proposed demand analysis of monitoring video intelligent recognition system and described function of all modules. According to result of demand analysis, we proposed the all design of system, introduced work flow of all functional modules. At last, with Visual C++ and DirectShow software, we implemented the system.

Keywords: Video Abstraction; Key Frame Extracting; Intelligent Monitoring; Intelligent Recognition.

1. Introduction

In recent years, with the rapid economic development of country and the comprehensive national strength increased significantly, all steps of life is also growing safety consciousness, so the demand of the alarm system becomes more and more important, which leads to the video surveillance has been widely used in the production life. Although airport, security check, railway stations and shopping malls and important the social security public all have installed monitoring and control system, the task people need to complete is still a lotion the actual monitoring, the current main video monitoring function is simple. For timely and effectively analysis, tracking and monitoring video content can

* Corresponding author.

timely prompt, report to relevant departments when needed, and provide the security department and the public security organ help to solve cases, which realized 'intelligent' video monitoring. Intelligent video monitoring mainly make use of special function of the computer without human intervention to analysis, tracking and identify the video content and signal. The Intelligent Video Surveillance and Recognition System Based on Summarization Video, mainly aim at extracting from surveillance video and analyzing each event further, in order to identify those kinds of events. Video Abstract Technology is mainly used in monitoring the video analysis and extraction. As for a 24-hour surveillance video, applying video technology will extract people, vehicle or other moving objects enter the scene regarded as an event in the 24 hours from the video. Then the users can selectively choose a video to play when they see surveillance video. Further, we can also use video analysis technology to analyze and classify these abstract events; users only need to watch the event categories which they are interested in. The advantage of this way is that you don't need to playback the original 24-hour surveillance video from beginning to end, and it's convenient for users monitoring video and looking for targets fast and saving a lot of time.

2. Analyzing System Requirements

The Intelligent Video Surveillance and Recognition System Based on Summarization Video applies Smart Client Software Developed by C/S Architecture, which can be run on the server and personal PC at the same time. It has the following features:

1) The systems support all avis surveillance video coding format, other format video require format conversion;

2) The system support common image format on the network, including JPG, PNG, jpeg, BMP, GIF format;

3) The systems support the camera video and local video to input information;

4) In the process of extracting video, video extraction processing time depend on the video frame rate, resolution, and length, but the extraction time less than one fifth of the entire length of the video;

Video abstract module can handle at least 150 video abstract events in the meanwhile, and the length of every video in all the events generated needed to be in 2 minutes;

6) Fireworks detection module can alert in 1 seconds when finding suspected smoke marks and suspected flame in real-time monitoring video, and returns the time of suspected fire in the real-time monitoring.

7) Fireworks detection module identification rate for ordinary flame above 90%, for large fire recognition rate is above 95%;

8) Sensitive information detection module retrieve speed can close to computing environment text database retrieval speed as far as possible, namely video content retrieval in text retrieval speed (using database);

9) Sensitive information detection module can input and store sensitive information (manual or automated), training and storage of determination;

10) Sensitive information detection module can be carried out in accordance with the logo color fast recognition and detection, and detection accuracy is above 90%, which can retrieve the check for a quick;

11) Sensitive information detection module can quickly identify and retrieve according to sensitive information content, retrieval rate more than 90%, retrieve the check for a quick;

12) Banner identification module can alert in 1 second when meet suspected banner events in real-time monitoring video, and returns the real-time monitoring of a suspected banner in time;

13) Banner recognition module for banner event recognition rate is above 90%.

System function module mainly include: video module, fireworks recognition module, banner identification module, the sensitive information retrieval module.

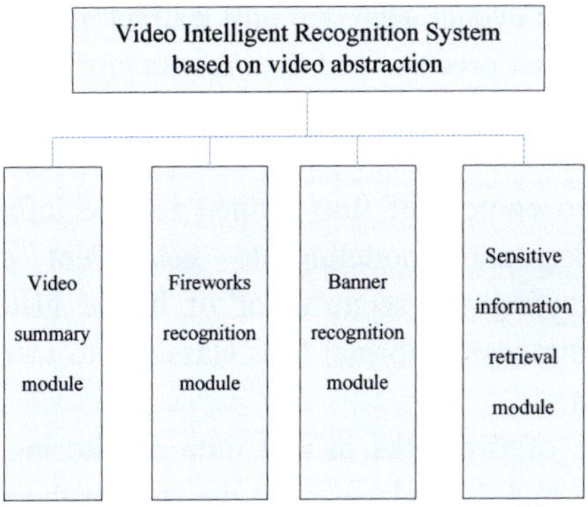

Fig. 1. Function diagram of video intelligent recognition system.

3. System Overall Design

The goals of the system mainly realize the following functions:

1) Moving object detection and tracking in the video. It is a prerequisite for video abstract extraction. Video abstract event refers to the movement of an object in video. In order to the extracted video abstract event from video must first carried on the detection and tracking, record its contour information such as size, location, and movement of the track, then according to these information, extracting it from the video. Frame differential method is used for video moving object detection and tracking, the algorithm will be used as the core program of video extraction technology.

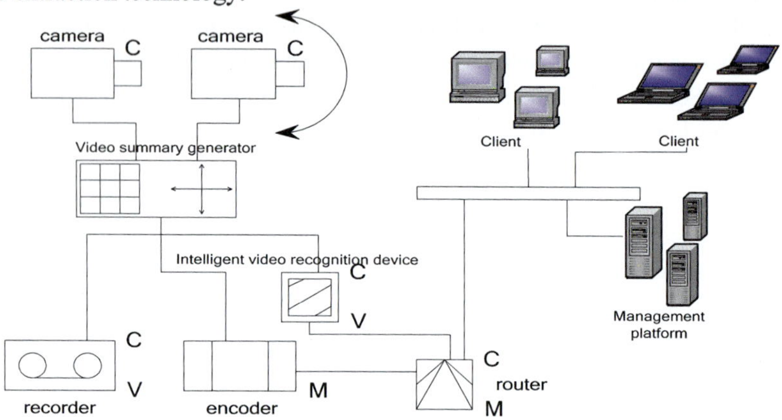

Fig. 2. Topology diagram of system.

2) The extraction of video summarization. After detecting and tracking moving object on video, the system extracts these events on the basis of saving the details of the sports events. Here are only extract minimal contours of sports events and its movement process, and then restructure these events to set in the background of the source video, and semi-transparent processing and label it, in order to identify.

3) Get the video content of fine-grained feature information. The use of mixed Gauss background modeling to get event contour and shape characteristics; Extracting the scene color of image histogram method; The design of the multi dimension normal Bias classifier to identify and classify the behavior of the event.

4) Identification of fireworks in real time monitoring. It mainly refers to real-time monitoring in public places or at the start of the forest fire can timely identification of fire abnormal, then immediately alerts, so that relevant personnel to take timely measures, fire will be nipped in the bud stage. The basic principle of fire detection is to use the burning flames and smoke on the dynamic characteristics and color characteristics were detected. In moving

object detection algorithm combined with the characteristics of flame and smoke finished smoke detection.

5) Banner event recognition in real time monitoring. The scene in the real-time monitoring of public places, monitoring personnel often can't monitor mass, through the use of a banner recognition technology can help monitoring personnel to identify illegal protest banners to sound the alarm and allow monitoring personnel can immediately take corresponding measures, to prevent the illegal demonstrations in public life caused irreparable consequences. Banner recognition basic principle is identified firstly banner specific colors, such as red, white, and yellow, and then to judge whether this is a color region shape to identify banner.

4. System Implementation

This paper mainly expounds the video access to extraction module implementation process, which is realized based on SDI application types. In order to make the interface easier to operate, so let the original window into two, the system used is Mr. Christi posed, we achieve a word to the left to achieve the directory tree in relation to CSizing Control Bar class, which used in the system is called CCoolTabCtrl wrapper class. Directory entry click on the left, then appears corresponding so we can play video. The main algorithm code system is implemented in a dialog box class: CVCKBASEDlg, also in order to facilitate the operation. The realization of human-computer interaction, creating a menu item, used to open, close, pause, stop the video file to be processed, but also add the right-click shortcut functions. Video abstract extraction module interface is shown in figure 3.

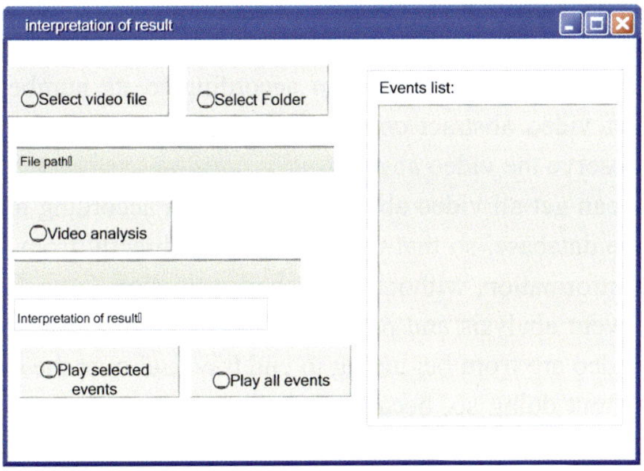

Fig. 3. Main software interface of video abstract model.

Taking into account the AVI is a relatively large video files, in the low configuration of the experimental conditions, the speed of this file is relatively slow, possibly will have an impact on the course of the experiment. On the method of multi thread technology can solve this problem. When click on the "key frame extraction" button, vc6.0 afxbeginthread function will be triggered, a linear equation was constructed by createthread function, so that we can image frames of a shot boundary detection and extraction of shot boundary: afxbeginthread; the ConvertAVIToBMPThread is the calling thread function. Mainly used for extracting the image frame of shot boundary. Shot detection of AVI video files, the algorithm based on mutual information. In this algorithm, it is necessary to calculate the amount of information interaction between two adjacent to each video frame, and in the video processing is to each frame image are analyzed, so to save the physical image information of the frame.

5. Conclusion

The video abstract technology as the foundation, developed a set of surveillance video intelligent identification system, which can be used in firework identification, banner recognition and sensitive information retrieval and so on. The main advantages of the system developed in this paper are:

1) using the video abstract technology, extraction of video abstract advantage, analysis video when system does not enter the event specific features (color, shape, etc.), but no difference of the video in all sports events as a video abstract is extracted, and then integrated into a video to play, so that the user can within 1 minute visiting hours of surveillance video summary.

2) all the video abstract events are corresponding to the number, so the user in watching all the video abstract event process, if it is found that they are interested in video abstract events can according to its numbers, then in the event list select video abstract corresponding to the single player, in order to more clearly observe the video abstract.

3) system can get all video abstract information according to certain format to store into the database, so that users can read it directly from the database of video abstract information, without the second extraction.

4) video event analysis and recognition technology of existing if the event analysis of a video are from beginning to end traversal over the entire video, but the system without doing so, because from the beginning to the end to a video analysis, time is very consuming. After extracting of video abstract, then event analysis technology of the video abstract is used, which saves a lot of time.

References

1. LI Ying. An OverView of Video Abstraction Techniques[R].Image Systems Laboratory, HP Laboratory Palo Alto, HPL -2001-191, 2001.
2. OUYANG Jian-quan, LI Jin-tao, ZHANG Yong-dong. Survey on video abstraction technolgy [J]. Computer Engineering, 2005, 30(10): 7-9.
3. WANG Chen, LIU Gui-qing, LAO Song-yang, et al. Event oriented Film Abstract [J]. Journal of Image and Graphics, 2005, 10(5): 642-649.
4. LIU Gui-qing, LI Jian-cheng, XIAO Peng, et al. Video summarization based on the "Entities-Description-Utilities" Model. Computer Engineering & Science 2005,27(10):31-34.
5. GUO Wei-xuan, GUO Bao-long. Design and realization of video processing system based on DM642. Electronic Design Engineering, 2009, 17(1): 79-81.
6. Zhang H J, Jianhua Wu. An integrated system for content based video retrieval and browsing [J]. Pattern Recognition, 1997, 30(4): 643-657.

A smart video surveillance system based on moving target recognition

Wan-Jia Qu* and Zhong-Lin Xu
Aviation University of Air Force,
Changchun, Jilin 130022, China
*E-mail: *1259737654@qq.com*

Yu-Wei Yuan,
Naval Aeronautical and Astronautical University,
Yantai, Shandong 264001, China
E-mail: yyw57156@hotmail.com

Due to the increasing number of accidents happening when flying target landing in the weapon testing field, a smart video surveillance system based on moving target recognition was designed. The system adopts the capturing front-server-decision model. In our paper, a method for detecting moving targets images using background difference method and frame difference method is first introduced. Secondly, target recognition is studied by the technology of contour extraction and edge detection. Finally, characteristic parameters of target are extracted by feature algorithm. Based on it, key technologies involved in the system are described in detail. Additionally, corresponding algorithm is designed using OpenCV in Visual C++ 6.0, and part of the key codes are given. Simulation results shows that the system designed can meet the needs of monitor and control of flying target, and also verify the effectiveness of the algorithm.

Keywords: Smart Video Surveillance System; Target Recognition; Detecting Moving Targets; Characteristic Parameters; Opencv.

1. Introduction

A Smart Video Surveillance System possess the advantages of non-involvement, high accuracy and good real-time performance and has been commonly applied in the fields of everyday life. The developed countries started research of smart surveillance earlier and has obtained much of a progress in military, aviation and other advanced fields. In this system, the demands of clients are to realize the function of safe landing of targets on the scene, meanwhile, the state parameters should be shown on the screen, in order to evaluate the potential threat to a safe

---

* Corresponding author.

landing. If the information is judged to be a threat, the system will provide the flight crew with the real-time data which would be important to flight attitude adjustment. To meet with this requirement, writer has designed a moving targets recognition based intelligent video control system.

2. System Component

2.1. *Video-capture-front-end*

Cameras are the main device to collect target statement data during landing, it takes CCD image sensor as the core, including synchronizing signal circuit, video signal processing circuit and battery. Besides basic parameters, cameras are required to be equipped with soft trigger, time delay exposure, white balance and anti flicker. It is located outside the observing spots around the landing area. Each spot equipped with a camera and lens, if necessary, a cloud platform should be considered. They transfer the signal to PC through the interface of each camera and then decode compress and further process.

2.2. *Server side*

The function includes video code and decodes module, video processing module and windows surface module. The video code and decodes module adopts MPEG-4 video compressing technology standard. It compresses and code video signal transformed by the video capture front end save the video on the hard disk as documents and files. The video decoding module is in charge of decoding and replaying. This module would observe and recognize the moving objects in the scene, and judge if the present state would threat a safe landing and make decisions. The window surface module provides the real time images of landing process, and simultaneously supply with the mutual interface between staff and aircraft.

2.3. *Behavior understanding and decision*

Images processed by server would require further analysis of the target characters and the understanding and explaining of both target movement and objective scene in order to categorize operations. This module consists of behavior understanding, threat evaluation and decision deduction. The first part mainly compares the statement data of altitude, position, velocity and attitude of target from processed images. Threat evaluation compares parameters data and setting standard to judge, if the aircraft can carry out a safe landing upon those data. Decision deduction takes the result of threat evaluation as the input information transforming to pilot for timely adjustment for flying state, in order to avoid aerial accident effectively.

3. Key Technologies

System needs to recognize whether there is a security threat in the monitoring scene, thus the technology of moving target identification is the key to implement this system. For dynamic image analysis and identification of moving targets, the basic principle is to put the sequence of image signal into the computer made by camera, through the image preprocessing, moving target detection, feature extraction, the methods of fuzzy clustering, feature matching so that we can implement target search, identification of the object, even for real-time target tracking. In the system, the main work includes moving target detection, feature extraction and target recognition, the author will focus on the introduction.

3.1. *Moving target detection*

For the question of the conventional motion target detection method is not suitable for weapon test field extraction of the target in complex background, using a Gaussian modeling background difference method and interframe difference method. Firstly, adopt the method of Gaussian model to accept the background image from a certain number of video sequences. In the subsequent processing, first using the interframe difference method to detect the outline of the moving targets for each frame video image, and then use the outside frame image motion target to go on the statistical filtering updating for the background model. Not only can it reduce the prospect of moving targets on the influence of background model updating, ensure the real-time performance of the system at the same time, but also can get a complete target image.

Open source computer vision library (OpenCV) is a library made by Intel company located in Russia's research laboratory developed a set of freely available, and composed of some function C and C++ class that is used to implement some algorithms for image processing and computer vision.

3.2. *Moving target identification*

Find a difference in the image contour and get its external rectangle. Get rid of small rectangle as noise area, and the rest as a moving target. At the same time set ROI of the current frame pFrame and background pBkImg for the external shape of moving targets. Firstly, set the current frame of ROI 0, then to the current frame and background of ROI for logic or operation. For moving target area, the current frame and background image information is the same, so it achieves the purpose of the elimination of the influence of moving targets on background updating.

After the identification of image edge detection and extraction. As Canny operator is a good method of edge detection. It has many advantages, such as no residual real edge, good positioning performance and only one response for a single edge point etc, and the algorithm of function cvCanny using Canny algorithm to detect the edge of the input image, then output the edge.

3.3. *The video frequency target comprehend*

If more than 2 cameras are installed on the monitoring site, the coordinates and azimuth angle α0 and pitching angle β0 of each camera are measured in advance. Take a flight target from the different direction, several cameras to have synchronously in the meantime with time record device, to same time target up of together point can adopt to hand over to remit method resolution space of 3D sit a mark.

Such as figure 1 shows: O1, O2 is 2 set cameras, T, T ′ is fly target diagraph to order and it is on the horizontal plane of projection. T space sits a mark for (X, Y, Z). (XO1, YO1, ZO1) is opposite for the camera O1 in sit a mark to fasten a sit of O-XYZ mark, (XO2, YO2, ZO2) is opposite for the camera O2 in sit a mark to fasten sit of O-XYZ mark. (Δ X1, Δ Y1, Δ Z1) is opposite for target T in sit a mark to fasten a sit of O1-XYZ mark, (Δ X2, Δ Y2, Δ Z2) is opposite for target T in sit a mark to fasten sit of O2-XYZ mark. The diagraph orders T is sit a mark to fasten O1-XYZ and the azimuth angle in the O2-XYZ, pitch angle of elevation is a α1, β1, α2, β2.

For analyzing convenience, cast target shadow at the XOY flat surface, such as figure 2 shows. So X, Y, Z can be achieved:

$$X=(XO1+XO2+\Delta X1+\Delta X2)/2 \quad (1)$$

$$Z=(YO1+YO2+\Delta X1\tan\alpha 1+\Delta X2\tan\alpha 2)/2 \quad (2)$$

$$Y=(YO1+YO2+\Delta X1\sec\alpha 1\tan\beta 1+\Delta X2\tan\alpha 2\tan\beta 2)/2 \quad (3)$$

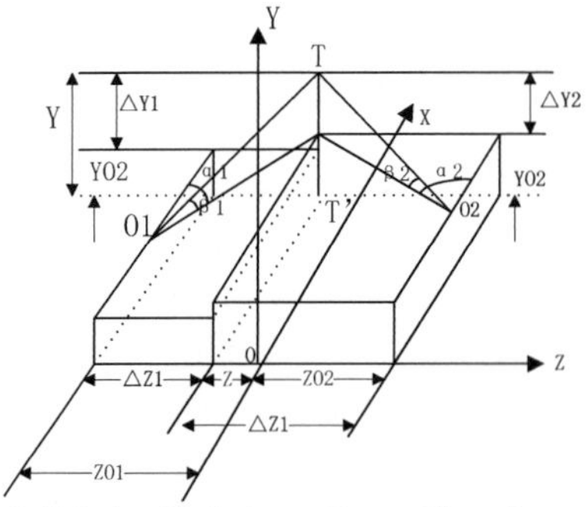

Fig. 1. Relationship between the position of camera and target in the intersection measurement.

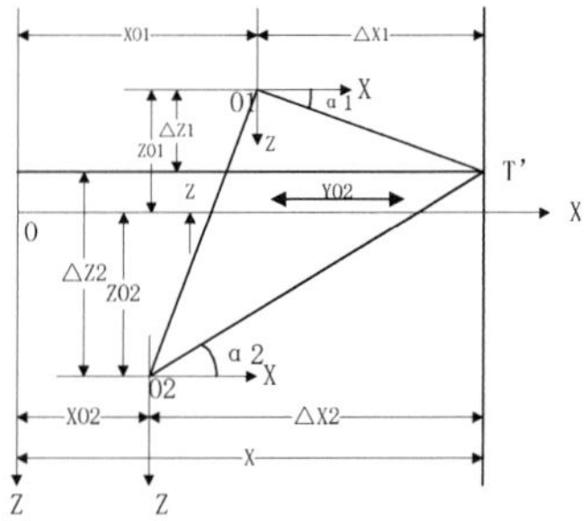

Fig.2. Projection of the target T on XOZ plane.

4. Experiment Test and Result Analysis

Test environment in the CPU 2.31GHzs, is the PC of the memory 960 MBs on board and use VC++6.0 carry out an intelligence video frequency to exercise in the supervision the examination of target with OpenCV1.0. Adopt that calculate way 320*240 imitate the landings video frequency of the flight target to carry on examination, results such as figure 4 show. Through the calculation, the speed of that calculate way processing picture is 12 frames/s, then processing each picture is about 16 ms, which can satisfy the request of system.

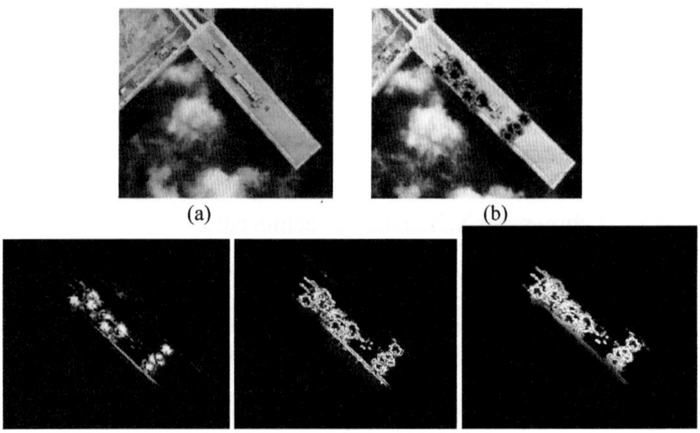

(a) (b)

(c) (d) (e)

Fig. 3. Experimental results of test.

Diagram 3(a) is to appear sport target current of the place a picture, figure 3(b) is a Gauss background under the model of current background picture, figure 3(c) is that the background which sets up a mold with the Gauss is bad divide method and bad cent the method combine together the result picture that the method examination come out, figure 3(d) checked to find out from the examination badly cent the picture of sport target outline picture, figure 3(e) is use Canny calculate carried on an edge examination and withdraw behind of picture.

From comparison of figure 3(a), figure 3(b) with figure 3(c) we can know, this system not only can clearly show the simulation test scene monitoring pictures, but also the moving object can be detected accurately, which shows the moving object detection method can be well established and update the background model, carrying out a foreground target and background picture in the meantime of complete partition. Compared figure 3(a), figure 3(d) with figure 3(e) we can know, the system can satisfy a request in identify moving targets. By finding the contour of the difference image, the moving object is extracted from the video frame.

5. Conclusions

From reading the video frequency pictures, this system succeeds in catching sequence, the Smart Video Surveillance System can actually examination and identify a moving target. At the same time, the position of the moving object on the target surface is accurately displayed, making the target parameter information more accurate, also for guided decision to provide importance basis. Now uses simulated flying target landing video experiment, next step, according to the situation in the target's flight landing video makes a further analysis and study.

References

1. LI Ying. An OverView of Video Abstraction Techniques[R].Image Systems Laboratory, HP Laboratory Palo Alto, HPL -2001-191, 2001.
2. OUYANG Jian-quan, LI Jin-tao, ZHANG Yong-dong. Survey on video abstraction technolgy [J]. Computer Engineering, 2005, 30(10): 7-9.
3. WANG Chen, LIU Gui-qing, LAO Song-yang, et al. Event oriented Film Abstract [J]. Journal of Image and Graphics, 2005, 10(5):642-649.
4. LIU Gui-qing, LI Jian-cheng, XIAO Peng, et al. Video summarization based on the "Entities-Description-Utilities" Model. Computer Engineering & Science, 2005, 27(10):31-34.
5. GUO Wei-xuan, GUO Bao-long. Design and realization of video processing system based on DM642. Electronic Design Engineering, 2009, 17(1):79-81.
6. Zhang H J, Jianhua Wu. An integrated system for content based video retrieval and browsing [J]. Pattern Recognition, 1997, 30(4): 643-657.

Design of MUAV target recognition system based on binary descriptor of Brisk

Ya-Wei Liu* and Xiao-Min Li†
University Ordnance Engineering College,
Shijiazhuang, China
*E-mail: *yaweiliu0211@163.com, †1217317854@qq.com*

Targeting the application of intelligence reconnaissance of MUAV in model airplanes, the reconnaissance intelligence automatic identification system is designed to generate fast and exact interpretations of targets of MUAV. The system contains two main modules: the stable platform control module and the target recognition. In this paper, we used BRISK and hamming distance match keypoints and introduced a new way to detect keypoints with adaptive threshold vale method, as well as screened keypoints with the RANSAC algorithm. The experimental results show that it has a good robustness to scale transformation, lean transformation and is able to prevent problems of MUAV. Besides, the BRISK algorithm is nearly 20 times faster than SURF algorithm.

Keywords: Muav; Stable Platform Control; Target Recognition; Brisk Algorithm.

1. Introduction

With the modern battlefield becoming more and more complex, getting information and the capture of the aircrafts has become the primary task of war. It is the important factors to the implementation of decisions and win the war that investigate the battlefield information and obtain and process and distribute the information reliably and accurately. With the battlefield shift to urbanization gradually, the application of the large unmanned aerial vehicle (UAV) severely restricted. However, due to its small volume, low cost, low casualties and strong ability to survive, Small UAV has been widely concerned in the world. But because of various limitations, the development of the Reconnaissance used to small UAV is not good. For example, during the reconnaissance of small UAV, the reconnaissance equipment is installed in the body of small UAV. So the quality of the reconnaissance images decrease under the shaking of the UAV. What's worse, because that the small UAV is often used for individual operations or small scope of operations and there is not a perfect automatic

* Corresponding author.

target recognition system, it is necessary to arrange a number of operational staff to observe . As a result , it can weaken the combat effectiveness of other issues. It will make a seriously bad influence on the effect and application of small UAV reconnaissance.

In this paper, Aiming at the features of small UAV reconnaissance system, an automatic target recognition system based on the binary descriptor of BRISK is proposed. The system can improve the intelligence of the system that based on the adaptive threshold it can choose different number of feature points during the feature detection in the target recognition. Besides, the system can effectively improve the accuracy of feature recognition because that it screens through the bidirectional matching and the method of double screening of RANSAC. Through repeated experiments, the system is in practical value.

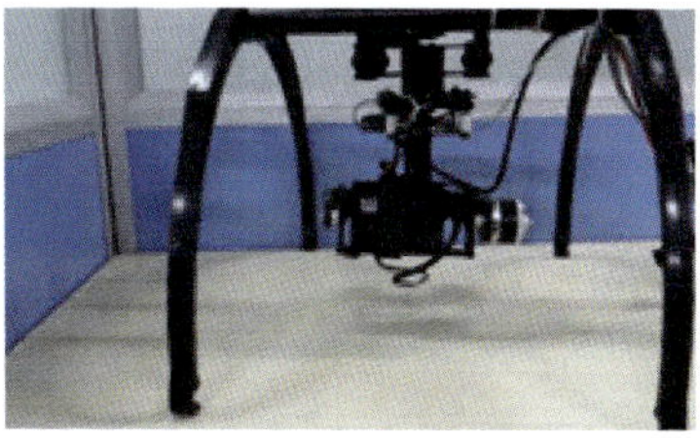

Fig. 1. The holder.

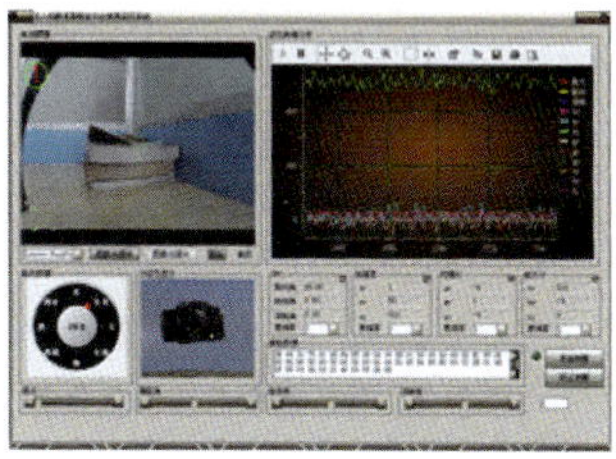

Fig. 2. The ground station.

2. Holder Control

Holder can increase the clarity of reconnaissance images by isolating the camera with the body of plane and have advantages in target reconnaissance. The structure of the holder in this paper is shown in figure 1.

The holder has the function of motor driving, servo controlling and camera adjusting. The range of pitching angle of the holder is from -90 degrees to +90 degrees and the range of yaw angle is from -180 degrees to +180 degrees. The cover of the holder meets the requirements of target reconnaissance. The holder is controlled by the ground station. The ground station is developed by VC++ and OpenCV. As shown in figure 2.

The ground station can control the raising angle and the roll angle and the yaw angle of the camera through sending data by RS232 port. What's more, there is a sensor on the Holder by which we can get the current angle of the camera on the holder in real-time. And A HERO camera and a T510 video acquisition card are used to collect the data.

The control system of the holder is shown in figure 3. The balance system of the holder is consist in two infrared sensors. One of the infrared sensors which is installed in the horizontal axis of the holder is used to get the roll angle of the holder. The other installed in the horizontal axis of the holder is used to get the raising angle of the holder. When the outputs of the two sensors are the same, the holder is in balance. Otherwise, if the holder is not in balance, the controller will adjust the holder according to the difference of the outputs.

When the holder is balance, the output of the two infrared is equal, what is the standard of the balance of holder. If the output are different, adjusting the holder according to the output. In order to keep the holder balance, PID control has been introduced to control the motors. As shown in figure 4.

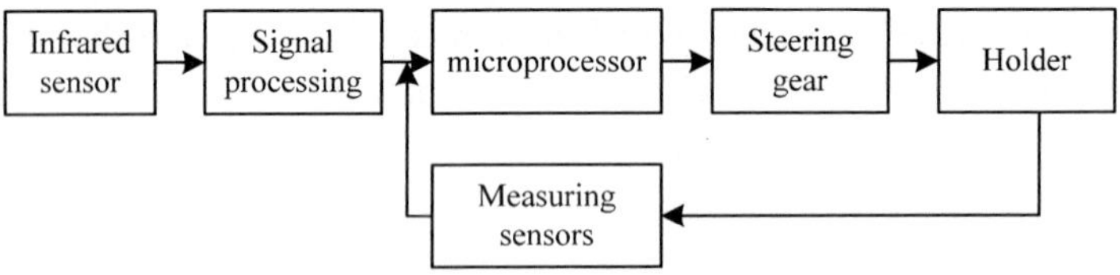

Fig. 3. The control of holder.

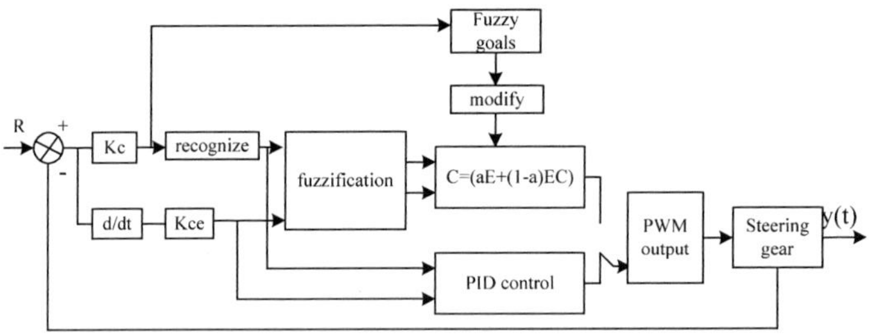

Fig. 4. The adaptive fuzzy PID.

3. Target Recognition

The algorithms of SIFT and SURF which have been widely used in target recognition perform stably. However, the common problems are time-consuming and the memory of system. The main factor to the problems is the form of character description that use multidimensional vector. This paper put forward a target recognition algorithm based on BRISK which use binary characters as the character descriptor and match features in Hamming distance. The algorithm speeds up the recognition and reduces the occupancy of memory. In order to improve the accuracy of the algorithm, the method of adaptive threshold value and the dual screening of characters with bidirectional matching and RANSAC have been proposed.

The process of the system are as follows: Firstly, collect a real-time image and process. Secondly, detect and extract the feature points with the method of adaptive threshold value. In the end, match the feature points. As is shown in figure 5, error matching points have been eliminated in the method of bidirectional matching and RANSAC.

3.1. *The method of adaptive threshold value*

During the process of target recognition based on features matching, The adaptive threshold values are changing with the complexity of the target. The algorithm of BRISK has strong robustness, high precision and efficiency, but the threshold value must been set manually. So in this paper the method of adaptive threshold value of BRISK is proposed.

The feature points should be increased along with the increase of the complexity of the image ideally. The energy J, information E and contrast C are the main parameters to measure the complexity of images. The formula of the complexity is as follows:

$$\text{Complexity}=E+C-J \tag{1}$$

In the process of calculation, in order to eliminate the influence of dimensional, normalization processing is required, and grade to the threshold.

3.2. *The method of the screen of features*

In this paper, The methods of bidirectional matching and adaptive threshold value have been used in the screen of features. In order to eliminate the error matching points such as “one to more” and “more to one” ,the methods of bidirectional matching have been used. Besides, In order to improve the accuracy of feature matching, RANSAC has been used in this paper to realize further screening. As a result, the accuracy of target recognition is improved.

4. The Experimental Results

The experiments are developed using VS2010. We program in C++ based on OpenCV. The hardware system is as follows: the processor is Inter (R) Core (TM) i3 CPU M390@2.67GHz; the memory is 4.00GB; the hard disk is 500G; the operating system is Windows 7.

4.1. *The validation of double screening*

The performance of the feature points screening algorithm is tested by using two pictures which have a few differences on scale and rotation and gradient of images. The test result is shown in figure 6.

Fig. 6-1. The coarse matching results.

Fig. 6-2. The matching results after screening.

We can see that from the experiment the number of coarse matching is 342, and the test result is shown in figure 6-1. As a comparison , the number of exact matching witch has used to bidirectional match and RANSAC is 127, and the test result is shown in figure 6-2. By comparing of coarse matching and exact matching, the matching result by screening is more accurate.

4.2. *Robustness testing of BRISK*

In the experiment, robustness of the algorithm has been tested in the condition of scale, rotation, gradient and shelter changes. The test result is shown in figure 7.

Template Target

Fig. 7-1.The test picture of scale changes.

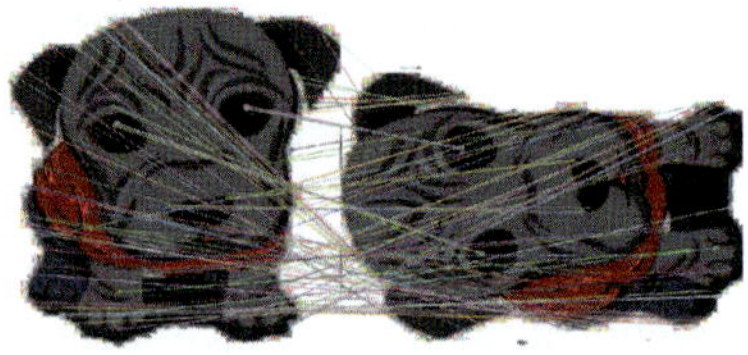

Temple Target

Fig. 7-2. The test picture of rotation changes.

Template Target

Fig. 7-3. The test picture of gradient changes.

Template Target

Fig. 7-4. The test picture of shelter changes.

The test results under the condition of scale changes have been shown in figure 7-1.We can see from the experiment that BRISK can also perform well when the target down to 0.7 times. The test results under the condition of rotation changes have been shown in figure 7-2. The experiment results show that the algorithm of BRISK can identify the target easily when the target rotate 90 degrees. In figure 7-3, the test results show that the number of matching points is small when the target leans. In figure 7-4, the test results show that the algorithm of BRISK keeps good performance under the condition of shelter changes.

Table 1. Robustness testing of BRISK.

Transform / classes	scale changes	rotations changes	gradient changes	shelter changes
Number of coarse matching	154	137	101	134
Number of exact matching	104	105	28	96
Time-consuming of BRISK(s)	0.063	0.092	0.082	0.081
Time-consuming of SURF(s)	1.438	1.695	1.620	1.423

The number of coarse matching points and exact matching points and the contrast of time-consuming of BRISK and SURF are show in table 1. The test shows that the algorithm of BRISK is 20 times faster than SURF.

In conclusion, the algorithm of Brisk for target recognition has better robustness when the target changes with the change of environment and BRISK is faster than SURF.

5. Conclusion

In this paper the target recognition system of MUAV is mainly studied. The algorithm of BRISK has been applied and the algorithm of adaptive threshold value and the combination bidirectional matching and RANSAC have been proposed. Several experiences have been conducted to test the robustness of the proposed algorithm in case of the changes of scale, rotation, gradient and shelter and compare with the algorithm of SURF. From the experimental results, we can see that the algorithm of BRISK has good robustness to the changes of scale, rotation and shelter and has poor robustness to the change of gradient. The algorithm of BRISK is 20 times faster than SURF, which meet the real-time demand in target recognition of MUAV.

References

1. Karpushin, M.,Valenzise, G., Dufaux, F.,Improving distinctiveness of brisk features using depth maps, in Image Processing (ICIP), 2015 IEEE International Conference on , vol., no., pp.2399-2403, 27-30 Sept. 2015.

2. Chongyu Pan, Tianjiang Hu, Lincheng Shen, BRISK based target localization for fixed-wing UAV's vision-based autonomous landing, in Information and Automation, 2015 IEEE International Conference on , vol., no., pp.2499-2503, 8-10 Aug. 2015.
3. SuGil Choi, Seungwan Han, New binary descriptors based on BRISK sampling pattern for image retrieval, in Information and Communication Technology Convergence (ICTC), 2014 International Conference on , vol., no., pp.575-576, 22-24 Oct. 2014.
4. Mohamed, N.J., Target course recognition using nonsinusoidal look-down radars, in Electromagnetic Compatibility, IEEE Transactions on , vol.36, no.2, pp.117-127, May 1994.
5. Xiao Tengjiao, Zhao Danpei, Shi Jun; Lu Ming, High-speed Recognition Algorithm Based on BRISK and Saliency Detection for Aerial Images, in Intelligent System Design and Engineering Applications (ISDEA), 2013 Third International Conference on , vol., no., pp.816-819, 16-18 Jan. 2013.
6. Zhihui Zheng, Hanxizi Zhang, Bo Wang, Zhifeng Gao, "Robust traffic sign recognition and tracking for Advanced Driver Assistance Systems," in Intelligent Transportation Systems (ITSC), 2012 15th International IEEE Conference on , vol., no., pp.704-709, 16-19 Sept. 2012.
7. Singh, S., Rao, D.V., Recognition and identification of target images using feature based retrieval in UAV missions," in Computer Vision, Pattern Recognition, Image Processing and Graphics (NCVPRIPG), 2013 Fourth National Conference on , vol., no., pp.1-4, 18-21 Dec. 2013.
8. Xuqiang Zhao; Qing Fei; Qingbo Geng, "Vision based ground target tracking for rotor UAV," in Control and Automation (ICCA), 2013 10th IEEE International Conference on , vol., no., pp.1907-1911, 12-14 June 2013.
9. Zeng Fucen; Shi Haiqing; Wang Hong, "The object recognition and adaptive threshold selection in the vision system for landing an Unmanned Aerial Vehicle," inInformation and Automation, 2009. ICIA '09. International Conference on , vol., no., pp.117-122, 22-24 June 2009.
10. Bekele, D.; Teutsch, M.; Schuchert, T., "Evaluation of binary keypoint descriptors," in Image Processing (ICIP), 2013 20th IEEE International Conference on , vol., no., pp.3652-3656, 15-18 Sept. 2013.

Time difference based highly accurate anemometer: architecture and practice

Li-Wei Zhang*

Tianjin Navigation Instruments Research Institute,
Tian Jin, 300131, China
E-mail: zlw170@163.com

Ying Yao[a], Long Xiao[b] and Shi-Bo Yin[c]

Systems Engineering Research Institute,
Beijing, 100036, China
E-mail: [a]baeng@163.com, [b]seridhs@163.com, [c]expcty@163.com

Anemometer is widely used as an important measuring instrument in modern ground, aerial and aquatic vehicles. The paper presents a novel type of anemometer based on time difference. There are three key contributions as follows. Firstly, the key principle for designing an ultrasonic anemometer is described as an novel approach by using time difference. Secondly, the architecture design, electro circuit design and the filter arithmetic are given in detail to show the feasibility of our design. Thirdly, the implementation is measured in a real scenario to determine the accuracy of this approach. Final results show that our system can measure the wind speed and direction with higher accuracy and reliability, and with stronger anti-jamming properties than other known instruments.

Keywords: Anemometer; Time Difference; Accuracy.

1. Introduction

Winds influence a lot on the security of shipboard aircrafts when they enter or leave the board. So accurate measurement for wind means a lot to flight safety. Ultrasonic based anemographs, have some obvious advantages comparing with the traditional ones :no rotatable components which means no abrasion or aging problems causing by rotation, high sensitivity, zero start-up-wind-speed, high measuring speed, no special maintaining, etc.[3]

* Corresponding author.

2. Principle of Time Difference Based Anemometer

2.1. *Model for unidimensional ultrasonic anemograph*

Assuming the distance between two transmit-receive ultrasonic sensors is d, time needed for ultrasonic spread from probe 1 to probe 2 is t_{12} when following the wind, the wind speed is V_w and the ultrasonic speed is V_s , following can be deduced from the mode shown in picture 1:

$$\frac{d}{t_{12}} = V_S + V_W \qquad \frac{d}{t_{21}} = V_S - V_W$$

Formula 1

And the wind speed at that moment can be deduced as:

$$V_W = \frac{d}{2}\left|\frac{1}{t_{21}} - \frac{1}{t_{12}}\right|$$

Formula 2

2.2. *Principle of planar ultrasonic anemograph*

Our design , based on the basic principle of time difference and combines parameters of ultrasonic sensor like sensitivity and directivity etc., builds a system using two pairs of orthotropic ultrasonic sensor that can both transmit and receive . As is shown in picture 1 :

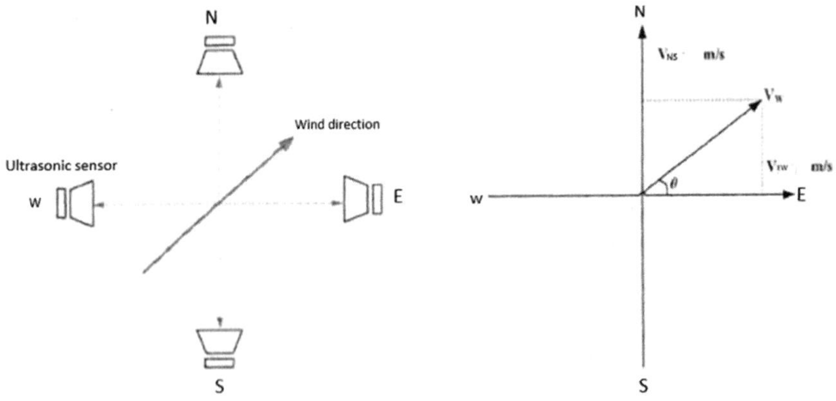

Picture 1

Picture 1 shows the sensors generate ultrasonic signals at the same rate and we measure the transmission time at both forward and reverse direction. So, we can get the horizontal speed and direction of the wind from calculation as picture 1 shows. Assuming the vector speed of the wind is V_w whose component along the east-west direction is V_{ew} and the one along the south-north direction is V_{sn}. So, the vector speed of the wind and the angle can be deduced as:

$$V_W = \frac{L}{2}\sqrt{\left(\frac{1}{t_{ew}} - \frac{1}{t_{we}}\right)^2 + \left(\frac{1}{t_{ns}} - \frac{1}{t_{sn}}\right)^2} \qquad \theta_0 = \arcsin\frac{\left|\frac{1}{t_{ns}} - \frac{1}{t_{sn}}\right|}{\sqrt{\left(\frac{1}{t_{ew}} - \frac{1}{t_{we}}\right)^2 + \left(\frac{1}{t_{ns}} - \frac{1}{t_{sn}}\right)^2}}$$

Formula 3

The only thing needed to measure the speed and direction of the wind is t_{ew}, t_{we}, t_{sn} and t_{ns}, which is independent on the spreading speed of the ultrasonic, so this feature means our design eliminates the interference of temperature.

3. The Principle Diagram of the Measurement System

3.1. *The system diagram*

The system model is designed as micro-controller+CPLD using Altera ATMEGA128 and EPM570T100C5N logical hips as the central processor ,control and driver circuit for ultrasonic sensors, amplifying and receiving circuit for ultrasonic signals, system time circuit, processing circuit and peripheral interface circuit.

3.2. *Overall design*

According to the overall design requirements, the anemometer is orthotropic placed, the distance between two probes which is made of 302L stainless steel is controlled around 200mm, using high frequency transducer that can both transmit and receive. Using the rubber sealing ring of vibration isolation at the joint to enhance the level of protection.

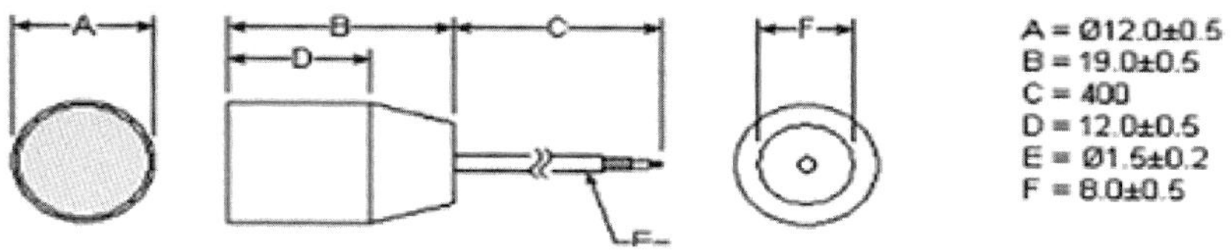

Picture 2. The probe of the Anemometer.

Description of the signal

Signals at the transmit side are square wave signals whose frequency is 200kHZ and peak is 400V. At the receive side, after amplification signals are restored as the same as that at the transmit side.

Recording mode of the propagation delay

In our experiment, we used 0.025us as the statistical time unit in CPLD. we can judge the effectiveness of the signal received by analyzing the time interval between pulse signals.

4. Circuit Designing

This anemometer uses high performance processor as its controlling and calculating core, samples the input and output waves of the logic controlling driver, and finally outputs the calculating result in time.

4.1. *Logic control waveform output*

The controller controls the logic devices output as waves, logic devices output square waves whose frequency is 200kHZ, after rising voltage by s4427, the waves then are sent to driver circuits of the probes.

4.2. *Booster output driver circuit*

Different from Single-phase transformer isolation amplifier using in most anemometer whose output is 0-400v waves, in our designing we use symmetric positive and negative transformer isolation amplifier, it outputs signals to symmetric waves peak at ±200V by using driver IRFR220 of MOS tube. In this way, we can reduce the delay during the waves going up and down thus rise the accuracy of the measurement.

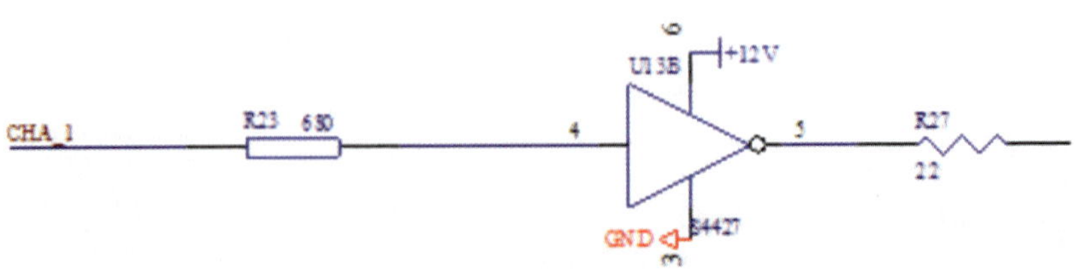

Picture 3. Wave-output circuit of the driver.

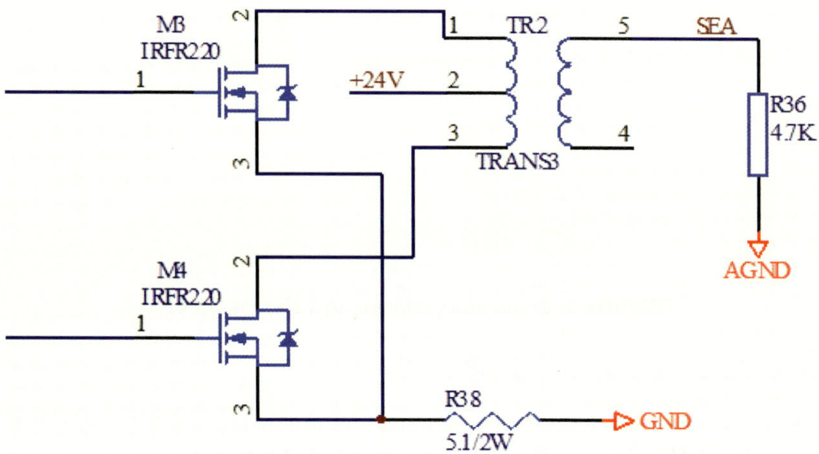

Picture 4. Drive waveform booster circuit.

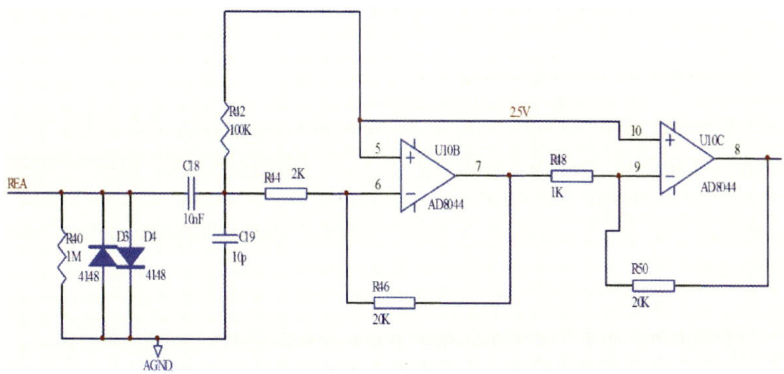

Picture 5. Echo amplification filter circuit.

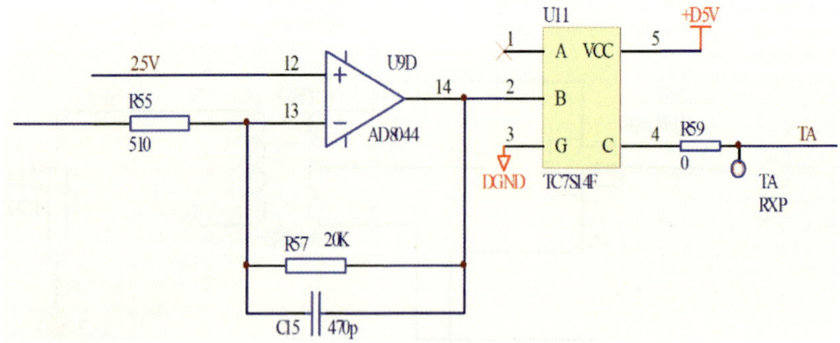

Picture 6. Echo comparison and Restoring circuit.

4.3. *Echo sampling amplification filter*

As is shown in picture 6, firstly, the input ultrasonic probe senses the sinusoidal echo 200KHz signal influenced by the wind, then rc high-pass filters after limiting voltage to eliminate coupled DC component; Secondly, amplifies the signal to 20 times by the second level AD8044 subtraction amplification circuit, and gets the 0-5v sinusoidal waves. Comparing with the one-level amplification circuit , Using two-level amplification circuit can effectively reduce interference thus rise the accuracy of the signal restored.

4.4. *Comparison and restoring*

After filtering and amplification, signals are passed to AD8044 subtraction comparing circuit and are restored to 0-5v square waves, then eliminate the shake of waveform lag back by TC7S14F, and outputs the accurate square waves to the logic devices which execute the following filter waveform calculation.

4.5. *Designing and Experiment for electromagnetism compatibility*

To eliminate the negative natural effects, we build some protection measurements: for the power protection and the communication part, in the first level we use gas discharge tube to absorb most extra energy generated by thunders or surges and designed different model protection and common model protection separately. In the second level we use piezo resistor to absorb the rest of the extra energy in power power protection and .PPTC in communication protection. In the third level, we use TVS (Transient Voltage Suppressor) to rise the peak instant absorption power to 200w to 2000w and above, and to reduce the response time to the order of picosecond.[4]

5. Data Processing

5.1. *Hardware filter processing*

We modulate the driving pulse and settle phase position reversal point so as to locate the reversal point accurately, and get high time difference value after software filtering.

5.2. *Diving filtering*

As we can judge the effectiveness of the signal received by analyzing the time interval between pulse signals in the signal series received. We can also exclude the instance spike signals in this way. And by measuring the speed of wind 10-20 times per second along with the diving filtering algorism, we can further the stability and consistency of the measurement data.

5.3. *Least square fit*

In the experiment we did, the environment data is temperature: 24.7°C, humidity: 72%RH, atmospheric pressure: 948.4hPa, technical stander we followed is JJG431-86 Portable 3-Cup Anemometer Type DEM6.

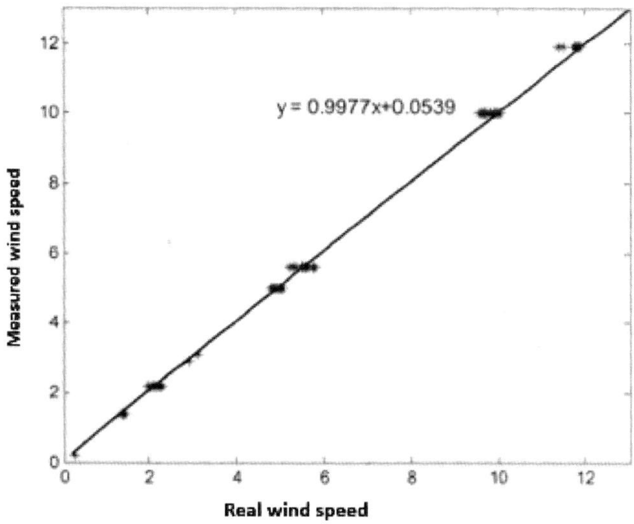

Picture 7. Diagram of Least Square Fit Data.

In this experiment, we tested 11 groups of winds with different speed, in each group we measured 18 times. As is shown below, the degree of linearity of our anemometer is good enough to realize accurate measurement.

5.4. *Measurement for the wind critical angle*

To deal with the critical angle problems, firstly we sort the N groups of direction angle data which is going to be averaged, then we can get the difference value between the maximum and the minimum, if this value is larger than the threshold value settled ahead (for example 300°),then we plus 360° to the minimum and sort again. At last, average the new group of values and get the new median, if this median is larger than 360°, then subtract 360° and we will get the correct value.

6. Conclusion

Our system is based on the time difference algorism, and analyzed the impact factors of accuracy of our anemometer qualitatively and quantitatively. The experiment result shows that our anemometer reach the accuracy of the same kind abroad, and has advantages like higher measuring speed , smaller size and greater anti-interference ability etc., which has high practical value.

References

1. WU Zong-ling, YAN Lian-shan, ZENG De-bing, CHEN Juan-zi,. Design of ultrasonic wave anemometer with high precision based on ARM + CPLD. Transducer and Microsystem Technologies.2012.
2. CAO Ke-jin, CUI Guo-heng, ZHU Yin-bing. Ultrosonic Anemometer Theory : Modeling and Analysis ACOUSTICS AND ELECTRONICS ENGINEERING.2010(1)
3. JIN Jing,TANG Hui-qiang.Design of Ultrasonic Wind Velocity Measurement System Based on ARM. INSTRUMENT TECHNIQUE AND SENSOR.2009.06
4. I. Saito, K. Mizutani, N. Wakatsuki, et aL Two-axis anemometer with acoustic reflectorusing single pair of loudspeaker and microphone [J].Jpn.J.Appl.Phys.,2007.

A three-layer framework for correlated multi-label classification with respect to package recommendation

Jing-Kang Yang*, Jie Shi†, Rui Xuan‡ and Regina Wang§

International School Queen Mary University of London,
London, UK
*E-mail: *j.yang@se13.qmul.ac.uk, †j.shi@se14.qmul.ac.uk,*
‡r.xuan@se13.qmul.ac.uk, and §t.wang@se13.qmul.ac.uk

Zhao-Jing Zhang¶ and Liang Dong||

Key Laboratory of Universal Wireless Communication,
Ministry of Education,
Beijing University of Posts and Telecommunications,
Beijing, 100876, China
E-mail: ¶buptzzj@163.com, ||liangdong@bupt.edu.cn

Recommendation in APP becomes increasingly popular as mobile Internet develops rapidly nowadays. Different from music recommendation, tariff package recommendation in telecoms operators' APP is able to use customers' recorded information for better performance. Consequently, multi-label learning can be applied for recommendation in that case. During the past decade, multi-label learning has a wide application in real work, such as automatic annotation for multimedia contents, web mining, tag recommendation, etc, and many famous frameworks, such as Binary Relevance(BR). It can raise and solve large amounts of multi-label classification problems successfully. In this paper, a new framework is presented which aims at solving the problem of correlated multi-label classification (CMC), which can be applied in tariff package recommendation. The new framework has three layers. On the first layer, all labels are transformed into several new-generated groups using adapted Principal Components Analysis. Each group is scored on the middle layer and the score will be disperse to original labels on the last layer. Furthermore, we analyze the performance of the framework on tariff package data set. The outcome shows that our framework can obtain more profits with a low time complexity compared to traditional methods.

Keywords: Recommendation; Multi-Label Classification Correlated Label; Data Mining.

1. Introduction

With the rapid development of mobile Internet, recommendation in APP becomes increasingly popular. For instance, music recommendation in music

* Corresponding author.

APP can use the record of users' history or music list to predict users' preference[2]. It is said that if the information gathered is just the object users once chosen without more essential clues like users' age, salary, etc, classic association algorithms such as Apriori are useful.[6] However, if more information is provided, just as operator like China Mobile, not only knowing the package users subscribed before but also users' private data including age, average call duration, even users' location, to make the best use of those data, multi-label classification is oriented for tariff package recommendation.

Multi-label classification is a classic problem of machine learning. Classification mainly deals with the problem when data is under the space which combines feature space X and label space Y . Specially, label space in multi-label classification problem is more than two dimensions, which draws a main difference from binary classification. Recent years, multi-label classification has received extensive concern[1]. Many people have obtained some achievement in this area, for instance, early researches have achieve much success on the problem of multi-label text categorization, and multi-label learning also attach much attention with automatic annotation for multimedia contents including image[7], bioinformatics[4][5], web mining[8][9], etc. Nevertheless, in real work, when using multi-label classification to solve problems, the data dimension is so high that this work cannot achieve ideal results. Usually, high data dimension leads low density of information and complex arithmetic. It even brings a series of problems like over-fitting. In history, some people have tried to use different methods to solve this problem. Argyriou raised a convex multi-task feature learning method to solve sparse representations problems, and Boutell introduce the famous Binary Relevance method to simplify the complex multi-label problem into binary classification[10]. A great amount of other methodologies are proposed, aiming at the specific multi-learning problem.

However, few people keep eyes on thinking over leveraging correlation between labels and how to use correlation to reduce data dimension and simplify the solution. Correlation between labels is so important in real world application that we cannot ignore it. In image recognition field, a picture containing many items can be marked by several labels, and there is no doubt that the correlation between labels is a critical factor to consider, since the items in a picture may often have strong relationship[7]. Also, in the case of tariff package, if we know two packages always has a strong connection, we would better tie up the package together. All in all, the correlation among labels is very important but still puzzled for us to use the correlation analysis to handle problems in real life.

As a result, in this paper, we put forward a new framework to solve the problem of multi-correlated-label classification (MCLC). Through studying multi-correlated labels, we transfer the multi-dimension problem to an easier

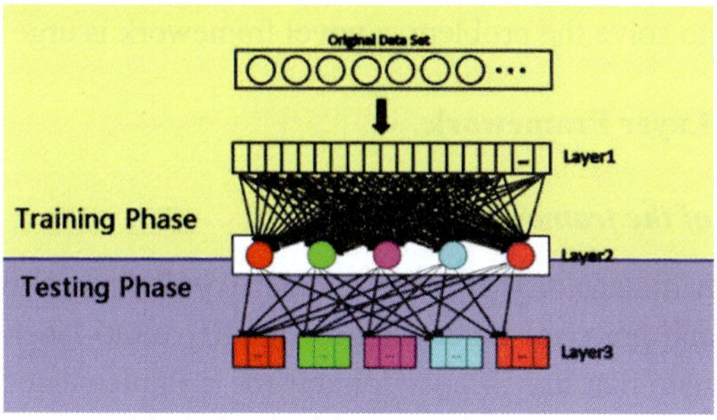

Fig. 1. Graph of three-layer framework

problem in less dimension. This method can simplify operation greatly. Moreover, we can solve extremely complicated problems in relatively easier ways, which enables quantitative cost saving and reasonable decisions.

Primarily, a three-layer framework is formulated for realizing dimension reduction. And we use Principal Component Analysis (PCA) with rotation matrix to finish reducing dimension. And then, a new matrix is created which shows the relations between different features and different labels. Furthermore, for every feature, it can be transferred a continuity problem by analysis of regression. Finally, utilizing the inverse transformation, we get ultimate result. Obviously, if meeting quite big data about individuals and labels, this way may reduce the amount of calculation and save much time. Aimed at the profit problem, we can realize benefit maximization through finding out the highest point of each feature. In actual production and real life, this method is very serviceable and meritorious. It can also assist all walks of life to work out their concrete issues.

The whole paper is planned to divide mainly three parts except introduction. In the first part, we give a problem statement to present the drawback of classic methods and the necessity of the new method in a thorough way. Then, we illustrate the framework of the multi-label algorithm in detail. Finally a experiment is carried out to verify the correction of this method with the tariff package dataset form China Mobile in the second part.

2. Problem Statement

The problem of recommendation can be transformed as multi-label classification, however, as for package recommendation, there are 2000 different kinds of tariff services, and telecom operator often sets a limit on the number of

packages recommended for users at a period in order not to disturb users. Thus, there are
1,331,334,000 possible combinations if the limit is 3. If we implement the classic multi-label classification like BR, the training data for every combination is sparse. Thus, to solve the problem, a novel framework is urge needed.

3. The Three-Layer Framework

3.1. *Overview of the framework*

Because of high dimensional label space and correlations between labels, we design a novel and practical framework to simplify multi-label classification. It is necessary to note that this 3-layer framework is implemented on training set, while for test set, we only need to apply last two layers. The framework is illustrated in the *figure* 1. Input of the framework is the whole data in training set $S = (x_1,y_1),(x_2,y_2),...,(x_n,y_n)$, including those in instance space X and label space L.

3.2. *Connection between layer 1 and layer 2*

As is illustrated in the figure 1, the main mission of layer 1 is to restore label space L, which is spanned by vector $y = [y^{(1)},y^{(2)},...,y^{(m)}]$. However, if we implement a multi-label learning algorithm to allocate instances into m labels directly, the computational complexity can be in an extremely high level. For example, one of the classic methods of multi-label learning is One-versus-All method, which need to generate nearly n^2 classifiers to combine them together for prediction. It can be found easily that the number of classifiers grows in square ratio as labels increase. In that case, to alleviate computational complexity, dimensionality reduction is needed to form a few super-labels, with one of which can represent a certain number of origin labels. Thus, principal component analysis (PCA) is implemented for dimensionality reduction. In this part, we will explain how the connection between Layer 1 and Layer 2 is formed, i.e. how labels in layer 1 consist into the groups in layer 2. The whole process is divided into three steps as follow.

3.2.1. *Detect the correlation matrix*

In order to implement principal component analysis, the first step is to detect the correlation matrix of labels, which is in a high dimension and needed to be shrunk into a low dimensional space. The samples in label space can be denoted as a matrix M, which is calculated by equation(1). For convenience, the label values of the n-th instance can be denoted as the vector y_n, as equation(2), while every instance value of the $m-th$ label is restored in the vector $y^{(m)}$, shown in equation(3).

$$M = \begin{pmatrix} y_1^{(1)} & \cdots & y_1^{(m)} \\ \vdots & \ddots & \vdots \\ y_n^{(1)} & \cdots & y_n^{(n)} \end{pmatrix} \tag{1}$$

$$\boldsymbol{y_n} = (y_n^{(1)}, y_n^{(2)}, \ldots, y_n^{(m)})^T \tag{2}$$

$$\boldsymbol{y^{(m)}} = (y_1^{(m)}, y_2^{(m)}, \ldots, y_n^{(m)})^T \tag{3}$$

The correlation matrix of labels S can be calculated as follow. Since then, we successfully detect the coefficient of i-th label and j-th label easily by referring either s_{ij} or s_{ji} in correlation matrix S, which is shown in equation(4).

$$S = \frac{1}{n}\sum_{i=1}^{n}(\boldsymbol{y}_i - \overline{\boldsymbol{\mu}})(\boldsymbol{y}_i - \overline{\boldsymbol{\mu}})^T$$

$$where\ \boldsymbol{\mu} = (\boldsymbol{\mu}_1, \boldsymbol{\mu}_2, \ldots, \boldsymbol{\mu}_m)^T \tag{4}$$

$$\boldsymbol{\mu_m} = \frac{1}{n}\sum_{i=1}^{n} y_i^{(m)}$$

3.2.2. *Use principal component analysis*

In order to find the principal component, a projection calculation is needed for mapping all the original data in the space spanned by labels on that spanned by new components. Here we call the new components group. By the method of maximum variance formulation, the variance of data is needed to be derived in new space. Let u_1 be the most important direction of new space, in that case the variance of the data transformed is S^0, where S means the covariance matrix of former data in origin label space. The equation to derive S^0 is shown in (5).

$$\boldsymbol{S'} = \frac{1}{n}\sum_{i=1}^{n}(\boldsymbol{u}_1^T \boldsymbol{\mu}_i - \boldsymbol{u}_1^T \overline{\boldsymbol{\mu}})^2 \tag{5}$$

To maximize the variance of new data set, constraint is set so that Lagrange multiplier can be used to find an unconstrained maximization of the following equation(6).

$$\boldsymbol{u}_1^T \boldsymbol{S u}_1 + \lambda_1(\boldsymbol{1} - \boldsymbol{u}_1^T \boldsymbol{u}_1) \tag{6}$$

Solving it by derivation, the variance can achieve maximum when λ_1 is the largest eigenvalue of covariance matrix *S*, and u_1 must be an eigenvector corresponding to λ_1. Similarly, all the other orthogonal component can be solved in this way. Consequently, to replacing the formal labels by *m* new principal, the optimal linear projection is defined by first *m* eigenvectors of *S*, i.e. $u_1, u_2, \ldots, u_m$, ranked by their corresponding eigenvalues.Regularly, u_m can be selected as principal component when the corresponding $\lambda_m \geq 1$. Other reasonable choice to determine u_m is also acceptable.Consequently, the new group is equation(7).

$$\begin{aligned} &\boldsymbol{G} = \boldsymbol{M} \cdot \boldsymbol{U} \\ &\quad \textit{where} \quad \boldsymbol{U} = (\boldsymbol{u}_1, \boldsymbol{u}_2, \ldots, \boldsymbol{u}_m) \end{aligned} \tag{7}$$

3.2.3. *Rotation matrix*

After Principal Component Analysis, the former *m* labels are transformed into T new groups, which can reflect the situation which *m* labels represent successfully, so much computational quantity is saved due to dimensionality reduction. At the same time, each two new groups are orthogonal, which maintains that the correlation coefficients among all groups are zero. However, the new groups which PCA provides have explicit hierarchy, with a common phenomenon that the captivity of comprehension of first group may be several times greater than the third or fourth one. Briefly, all the new components do not share the same weight, which is not suitable for the following scoring process.

To alleviate the unbalanced situation of new groups, one practical method is rotation because with a rotation matrix, M groups can be transformed into *M* balanced groups, sharing a relatively equal captivity of comprehension. The information M balanced components provide equates that *M* components derived by principal components analysis.

There are many methods to do the rotation, such as R. With the rotation, we can derive a new balanced groups space to replace the group space T merely derived by PCA. As the equation(8), the new transforming matrix is *H*. The new matrix *G* in space T can be obtained by the equation(9).

$$H = U \cdot R \tag{8}$$

$$G_{new} = M \cdot H \tag{9}$$

3.3. *Works on layer 2*

Layer 2 is an important section of the framework as it is both the terminal of training phase and the starting point of test phase. Thus, in this part, we introduce the layer for training phase first, and for test set afterwards.

As for training phase, m labels in original label space have been transformed into t groups in the new-generated group space. For every instance, there are t values corresponding to every group. It is needed to note that the value in the new space is considered to be numerical because the coefficients in principal component matrix are most likely a continuous value. Thus, the former label space with binary value is shrunk into a much lower-dimensional group space containing numerical values. Accordingly, the former problem of large-amount-label classification is transferred into the problem of multiple regression, which is much easier to solve.

To illustrate it more clearly, we assume that we have M labels originally, with each instance having several labels marked positive and others marked negative. Through principal component analysis and rotation matrix, we obtain t groups replacing the former labels, with every group value numerical. Now, we couple the feature space X with the group space T by the unification of instance to construct the training set. Thus we successfully switch the former m-label classification into t independent regression problems.

There are many mature regression algorithms available such as linear regression and SVM. (Reference) Consequently, the training phase is finished with an outcome of t rules, generated by regression algorithms. The set of rules is denoted as H, where H = $\{h_1, h_2, \ldots, h_t\}$. As for test phase, the input of layer 2 is the test set which only have feature space. With the rules generated in training phase, we can obtain the group space by H, where H : X $\rightarrow$ G. Thus, as the test set enter in the layer 2, every instance will have their corresponding t group values.

3.4. *Connection between layer 2 and layer 3*

This part is totally in a phase of performing prediction for test set. As the rule set H derived in layer 2, every instance will have their corresponding T group values. However, the ultimate task of multi-label learning is to predict suitable labels for each instance, rather than doing the group prediction. Thus, the layer 2 is served as a latent layer and what we need to do now is giving the scores back to labels. We have two steps to reach the target.

3.4.1. *Determine the labels belonging to each group*

After principal component analysis with rotation, information reveals by each group can reach a balance and correlation between labels is still in a low rate. Thus, we can treat it approximately that each group reflects a certain amount of labels, and labels under the same group are in a strong connection. Taking the correlation into account, we can alleviate the correlations among labels we finally choose by selecting a single label in every group. Since the groups are relatively independent, labels in different groups will be in a low rate of correlation correspondingly. To allocate labels into correct groups, we need to exploit the property of matrix *U*. For each label, their contribution to a certain group is noted in the matrix. Thus, the corresponding group of the maximum contribution value one label has is the group that label belonged to.

3.4.2. *Using reconstructing matrix to return the score back*

In the phase of testing, we obtain the scores of each group. The scores of each group is restore in the matrix G_{test}. To return the score back to labels, a reverse transform is applied as equation(10). For the matrix M_{test}, it restores the scores of labels one instance has in row.

$$Mtest = Hinv \cdot Gtest \tag{10}$$

3.5. ***Output derived by layer 3***

In real world application, it is natural that labels have their social meaning. For example, when we tackle with the tariff package recommendation, the suggested packet for customers should not only be the suitable one but the profitable one. In this part, a profit model is applied for making a better decision. We determine the profit value of each label in the vector *p*. So we can get the matrix representing the expectation of profits if one choose one label[3].

Consequently, we can firstly select the group with max score as the aiming group. As is illustrated above, each group will possess several labels, so the target labels are those who have maximum score in the target group. If we want to choose *n* recommended labels, we need to find first *n* score in target group. If the number of labels one group has is limited, we need to turn to the group with the next highest score.

4. Experimental Results

In this section, we show the experimental results of our proposed framework on the data set of tariff package purchase situation in comparison with the classic multi-label method, Binary Relevance method.

	June		July		August	
Classic Method (BR)	hit rate	0.6357	hit rate	0.5190	hit rate	0.6142
	cover rate	0.3185	cover rate	0.2613	cover rate	0.3053
3-LAYER Framework	hit rate	0.6473	hit rate	0.5219	hit rate	0.6846
	cover rate	0.3445	cover rate	0.3026	cover rate	0.4978

Fig. 2. The experimental result showing the framework's better performance.

4.1. *Information of data sets*

The dataset we used is information of mobile users, who once has subscribed some tariff package. The dataset has 0.32 million instances and up to 60 features, including telephone expenses in six months, average call duration, monthly data usage, etc. Most importantly, the dataset contains the purchase situation of 20 different tariff packages. Our goal is to verify the advantage of the three-layer framework.

4.2. *Specification of framework*

There are several alternative points in the framework. On layer 1, when we conduct rotation after principal component analysis, several methods can be implemented, such as varimax, quartimax and equamax. Meanwhile, on the layer 2, we successfully convert the multi-label classification into few regression problems. In the experimental section, we use linear regression to derive the score matrix.

4.3. *Outcome of experiment*

We have compare the outcome of the 3-layer framework with that of Binary Relevance. We used three different data sets which are extracted in different months in order to provide a thorough comparison. The outcome is shown in the Figure 2, which indicates that with the three-layer framework, the accuracy of package recommendation can achieve an improvement. It is necessary to mention that the standard of evaluation uses both average hit rate and average covering rate. For example, for a single instance, it may have 5 correct choices in reality but a learner is set to allocate 3 choice at a time, while the result shows that the learner does 2 correct answers and 1 mistake. Thus, the hit rate is $\frac{2}{3}$ and the covering rate is $\frac{2}{5}$. The term 'average' means those two index is supposed to show the performance of multi-label learner through the outcome.

$$hit\ rate = \frac{number\ of\ correct\ labels}{number\ of\ predicted\ labels}$$
$$cover\ rate = \frac{number\ of\ correct\ labels}{number\ of\ observed\ labels}$$

5. Conclusion

In this paper, the problem of multi-label learning is addressed by taking into account the fact that classic method will have large computational complexity and correlation between labels is common in real-world application. Accordingly, a novel three-layer multi-label learning framework is proposed, which works by leveraging principal component analysis to simplify the complex multi-label classification into few regression problems. Applying the orthogonality of PCA, we finally obtain target labels for every instance, which take the correlation between each other into consideration. The simulation outcome shows that our framework can obtain more profits with a low time complexity comparing to classic methods. The new framework can solve the problem of tariff package recommendation in a fine way.

Acknowledgment

This work is supported by Research Innovation Fund for College Students of Beijing University of Posts and Telecommunications (Project No.243).

References

1. M.Zhang and Z.Zhou, *A review on multi-label learning algorithms*,in IEEE Transactions on Knowledge and Data Engineering, VOL.26,NO.8, August 2014.
2. Eerola, T., Lartillot, O., Toiviainen, *Prediction of multidimension alemotional ratings in music from audio using multivariate regression models*, In: Proceedings of the 10th International Conference on Music Information Retrieval, pp. 621626 (2009)
3. Z.Zhang, Regina Wang, *Profit maximization analysis based on data mining and the exponential retention model assumption with respect to customer churn problem*, in ICDM 2015, Atlantic city, NJ. 2015.

4. Clare and R.D King, *Knowledge discovery in multi-label phenotype data*, in Computer Science 2168, L. De Raedt and A. Siebes, Eds.Berlin, Germany: Springer, 2001, pp42-53.
5. Elisseeff and J.Weston, *A kernel method for multi-labelled classication*, in Advances in Neural Information Processing Systems 14, T. G. Dietterich, S. Becker, and Z. Ghahramani, Eds. Cambridge, MA: MIT Press, 2002, pp. 681687.
6. Stanii P, Tomovi S. *Apriori multiple algorithm for mining association rules*, Information Technology And Control, 2015, 37(4).
7. Yu-Kun Li, et.al. *Leveraging Implicit Relative Labeling-Importance Information for Effective Multi-Label Learning*, 2015 IEEE International Conference on Data Mining, Atlantic city, NJ. 2015.
8. H. Kazawa, T. Izumitani, H. Taira, and E. Maeda, *Maximal margin labeling for multitopic text categorization*, in Advances in Neural Information Processing Systems 17, L. K. Saul, Y. Weiss, and L. Bottou, Eds. Cambridge, MA: MIT Press, 2005, pp. 649656.
9. L. Tang, S. Rajan, and V. K. Narayanan, *Large scale multi-label classication via metalabeler*,in Proceedings of the 19-th International Conference on World Wide Web, Madrid, Spain, 2009, pp. 211220.
10. Argyriou A, Evgeniou T, Pontil M. *Convex multi-task feature learning* [J]. Machine Learning, 2008, 73(3): 243-272.

Applying improved split Bregman method to fluorescence microscopic image restoration

Chang-Chun Zhang, Yu Wang* and Hong-Bing Xiao

School of Computer and Information Engineering,
Beijing Technology and Business University,
Beijing, 100048, China
**E-mail: wangyu@btbu.edu.cn*

Fluorescence microscopic image restoration is of great importance, which can be applied to many fields, such as astronomical imaging, electronic microscopy, single particle emission computed tomography (SPECT) and positron emission tomography (PET). Traditional image restoration based on split Bregman (SB) algorithm can preserve sharp edges, and save the image texture. However, serious staircase effect phenomena usually occurs with traditional image restoration. Therefore, an improved image restoration algorithm is proposed in this paper, which considers two aspects. One is that the total variation (TV) regularization is used, which is an effective tool to recover blurred images. The other is that the weight function of the total variation is involved, which can not only suppress the staircase effect, but also preserve the image texture information. The experimental results show superior performance in terms of both objective criteria and subjective human vision via processing simulated and real fluorescence microscopic degraded images, compared with the traditional restoration image methods.

Keywords: Image Restoration; Split Bregman; Weighted; Total Variation.

1. Introduction

Fluorescence microscopy is an emerging and effective microscopy technique, which uses the phenomenon of fluorescence to study properties of the living specimen in biomedical research [1]. However, it is impossible to capture a perfect image because of unavoidable constraints during the imaging: noise and blurring. The image degradation from two factors: First, the physical limitations of microscopic imaging system causes blurring, which is characterized by the point spread function (PSF) [2]. Second, most of the measurements are conducted at low photon flux of conduct that will produce Poisson noise. For

*Corresponding author.

inherent optical limitations, as well as demand for the high-quality image, it is a more than meaningful work for image restoration by digital image processing.

Currently, the split Bregman algorithm [3] is one of the most effective methods to solve norm optimization problems. It was first used to image denoising based on a Bregman distance iterative regularization algorithm proposed by Rudin, Osher, and Fatemi [4]. Subsequently, the image restoration model with split Bregman iterative was proposed by Goldstein [5].

Traditional image restoration based on split Bregman (SB) algorithm can preserve sharp edges, and save the information of image texture. Serious staircase effect phenomena [6-7], however, is generally accompanied. Therefore, an improved image restoration algorithm based on SB algorithm is proposed in this paper, which considers two aspects. One is that the TV regularization is used, which is an effect tool to recover blurred images [8]. The other is that the weight function of the total variation is involved, which can not only suppress the staircase effect but also save the image texture information.

The outline of the rest of the paper is as follows. In section 2, we briefly start with reviewing on basic split Bregman algorithm. A fluorescence microscopic image restoration based on improved split Bregman method in section 3. Then, showing the experimental results of the method proposed by this paper and the other two compared methods via processing simulated and real fluorescence microscopic degraded images in section 4. Finally, a conclusion and future work is drawn in section 5.

2. Basic Split Bregman Algorithm

Supposing $W(u)$ and $f(u)$ are convex function about variable *u,* the unconstrained problems about variable *u* can be described as:

$$\min_{u}\|W(u)\|_1 + f(u), \tag{1}$$

where the symbol $\|\bullet\|_i$ denotes the *i*-norm.

By introducing an auxiliary variable *d,* the formulation (1) was equivalently transformed to the following constrained optimization problem:

$$\min_{u}\|d\|_1 + f(u) \quad s.t. \quad d = W(u) \tag{2}$$

According to the variable splitting theory and SB method to solve the (2), we can obtain iterative formula as follows:

$$u^{k+1} = \min_u \left(\frac{\lambda}{2} \left\| d - W(u) - \gamma^k \right\|_2^2 + f(u) \right), \tag{3}$$

$$d^{k+1} = \min_d \left(\|d\|_1 + \frac{\lambda}{2} \left\| d - W(u) - \gamma^k \right\|_2^2 \right), \tag{4}$$

$$\gamma^{k+1} = \gamma^k + \left(W\left(u^{k+1}\right) - d^{k+1} \right). \tag{5}$$

3. Proposed Method

Image restoration is an inverse problem [9-10]based on some prior knowledge such as blur kernel. Fluorescence microscopic image quality is limited by two factors: the one is a blur modeled by the point spread function (PSF) indicated here by K, and the other is Poisson statistics owing to low photon flow under detection. Mathematically, a blurry image f can be described as [8]:

$$f = K * u + N, \tag{6}$$

where K is a blur kernel whose support is small compared to the image size, u is the latent sharp image, N is additive noise, and the symbol $*$ stand for the convolution operation.

In this article, we proposed an method for solving (6) for the weighted TV regularization. A fluorescence microscopic image restoration objective function can be expressed as:

$$\min_u \sum_i^{n \times n} \alpha_i \|D_i u\|_1 + \frac{\mu}{2} \|Ku - f\|_2^2, \tag{7}$$

where $u \in R^{n \times n}$, and $D_i u \in R^2$ is the forward finite difference at pixel i for u. The fidelity term $\|Ku - f\|_2^2$ measures the difference between the original real image f and the restored image u, and μ is regularization parameter which is used to balance the two terms for minimization. α is the weighted function, which is described as:

$$\alpha_i = \frac{n^2 \chi_i}{\sum_i^{n \times n} \chi_i}, \quad \chi_i = \frac{1}{1+\sigma\|D_i \tilde{u}\|}, \tag{8}$$

where $\tilde{u}$ is an estimation of the original image, and χ is the intermediate parameter and parameter σ which is step length to adjust weight coefficient α_i, and σ is determined based on the situation of experiments. In this paper, we set it as 8 according to the practical experience.

Total variation image deblurring problems can be considered as an unconstrained optimization problems. Therefore, by introducing auxiliary variable y. The formulation(7) can be converted into constrained model about restored image u and auxiliary variable y, which can be expressed as:

$$\min_u \sum_i^{n \times n} \alpha_i \|y\|_1 + \frac{\mu}{2}\|Ku - f\|_2^2 \ s.t.\ y = D_i u. \tag{9}$$

We use the variable splitting theory and SB method to solve the (9), the algorithm flow of the proposed improve the SB method are depicted in Table 1.

Table 1. The flow of the improved image restoration algorithm.

1: Set $k = 0$, choose λ, μ, and σ.

2: Initialize u^0, y^0, and γ^0.

3:**while** stopping criterion is not satisfied, **do**

4: $u^{k+1} = \arg\min_u \left(\frac{\mu}{2}\|Ku - f\|_2^2 + \frac{\lambda}{2}\left\|y^k - D_i u^k - \gamma^k\right\|_2^2 \right)$;

5: $y^{k+1} = \arg\min_y \left(\alpha_i \left\|y^k\right\|_1 + \frac{\lambda}{2}\left\|y^k - D_i u^{k+1} - \gamma^k\right\|_2^2 \right)$;

6: $\gamma^{k+1} = \gamma^k + \left(D_i u^{k+1} - y^{k+1}\right)$;

7: **end while**

8: **return** u^{k+1}.

4. Experimental Results

In this section, we describe some numerical experiments to illustrate the potential advantages of the proposed method discussed in this paper. We show the results of our method and other comparative methods including FTVd-FM [8] and traditional split Bregman [3] methods. In our experiments, by means of convolving a Gaussian PSF(size 5×5, standard deviation $\sigma = 1.5$) add the Poisson noise. We choose the signal-to-noise Ratio (*SNR*) [11]method as a standard, the restoration results of the synthetic image shown in Figure 1.

In order to evaluate the practical relevance of our method, we also show the reconstruction results on a real fluorescence microscopic image in Figure. 2.In real image stack example, we verify that the proposed method avoid oil-painting artifacts and preserve line-like.

Figure. 3(a) is the local close-up of Figure. 2(a) whose position is shown using a black box, and Figure. 3(b)–(d) is the corresponding close-ups of Figure. 2(b)–(d) in the same region.

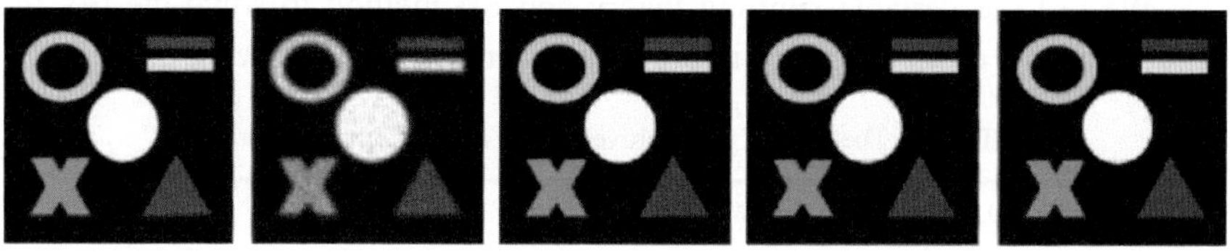

(a) (b) (c) (d) (e)

Fig. 1. Restoration results on the synthetic image of size 130×130 pixels.(a) original synthetic image, and (b) corresponding degraded image , (c) FTVd-FM method restoration (*SNR*=45.50dB), (d) traditional split Bregman method restoration(*SNR*=49.41dB), and (e) proposed method restoration (*SNR*=49.68dB).

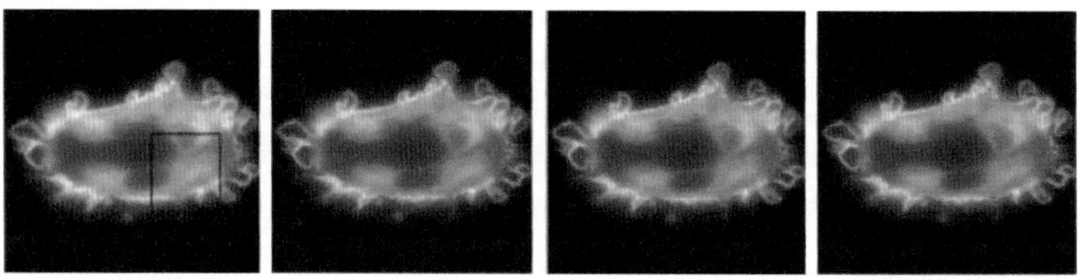

(a) (b) (c) (d)

Fig. 2. Restoration results on a real fluorescence microscopic image of size 512×512 pixels from the Internet in [12]. (a) original image (b) FTVd-FM method restoration , (c) traditional split Bregman method restoration, and (d) proposed method restoration.

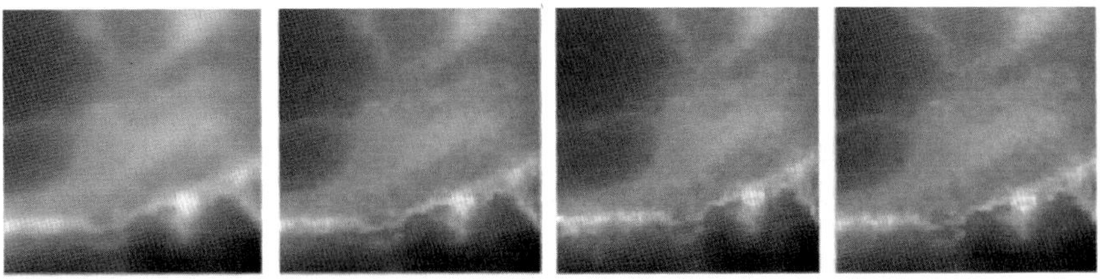

(a) (b) (c) (d)

Fig. 3. (a) A close-up of Fig. 2(a), the corresponding close-up of (b) Fig. 2(b) , (c) Fig. 2(c), and (d) Fig. 2(d).

It can be from Fig. 1, Figure. 2, and Figure. 3 that the restoration results of proposed method is superior to the FTVd-FM and traditional split Bregman method to a certain extent.

5. Conclusion

In this paper, we proposed an image restoration algorithm based on traditional split Bregman for fluorescence microscopic image restoration, which considers two aspects. On one hand, the TV regularization is used to recover blurred images. On the other hand, the weight function of the total variation is involved, which can not only suppress the staircase effect, but also preserve the image texture information. The experimental results show the feasibility and effectiveness by processing simulated and real fluorescence microscopic degraded images, compared with the traditional restoration image methods.

Acknowledgments

This work is supported by Beijing Talents Fund (No. 2014000026833ZK14), the Importation and Development of High-Caliber Talents Project of Beijing Municipal Institutions (No. CIT&TCD201504010), and excellent talents training foundation of Beijing (No. 2012D005003000010).

References

1. S. Lefkimmiatis, A. Bourquard, and M. Unser, Hessian-based regularization for 3D microscopy image restoration, in *Proc. IEEE Int. Symposium on Biomedical Imaging*, (Barcelona, Spain, 2012).

2. Y. Wang, Q. H. Dai, P. Y. Guo, Z.W. Liu, Blind deconvolution subject tosparse representation for fluorescence microscopy, *Optics Communications*, vol. 286(2013), pp.60-68.

3. J. Wang, K. Lv, N. He, and Q. Wang, Total variant image deblurring based on split Bregman method, 40(8), (2012), pp. 1503-1508.

4. L. Rudin, S. Osher, Nonliner total variation based noise removel algorithm, *Physica D*, 60(8), (2012), pp. 1503-1508.

5. T. Goldstein, S. Osher, The split Bregman algorithm for L1 regularized problem, SIAM Journal Imaging Sciences, 2(2), pp. 323-343, (2009).

6. S. Brette, J. Ider, Optimized single site update algorithms for image deblurring, in *Proc. IEEE Int. Conference on Image Processing*, (Lausanne, Switzerland, pp. 65-68, 1996).

7. W.T. Yin, J.F. Yang, Alternating direction algorithms for total variation deconvolution in image reconstruction, http://www.optimization-online.org/DR-HTML /2009/11/2463.html.

8. J. F. Yang, W.T. Yin, Y. Zhang, and Y.L. Wang, A fast algorithm for edge-preserving variational multichannel image restoration, SIAM Journal Imaging Sciences, 2(2), pp. 569-592, (2009).

9. Y. F. Gu, B. Yan, L. Li, F. Wei, Y. Han, J. Chen, Image reconstruction based on total variation minimization and alternating direction method for Compton scatter tomography, *Acta Physica Sinica*, 63(1), (2014).

10. A.N. Tikhonov, V.Y. Arsenin, Solutions of ill-posed problems, (Winston Washington, DC:VH. Winston&Sons, 1997).

11. Y.L. Wang, J.F. Yang, W.T. Yin, Y. Zhang, A new alternating minimization algorithm for total variation image reconstruction, SIAM Journal Imaging Sciences, vol. 1(2008), pp. 948-951.

12. http://www.mediacy.com/.

A new image enhancement method based on high-dimensional information geometry

Jun-An Wu* and Wen-Ming Cao
College of Information Engineering, Shenzhen University,
Shenzhen, 518060, China
**E-mail: wujunan@email.szu.edu.cn*

A blurring image enhancement method based on high-dimensional space information geometry is proposed in this paper. In this novel image enhancement method, every image is considered as a point in the high-dimensional space. With the "blurring-blurring-deblurring" thought, this method can get ideal enhanced images by comparing the corresponding locations of various images with simple geometry analysis. The algorithm is implemented using the MATLAB tool in this paper, and the experiment results show that the effect of this algorithm is good. Also, compared to traditional image enhancement methods, this method is simple and efficient.

Keywords: Image Enhancement; High-Dimensional Information Geometry; Blurring-Blurring-Deblurring.

1. Introduction

There are many factors which affects the clarity of images, such as uneven outdoor light can cause the gray level over concentration. Besides, noise pollution will be produced in the image transmission which inevitably decreased the quality of the image. Take the above into consideration, we need to improve the image quality before we analyze the images. The commonly used methods of image enhancement can be divided into two categories: space domain enhancement method and frequency enhancement method [1]. No matter space domain enhancement method or frequency enhancement method, both of them are around restraining the image noise and strengthening the image's edges or local details to improve the image contrast.

A blurring image enhancement method based on high-dimensional space information geometry is proposed in this paper. In this novel image enhancement method, every image is considered as a point in the

* Corresponding author

high-dimensional space [2,3]. With the "blurring-blurring-deblurring" thought, this method can get ideal enhanced images by comparing the corresponding locations of various images with simple geometry analysis.

2. Blurring Image Enhancement Based on High-dimensional Space Information Geometry

Shou-jue wang, academician of Chinese academy of science, makes a lot of supplements to multi-dimensional space geometry theory based on the progress of the predecessors, and widely applies it in information science which has achieved great results [4,5,6].

2.1. *Basic contents of high-dimensional space information geometry*

Summarily, three main features of high-dimensional geometry methods are introduced as following:

(1)All the theory methods and computing procedures are derived from geometry geometric concepts in high-dimensional space.

(2)During the computing procedures, all subspaces should be described by points in the high-dimensional space.

(3)Complicated computational problems in high-dimensional space are solved with combinations or iterations of basic operation in 2D planes.

Academician Wang makes full use of geometry's characteristics of intuitive and practical to found the point distribution theory in high-dimensional space and uses descriptive geometry as a tool in high dimensional space to map the multi-dimensional information to the points in high-dimensional space. Then, the relations between multi-dimensional information can be analyzed by the points distribution or the properties of geometry in high-dimensional space.

2.2. *The principle of homology-continuity (PHC)*

In this paper, every image is considered as a point in high-dimensional space. As a common sense of human being, that the points belong to the same class should be located in feature space with continuity. These points are homologous. (The homologous things are defined as: these things belong to one same class but not completely identical. Namely that the difference between them is gradual or non-quantifiable, at least one gradual process must be existed between them. So the objects in this gradual changing process belong to the homologous things.) The high-dimensional space geometry theory is based on the principle of homology-continuity (PHC). Therefore the problem of image enhancement becomes the process of finding another point (the deblurring image) which is the homologous point about the observed image (the blurring image).

3. The Blurring Image Enhancement Algorithm

3.1. *Image enhancement principle*

For the sake of easy understanding, we use images in 2D planes to explain the image enhancement principle. Suppose the original image we need to enhance is A, it changes to be B after first blurring. And vector $\vec{A}$ and vector $\vec{B}$ is the vector expression of image A and image B respectively. Compared with point B, point A is an enhanced point. So, if we connect point B and point A, we could make a conclusion that the direction of vector $\overrightarrow{AB}$ is the general direction in which we can make image enhanced. We can get image C after making the second blurring to B. Similar to the above analysis, we could get point C which is the mapping to image C in the high-dimensional space, and vector $\vec{C}$ is the vector expression of image C and the direction of vector $\overrightarrow{BC}$ is also the direction in which image can be enhanced.

Now, we have got three vector $\vec{A}$, $\vec{B}$, $\vec{C}$ in high-dimensional geometry which meets the minimum condition for operation, that is to connect point C-B-A. The homologous images have the properties of continuity. Hence, in fact, there are many continuous points between A~C which could form a hyper-curve. We need to do next is to find the point in the hyper-curve. Theoretically, it is free to choose a vector for calculation here, and we choose a group which is relatively simple. We may as well set the enhanced image mapping to the vector $H = \alpha A - \beta(B - A)\gamma(C - A)$ in high-dimensional geometry. We could represent it in the following figure.

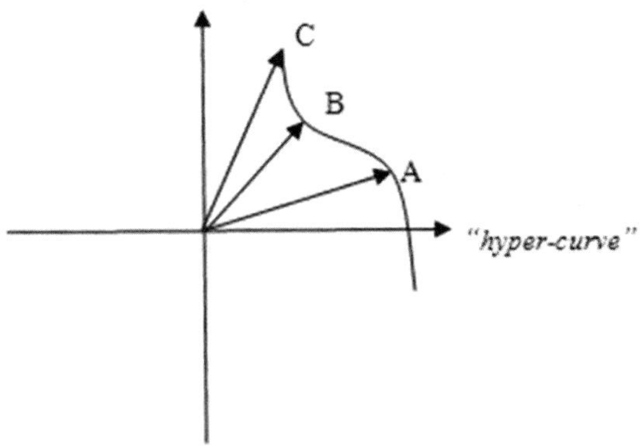

Fig. 1 .Point A corresponding to the original image, point B corresponding to the once blurring image, point C corresponding to the twice blurring image, the hyper-curve based on them.

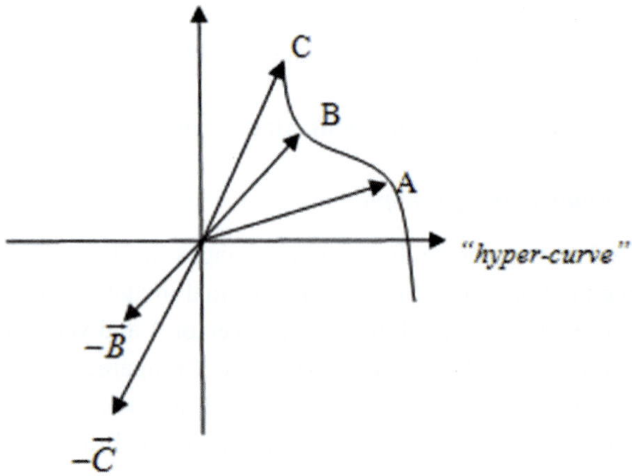

Fig. 2. Making $-\vec{B}$ and $-\vec{C}$ to prepare for the next operation.

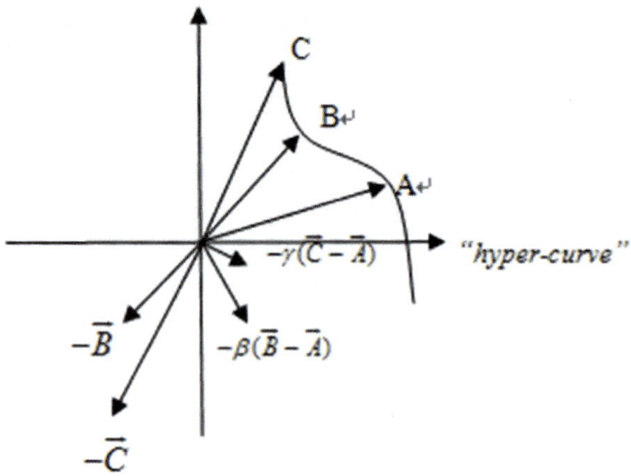

Fig. 3. Using partial parameters to respectively get $-\beta(\vec{B}-\vec{A})$ and $-\gamma(\vec{C}-\vec{A})$.

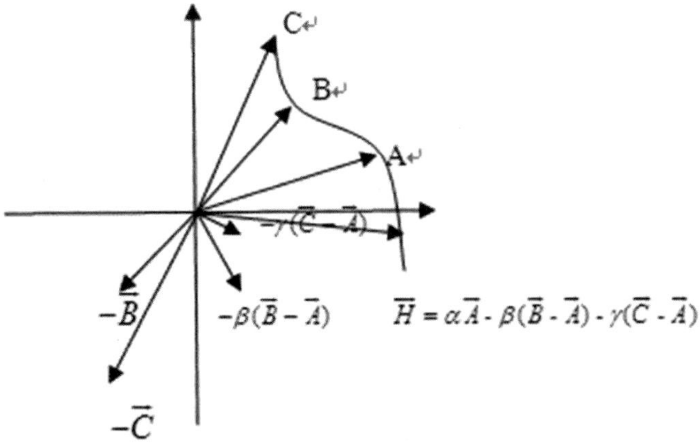

Fig. 4. Adjusting parameters to enable the end point of vector $\vec{H}$ which maps to the enhanced image fall on the hyper-curve.

Next, we are targeted to move the enhanced image to that direction. This gradually approaching method reflects the superiority of high-dimensional space geometry as we can easily work out by simple geometric operation.

3.2. *Algorithm description*

i. Read the original image A, RGB value (x,y,z) of the pixel. Calculate the mode a_i of the pixel according to $A = \sqrt{x^2 + y^2 + z^2}$. Calculate the whole pixels a_1 ... a_n of this image by loop command respectively. As a consequence, we will get the coordinates of the image A=($a_1, a_2, \ldots a_n$).

ii. Similarly, calculate the coordinates of the blurred image B and re-blurred image C. Map them to three points in high-dimensional space.

iii. Adjust H to the specific location as the above figure according to the formula H=A+p(A-B)+q(A-C), choose p q flexibly according with particular situations. The principle is to make H in the first quadrant and close to positive half axis.

iv. Observe the quality of the output image. Go back to step 2 if not satisfied. Set A=H at this moment, and do the loop operation until the output result is satisfied.

4. Experiment Results

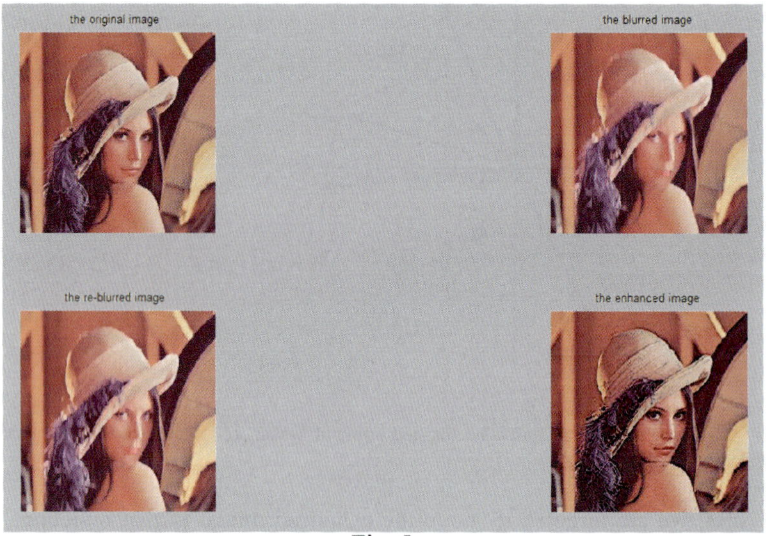

Fig. 5

To verify the validity and efficiency of this algorithm, we randomly choose an image from the MATLAB library, and get the experiment image through corrosion. The output effects according to the algorithm code are successively: the original image, the blurred image, the re-blurred image, the enhanced image. We can get more ideal effect through multiple calculations. From figure 5, we can find that the visual effect of the image is great improved by the enhancement.

5. Conclusion

This paper applies the high-dimensional space geometry theory to the image enhancement field, making the image enhancement algorithm based on high-dimensional space geometry theory verbalized. This method maps images to points in the high-dimensional space, and gets ideal image enhancement effect by study relationships between the direction of vectors and the location of points. Compared to traditional image enhancement methods, this method is obviously simple and efficient, and should get more extensive research and application.

Acknowledgments

The work is supported by National Natural Science Fund (Grant No. 61375055).

References

1. Hamid R. Tizhoosh. *Fuzzy Image Processing* [M]. Springer-Verlag.1997.
2. WANG Shoujue, "Biomimetic Pattern Recognition", Neural Networks Society (INNS, ENNS, JNNS) Newsletter, Vol.1, No.1, pp.3-5, Mar.2003.
3. Wang Shoujue and Zhao Xingtao, "Biomimetic Pattern Recognition Theory and Its Applications", Chinese Journal of Electronics, Vol.13, No.3, July 2004.
4. Wang Shoujue, XUJian, Wang Xianbao, Qin Hong, "Multi-camera Human-face Personal Identification System Based on the Biomimetic Pattern Recognition" (in Chinese), Acta Electronica Sinica, Vol.3 1, No.1, pp: 1-4, Jan 2003.
5. Wang Shoujue, Qu Yanfeng, Li Weijun, Qin Hong, "Face Recognition: Biomimetic Pattern Recognition vs. Traditional Pattern Recognition", (in Chinese), Acta Electronica Sinica, Vol.32, No.7, pp 1057-1061, July.2004.
6. Wenming Cao, Xiaoxia Pan, Shoujue Wang, "Continuous Speech Research Based on Two-Weight Neural Network", Advances in Neural Networks - ISNN 2005: Second International Symposium on Neural Networks, LNCS 3497, pp. 345-350, 2005.

Interactive rendering of approximate soft shadows using ray tracing with visibility filtering

Chun-Yi Chen*, Hua-Min Yang and Hua Li

School of Computer Science and Technology,
Changchun University of Science and Technology,
Changchun, 130022, China
**E-mail: chenchunyi@hotmail.com*
www.cust.edu.cn

A visibility-filtering-based algorithm for rendering approximate soft shadows is proposed, which is well compatible with the framework of ray tracing. The luminaire in a 3D scene is treated as an ideal point light source and a binary light-visibility map for the visible region of the 3D scene is subsequently created by casting shadow rays. The fractional light visibility for visible points in penumbrae is computed by smoothing out the abrupt changes in the binary light-visibility map with a box filter. The light leaving visible points into the eye resulting from direct illumination is obtained by multiplying the one without considering possible occluders by the corresponding light-visibility value achieved by filtering the binary light-visibility map. Our algorithm can render visually pleasing soft shadows with only a small amount of overhead.

Keywords: Visibility; Soft Shadow; Ray Tracing; Box Filter; Penumbra.

1. Introduction

Shadows are important to an image of a 3D scene generated by using computer algorithms, because they provide some visual cues that can help us to understand the spatial relationships between various geometric objects in the scene. When point light sources exist in a 3D scene, the shadows therein may feature sharp edges, which are generally called hard shadows. However, strictly speaking, ideal point light sources are very rare in the "real world". The luminaires that exist in our surroundings are commonly nonpoint light sources with a finite size, e.g., a lamp may be regarded as a nonpoint light source. Unlike point light sources, nonpoint light sources can generally cause soft shadows with blunt edges in a 3D scene. The rendering of soft shadows has attracted much attention over the years in the computer graphics community.

---

* Corresponding author.

In the literature [1-3], numerous variants of the shadow mapping or shadow volume algorithm have been proposed for rendering soft shadows. These algorithms can be readily integrated into existing rendering systems based on rasterization methods. Of course, generally speaking, they can only produce approximate soft shadows. However, they are suitable for interactive applications. On the other hand, ray tracing has been widely employed in realistic rendering that requires the computation of global illumination. It is traditionally thought of as a time-consuming algorithm and hence is generally used in off-line applications, for instance, movie rendering. Nevertheless, parallel ray tracing based on graphics processor units (GPUs) has recently been under intensive study, and many advances in rendering speed have been made. Horn *et al.* [4] developed an interactive ray tracing system based on k-D tree which takes advantage of GPU's parallelism; however, their system can only generate hard shadows. Karras and Aila [5] designed a parallel algorithm running on GPUs for creating bounding volume hierarchies (BVHs) used in ray tracing. Doyle *et al.* [6] presented a dedicated microarchitecture for building binned BVHs based on surface area heuristic (SAH). Carcía *et al.* [7] devised a parallel method for constructing linear BVHs used to speed up ray tracing. Mehta *et al.* [8] employed ray-tracing-based occlusion calculations and adaptive image-space filtering to produce soft shadows from planar area light sources. Regarding generating soft shadows using ray tracing, despite the advances mentioned above, achieving interactive rendering speed for a 3D scene containing a large amount of geometric primitives is still somewhat difficult when common personal computers are used. As a result, new methods that can accelerate the ray-tracing-based rendering of soft shadows are useful for practical applications related to realistic rendering.

In this paper, we propose a method of rendering approximate soft shadows by using ray tracing. Our method does not accurately calculate the fractional visibility between a point in penumbrae and a luminaire. Our method can be readily integrated into existing ray tracing systems and can render approximate soft shadows with a small amount of time and space overhead.

2. Algorithm Overview

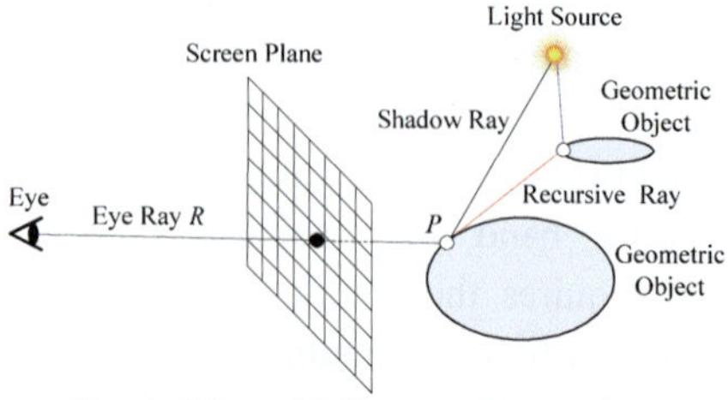

Fig. 1. Schematic diagram of ray tracing.

The basic geometry with respect to ray tracing is illustrated by Figure 1. The key to our algorithm is to obtain an approximate fractional visibility between a point P in the 3D scene and the light source. Algorithm 1 shows an outline for rendering soft shadows based on ray tracing with visibility filtering, where it is assumed that there are M rows and N columns of pixels in the screen plane.

Algorithm 1. Rendering approximate soft shadows based on ray tracing

1. Create three 2D buffers $Vis[M \times N]$, $DirI[M \times N]$ and $IndI[M \times N]$. Each element of the 2D buffer Vis is initialized to 1. Each element of the 2D buffers $DirI$ and $IndI$ is initialized to 0.
2. Cast an eye ray R from the eye through each screen pixel into a 3D scene and determine whether the eye ray R intersects a geometric object in the 3D scene.

If intersections exist **then**

(1) Find the intersection point P that is nearest to the eye.

(2) Determine the binary visibility V between point P and the point light source by casting a shadow ray.

(3) Let $Vis[R] \leftarrow V$;

(4) Calculate the light L_d leaving point P to the eye as a result of the direct illumination without considering possible occluders between point P and the point light source. Let $DirI[R] \leftarrow L_d$.

(5) Based on recursive ray tracing, calculate the light L_i leaving point P to the eye as a result of the indirect illumination. Let $IndI[R] \leftarrow L_i$.

else

$DirI[R] \leftarrow$ background color.

end if

3. Smooth out the abrupt changes in the binary light-visibility map Vis by using a box filter.
4. For each screen pixel associated with eye ray R, compute the total light arriving at it by employing the formula: $T_R = Vis[R] \times DirI[R] + IndI[R]$.

In the above algorithm, for succinctness, the eye ray R is used as a buffer index, because each eye ray uniquely corresponds to an element in a buffer.

3. Fractional Light Visibility based on Smoothing Filtering

In algorithm 1, we have employed a box filter [9] to smooth out the abrupt changes in the binary light-visibility map. The basic principle of visibility filtering is depicted by Figure 2. The value of the elements in a binary light-visibility map is either 0 or 1. An element in the binary light-visibility map with a value of zero means there is an occluder between the corresponding visible point P and the point light source. Abrupt changes in a binary light-visibility map occur in the region in which the binary visibility varies abruptly from 0 to 1. Intuitively, penumbrae casted by a light source in a 3D scene are likely to take place in this region. The fundamental idea behind our algorithm is that binary light visibilities in the aforementioned region can be artificially transformed into fractional ones by making use of smoothing filtering operation. With fractional light visibilities, approximate penumbrae can be readily rendered.

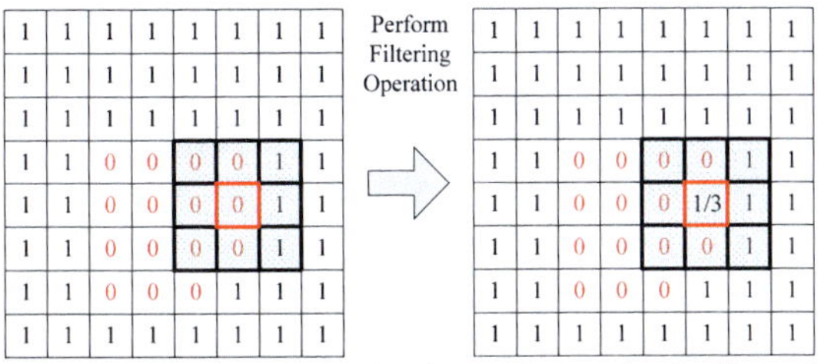

Fig. 2. Schematic diagram of smoothing out the abrupt changes in the binary light-visibility map with a box filter. Here, a 3×3 box filter is displayed as an example.

It needs to be pointed out that although a 3×3 box filter is shown as an example in Figure. 2, other box filters with a wider size should be used if one wants to generate smooth enough shadow edges. Moreover, the smoothing filtering described above is indeed a spatial convolution operation. To better utilize massively parallel hardware available in today's personal computers, we have designed a parallel algorithm running on GPUs for calculating spatial convolution.

4. Implementation and Results

Our algorithm presented above has been implemented in a ray tracer running on GPUs developed based on NVidia's OptiX SDK 3.7.0. The computer system on which the ray tracer runs includes an Intel(R) Xeon(R) CPU E3-1225 v3 @ 3.20GHz together with 8 GB RAM and an NVidia Quadro K2000 graphics card.

All the images shown later are generated at a resolution of 1024(width) × 768(height). Four different 3D scenes have been considered; the first one including a bunny has 69452 faces; the second one including a cow has 5805 faces; the third one including a dragon has 869929 faces; the fourth one including a fairy has 173923 faces. The material associated with each face contains both specular and diffuse components, resulting in that mirrored objects appearing on some faces can be seen clearly in the rendered images. In addition, BVHs have been employed as the acceleration data structures for ray tracing, and a 21×21 box filter has been used to smooth out the binary light-visibility map.

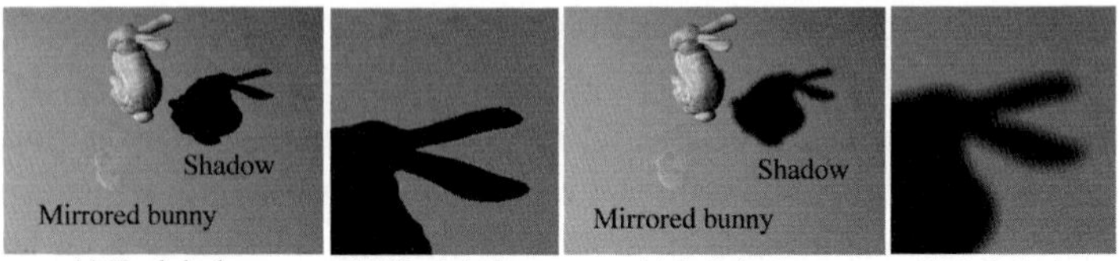

(a) Hard shadow (b) Hard shadow (c) Soft shadow (d) Soft shadow

Fig. 3. Rendered images with hard or approximate soft shadows.

(a) Binary map (b) Fractional map

Fig. 4. Binary and fractional light-visibility maps.

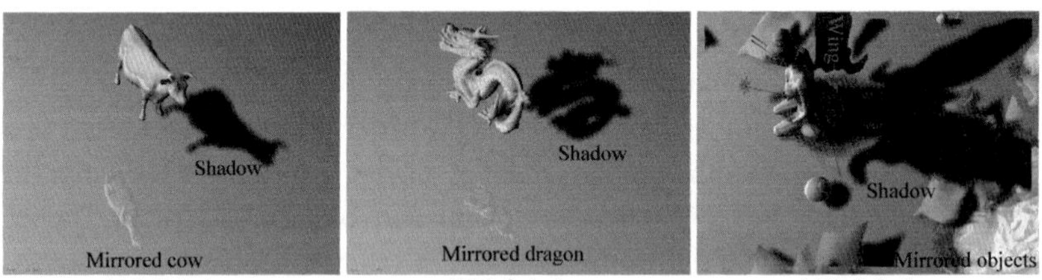

(a) Soft shadow (b) Soft shadow (c) Soft shadow

Fig. 5. Rendered images with approximate soft shadows.

Figure 3 shows the rendered images with hard or approximate soft shadows of the 3D scene including a bunny; the corresponding binary and fractional light-visibility maps are given by Figure 4. By comparing Figure 3(b) and 3(d), it is found that the use of our algorithm does significantly improve the visual reality of the rendered shadows. Note that, the mirrored bunny is rendered correctly in the results shown by Figure 3. Hence, our algorithm is well compatible with the computation of global illumination based on ray tracing; the basic idea behind our algorithm can be easily understood by making a comparison of Figure 4(a) with 4(b); i.e., we use a box filter to transform the binary light-visibility map into a fractional one. The rendered images of other three 3D scenes are illustrated by Figure 5. Timings show that it takes about 89.1 ms, 86.4 ms, 277.8 ms and 263.2 ms to render the images of the 3D scenes containing bunny, cow, dragon and fairy, respectively. It is apparent that the rendering of 3D scenes with more faces is more time-consuming.

5. Conclusions

In this paper, a novel algorithm for rendering approximate soft shadows by using ray tracing running on GPUs has been presented. It first treats the luminaire as an ideal point light source and creates a binary light-visibility map for the visible region of the 3D scene, and then finds an approximate fractional light visibility for visible points in penumbrae by smoothing out the binary light-visibility map with a box filter. Our algorithm can produce visually pleasing soft shadows and does not incur a significant amount of overhead.

Acknowledgments

This work was supported by Jilin Provincial Science & Technology Program of China (20140204009GX) and Changchun municipal Science & Technology Program of China (14KG008).

References

1. B. Yang, Z. Dong, J. Feng, H. Seidel, and J. Kautz, Variance Soft Shadow Mapping, *Comput. Graph. Forum* **29**, 2127 (2010).
2. U. Assarsson and T. Akenine-Möller, A Geometry-based Soft Shadow Volume Algorithm using Graphics Hardware, *ACM Trans. Graph.* **22**, 511 (2003).
3. J. -M. Hasenfratz, M. Lapierre, N. Holzschuch, and F. Sillion, A Survey of Real-time Soft Shadows Algorithms. *Comput. Graph. Forum* **22**, 753 (2004).
4. D. R. Horn, J. Sugerman, M. Houston, and P. Hanrahan, Interactive k-D Tree GPU Raytracing, *in Proc. of Symposium on Interactive 3D Graphics and Games (I3D'07)*, (Seattle, USA, 2007).

5. T. Karras and T. Aila, Fast parallel construction of high-quality bounding volume hierarchies, *in Proc. of High-Performance Graphics Conference (HPG'13)*, (Anaheim, USA, 2013).
6. M. J. Doyle, C. Fowler, and M. Manzke, A Hardware Unit for Fast SAH-Optimised BVH Construction, *ACM Trans. Graph.* **32**, 139 (2013).
7. A Carcía, S. Murguia, U. Olivares, and F. F. Ramos, Fast Parallel Construction of Stack-less Complete LBVH Trees with Efficient Bit-trail Traversal for Ray Tracing, *in Proc. of ACM SIGGRAPH International Conference on Virtual-Reality Continuum and Its Applications in Industry (VRCAI'14)*, (Shenzhen, China, 2014).
8. S. U. Mehta, B. Wang, and R. Ramamoorthi, Axis-Aligned Filtering for Interactive Sampled Soft Shadows, *ACM Trans. Graph.* **31**, 163 (2012).
9. J. F. Hughes, A. V. Dam, M. McGuire, *et al.*, *Computer Graphics: Principles and Practice, 3rd edition*. (Addison-Wesley, 2014).

Design of global cooperative signaling system for train control

Wan-Ting Zhao*, Xiang-Xian Chen†, Zhu-Jun Ling[1] and Zhen-Peng Nie[1]
College of Biomedical Engineering & Instrumental Science, Zhejiang University,
[1]Zhejiang United Science & Technology Co., Ltd.
*Email; *wtzhao@zju.edu.cn, †xxchen99@zju.edu.cn*

Wen-Bin Tang[1], Guo-Dong Teng[2] and Lan-Qing Hu[3]
[1]College of Institute of Information Technology, China Jiliang University
[2]Department of Computer Science, Hangzhou Normal University
[3]College of Computer Science and Technology, Zhejiang University,
Hangzhou, 310000, China

The current Communication-based Train Operation Control (CBTC) system adopts the adhesion, closed system architecture and autonomous safety model. It only makes the brakes respond to the train operation speed, train position, the input of the front-back train relationship and other input states. With the rapid development of urban rail transit, train operations are becoming more and more complex, and operators need to respond to more external events. However, since the current data acquisition equipment of CBTC system only has basic functions but has neither data acquisition nor data interface for other emergencies, the system has no response stored or is unable to respond in time for exogenous emergencies. Therefore, based on the current CBTC system, this article presents a new generation of global cooperative signaling for train control. Through increasing acquisition of external environment data acquisition equipment and interface, the safety cooperative function of wayside subsystem, carborne subsystem and other subsystem can be improved. Also, the new generation of global cooperative signaling for train control is able to have a timely and efficient response to fail-safe information. Improving the safety and reliability of train operation system brings about the fail-safe of the whole rail transit system.

Keywords: Global; Cooperative; Train Control; Operation; Fail-Safe.

1. Introduction

The basic task of urban rail transit is transporting passengers in a safety, punctuality, high efficiency and high density way. In recent years, with the rapid development of Chinese economy and the acceleration of urbanization, urban rail transit has entered a period of rapid development[1]. With the leap development of computer technology, communications technology and control

*Corresponding author.

technology, comprehensive the utilization of 3C Technology, CBTC system has become the development direction of the word advanced train control operation system[2]. The CBTC system has advanced train positioning technology, safety processor technology and wireless communication technology, getting rid of the restraint of the train occupancy and the information transmission by the ground rail circuit equipment, achieving move block, and breaks the limit of fixed block to rail interval, greatly shorten the train railing interval, transit density has also been improved, thus improving the efficiency of transportation[3].

Traditional CBTC system uses the safety autonomous technology framework based on the principle of fail-safe. In order to ensure the fail-safe principle for the system, current CBTC system usually uses the traditional adhesion and closed system architecture and safety autonomous model. Except to the responder equipment set up for collecting data of location, speed, radar device, speed sensor, and beacon reader, it lacks the detecting equipment such as smoke sensors, humidity sensors for supervise flood, infrared detection device used for detecting the foreign stuff dropped off the rail[4]. All of these deficiencies lead to a delayed response and safety alternative solution of the emergence of environmental anomalies and other unexpected events in the train operation process, such as explosion, poison gas, fire, and water, eventually lead to serious consequences. For example, the Daegu subway arson in South Korea, February 18, 2003, due to improper operation of the staff and the imperfect emergency system, eventually resulted in a disastrous consequence with 126 deaths, 146 injured, 318 missing, and up to 46 billion losses of property[5].

The new generation system of the global cooperative signaling train control system has make the breakthrough of the current control system's autonomous control model. Combining the passive reaction of the traditional control system, the new system established a new type of active prevention, active warning, and global safety cooperative control model. Based on traditional CBTC system, the new system increases the equipment which can collect and transmit the real-time date and detect state for vehicle's external environment and inside state. Through real-time data acquisition and analysis of train structure, traction power supply system, rail line, train staff's status, line environment etc. To ensure the train can monitor the state of the train body and rail line in real-time during the train operation process, if there is a system fault or abnormal environment or other emergency situations, the train can make timely appropriate emergency treatment, avoiding serious accidents in maximum, improving the safety and reliability of train operation, enhancing the operating efficiency. The global cooperative signaling train control system combines many kinds of faults and accidents supervision measures and the treatment of failed integrated safety, to achieve the principle of global cooperative rail transit system—fail-safe.

2. System Architecture

Traditional CBTC system usually adopt the closed system architecture and safety autonomous control mode, which only equipped with the transponder device for train position acquisition, speed measurement, radar speed measuring device, speed sensor, beacon reader and other basic infrastructure. On the basis of the current CBTC system, the new system adds on the interface of real-time data acquisition and transmission to the external environment, such as temperature sensor for fire detection, humidity sensor for flood detection, smoke detector for smoke detection, infrared detector for rail foreign stuff dropped off detection, etc. Through the detecting equipment and data interface, it is possible to realize a real-time data acquisition and analysis of rail state of natural environmental disaster, fire or other safety condition, train internal environment and train body structure, train traction power supply and braking system fault and staff condition. The new system establishes an active prevention, and warning system and a timely response system to realize the global cooperative safe and autonomous train control system.

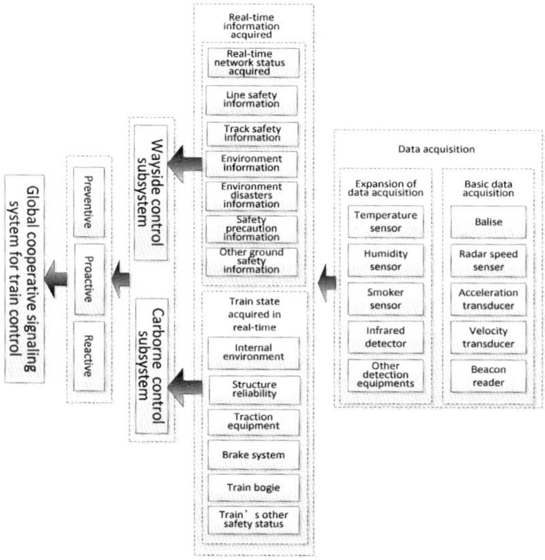

Fig. 1. Global cooperative signaling train control system frame.

2.1. *Wayside control subsystem*

Wayside control subsystem includes Zone Controller (ZC) and Computer Based Interlocking (CBI). It exchanges database through Data Communication System(DCS) to other CBTC subsystem such as wayside-carborne subsystem and automatic train Supervision (ATS) system. To complete the interlock logic control between route and wayside signaling equipment, receives the location and operation information from each train in the control area, locates all trains in the area according to the position and condition of the known obstacle along the line, and relay interfaces the implementation of wayside signal equipment safety detection and control. Then based on the position and status of all known obstacles on the line it determines the operating rights of all the trains in the region, and through the relay interface to achieve the safety inspection and control of the rail signal equipment. In view of the shortcomings of the current rail transit control subsystem, the external environment and the external system cannot be processing timely or response appropriately, the new generation of global cooperative signaling wayside control subsystem expanded line network real-time data acquisition system interface, through the acquisition of rail route safety information, natural environmental information, environmental disaster information and other ground safety information data, to achieve the rail transit system fail-safe principle. As shown in Figure 2, the wayside control subsystem consists of interlocking application logic, line data management base, logic controller, zone controller, maintenance support workstation, local workstation and safety computer platform. The relay interface is a bridge that connects the safety computer platform and the external equipment, and the safety computer platform collects the signal through the interface of state of the external device and the output drive. Interlocking application logic provides interlock file. Logic controller is interlock logic execution software, which reads the logical file, carries out the logic operation. Based on controller area wayside equipment status and data lines on the train, it commands and dispatches. And it runs on two by two second safety computer platform. According to the CBTC architecture, the control subsystem of the rail is transmitted by the relay interface and the rail signal equipment; Through Ethernet and shielding door system, carborne control system, ATS system and adjacent rail transmission control subsystem are connected to realize the signal transmission, supervision and control of equipment status.

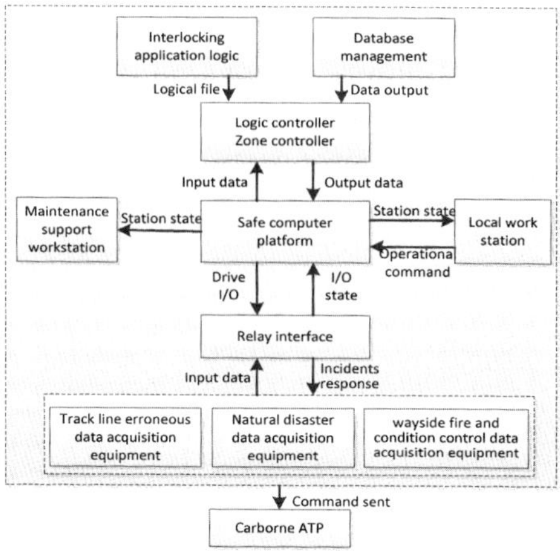

Fig. 2. Global cooperative signaling of wayside control system frame.

2.2. *Carborne control subsystem*

Current CBTC carborne subsystem usually adopts the closed system architecture and safety autonomous control mode. The external sensing equipment is limited to speed sensor, acceleration sensor, radar speed measuring device and beacon reader. Due to the lack of other data acquisition equipment and acquisition interface, it can only determine the speed and the initial position of the train, relocating the train and realize some other basic functions. A new generation of the global cooperative signaling train control system increases train supervision structure and its internal environment data acquisition device, adds on an acquisition interface to achieve a more accurate train positioning safety and safe speed measurement. A more accurate speed range curve is used to calculate the driving mode, which can bring out the automatic unmanned reentry function of the train, and guarantee the necessary braking sequence, completing the database's load, update and check, improving the exchange of data and information with other subsystems, and protects the principle of rail transit system—fail-safe.

The new generation of global cooperative signaling train control system is divided into a Carborne Controller (CC) and peripheral equipment. CC includes the interface of the onboard equipment and the train subsystems, and is connected with the interface of ZC through the data communication system. Train controller is responsible for train positioning, the implementation of the speed and mobile license, and other related functions, to achieve the function of Automatic Train Protection (ATP) and Automatic Train Operation (ATO). The main function of the carborne ATP system is guarantee safe speed and safe position of the train. In addition, it includes the version of database's uploading, upgrade and error calibration, train control mode management, mobile license management, speed range curve calculation, train control and CC safety supervision[7]. The carborne subsystem peripheral equipment includes detection proximity sensor of platform which close to the disc of wayside, and use antenna to detect line location beacon response query, measuring the speed of train speed sensor, radar, and operation unit display for the driver.

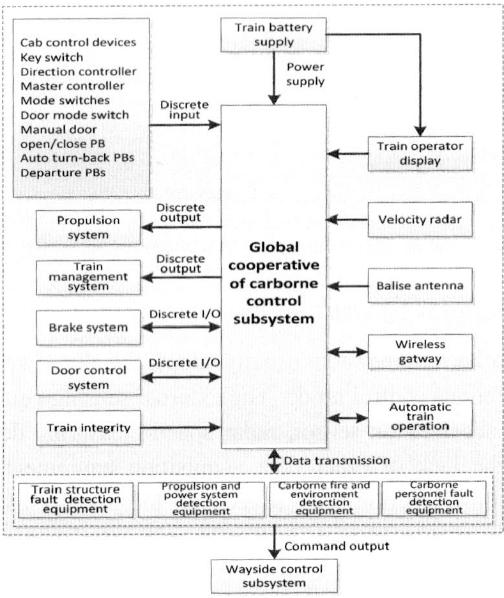

Fig. 3. Global cooperative signaling of carborne subsystem frame.

3. Train Operation Fault Analysis

For the fault analysis of train operation, it should start from the source. Finding the faulty source of the train operation, then collecting and analyzing the data of the faulty source, which focus on the different types and specific operative environment, finally determining the mode to response every single fault. The fault sources in the train operation can be broadly divided into the carborne fault source and the wayside fault source. The carborne fault source can be divided into the fault as train structure, traction power supply system and station staff system. Wayside fault source can be broadly divided into rail line fault, natural disaster and fire control fault. The following charts are for some specific fail-safe measures:

Table 1. Vehicle body structure fail-safe measure.

Faulty unit	Subcomponent fault	Faulty content	Train control system response (carborne ATP)
Train body	Roof/sidewall/emergency exit etc.	Badly damaged	Speed limit, detain the train in next station
	Fire/flood etc.	Environmental abnormally	Emergency braking(EB)
	Explosion/toxic gas	Terrorist attack	EB
Draw gear buffer device	Draw gear head	Damage or looseness	Stop and detained in next station
	cable	Leave over	EB
		Drop out	Stop and detained in next station
Bogie	Wheelset (wheel)	Overwear/crack/geometric shape abnormity	Limit speed, detain the train in next station
	Axle	Crack/peeling	Limit speed, detain the train in next station
	Rolling bearing and axial	Fracture/sudden heating	Common or EB
		Crack/deformation/high temperature/vibration abnormity	Limit speed, detain the train in next station
	Flexibility suspension device	Crack/deformation	Limit speed, detain the train in next station
	Traction device	Tail hook or buffer bottom crack/deformation	Limit speed, detain the train in next station
Breaking system	Shoe	Heat crack/glissade crake	Limit speed
	Magnetic rail/regenerative/resistor breaking and breaking control system	fault	Limit speed

4. Safety Cooperative and Comprehensive Detection Technology

4.1. *External environment safety cooperative technology*

Besides the traditional CBTC system's external sensing devices, the new generation of global cooperative signaling cooperative train control system expands the detection device of fire, flood, explosion, smoke which are for the fire protection and environmental control, and the infrared detection device for the protection of stuff dropped off the rail. When the fault and accidents happened, the train control system can use these detection devices to check the specific location whether has the hazards above or not. Wayside control subsystem collects data through real-time data acquisition sensor, if detects hidden dangers, the system will withdraw MAL to the entrance area, and then send the information to CC, when CC received the information, it will take the breaking control if necessary to ensure the train cannot enter the safety-risk area.

If staff felled off rail, but the train has already entered the area, CC will implement an emergency brake to ensure the safety of the passengers.

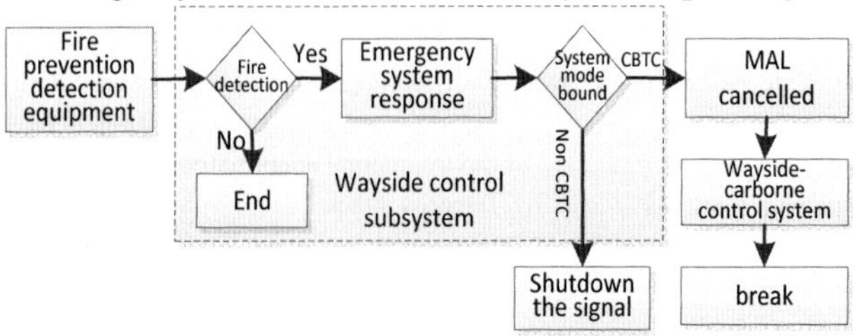

Fig. 4. Fire prevention technology framework.

4.2. *ATS safety cooperative technology*

Under the condition of knowing the faults happened, the dispatcher can send a command to shut down the wayside control subsystem through the ATS system in time. Therefore, it can prevent serious accidents from occurring, achieving the safety cooperative effect of ATS. Driver can also send a command to shut down the wayside control subsystem through CC system when knows the fault happened. Once the system receives the command from ATS system, it will orient to safety side and shut down the wayside control subsystem, and then withdraw MAL to the entrance area, then send the information to CC, when CC received the information, it will take the breaking control if necessary to ensure the train cannot enter the safety-risk area. If the train has already entered the area while the fault happened, CC will implement an emergency brake to ensure the safety of the train.

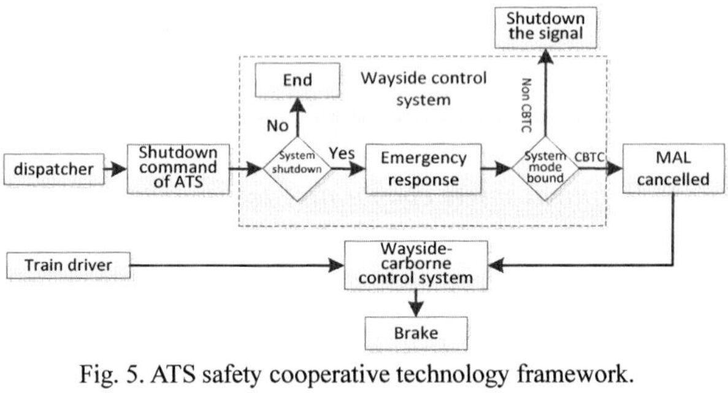

Fig. 5. ATS safety cooperative technology framework.

5. Summary

The new generation of global cooperative signaling train control system makes the breakthrough of the partial safety control mode of the current CBTC system. It establishes an active prevention, proactive warning, and a combination of traditional passive reactive control system. Based on the current CBTC system, the new system increases the real-time data acquisition and detection to supervise the external environment and the train inside state. For example, the temperature sensor for fire detection, humidity sensor for flood detection, smoke detector for smoke detection, infrared detector for rail foreign stuff dropped off detection, etc. Through the acquisition of rail route safety information, natural environmental information, environmental disaster information and other safe information data, the new system can make a real-time supervision for the train. When the fault happened, the train system can make a timely and appropriate emergency disposal, which maximum extent avoid serious accidents occurs, improve the reliability and safety of the train operation. The new generation of global cooperative signaling train control system realizes the principle of the rail transit system—fail-safe.

Acknowledgments

This work is supported by the funding from the National Science and Technology Infrastructure Program of China under Grant 2015BAG19B03.

References

1. Panfeng Liu, Xixi Qiu, Xiangxian Chen. Urban Rail Transit Train Automatic Protection Operation System Software Design[J].Computer Engineering.2012,38(19): 1-5
2. Xiaoqing Zeng, Changlin Wang, Shujing Zhang. Communication-based Train Operation Control for Rail Transit [M]. Shanghai: Tongji University Press, 2007. 12-22
3. Rail Transit Vehicle Interface Standards Committee of the IEEE Vehicular Society. IEEE Standard for Communications— Based Train Control (CBTC) Performance and Functional Requirements [M]. New York: IEEE Publisher, 1999:5-8
4. Yuehui Yang, Real-time Performance Research for Urban Rail Transit Sensor Network Data Transmission [D]. Beijing: Beijing Jiaotong University, 2014
5. LI Wei-wei, TANG Zhen-min Prof. Analysis on Metro Operation Accidents and Study on Countermeasures [J]. China Safety Science Journal, 2004, (14)
6. IEEE.1474.1-2004 - IEEE Standard for Communications-Based Train Control (CBTC) Performance and Functional Requirements[S].USA: IEEE, 2004

The liver nonstandard segment based on relationship of vascular blood supply

Yuan Gao
Educational Technology Center, The Chinese PLA General Hospital, Beijing, PR China
E-mail: Gaoyuan301@126.com

Ke Tan*
Educational Technology Center, The Chinese PLA General Hospital, Beijing, PR China
E-mail: Tanke123@sina.com

A liver is divided into 8 segments. Sometimes patients only need a small portion of the liver, not the whole segment, thus we brought up a new idea that surgeons could make a more accurate operation plan according to freedom grouping of vessels. We develop a new nonstandard liver segments method to ensure the maximum retention quality of affected healthy tissues and patients function. Clinicians could make a more accurate operation plan according to freedom grouping of vessels. The resection portion based on conventional segments method is more than our nonstandard resection volume. Therefore, it could ensure the maximum quality of affected healthy tissues and patients function.

Keywords: Nearest Neighbor Segment Approximation; 3d Reconstruction; Vessel Tree Reconstruction.

1. Introduction

Primary hepatocellular carcinoma (HCC) and metastatic liver cancer are refractory malignancies. The alternative treatment mainly includes surgical operation, minimally invasive surgery and living donated liver transplantation (LDLT). Surgical planning of liver tumor resections requires detailed three-dimensional (3D) understanding of the liver resection volume, area around the tumor and relationship of intrahepatic vessels, complex arrangement of vasculature, liver segments and tumors.

---

* Corresponding author.

Advances in radiological techniques have allowed three-dimensional (3D) hepatic reconstruction from multiple 2D slices of CT or MRI images, which can provide the complex anatomy present in slices and volumetric information [1-4]. 2D medical images may require special training, and although radiologists have these skills, they must often communicate their interpretations to the referring physicians, who sometimes have difficulty visualizing the 3D anatomy. Therefore, 3D reconstruction is useful to understand tumor localization and to make a resection proposal.

Knowledge about location and sizes of liver segments is important for choosing an optimal surgical resection approach h and predicting postoperative residual liver capacity. Clinicians often need the information of liver segments based on standard liver segment algorithm. However, the liver vascular variations are more likely to happen, especially in the presence of malignancy. Sometimes the volume of tumor does typically not require completely removed the liver segment. Therefore, we developed a nonstandard liver segment tool, and refined the existing segment method for maximum effect of surgery.

2. Materials and Methods

2.1. *Medical background*

Liver surgery is composed primary of Split-Liver Transplants and Oncologic Resections[5,6]. The planning of both surgeries can benefit from volumetric estimations which refer to the portion of the liver.

A liver segments[7] is defined by the supplied territory of a third order branch of the portal vein, thus determining 8 segments. Since the 8 segments are relatively independent. Therefore resection one of these segments will not affect the function of the liver. But the conventional segments method is not very precise, sometimes patients only need the small portion of the liver, not the whole segment, thus we brought up a new idea that surgeons could make a more accurate operation plan according to freedom grouping of vessels.

2.2. *Preparation work*

The section mainly includes the vessel segmentation and centerline extraction model. Despite the transformation of gray level could highlight the vessels, differences of vessel diameter and concentration of contrast agent give rise to higher gray value of larger vessel and lower gray value of tiny blood vessel in the image. The result is difference of surrounding tissue is hard to distinguish and influence the segmentation effect. The difficulty of segmentation is how to get the low gray value vessel and distinguish the surrounding similar vessels. In this novel, we enhance the vessels pixel value based on the Hessian matrix eigenvalue algorithm. Then we extract the vessels using the region growing algorithm. After the range conversion on mask data, we extract the centerline of vessels and build the tree structure.

2.3. *Standard liver segments*

The conventional classification of liver anatomy divides the liver into eight functionally independent segments.

Each segment has its own vascular inflow, outflow and biliary drainage. In the centre of each segment there is a branch of the portal vein, hepatic artery and bile duct. In the periphery of each segment there is vascular outflow through the hepatic veins. Right hepatic vein divides the right lobe into anterior and posterior segments. Middle hepatic vein divides the liver into right and left lobes (or right and left hemi liver). This plane runs from the inferior vena cava to the gallbladder fossa.

The Falci form ligament divides the left lobe into a medial- segment IV and a lateral part - segment II and III. The portal vein divides the liver into upper and lower segments. The left and right portal veins branch superiorly and inferiorly to project into the center of each segment (Figure 1).

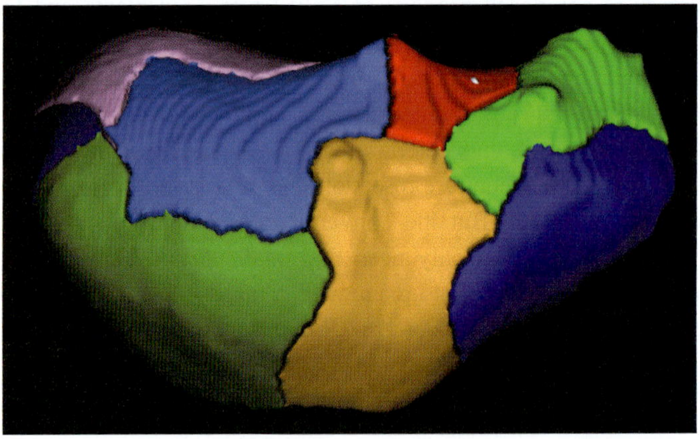

Fig. 1. Standard liver segments.

2.4. *Nonstandard liver segments*

The segmentation of liver data and extraction of vessel centerline has been completed, therefore dendritical structure is structured.

2.4.1. *Display of hepatic vascular*

Rendering of the vessel centerline is done in 3d environment for analysis purpose (Figure 2).

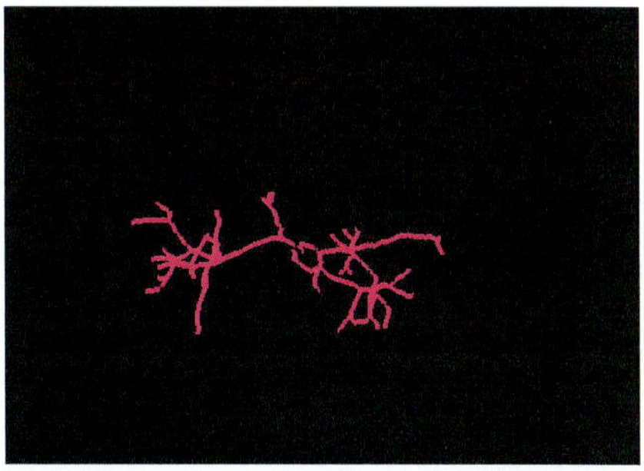

Fig. 2. Vessel centerline is extracted.

2.4.2. *Interaction mode of liver vessels*

The method of the interaction mode is to choose the centerline of vessels randomly. Based on the Nearest Neighbor Segment Approximation [8](NNSA), we move the mouse and pick up the blood vessel and branches when the distance of projection line to the vessel is 2mm (Figure 3). Holding down the mouse means confirm the pick-up action.

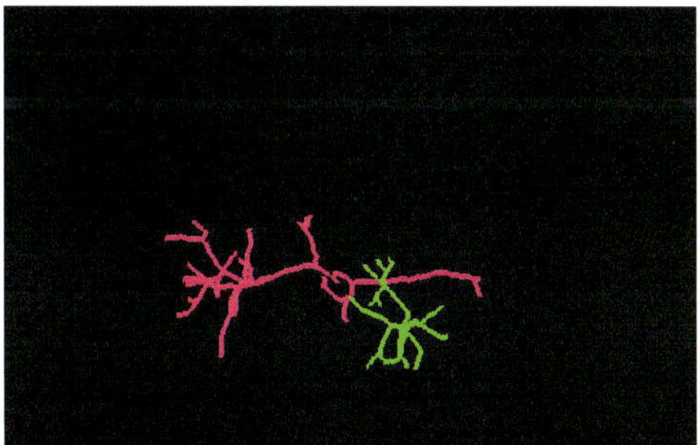

Fig. 3. Vessel and branches are picked up.

2.4.3. *Nonstandard liver segments method*

According to the Nearest Neighbor Segment Approximation (NNSA) algorithm, we can calculate the liver segments and present the outcome in the form of table. (Table 1).This method is based on the Euclidean distance function.

$$d_i(v) = \min_{P' \in B_i} \|p - p'\|_2 \quad p \in L$$

B_i (i = 1,.....n) is one of vessel branch. liver voxels $p \in L$.

$$Gnnsa(p) = \min(P_1, P_2....P_n)$$

$$P_n = \sqrt{(pax - pbx)^\wedge 2 + (pay - pby)^\wedge 2 + (paz - pbz)^\wedge 2}$$

In other words, di(v) is the Euclidean distance of the voxel v to the nearest voxel belonging to the branch Bi. Pix is the pixel corresponding to the points in image.

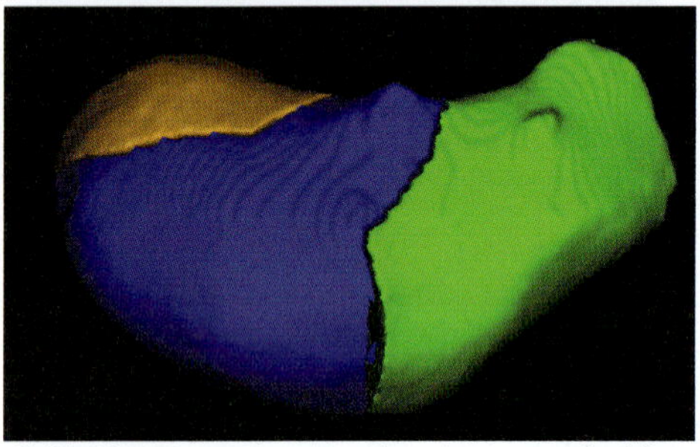

Fig. 4. The liver is divided into 3 segments based on the nonstandard liver segments method.

Table 1. Comparison of conventional resection portion with nonstandard resection volume.

Case	Liver volume (ml)	Resection portion (conventional method)(ml)	Resection portion (Nonstandard method) (ml)	C/V(ratio of conventional resection to liver volume)	N/V (ratio of nonstandard resection to liver volume)
1	1667	572	482	0.343	0.289
2	1714	751	621	0.438	0.362
3	1665	585	512	0.351	0.308
4	1785	664	572	0.372	0.320
5	1587	611	542	0.385	0.342
6	1618	352	287	0.218	0.177

3. Results

Figure 2 is the extracted centerline, figure 3 showed an interactive mode. From the figure 4, the liver is divided into 3 segments based on the nonstandard liver segments method. By using the nonstandard liver segments method, we can avoid the maximum number of destruction healthy tissues.

According to the Nearest Neighbor Segment Approximation (NNSA) algorithm, the outcome is calculated and present in table 1. We take case 1 as an example, the liver volume is 1667ml, and resection portion is 572ml bases on the Split-Liver Transplants and Oncologic Resection algorithm. According to our nonstandard resection method, the resection volume is 482ml. The ratio of the conventional method to original liver volume is 0.343, the ratio of nonstandard resection volume to whole liver volume is 0.289. Obviously, the nonstandard resection volume is smaller than conventional method. In other words, the reserved liver volume of nonstandard method is more than conventional reservation. From the remaining 5 cases, the results are the same.

4. Conclusion

As we know, the liver segments are defined by the supplied territory of a third order branch of the portal vein, thus determining 8 segments. Since the 8 segments are relatively independent. Therefore resection one of these segments will not affect the function of the liver. Sometimes the volume of tumor does typically not require completely removed the liver segment. Therefore, we developed a nonstandard liver segment tool, and refined the existing segment method for maximum effect of surgery.

From the results, we take case 1 as an example. The liver volume is 1667ml, and resection portion is 572ml bases on the Split-Liver Transplants and Oncologic Resection algorithm. According to our nonstandard resection method, the resection volume is 482ml. Obviously, the nonstandard resection volume is smaller than conventional method. In other words, the reserved liver volume of nonstandard method is more than conventional reservation. From the remaining 5 cases, the results are the same. Clinicians could make a more accurate operation plan according to freedom grouping of vessels. The resection portion based on conventional segments method is more than our nonstandard resection volume. Therefore, it could ensure the maximum quality of affected healthy tissues and patients function.

Acknowledgments

This work is supported by Beijing Natural Science Foundation(4132080).

References

1. Ming-ke Gao, Yi-min Chen, Quan Liu, Chen Huang, Ze-yu Li, Dian-hua Zhang:Three-Dimensional Path Planning and Guidance of Leg Vascular

Based on Improved Ant Colony Algorithm in Augmented Reality. [J].Medical Systems 39(11): 133:1-133:11 (2015)

2. Newe A, Becker L, Schenk A. Application and Evaluation of Interactive 3D PDF for Presenting and Sharing Planning Results for Liver Surgery in Clinical Routine (vol 9, e115697, 2014) . [J]. PLOS ONE, 2015, 10(3).
3. KHAMBAY, BNEBEL, JCBOWMAN, et al. 3D stereophotogrammetric image superimposition onto 3D CT scan images: the future of orthognathic surgery. a pilot study.[J]. The International Journal of Adult Orthodontics and Orthognathic Surgery, 2002, 17(4): 331-341.
4. HEINZ K H, RALPH B. Shading 3D-Images from CT using Gray-Level gradients. [J]. Medical Imaging, IEEE Transactions, 5(1): 45-47.
5. BERNHARD R, ALEXANDER B, BEICHEL R, et al. Liver surgery planning using virtual reality. [J]. IEEE Computer Graphics and Applications, 2006, 26(6): 36-47.
6. GANE D, MDA, SALIBAB M D, et al. Randomised trial of efficacy and safety of oral ganciclovir in the prevention of cytomegalovirus disease in liver-transplant recipients. [J]. The Lancet, 1997, 350(9093): 1718-1719.
7. LEELAUDOMLIPI S, SUGAWARA M D, KANEKO J, et al. Volumetric analysis of liver segments in 155 living donors. [J]. Liver Transplantation, 2002, 8(7): 612-614.
8. Ronen Basri, Tal Hassner, and Lihi Zelnik-Manor. Approximate nearest subspace search with applications to pattern recognition. In Computer Vision and Pattern Recognition (CVPR'07), pages 1–8, June 2007.

Image content authentication algorithm based on finite ridgelet transform

Rui-Ling Zhu*

Department of Information Engineering, Shandong Communication & Media College, Jinan 250200, China
E-mail: rlzhu@163.com

Zhuo-Ran Han

School of Electronic Information and Electrical Engineering, Shanghai Jiao Tong University, Shanghai 200240, China
E-mail: hzrtom1996@163.com

In this paper, we propose a novel content-based image semi-fragile watermarking algorithm for authentication in finite ridgelet transform (FRIT) domain. The authentication watermark was generated according to the image feature, which was the mean value of an image extracted from FRIT. It was embedded in the FRIT coefficients, which were selected by the minimizing Bayesian risk estimate. Experimental results show the algorithm is simple and effective, providing very good classification from malicious tampering to incidental modification. Besides, it has the ability to localize the malicious tampering at a high resolution.

Keywords: Finite Ridgelet Transform, Authentication, Bayesian Estimate.

1. Introduction

The increasingly easy access to digital multimedia via the Internet and the increasingly powerful tools for editing digital media editing tools have made protection of the intellectual property rights and authentication of digital

* Corresponding author: Rui-Ling Zhu received the PhD. degree at the Shandong University,and now she is professor at the Department of Information Engineering, Shandong Communication & Media College. Her research interests include digital image processing and computer vision.

multimedia a very important issue[1]. One way to solve this problem is the digital watermarking. Image authentication technology is mainly divided into complete authentication and content authentication. Certification completely focuses on the overall integrity of the image data, which does not allow the image content has any change[2], while content authentication is capable of tolerating the change towards the watermarked image to some extent[3], such as the lossy-compression. Content authentication uses a method based on image characteristics in general, which provides classification of malicious tampering and incidental modification, and it is able to localize tampering.

A content-based image signature for authentication using the ridgelet transform was proposed in Ref.[3], however, the inverse transform is ill-conditioned in the presence of noise and requires an iterative approximation algorithm. Finite ridgelet transform can achieve both invertibility and nonredundancy[5]. We propose a novel content-based image semi-fragile watermarking algorithm for authentication in finite ridgelet transform (FRIT) domain. The authentication watermark is generated according to the image feature, which is the mean of an image extracted from FRIT. It is embedded in the FRIT coefficients, which are selected by minimizing Bayesian risk estimate[6].

2. Finite Ridgelet Transform

To overcome the weakness of wavelets in higher dimensions, Candès and Donoho pioneered a new system of representations named ridgelet, which deals effectively with line singularities in 2-D. The idea of ridgelet transform (RT) is to map a line singularity into a point singularity using the radon transform. Then, the wavelet transform can be used to handle the point singularity in the radon domain.

Due to the radial nature of the ridgelet, straightforward implementations of discretion with continuous formulae would require interpolation in polar coordinates, from which the results would either be redundant or fail to be fully reconstructed. So, a better solution, FRIT was introduced[5] to perform 1-D transform in finite Radon transform (FRAT) domain.

The FRAT is defined as summations of image pixels over a certain set of "lines". The FRAT of a real function f defined on a finite grid Z_p^2 ($Z = \{0,1,2,\cdots,p-1\}$), where p is a prime number, is

$$r_k[l] = FRAT(k,l) = \frac{1}{\sqrt{P}} \sum_{(i,j \in L_{k,l})} f[i,j], k \in Z_p^*, l \in Z_p \tag{1}$$

where $Z_p^* = \{0,1,\cdots,p-1,p\}$. $L_{k,l}$ defines the set of points that form a line on the lattice Z_p^2, specifically

$$\begin{cases} L_{k,l} = \{(i,j): \ j = ki + l(\bmod p), i \in Z_p\}, 0 \le k < p \\ L_{p,l} = \{(l,j): j \in Z_p\} \end{cases} \tag{2}$$

where k is the line direction and l is its intercept. FRAT is invertible by the finite back-projection (FBP), thus provides a representation for a generic image. The FBP operator is defined as the sum of radon coefficients of all the lines that go through a given point, that is

$$FBP_r(i,j) = \frac{1}{p} \sum_{(k,l \in P_{i,j})} r_k[l], (i,j) \in Z_P^2 \tag{3}$$

where $P_{i,j}$ denotes the set of indices of all the lines that go through a point $(i,j) \in Z_p^2$. More specially, using Eq. 2 we can write

$$P_{i,j} = \{ (k,l) : l = j - ki \, (\bmod p), k \in Z_p \} \cup \{ (p,i) \} \tag{4}$$

3. Authentication Algorithm Based on the Content

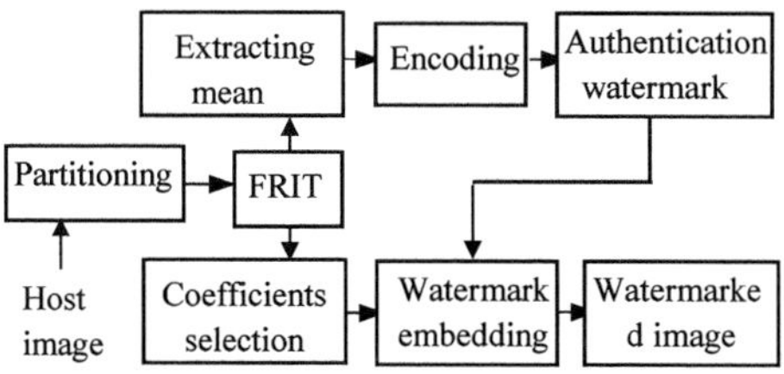

Fig. 1. The diagram of watermark generating and embedding.

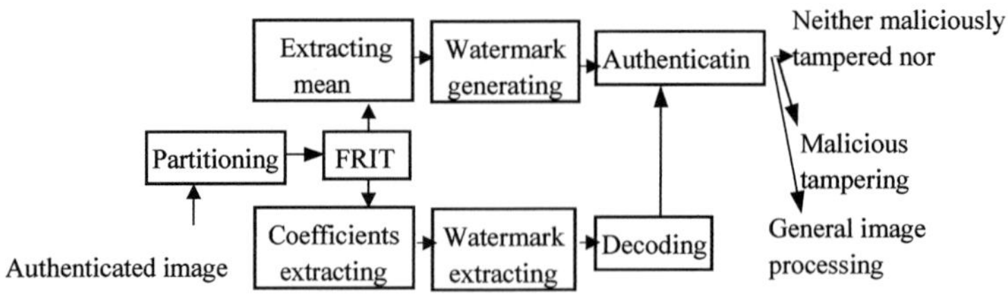

Fig. 2. The diagram of watermark extraction and authentication.

(a) Watermarked image

(b) Authenticated image

(c) Tamper detection for the Fig.(b)

Fig. 3. Tamper detection for the thenticated image.

Just as shown in Fig. 1 and Fig. 2, the whole algorithm is divided into, authentication watermark generation, embedding, extracting, and authentication, four major processes, the first two process performed by Fig. 1 and second two process performed by Fig. 2

4. Watermark Embedding

4.1. *The significant FRIT coefficients selection*

In order to obtain the distribution characteristic, and then to extract the perceptually significant coefficients for embedding watermark, we have analyzed a number of images, and gotten the statistical characteristic of the FRIT coefficients, shown in Fig. 4. Let us call I_b^a ($b = 0,1,2,3$, $a = 0,1$) sub-band at resolution level $b = 0,1,2,3$, detail bands when $a = 0$ and approximate band when $a = 1$. Fig. 4 demonstrates the oscillation wave about the FRIT coefficients of each sub-band along each direction. It shows that there is a part of coefficients containing the most energy in each sub-band, which are the perceptually most significant components of the data in several directions. Insertion of a watermark into these bigger coefficients makes the watermark robust to signal processing operation. Thus, we intended to extract those perceptually significant components, and embed watermark into them.

Watermark was embedded in the FRIT coefficients, which were selected by the minimizing Bayesian risk estimate. After many standard and generic images were analyzed statistically, the position sequence of selected coefficients using the method presented above may be obtained. It was then stored as sequence R.

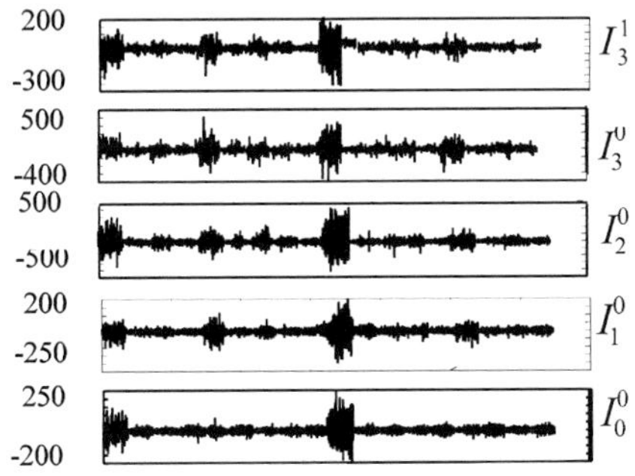

Fig.4. The oscillation wave about the FRIT coefficients.

4.2. *Authentication watermark generation and embedding*

Given a possibly corrupted image, represented as I with the size $M \times N$, it was partitioned into blocks of dimension $P \times P$ of pixels, and the FRIT was performed for each block. The authentication watermark is generated according to the image feature, which was the mean of an image extracted from FRIT. The results presented in Table 1 were carried out on the Lena image, where W_Z denotes watermark, and ΔW_Z denotes the change of watermark after image processing. Embedding generated watermark into host image according to [7].

4.3. *Image authentication*

Let variable D_z represent the differential matrix, which can be expressed as

$$D_z = I_z' - I_z^* \tag{5}$$

where I_z' is original watermark, and I_z^* is the watermark extracted from the authenticated image. Variable D_z may demonstrate directly the change of watermark, namely it shows that whether or not the watermarked image is tampered with.

Table 1. Difference of the authentication watermark after several image processing.

Image processing	JPEG compression (70%)	JPEG compression (10%)	Gaussion noise (σ^2=20))	Median filtering (5×5)
$\Delta W_z / W_z$ (this paper)	0.0071	0.0459	0.0160	0.0100
$\Delta W_z / W_z$ (Ref. [2]	0.0235	0.1814	0.0320	0.0815

Let variable TAF_z represent the tampering function which can be expressed as:

$$TAF_z(i,j) = \frac{1}{3\times 3}\sum_{u=i}^{i+2}\sum_{v=j}^{j+2} D_z^{\ 2}(u,v) - \left(\frac{1}{3\times 3}\sum_{u=1}^{i+2}\sum_{v=1}^{j+2} D_z(u,v)\right)^2 \tag{6}$$

Where $1 \le i \le M_z - 2, 1 \le j \le N_z - 2, M_Z = floor(M/p), N_Z = floor(N/p)$.

Given the specified threshold τ, only three cases are possible:

Case A: $(TAF_z)_{MAX} \approx 0$, image is not temped. Case B: $(TAF_z)_{MAX} < \tau$, image is temped to some degree, which is considered as incidental and negligible. Case C: $(TAF_z)_{MAX} >= \tau$, image is temped. The greater the TAF_z is, the severer the tempering is. According to variable D'_z, the temped location is determined. The value of τ is determined by actual demand.

5. Experimental Results

For the purposes of checking the robustness and fragility of this algorithm, we performed several attacks on the watermarked image: median filtering, adding Gaussian white noise, Pepper& salt, Enhancing, low filtering, and JPEG compression. Some results are shown in Fig. 3, and Table 2. The results presented permit us to appreciate that the algorithm is simple and effective, providing remarkable classification of malicious tampering and incidental modification. Moreover, it has the ability to localize tampering at a high resolution.

Table 2. Image authentication results corresponding to some image processing.

Image processing	Gaussian noise (σ^2=15)	Gaussian noise (σ^2=30)	Pepper & salt (1%)	Low filtering (3×3	Enhan-cing(2%)	Median filtering (7×7)	JPEG compression (QF=70)	JPEG compression (QF=50)
$TAF_z)_{MAX}$	0.691	1.432	2.395	1.670	1.516	1.525	0.172	2.889
Authentication results in this paper	Y	Y	Y	Y	Y	Y	Y	Y
Authentication results by Xie	N	N	N	N	N	N	N	N

6. Conclusions

We proposed a novel content-based image semi-fragile watermarking algorithm for authentication in finite ridgelet transform (FRIT) domain, which has the following characteristics:

(1) The authentication watermark was generated according to the image feature, which was the mean of an image extracted from FRIT.(2) The watermark was embedded in the FRIT coefficients, which were selected by the minimizing Bayesian risk estimate.(3) The algorithm provides remarkable classification of malicious tampering and incidental modification, as well as the ability to localize tampering at a high resolution. Meanwhile, the authentication algorithm can also be used for copyright protection.

Acknowledgments

This work is supported by A Project of Shandong Province Higher Educational Science and Technology Program (J12LN89,J15LN53)

References

1. Huang Qing-long, Liu Jian-lan, Chen Jin. Blind digital watermarking technique based on Fresnel diffraction transform and phase encryption mask, J. Journal of Optoelectronics Laser, 2008, 19 (5) 667-670.

2. HuYu-ping, Chen Zhi-gang. Wavelet domain semi-fragile watermarking algorithm for image authentication, J. Acta Electronica Sinica, 2006, 34 (4) 653-657.

3. Zhen Yao, Nusir Rujpoot. Radon/Ridgelet signature for image authentication, C. Proceeding of International Conference on Image Processing, Singapore, 2004, 43-46.

4. Liehua Xie, Gonzalo R. Arce. A class of authentication digital watermarks for secure multimedia communication, J. IEEE Transactions on Image Processing , 2001, 10 (11) 1754-1764.

5. Minh N. Do, Martin Vetterli. The finite ridgelet transform for image representation, J. IEEE Transactions on Image Processing, 2003, 12 (1) 16-28.

6. S.Grace. Chang, Bin Yu, and Martin Veterli. Adaptive wavelet thresholding for image denoising and compression, J. IEEE Transactions on Image Processing, 2000, 9 (9) 1532-1546.

7. Brian Chen, Gregory W. Wornell. Digital watermarking and information embedding using dither modulation, C. 1998 IEEE Second Workshop on Multimedia Signal Processing, California: IEEE, 1998, 273-278.

An resolution impact analysis for the catadioptric system

Mai Jiang*, Ming-Ming Qiao and Li-Min Zhang
Criminal Investigation Department, National Police University of China, Shenyang, Liaoning Province, 110035, China
**E-mail: jiangmai@npuc.edu.cn*
www.npuc.edu.cn

This paper analyses the hyperoloidal parameters impact on the spatial resolution. The most important part of an omnidirectional vision system is the catadioptric mirrors. It is important to note that single-viewpoint property is suitable for image analysis and processing because we can easily generate any desired image through projecting the original circle image on any designated image plane. Thus, we briefly introduced the single-viewpoint general constraint equations. We then analyzed the hyperbola parameters influence to the system structure because perfect curvature of the hyperboloid mirror is the only option for a clear picture. The catadioptric system's spatial resolution distance impact on the vertical and horizontal object was subsequently discussed. Lastly, we concluded that we can exploit a flexibility hyperboloidal mirror by choosing the curvature of the mirror for different objects identification.

Keywords: Catadioptic mirrors; Single-Viewpoint Constraint; Hyperbola Parameters; Spatial Resolution.

1. Introduction

In recent years, the omnidirectional vision system has been widely used for many applications, such as robot navigation, virtual reality, sensitive area monitoring, video conference telepresence and remote surveillance [1, 2, 3, 4].

There are several different ways to acquire an omnidirectional image of the entire scene, which including four main types in term of structure: using a rotating camera which consists of a conventional camera and a mechanic system, multi-camera by combining conventional cameras pointing to different directions, a dioptric (means refraction) fisheye lens which use wide angle lens, and finally, convex mirrors combing with conventional sensors which belong to the catadioptric (refraction and reflection) system [5, 6, 7]. However, the single rotating camera and poly-camera methods bring in mechanical and cost problems which cause accurate positioning troubles; the non-central system can't satisfy the single viewpoint demand which means lead to a blurring image.

---

* Corresponding author.

Compared with other convex mirrors, the hyperboloidal catadioptric vision system accord with the single viewpoint constraint which allow us easily calculate the incident position of light to reconstruct the perspective images [8], and also have been developed to observe 360 degree environment information in the horizontal direction and nearly 90 degree in the vertical direction separately. In order to obtain a clear picture, we need the vision system can cover both hyperboloidal mirror and camera optic lens focus points. That means the hyperboloidal mirror should placed up and facing to the camera lens below, meanwhile, the mirror and the lens focus should be aligned with the same axis, in this way, all the entire space reflect lights around the mirror will gathered in the camera sensor through the lens focus.

Early research mainly focus on the image transform algorithm, but a proper hyperboloidal mirror shape can significant improve the catadioptric system sight, meanwhile, we can also control the vision system's parameter relationships through simulation for cost savings, and in accordance with our need, we design the mirror curvature based on the image resolution that we want to get from the camera.

2. The Single-Viewpoint Constraint

The catadioptric omnidirectional vision system generally consists of two main sections, the mirror and camera. In the single-viewpoint system, there is a one-to-one correspondence between the object and its own imaging, so we can get the perspective image which is consistent with the human visual via inversion algorithm on the original image. The single-viewpoint constraint guarantee each lights through the lens center based on the perspective principles, each pixel corresponds to an incoming light, therefore, the omnidirectional image can be projected onto a plane at any distance, so this constraint allows us transform the image onto any projection image plane and offer a wider FOV at the same time, after that, we can use the general methods in image analysis and processing. From early research work, we get the final universal Equations solutions:

$$\left(z-\frac{s}{2}\right)^2-r^2\left(\frac{k}{2}-1\right)=\frac{s^2}{4}\left(\frac{k-2}{k}\right) \qquad (k\geq 2) \tag{1}$$

$$\left(z-\frac{s}{2}\right)^2+r^2\left(1+\frac{s^2}{2k}\right)=\left(\frac{s^2+2k}{4}\right) \qquad (k>0) \tag{2}$$

These two Equations (1) and (2) above defined all the mirror curves which satisfied with the single-viewpoint constraint. The parameters s and k values are possessed of certain geophysical significance and if they fail to meet the $c>0$, $k>0$ limitations, there will be identified as the regressive solutions which not use to framed the single-viewpoint imaging system. If the parameter meet $k > 2$, $c > 0$, we can get the hyperboloid mirrors.

The hyperboloidal mirror which is a useful enhancement way for FOV, also it satisfy the single-viewpoint constraint, which makes easy to generate any desired perspective image from the original image.

3. The Parameters Impact Analysis

To generate omnidirectional images, perfect curvature of the hyperboloid mirror is the only candidates for a clear picture. We can get the mirror's shape by simulation method.

The omnidirectional vision system's spatial resolution is distribute of nonlinear style around the hyperboloidal mirror axis center. In order to facilitate the analysis we established the coordinate system as shown in Fig. 1. For an vertical object $E1E2$ and assume the spatial vertical distance $|E_1E_2| = |X_1 - X_2| = L$, then the corresponding image pixels on the camera sensor will gradually decrease while the object far away from the hyperboloidal mirror. At this point, take the L length as the vision system's minimum resolution in its spatial region when the image distance equal to one pixel.

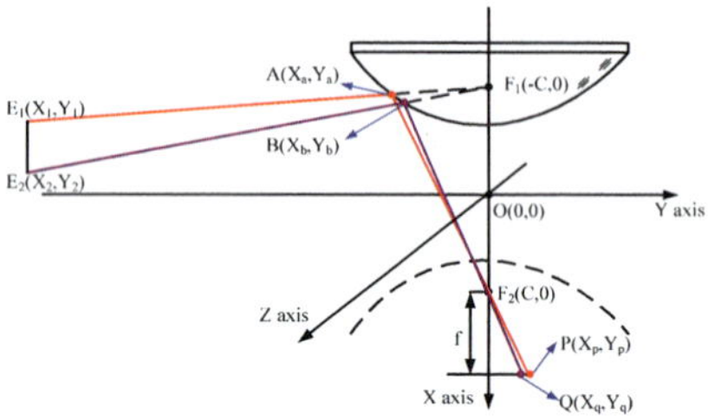

Fig. 1. The spatial resolution for the vertical object.

In Figure 1, every symbol is expressed as follows: E_1 and E_2---the vertex of the object, coordinates are (X_1, Y_1) and (X_2, Y_2) separately; A and B --- the intersection points of the incoming light $\overrightarrow{E_1A}$ and $\overrightarrow{E_2B}$ and the hyperboloidal mirror, coordinates are (X_a, Y_a) and (X_b, Y_b), respectively; F_1 and F_2---the hyperbola focus, coordinates are ($-C$, 0) and (C, 0) separately; P and Q ------ the imaging points, coordinates are (X_p, Y_p) and (X_q, Y_q) separately.

There exists a blind area on the CCD sensor surface by the features of the panoramic vision system, so eliminate the CCD sensor center image region in the simulations, in this scenario, sensor pixel size $\mu=7.4\times10^{-6}$, the number of sensor pixels 2048×2048, so we know that the range of the image region are approximate to 0.00037m~0.0075m by calculating the pixel numbers and size. The system parameter was setting as follows: a=84.63mm, b=91.99mm; mirror thickness h=100mm; hyperboloidal mirror upper surface radius D=0.1783m.

One pixel was the object imaging minimum constraints, so for a vertical object E_1E_2 with length was 0.1m shown in Fig. 1, which parallel to the *X-axis* and perpendicular to the *Y-axis* simultaneously, when the hyperboloidal mirror's major axis parameters a value changed, we can observe the corresponding position curve in Figure. 2-a, meanwhile, guaranteeing the image resolution under one pixel constraints. Then Fig. 2-b and Figure 2-c showed the target position curve with length of 3m and 5m, respectively, where L_1 is the vertical object length.

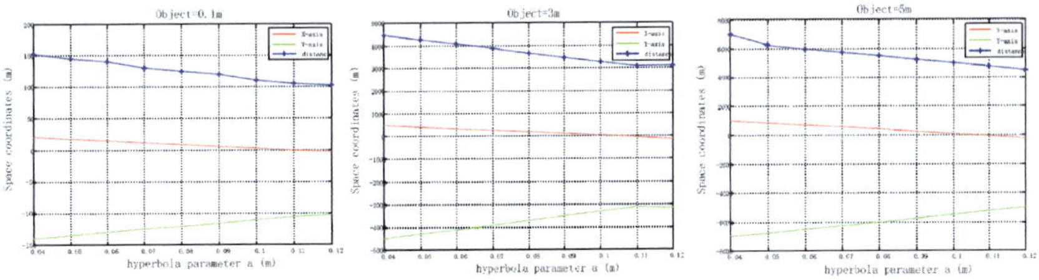

Fig. 2. The vertical object resolution ability when L1=0.1, 3 (from top to bottom) and 5m with parameter a.

In Figure 2, we can see the resolution distance is decrease with increase of the hyperbola parameter a, it is mainly because the hyperboloidal mirror became narrow while a increasing, therefore, the resolution distance for the vertical object will reduce. Similarly known, the hyperboloid mirror became flat while b increasing, therefore, the resolution distance for a vertical object will increase.

For a horizontal object E_3E_4 with length was 0.1m shown in Figure 3, which parallel to the *Z-axis* and perpendicular to the *X-Y* plane simultaneously.

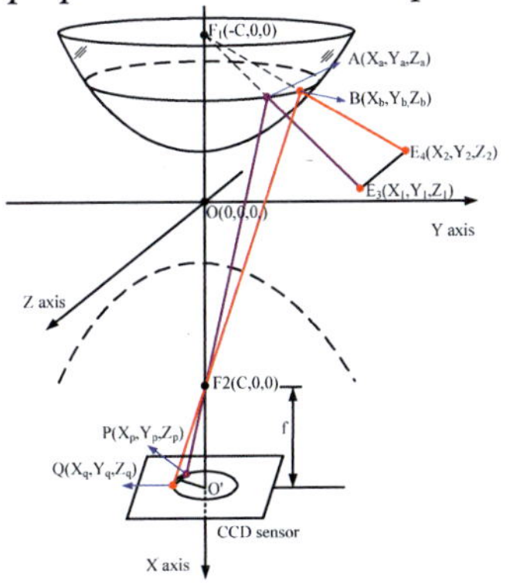

Fig. 3. The spatial resolution for the horizontal object.

For the sake of contrastive analysis, the hyperboloidal minor parameter influence to the spatial resolution for a horizontal object when *b* changing was illustrated in Figure 4.

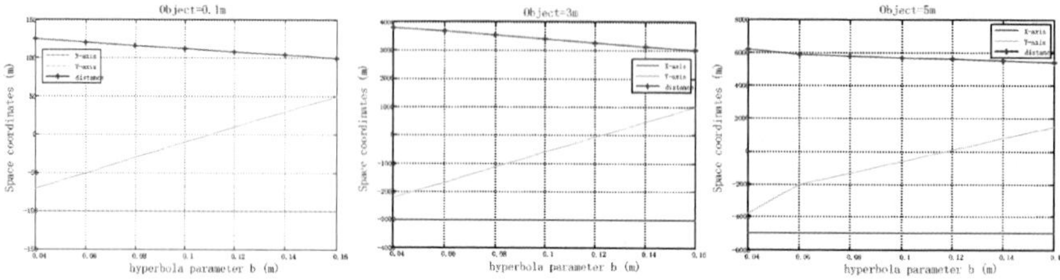

Fig. 4. The horizontal object resolution ability when L2=0.1, 3 (from top to bottom) and 5m with parameter *b*.

From Figure 4 above, the hyperboloidal mirror became flat while *b* increasing, therefore, the resolution distance for a horizontal object will decrease.

4. Conclusions

The catadioptric monidirectional vision system is widely used in surveillance, tracking objects, SLAM (simultaneous localization and mapping) and 3-D scene environment reconstruction and other fields. Perfect curvature of the hyperboloid mirror is the most for the catadioptric omnidirectional vision system, ishiguro has summarized mirror shapes and their differences in details, it has been proved that paraboloidal, ellipsoidal and hyperboloidal mirrors, combined with the conventional high-resolution cameras are the only methods that ensure the single-viewpoint property, the constraint equations provide that the mirror is positioned appropriately relative to the camera.

Figure 2 shown the spatial resolution distance will decrease with parameter *a* increasing for a vertical object, or increase when parameter *b* increasin). In the meantime, Fig. 4 gave the results for a horizontal object correspondingly, but the resolution distance trends were not so obvious. Based on the above results and actual demands, we can exploit a flexibility hyperboloidal mirror in which by choosing the curvature of the mirror, for different objects identification.

Acknowledgments

This work is supported by the PhD research startup foundation of National Police University of China (Video Investigation Simulation Design Based on the Omnidirectional Technology).

References

1. Y. Onoe, K. Yamazawa, H. Takemura, and N. Yokoya, Telepresence by real-time view-dependent image generation from omnidirectional video streams, Computer Vision and Image Understanding, 71(2): 154-165, 1998.
2. S. E. Chen, QuickTime VR-An image-based approach to virtual environment navigation, Proc. ACM Int. Conf. Computer Graphics, pp. 29-38, 1995.
3. M. Sugimoto, G. Kagotani, H. Nii, N. Shiroma, M. Inami, and F. Matsuno. Time Follower's Vision: A Teleoperation Interface with Past Images. In IEEE CG&A, Vol. 25, No.1, pp. 54-63, 2005.
4. T. Svodoba, T. Padjdla and V. Hlavac, Epipolar Geometry for Panoramic Cameras, In: Proceedings of the European Conference on Computer Vision, Bombay, India, 1998, pp. 299-303.
5. A. Maeder, H. Bistry, and J. Zhang. Intelligent Vision System for Robotic Applications. International journal of Information Acqusition (IJIA), 5(3): 259-267, September 2008.
6. S. H. Kasaei, S. M. Kasaei, S. A. Kasaei, M. Taheri, S. A. Monadjemi, Modeling and Implementation a Fully Autonomous Soccer Robot Based on Omni-directional Vision System, Industrial Robot: an International Journal, Emerald Group [ISSN 0143-991X] 37/3 (2010), pp. 279-286
7. J.W. Choi, R.E. Cury, G.H. Elkaim, Realtime obstacle avoiding pathplanning for mobile robots, in: Proceeding of the AIAA Guidance, Navigation and Control Conference, AIAA GNC 2010, Toronto, Ontario, Canada, 2010.
8. H. Ukida, N. Yamato, Y. Tanimoto, T. Sano, and H. Yamamoto, "Omni-directional 3D Measurement by Hyberbolic Mirror Cameras and Pattern Projection," Proc. 2008 IEEE Conf. on Instrumentation & Measurement Technol. Victoria, BC, Canada, May 2008, pp. 365-370.

Design of a colored QR code adaptive to logo contour

Dong Li*
Information engineering department, Engineering College of CAPF,
Xi'an, Shaanxi, China
E-mail: 563783912@qq.com

Hong-An Xie
Electronics department, Engineering College of CAPF,
Xi'an, Shaanxi, China
E-mail:15529332695@163.com

The information capacity of the black and white QR code with the logo is small and the logo will cover the encoding module that leads to a problem of high bit error rate. Thus, this paper designs a colored QR code which adapts to the size of the logo. Firstly, we do so by enhancing the information capacity of the color QR code by choosing a variety of colors such as encoding color, and selecting the greatest contrast color. Secondly, we extract the contour of the logo and place coding modules around the logo contour extracted. MATLAB simulation experiments and the bar code capacity analysis show that the scheme of the colored QR code can improve the capacity of the two-dimensional code and solve the problem of decreasing ability to correct errors caused by the covered encoding module.

Keywords: Color QR Two-Dimensional Code; The Contour Of The Logo; Date Capacity; Correction Capability; Multi-Band Data.

1. Introduction

A two-dimensional barcode, or a matrix code, is a type of graphic symbol representing certain data, and also a vital means of automatically storing and carrying information of products that can be read automatically by a machine[1-2]. In the mid 1980s, researchers in China begun their study on barcode technology, and some research institutes and universities made a certain number of achievements in the field; then the research results were applied in some aspects of our life. For instance, barcode technology was adopted by organizations like library and bureau of materials, etc. because they had a large number of data to be stored [3]. Since the 2D barcode is only in black and white, it looks relatively

* Corresponding author.

monotonous and not pleasing to the eye. Besides, its data capacity can still not meet the demand for storage in the information explosion of the society. To make it look more artistic, designers have invented the barcode with a Logo in its center, which to a certain degree has improved the appearance of barcode [4]. However, 2D barcodes designed in this way are of a poor capability in error correction. Meanwhile, to solve the problem that the 2 D barcode is of a low data capacity, basically two solutions have been put forward for enhancing its data capacity: increasing the density of 2D barcode or enlarging its size [5]. Through relevant analysis, it is found that both solutions are of certain limitations. First, to increase the density of 2D barcode requires a higher precision of code printers and code-reading equipment, which naturally leads to a higher cost for code printing and reading. Second, to enlarge the size of 2D barcode means the item to be attached with the code should also has a relatively large size; however, in the actual application of codes, it is often required that the code is relatively small so that it can be printed onto some products like electronic chips on which it will not be easy to print a code if the code is too large. Therefore, to find a solution for the poor data capacity of the black-and-white barcode, the colored 2D barcode is invented [6].

2. Design Concept of Colored QR Barcode

The black-and-white QR barcode widely used in our daily life is able to transform data into binary numbers and then represent the two digits, 0 and 1, with black and white respectively. Therefore, in order to increase the information density of a QR code, we think of the idea of employing a module to stand for N-nary numbers [7]. In this way, the information density of a QR code will be greater and a QR code in the same version will contain a larger amount of information.

Currently, for most black-and-white QR codes in advertisements, a logo is simply put in the center of the code, which will lead to a sharp decline in the error-correcting ability of a QR code. Moreover, with a logo that is too large, a QR code might not be read properly by machines. To solve that problem, it is proposed in this paper that we can first read the outline of the logo with a machine and then put the coding module around it so that the error-correcting ability of QR code can be enhanced.

3. Selection and Application of Colors

Color selection is essential to a colored QR code since colors selected can basically determined the effect of recognizing the code. In the paper, with consideration to the fact that QR code has a relatively high demand for the precision of reading equipment, we have selected colors with a high contrast among each other, so that during the reading process a certain color will not be mistakenly recognized as another similar color. At the mean time, selected colors should be able to be effectively read and recognized.

Reorganization of a colored QR code involves segmenting an image, and the segmentation of a color image is often carried out according to the RGB color space, which is a simple, direct and effective method [8]. As demonstrated in the RGB color model, colors that are most easily recognized include black, white, red, green, blue, magenta, cyan and yellow. In this paper, the data stream is in a quaternary system where the four digits, 0, 1, 2, and 3 respectively represent green, yellow, red and blue, and the corresponding relationship is show in Figure 1 as follows.

Corlor	Green	Yellow	Red	Blue
Quaternary	0	1	2	3
Color Display				

Fig. 1. Representation of colors by corresponding digits in quaternary numerical system.

4. Contour Extraction and Module Placement

4.1. *Extract of logo contour*

With regard to all black-and-white QR codes with a logo currently in our life, the code is firstly produced, and then according to certain category of error correction for the code, the logo is reduced to a size within the error-correcting ability of the code; after that, the logo is put amid the QR code. As a result, the logo will cover certain part of the code, and naturally the error-correcting ability of the code will sharply decline.

As put forward in this paper, to produce the colored QR code automatically adaptive to the logo size, we first extract the contour of a logo with image processing techniques and then encoding is made around the contour; therefore, no part of the code will be covered by the logo. The colored QR code produced in this way is of a strong error-correcting ability. In general we can get such a code through three steps: (1) scan the logo and determine the its size; (2) convert the logo to grayscale and obtain the binary image after determining the logo size;

(3) extract the edge contour of the logo.

After the logo contour is extracted, we are to deploy the coding module according to the contour.

4.2. *Placement of the coding module adaptive to logo size*

Since the logo in a colored QR code is changeable, the version number of the code is also undetermined. In placing the function pattern into a colored QR code adaptive to logo size, basically the same method of placing a function pattern into a black-and-white QR code can be used. First, we put three position detection patterns on three corners of the QR code, and then we add the finder pattern and the alignment pattern onto the code.

In this paper, the module is placed firstly at the bottom left corner of the matrix and then proceeds upward in a 4×2 array, when it reaches a function pattern, the module will be placed in a way as referred to in relevant literature [9]. If the contour of the logo covers an area where the module is to be placed, then we won't place the module there.

5. Encoding Process of Colored QR Codes

Encoding process of colored QR codes mainly includes three procedures, namely, information coding, generation of error-correcting code, and symbolization. Due to the limited length of this paper, placement of function patterns in colored QR codes will not be introduced in detail, and it can be referred to in relevant literature [10]. In general, the encoding process of colored QR codes can be described as follows:

(i)Convert data information into a stream of bits in quaternary numeral system;

(ii)Compose the stream of bits from Step 1 into a codeword, and obtain the data information codeword;

(iii)Work out the error correction codeword with the RS error-correction algorithm and add the obtained error correction codeword to the data information codeword from Step 2 to form the ultimate data codeword;

(iv)Symbolize the ultimate data codeword with corresponding color module to get the final coding module;

(v)Extract the contour of logo and place the coding module around the logo;

(vi)Place function patterns and codeword patterns into the matrix and obtain the final colored QR code through masking.

6. A Case of Encoding a Colored QR Code

In this part, the encoding process is illustrated with the case of a number sequence 20160309576432 (Symbol version 5-L). The information is converted into the ultimate stream of bits through the following steps.

The number is divided into groups by each three digits. Since 999 is converted into 33213 in quaternary numeral system, when the quaternary version of a group contains less than 5 digits, then 0 is put before it as the first digit.

(i) Convert groups of data into corresponding streams of bits in quaternary numeral system;item two,

(ii) Put together the obtained quaternary numbers into a sequence of bit streams;item four.

(iii) Convert the character count indicator into a quaternary stream of bits;

(iv) Add the character count indicator and the numeric mode indicator to the sequence of bit streams and obtain the final sequence of bit streams.

At the end, corresponding version information and format information are added into the matrix to generate a colored QR code automatically adaptive to logo contour, as shown in Figure 2

Fig. 2. Colored QR code obtained.

7. Analysis of Data Capacity and Error Correction Ability of Colored QR Code

In a black-and-white QR code, there are two possibilities for each module: to represent 0 or 1. When the number of module(s) is n, then the data capacity of a black-and-white QR code is . If we assume the number of color(s) selected for a colored QR code is k, by extending the symbolic representation of a black-and-white QR code, we use different colors to mark corresponding in the k-nary system and it is clear that a colored QR code with a total of n module(s) has a data capacity of . Therefore, as long as we select a total of more than 2 colors for encoding, the data capacity of a colored QR code is time(s) that of a black-and-white code.

By placing coding modules around the logo, we can effectively avoid the possible loss of modules in the area covered by the logo, and thus the error correction ability of the QR code is improved, as demonstrated in Table 1.

Table 1. Comparison of error correction ability between black-and-white QR code with a logo and colored QR code adaptive to logo size.

Error Correction Performance / QR Code	Codeword Covered by the Logo	Error Correction Ability
Black-and-white QR Code with a Logo	Parts of code words covered	Weak
Colored QR Code Adaptive to Logo Size	No codeword covered	Strong

8. Conclusion

The appearance of 2D barcode has greatly changed our life, yet the black-and-white 2D code is still of some defects. In view of the small data capacity of the black-and-white 2D code, this paper puts forward a method to design the colored QR code. Through an encoding process with multiple colors, we have managed to increase the coding density of information in the obtained QR code; besides, as proposed in the paper, modules are placed around the logo so that loss of modules can be avoided. In this way, the error correction ability of the colored code is enhanced.

References

1. ZHANG Li, DU Min, WU Baifeng. Sequence Two-dimension Symbolic Design and Mask Solution [J], Journal of Image and Graphics, 2009(07), 1426-1431.
2. CHEN Jiong. Study and Realization of Encoding and Decoding Technologies for QR Code [D], Xi'an: Xi'an University of Electronic Science and Technology, 2012.
3. LI Zhaohui. Study on Colored Two-dimensional Code Based on Image Processing [D], Tianjin: Tianjin Polytechnic University, 2009.
4. JIA Huaguo. Study and Application of Encoding and Decoding Algorithms for Colored Two-dimensional Code on Mobile Terminals [D], Hangzhou: Zhejiang University of Technology, 2009.
5. HUANG Cuicui. Design and Encoding Method of Central Region Visual Two-dimensional Barcode [D], Dalian: Liaoning Normal University, 2011.

6. HUANG Guoxiang.RGB Color Space and Its Application [D], Changsha: Central South University, 2002.
7. GB/T18284-2000[S]. National Standards for Quick Response Matrix Codes
8. LI Yunpeng. Study on Application and Hardware Implementation of RS Codes and TURBO Codes [D], Xi'an: Xi'an University of Electronic Science and Technology, 2003.
9. QR Code: Technology and Application [M], compiled by Article Numbering Center of China, Beijing: China Standards Press, 2002.
10. YU Yingzheng. Study on Relevant Technologies of QR Two-dimensional Codes [D], Beijing: Beijing Jiaotong University, 2014.

Detection of imagined and actual hand movements with EEG signals

Miguel Contreras, Javier Suárez and Xiao-Hua Yu*
Department of Electrical Engineering,
California Polytechnic State University, San Luis Obispo,
San Luis Obispo, CA 93407, USA
**E-mail: xhyu@calpoly.edu*

For an electroencephalograph (EEG)-based brain–computer interface (BCI) system, the detection of mental states from spontaneous EEG signals is very important. In this paper, an approach based on artificial neural network (ANN) and discrete wavelet transform (DWT) is developed. Computer simulation results on multiple subjects show that the proposed method can successfully classify imagined and actual hand movements using EEG signals.

Keywords: EEG Signals; Artificial Neural Networks; Wavelet Transform; Brain–Computer Interface.

1. Introduction

Electroencephalography (EEG) is the recording of electrical activity along the scalp which measures voltage fluctuations resulting from ionic current that flows within the neurons of the brain ([1]). In clinical studies, EEG signals are usually recorded from multiple electrodes placed on the scalp over a short period of time. It is often used to diagnose diseases related with mental activity disorders, such as epilepsy, sleep disorders, coma, encephalopathies, and brain death ([1]).

It is known that a simple movement of the arm or gesture with a hand can cause the brain to fire hundreds and sometimes thousands of neural signals. In [2], the feasibility of decoding human brain activity for physical movement is demonstrated by inferring voluntary natural movements from EEG signals. While more accurate results may be obtained by complicated invasive neural interface systems, the EEG based approach is noninvasive and only requires a fabric cap with sensors attached. It is not only easy to repair or replace, but also safer to users. However, studies show that much of the detailed EEG signals can be distorted or blocked by human skull; thus advanced signal processing and pattern recognition technologies must be developed to analyze EEG signals.

* Corresponding author.

Wavelet transform has recently been applied for EEG signal noise reduction and feature extraction ([3], [4]). In [3], statistical characteristics of wavelet transform coefficients, such as mean, standard deviation, the maximum and minimum value etc. are selected as features for epilepsy detection. In [5], EEG signal energy distribution is calculated using Parseval's theorem, which then becomes the input vector to an artificial neural network (ANN) for seizure detection.

In this paper, an approach based on ANN and discrete wavelet transform is investigated to detect imagined and actual hand movements. Unlike the epilepsy detection problem in [5] where single channel EEG signal is usually considered, EEG measurements are now obtained from a scalp with 64 electrodes which is placed on the primary motor cortex of a subject's brain. In other words, the EEG signal studied in this research is a 64 dimension vector, which is more difficult to process than a single channel signal. A multi-layer feedforward neural network is employed to identify the imagined and actual movements; the inputs of neural network is the information on energy distribution of the measured 64-channel EEG signals (i.e. percentage of the energy of a particular frequency band with respect to the total energy of signal) obtained using discrete wavelet transform (DWT). Computer simulation results on multiple subjects show that the proposed neural network model can successfully classify imagined and actual hand movements with a detection rate of 93.33%.

The rest of the paper is organized as follows: section 2 discusses the details of the proposed neural network model. Computer simulation results are presented in section 3. Finally, section 4 concludes the paper and gives direction for future works.

2. The Proposed Neural Network Model for EEG Signal Classification Using Wavelet Transform

The continuous wavelet transform of a signal *f(t)* can be written as:

$$T(\tau, s) = \frac{1}{\sqrt{s}} \int_{-\infty}^{\infty} f(t) \Psi\left(\frac{t-\tau}{s}\right) dt \qquad (1)$$

where $\psi(\cdot)$ is called the "mother wavelet" and ψ^* represents its complex conjugate; s is the scaling factor and τ is the shifting factor. In computer simulations, discrete wavelet transform is performed by selecting

$$s = 2^a, \ \tau = 2^b \qquad (2)$$

where *a* and *b* are positive integers.

Wavelet transform decomposes signals into multiple frequency bands, as shown in Fig. 1. That is, wavelet transform can be considered as passing a signal through a set of low-pass (LP) and high-pass filters (HP). Through this process, a signal can be decomposed into different levels. At each level, it contains a set of detail coefficients (D) and approximation coefficients (A), as shown in Fig. 2.

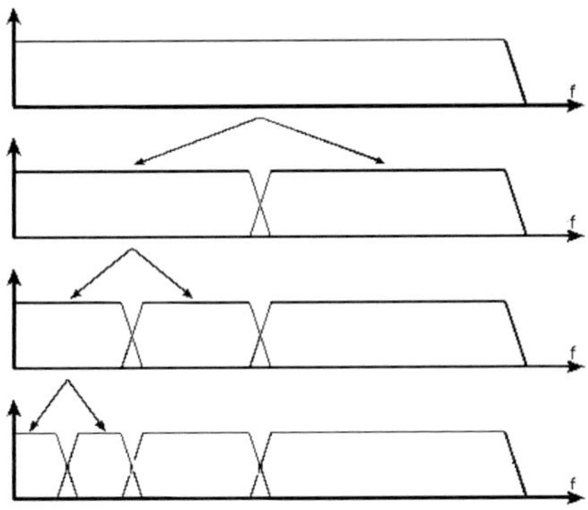

Fig. 1. The wavelet transform for signal decomposition [6].

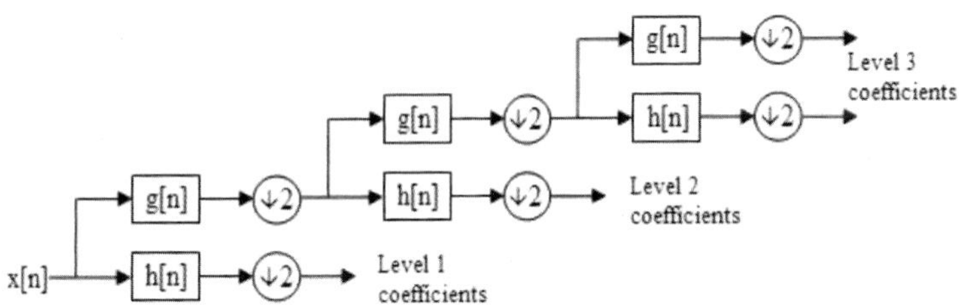

Fig. 2. The wavelet transform for signal decomposition [7].

Let E be the total energy of the signal, A_i be the approximation coefficients and D_i be the detail coefficients (i = 1, 2, ... L_1, where L_1 is the total number of approximation or detail coefficients of the first level decomposition), then:

$$E_T = \sum_{i=1}^{L_1} (A_i)^2 + \sum_{i=1}^{L_1} (D_i)^2 \tag{3}$$

Similarly, let C to be the approximation or detail coefficients of the specific decomposition level in interest (m). If it contains L_m coefficients, the specific energy E_s can be written as:

$$E_s = \sum_{i=1}^{L_m} (C_i)^2 \qquad (4)$$

Finally, the energy ratio E_r, represented by percentage, can be calculated by:

$$E_r = \frac{E_s}{E_T} \times 100\ (\%) \qquad (5)$$

The block diagram of the proposed approach is shown in Figure. 3.

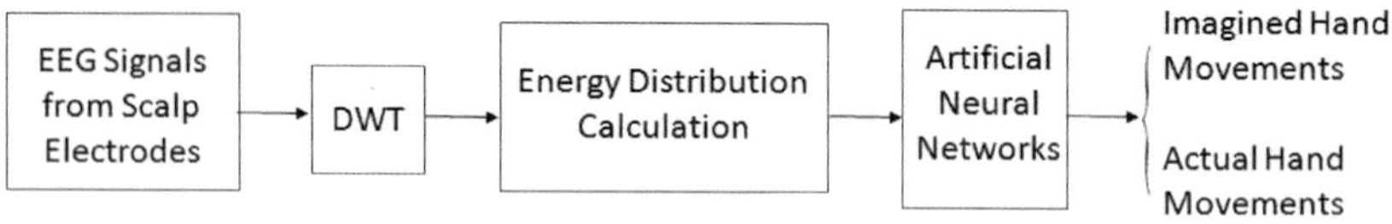

Fig. 3. The Proposed Algorithm Based on DWT and ANN.

3. Simulation Results

A typical electrode map to obtain EEG signals is shown in Fig. 4. The data used in simulation is from an online resource, PhysioNet. The data consist of one to two minute EEG recordings, obtained from 109 volunteers using a 64-channel EEG recorder (BCI2000) system ([8], [9]).

In this research, we consider the following two tasks performed by subjects:

Task 1: open and close left or right fist

Task 2: imagine opening and closing left or right fist

The signals obtained from 64 electrodes that are placed on the primary motor cortex of a subject's brain are sampled at 160 samples per second and recorded in EDF+ format (European Data Format). Obviously, it is very difficult to classify the two tasks by simply observing signal waveforms; thus feature extraction from signals must be performed. In this study, by trail-and-error, dB4 mother wavelet is employed and a total of five decomposition levels are considered. Unlike the approach proposed in [5] which considers the energy distribution at all wavelet decomposition levels for a single channel EEG signal, in this research, only the percentage of energy distribution in a particular frequency band is employed to reduce the dimensionality of input vector (64 channels) to the neural network. It is noticed that for the two classes (i.e. imagined and actual hand movements), the energy represented by A5 coefficients (approximation coefficients of level 5) show the most significant change among all the frequency bands (i.e. D1, D2, D3, D4, D5 and A5) and thus chosen to be the feature vectors. The decomposed signal in the frequency band of A5 is often referred as "delta" wave in EEG signals. Also, the influence of measurement noise can be minimized by not including D1 in the feature vector.

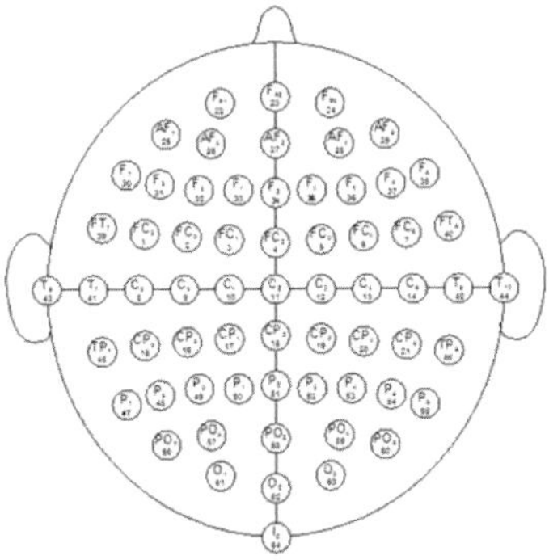

Fig. 4. EEG Electrode Map; 64 standard electrodes [9].

The artificial neural network employed in this research has 15 hidden neurons in its hidden layer. The nonlinear activation function of hidden neuron is hyperbolic tangent sigmoid (tansig):

$$g(x) = \frac{1-e^{-x}}{1+e^{-x}} \tag{6}$$

The weights of the neural network are initialized at random, and then updated to minimize the following objective function:

$$[e(n)]^2 = \frac{1}{2}[\hat{y}(n) - y_d(n)]^2 \tag{7}$$

where $\hat{y}(n)$ is the neural network output and $y_d(n)$ is the desired output (i.e. correct classification). The neural network is trained by Levenberg-Marquardt algorithm.

For neural network training and testing, we randomly selected 30 subjects from the database. Computer simulation results show that the neural network is able to identify the imaged hand movement with 100% successful rate. For actual hand movement, the detection rate is 93.33%.

4. Conclusions and Future Work

Electroencephalograph (EEG) signals record the electrical activities of biological neural networks in human brain, thus provide a non-invasive approach for building a brain–computer interface (BCI) system. In this research, artificial neural network is employed for the detection of imaged and actual hand movement. Feature extraction is performed by discrete wavelet transform and signal energy distribution, which then becomes the input vector of neural network. Computer simulation results show that detection rate is 93.33% for the proposed approach. Future works may include the addition of a pre-processing filter to reduce the noise in measurement data and perform more tests to fully study the performance of the proposed approach.

References

1. Electroencephalography, http://en.wikipedia.org/wiki/Electroencephalography.
2. Contreras-Vidal, J.L.; Presacco, A.; Agashe, H.; Paek, A., "Restoration of Whole Body Movement: Toward a Noninvasive Brain-Machine Interface System," Pulse, IEEE, vol.3, no.1, pp.34-37, Jan. 2012.
3. Kharat, P., Dudul, S., "Daubechies Wavelet Neural Network Classifier for the Diagnosis of Epilepsy", WSEAS Transactions on Biology and Biomedicine, 4(9): 103-113, Oct. 2012.
4. Zhang, X., Yin, L., Wang, W. "Wavelet Time-frequency Analysis of Electro-encephalogram (EEG) Processing", International Journal of Advanced Computer Science and Applications, 1(5), 2010.
5. Omerhodzic, I., Avdakovic, S., Nuhanovic, A., Dizdarevic, K. "Energy Distribution of EEG Signals: EEG Signal Wavelet-Neural Network Classifier". World Academy of Science, Engineering and Technology, (2010), http://arxiv.org/pdf/1307.7897.
6. Discrete wavelet transform, http://www.thepolygoners.com/tutorials/dwavelet/DWTTut.html
7. Discrete wavelet transform, http://en.wikipedia.org/wiki/Discrete_wavelet_transform.
8. Schalk, G., McFarland, D.J., Hinterberger, T., Birbaumer, N., Wolpaw, J.R. BCI2000: A General-Purpose Brain-Computer Interface (BCI) System. IEEE Transactions on Biomedical Engineering 51(6):1034-1043, 2004.
9. EEG Motor Movement/Imagery Dataset, http://physionet.org/pn4/eegmmidb/.

Target detection and tracking based on embedded multicore DSP system

Hao Chen*, Hong Zhang, Yi-Fan Yang, Lei Zhang and Yue-Ren Wang
Image Processing Center, Beihang University,
Beijing, China
**E-mail: chenhaoyiyou@163.com*

In this paper, a multicore DSP parallel implementation strategy for infrared target detection and tracking is proposed. Tracking-Learning-Detection is considered a highly efficient algorithm for tracking a single target. Although this algorithm can re-track a target when the target is occluded by other targets, many shortcomings still exist. This paper deals with the issue of target tracking by fusing local spectrum suppression based on human visual attention mechanism with tracking-learning-detection algorithm. Specifically, the area of the target is estimated to reduce the detection region and to increase the processing speed. Experimental results conducted on VIVID benchmark video library demonstrate that the proposed method can detect properly and track accurately infrared target in complex scenes.

Keywords: Visual Attention; Detection And Tracking; Local Spectrum Suppression; Tracking-Learning-Detection; Multicore DSP System.

1. Introduction

Conventional infrared target enhancement methods such as the median subtraction filter [1, 2], top-hat filter [3], and max-mean/ max-median filter [4] are widely used to reduce the background clutter. In recent years, researchers discovered that target detection by using human visual characteristics could often achieve more favorable results, because the selective attention property of the human vision can help people locate the salient target in the complex unknown scene quickly and precisely without being influenced by the complex background. In 1998, Itti firstly proposed a visual attention model [5] which attracted a wide-spread attention in the field of target detection and recognition. Subsequently, much effort has been devoted to detect the target from the visible image by the visual attention mechanism [6, 7, 8]. Recently, Spectrum Residual (SR) algorithm

* Corresponding author.

was proposed based on the Fourier Transform [9]. It caught many researchers' attention for its simple mechanism and efficiency. This work argued that the spectrum residual exactly corresponds to the image saliency.

Based on human visual attention mechanism, we propose a novel method based on the local spectrum suppression to detect infrared target. This paper deals with the issue of target tracking by fusing local spectrum suppression based on human visual attention mechanism with tracking-learning-detection algorithm [10].

2. TLD Algorithm and Improvements

2.1. *Local spectrum suppression*

Given a digital infrared image $f(x, y)$ of size $M \times N$, its transformed form in the frequency domain is:

$$F(u,v) = \frac{1}{MN}\sum_{x=0}^{M-1}\sum_{y=0}^{N-1} f(x,y)\exp(-j2\pi(ux/M+vy/N)) \tag{1}$$

As in the analysis of complex numbers, it is more convenient to express $F(u,v)$ in polar coordinates:

$$F(u,v) = |A(u,v)|\exp(-jP(u,v)) \tag{2}$$

Because the zero-frequency term dominates the values of the amplitude spectrum, the dynamic range of other intensities in the displayed image are compressed. To bring out those details, we perform a log transformation. The log amplitude spectrum is obtained by:

$$L(u,v) = log(A(u,v)) \tag{3}$$

In the second step, the sharp spikes are located in the log amplitude spectrum by using the cubic facet model. The cubic facet model assumes that in each neighborhood, the underlying grey-level intensity surface can be approximated by a cubic function. We define the 2D discrete orthogonal polynomials and show 9 sample masks for the 3×3 window.

In the third step, unlike recent approaches by using global spectral filtering, we define the concept of local spectrum suppression which adopts local spectrum filtering to smooth the spikes in the amplitude spectrum, thereby producing the pop-out of the infrared targets. In this work only the amplitudes are involved in the convolution while the phase information remains in the frequency domain [11].

2.2. *Improved TLD algorithm*

Three cascade classifiers are utilized to enhance the detection accuracy: (1) Variance classifier; (2) Random Forest classifier; (3) Nearest neighbor classifier. Image blocks not satisfying with preset constrains will be removed in each stage, and image blocks satisfying preset constrains will be sent into the next classifier. Such processing can not only improve the efficiency of classification, but also improve the stability of the classifier.

Although the TLD algorithm can achieve satisfactory detection results, however it needs for the detecting module to scan all scanning windows to detect targets. Furthermore, the TLD algorithm only provides the size of scanning windows with no information of the location. Therefore, it needs a large number of scanning windows, which occupy a lot of computing resources.

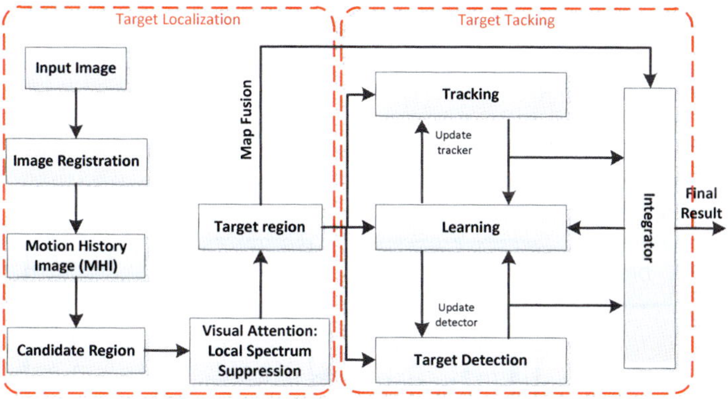

Fig. 1. Overall flowchart of the proposed method.

The main idea of the proposed algorithm, as shown in Figure 1, is describes as below:

(1) For aerial videos, the camera platform is moving, so firstly, the image registration is implemented based on the feature matching method before motion detection to compensate the camera motion. Then, the Motion History Image (MHI) [12] is used to get the candidate motion regions.

(2) Inspired by the property that regular patches in the spatial domain turn out to correspond to the spikes in the amplitude spectrum [13, 14, 15], we model regular patches which do not attract much attention by the visual system of our

human-beings. In this work, the log amplitude spectrum of the input infrared image is firstly computed. Then, we locate the amplitude spectral spikes by using the cubic facet model [16], and suppress the spikes by applying the local filtering. Finally, the target region in the spatial domain is obtained by reconstructing the 2D signal via the original phase and the filtered amplitude spectrum.

(3) Detecting module of the TLD algorithm detects moving targets by scanning window, where the parameters of scanning window are set as follows: the scaling factor is 1.2, the horizontal step is 10% of the width, vertical step is 10% of the height, and the size of scanning window is 10 pixels.

3. Implement on Multicore DSP System

Single core DSP encounters frequency and heat bottleneck, multi-tasking frequently breaks out the dense computing pipeline. Multicore DSP which is more tightly coupled compared to multiple DSP system considerably increases the multi-tasking efficiency. Algorithm performance was accelerated considerably through reasonably splitting up of the whole algorithm into many specific tasks which can be assigned to multiple cores.

3.1. *Hardware system*

Hardware system uses the new generation multicore DSP (TMS320C6678) from TI and FPGA. The block diagram shows as Figure 2.

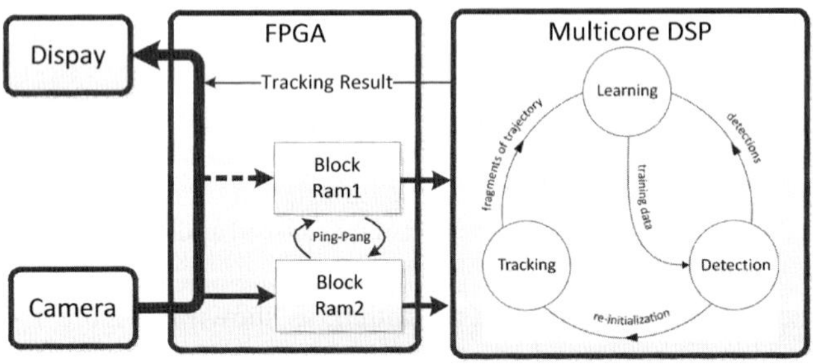

Fig. 2. Hardware system connection.

3.2. *CACHE optimization*

The access speed of external memory was 20 times slower contrast to internal memory. CACHE strategy effectively decreases the data access delay time by pre-fetch a whole cache line 128 bytes to CACHE memory for further processing.

The memory coherency problem was tough because each core has its own CACHE. There are approximately 2 types of inconsistency:

● CPU and DMA: DMA directly read from DDR, while CPU from CACHE.

● Two cores: Multiple copies of the same address in two cores.

The parallel processing with regard to multicore shared image frame requires manually CACHE coherency operation like Figure 3.

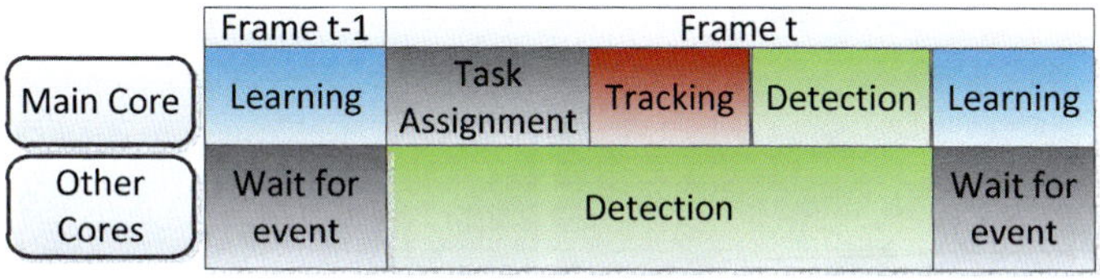

Fig. 3. CACHE consistency.

4. Experimental Results

VIVID database is utilized to verify the performance of the improved TLD algorithm. Experiment results demonstrate that the proposed algorithm can detect and track targets quickly and accurately under complex circumstances, such as disappear, turning, and mutual occlusion.

Figure 4 and Figure 5 are the experimental results under the case of object occlusion based on the TLD algorithm and proposed improved TLD algorithm respectively, where the red rectangle indicate the target location obtained by the TLD algorithm and the improved TLD algorithm.

It can be seen from Figure 4 (b) that the TLD algorithm is able to detect the moving target accurately despite the impact of a tree. It is known from Figure 4 (c) that the TLD algorithm is not able to track the same target after passing the tree. As can be seen from the results shown in Figure 5 (c), the improved TLD algorithm is still able to track the same target when passing the tree.

Fig. 4. Target tracking results via the TLD.

Fig. 5. Target tracking results via the improved TLD .

5. Conclusion

As a high performance single-target tracking method, the main characteristics of the TLD algorithm contain: it needs less prior information to achieve long-term target tracking with high computing speed and good real-time ability, and it can effectively track target when target disappear suddenly or under complex background. An improved TLD algorithm is here proposed to enhance the performance of the original TLD algorithm. Beside, multicore DSP parallel implementation strategy for infrared target detection and tracking was proposed and the whole system runs stable.

References

1. S. Qi, J. Ma, H, Li, S. Zhang, J. Tian, Infrared target enhancement via phase spectrum of Quaternion Fourier Transform, Infrared Physics & Technology . 62 (2014) 50-58.
2. J. Barnett, Statistical analysis of median subtraction filtering with application to point target detection in infrared backgrounds, Proceedings of the SPIE 1050 (1989)10–15.
3. V.T. Tom, Morphology-based algorithm for point target detection in infrared backgrounds, Proceedings of the SPIE 1954 (1993) 2–11.
4. S.D. Deshpande, Max-mean and max-median filters for detection of small-targets, Proceedings of the SPIE 3809 (1999) 74–83.
5. L. Itti, C. Koch, E. Niebur, A model of saliency-based visual attention for rapid scene analysis, IEEE Trans. Pattern Anal. Mach. Intell. 20 (1998) 1254-1259.
6. B.C. Ko, J. Nam, Object-of-interset image segmentation based on human attention and semantic region clustering, J. Opt. Soc. Am. A 23 (2006) 2462-2470.
7. Y. Xu, Y. Zhao, C. Jin, Z. Qu, L. Liu, X. Sun, Salient target detection based on pseudo-Wigner-Ville distribution and Renyi entropy, Opt. Lett. 35 (2010) 475-477.
8. W. Li, Saliency-based automatic target detection in forward looking infrared images, Proc. IEEE Conf. Image Process. (2009) 957-960.
9. X. Hou, L. Zhang, Saliency Detection: A Spectral Residual Approach, Proc. IEEE Conf. Computer Vision and Pattern Recognition, 2007.
10. Z. Kalal, J. Matas, and K. Mikolajczyk. P-N Learning: Bootstrap ping binary classifiers by structural constraints. IEEE Conference on Computer Vision and Pattern Recognition, 2010: 49-56.

11. J. Li, X. An, X. Xu, H. he, Visula saliency based on scale-space analysis in the frequency domain. IEEE Trans. Pattern Anal. Mach. Intell. 35(2013) 996-1010.
12. Z. Yin and R. Collins, "Moving object localization in thermal imagery by forward-backward MHI," in CVPR Workshop. IEEE, 2006.
13. M. Haralick, Digital step edges from zero crossing of second directional derivatives. IEEE Trans. Pattern Anal. Mach. Intell. 6(1984) 58-68.
14. L. Yang, J. Yang, K. Yang, Adaptive detection for infrared target under sea–sky complex background, Electron. Lett. 40 (17) (2004) 1083-1085.
15. H. Chen, H Zhang, Real-time automatic small infrared target detection using local spectral filtering in the frequency. Proceedings of SPIE - The International Society for Optical Engineering, 9273 (2014).
16. G.D. Wang, C.Y. Chen, X.B. Shen, Facet-based infrared target detection method, Electron. Lett. 41 (22) (2005) 1244-1246.

Dynamic properties and modified projective synchronization of two four-scroll hyperchaotic systems with non-identical fractional orders

Zhen-Wu Sun*

Institute of Applied Sciences, Shanghai Dianji University,
Shanghai, 201306, PR China
E-mail: sunzw@sdju.edu.cn

To speed up the application of chaos, the dynamic properties of two fractional-order hyperchaotic systems were investigated. It was found that both of them remain chaotic when their orders are in the range (0, 1]. Based on the stability condition of fractional-order differential systems, a controller was designed to realize modified projective synchronization of the two systems. Numerical simulation showed the effectiveness of the proposed controller.

Keywords: Hyperchaotic System; Modified Projective Synchronization; Fractional Order.

1. Introduction

In the investigation of chaos, two topics have attracted increasing interest in recent years. One is on hyperchaotic systems with multi-scroll attractors. It is believed that they can clearly improve the security of the communication scheme by generating complicated dynamics. However, most of hyperchaotic systems can only generate two-scroll attractors [1-2]. Although some systems with four-scroll attractors have been found [3], most of them are chaotic. Recently, Dadras *et al.* [4] and Cang *et al.* [5] presented one of hyperchaotic systems with four-scroll attractors respectively, called Dadras system and Cang system in the following. It is greatly possible that the two systems will be widely adopted in engineering and communication. The other is on fractional-order systems. The major merit of fractional calculus lies in the fact that it can offer a deeper insight into physical processes underlying a long range memory behavior, and has been proven to be a very suitable mathematical tool modeling the reality [6]. In addition, fractional-order chaotic systems have more adjustable variables than integer-order ones so that the former are superior to the latter in application [7]. Therefore, the synchronization of fractional-order chaotic systems has been

*Corresponding author.

paid much attention to. Among all kinds of synchronization, projective synchronization (PS) [8] is a noticeable one as the unpredictability of the scaling factor can enhance the security in digital communications [9]. Modified projective synchronization (MPS) is one kind of PS that the responses of the synchronized dynamical states synchronize up to a constant scaling matrix. The proportional feature of MPS can be used to extend binary digital to M-nary digital communications for achieving fast communication. MPS of fractional-order chaotic systems was already studied in some refs. [7, 9]. It is worth pointing out that the above refs. studying the synchronization of fractional-order systems were concerned with the ones with identical orders. However, the synchronization of ones with non-identical orders apparently has some advantages. One is that the orders of response systems are freed from the ones of drive systems, which provides more flexible mechanism in the selections of systems. Moreover, the orders are variable parameters, which can be used as secret keys in digital communication.

As is well known, exact solutions of fractional-order differential systems cannot be found easily [10]. Thus the systems are always studied by analytical and numerical methods, including Adomian decomposition method [11], Laplace decomposition method [12], and *so on*. Differential transform method (DTM) was proposed by Zhou [13] to solve linear and nonlinear problems. Arikoglu *et al.* [14] and Odibat *et al.* [15] extended the algorithm to solve fractional-order differential equations. However, DTM has a drawback that it only gives a good approximation to the true solution in a small region. To overcome the shortcoming, multi-step differential transform method (MSDTM) was presented in ref. [16]. The interval is divided into many sub-intervals and the series solution for each interval is obtained by DTM. MSDTM was used to study fractional-order chaotic systems recently, where it was called as multi-step fractional differential transform method (MSFDTM) [17].

Motivated by the above discussions, in this paper, we investigate the dynamical properties and MPS of fractional-order Dadras system and Cang system. The rest of this paper is organized as follows. Section 2 analyzes the dynamical properties of the two systems. Section 3 presents a sheme of MPS of fractional chaotic systems with non-identical orders. In section 4, MPS of the above two systems is designed and simulated by MSFDTM.

2. Dynamic Properties of Two Fractional-Order 4D Hyperchaotic System

2.1. *Necessary condition for fractional-order differential systems to remain chaotic*

Consider the following fractional-order differential system:

$$D^{\alpha}\mathbf{x} = \mathbf{f}(\mathbf{x}), \tag{1}$$

where $0 < \alpha \le 1$ and $\mathbf{x} \in R^n$. The equilibrium points of system (1) are calculated by solving equation $\mathbf{f}(\mathbf{x}) = 0$. These points are locally asymptotically stable if all the eigenvalues of the Jacobian matrix $A = \partial\mathbf{f}/\partial\mathbf{x}$ evaluated at the equilibrium points satisfy the condition [18]:

$$|\arg(\mathrm{eig}(A))| > \alpha\pi/2. \tag{2}$$

In a nonlinear system, a saddle point is an equilibrium point at which the equivalent linearized model has at least one eigenvalue in the stable region and one eigenvalue in the unstable region. A saddle point is called saddle point of index *n* if *n* eigenvalues of its ones are unstable. It has been proved that saddle points of index 1 in three-dimensional (3D) chaotic systems are responsible only for connecting scrolls [19]. Scrolls of attractors are generated around saddle points of index 2. Therefore, the necessary condition for a 3D system to remain chaotic is that at least one of its equilibrium points is a saddle point of index 2. If λ is an unstable eigenvalue of a saddle point, it is requested that

$$\tan(\alpha\frac{\pi}{2}) > \frac{|\mathrm{Im}(\lambda)|}{\mathrm{Re}(\lambda)} \quad\Rightarrow\quad \alpha > \frac{2}{\pi}\tan^{-1}\frac{|\mathrm{Im}(\lambda)|}{\mathrm{Re}(\lambda)}. \tag{3}$$

If the conclusion is extended to 4D nonlinear systems, the necessary condition for them to remain chaotic should be that at least one of its unstable equilibrium points is a saddle point of index 2 or 3.

2.2. *Property of fractional-order Dadras system*

Fractional-order Dadras system is described by

$$\begin{cases} d^{\alpha}x_1/dt^{\alpha} = ax_1 - x_2x_3, \\ d^{\alpha}x_2/\mathrm{d}t^{\alpha} = x_1x_3 - bx_2, \\ d^{\alpha}x_3/dt^{\alpha} = cx_1x_2 - dx_3 + gx_1x_4, \\ d^{\alpha}x_4/dt^{\alpha} = fx_4 - hx_2. \end{cases} \tag{4}$$

$0 < \alpha \leq 1$. a, b, c, d, f, g, and h are real constant parameters. In this paper, they are adopted as 8, 40, 2, 14, 0.05, 5, 0.2, respectively, so that system (4) is hyperchaotic when α is equal to 1. It has five equilibrium points: O(0, 0, 0, 0), E_1(-5.045, +2.256, -17.788, +9.025), E_2(+5.045, +2.256, +17.788, +9.025), E_3(+5.045, -2.256, -17.788, -9.025), E_4(-5.045, -2.256, +17.788, -9.025).

System (4) is linearized and the Jacobian matrix can be obtained as

$$J = \begin{bmatrix} a & -x_3 & -x_2 & 0 \\ x_3 & -b & x_1 & 0 \\ cx_2 + gx_4 & cx_1 & -d & gx_1 \\ 0 & -h & 0 & f \end{bmatrix}. \tag{5}$$

One can gain the eigenvalues of the linearized system by letting $|J - \lambda I| = 0$. The eigenvalues corresponding to O are obtained as 8, -40, -14 and 0.05. Therefore, the equilibrium is always an unstable saddle point of index 2 when $0 < \alpha \leq 1$. Doing the same calculation for the other equilibria, it can be verified that they all have the four eigenvalues, -39.398, 0.091 and -3.321$\pm$15.435i. E_1, E_2, E_3, and E_4 are unstable saddle points of index 1 by condition (2).

Simulation is performed for α = 0.9, 0.6, 0.3, respectively. The results are as shown in fig. 1, which demonstrate that chaos indeed exists in the three situations. One may easily consider the four scrolls in fig. 1 are generated around the four marginal equilibria, E_1, E_2, E_3, and E_4. In fact, the scrolls are around the origin as it is the sole unstable saddle point of index 2 of system (4). The interesting phenomenon was also reported in ref. [20].

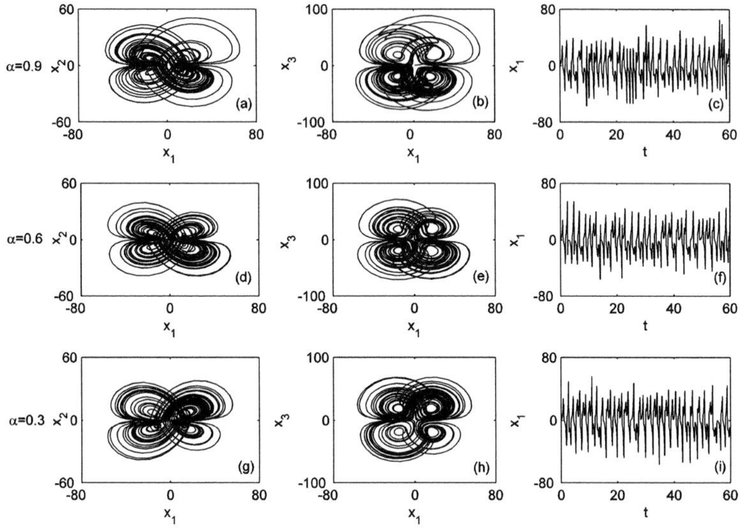

Fig. 1. Projections of the four-wing attractors of fractional-order Dadras system: (a)-(c) for α=0.9, (d)-(f) for α=0.6, (g)-(i) for α=0.3.

2.3. *Property of fractional-order Cang system*

Fractional-order Dadras system is described by

$$\begin{cases} d^{\beta} y_1 / dt^{\beta} = -jy_1 - sy_4 + y_2 y_3, \\ d^{\beta} y_2 / \mathrm{dt}^{\beta} = y_1 y_3 + my_2, \\ d^{\beta} y_3 / dt^{\beta} = ny_3 + qy_4 - y_1 y_2, \\ d^{\beta} y_4 / dt^{\beta} = py_4 - ry_3. \end{cases} \tag{6}$$

$0 < \beta \leq 1$. $j, m, n, p, q, r,$ and s are real constant parameters. In this paper, they are adopted as 50, -16, 10, 0.2, 16, 0.5, 10, respectively, so that system (6) is hyperchaotic when β is equal to 1. In the same way as above, the equilibria of system (6) are obtained as: O'(0, 0, 0, 0), E'_1(-28.284, -39.039, +22.084, +55.210), E'_2(+28.284, -39.039, -22.084, -55.210), E'_3(-28.284, +64.039,-36.226, -90.565), E'_4(+28.284, +64.039, +36.226, +90.565).

O' is an unstable saddle point of index 2, whose eigenvalues are -50, -16, 9.101 and 1.099. E'_1 and E'_2 are unstable saddle points of index 1, whose eigenvalues are 0.233, -55.864, -0.084$\pm$44.461i. The eigenvalues corresponding to E'_3 and E'_4 are 0.174, -60.522, 2.274$\pm$63.287i. Therefore, E'_3 and E'_4 are unstable saddle points of index 3 when β >0.977, and turn to unstable saddle points of index 1 when $\beta < 0.977$.

Numerical simulation is also for β = 0.98, 0.6, 0.3, respectively. The simulation results are as shown in fig. 2, showing that chaos still exists in the three situations. We cannot know whether the scrolls in fig. 2 (*a*) and (*b*) are generated around only O' or the three equilibria, O', E'_3 and E'_4 as all of them are unstable saddle points of index n ($n > 1$). However, it is confirmed that the scrolls in fig. 2 (*d*), (*e*), (*g*) and (*h*) are generated only around O' because it is the sole such unstable saddle point when β < 0.97. It is supported again that multi-scrolls in 4D chaotic systems can be generated around one equilibrium.

3. MPS Scheme of Two Chaotic Systems With Non-Identical Fractional Orders

Consider the following drive and response systems as

$$D^{\alpha}\mathbf{x} = \mathbf{f}(\mathbf{x}), \tag{7}$$

and

$$D^{\beta}\mathbf{y} = \mathbf{g}(\mathbf{y}) + \mathbf{U}(\mathbf{x}, \mathbf{y}). \tag{8}$$

Here $\mathbf{x}, \mathbf{y} \in R^n$ are n-dimensional state vectors of the drive system and the response system respectively. $\mathbf{f}, \mathbf{g} : R^n \to R^n$ are continuous nonlinear vector functions. α , β are the fractional orders of the two systems. $\mathbf{U}(\mathbf{x}, \mathbf{y})$ is a controller to be determined later.

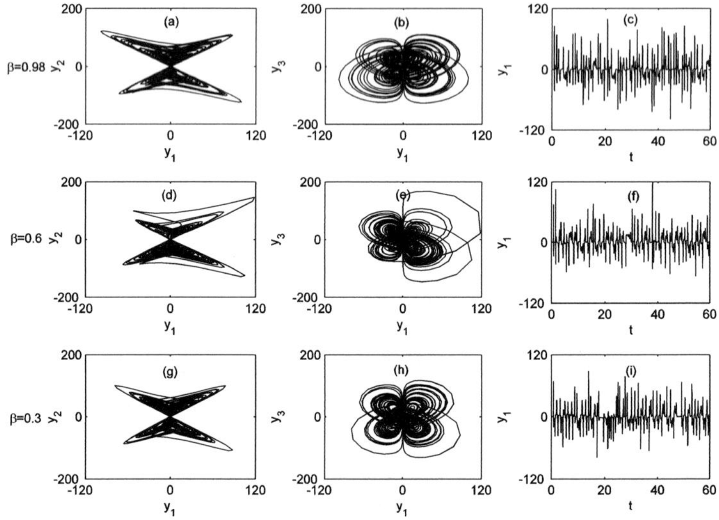

Fig. 2. Projections of the four-wing attractors of fractional-order Cang system: (a)-(c) for β=0.98, (d)-(f) for β=0.6, (g)-(i) for β=0.3.

In the definition of MPS, the synchronization error is defined as

$$\mathbf{e} = \mathbf{y} - \boldsymbol{\gamma}\mathbf{x}, \tag{9}$$

where $\boldsymbol{\gamma}$ is called the scaling matrix.

Definition 3.1. *For the drive system (7) and the response one (8), it is said they are MPS if there exists a nonzero matrix* $\boldsymbol{\gamma} = diag(\gamma_1, \gamma_2, \cdots, \gamma_n)$ *for arbitrary initial conditions* ***x****(0) and* ***y****(0) such that*

$$\lim_{t \to +\infty} \|\mathbf{e}(t)\| = 0, \tag{10}$$

where $\|\cdot\|$ *represents the matrix norm.*

To introduce conveniently, we decompose systems (7) and (8) as

$$D^{\alpha}\mathbf{x} = \mathbf{A}\mathbf{x} + \mathbf{F}(\mathbf{x}), \tag{11}$$

and

$$D^{\beta}\mathbf{y} = \mathbf{B}\mathbf{y} + \mathbf{G}(\mathbf{y}) + \mathbf{U}(\mathbf{x}, \mathbf{y}). \tag{12}$$

Here $\mathbf{A}$, $\mathbf{B}$ are the parameter matrices for the linear parts in the drive and response systems respectively, and $\mathbf{F}: R^n \to R^n$ and $\mathbf{G}: R^n \to R^n$ are their nonlinear parts. Then we separate the controller $\mathbf{U}(\mathbf{x}, \mathbf{y})$ into two sub-controllers $\mathbf{U}_1(\mathbf{x}, \mathbf{y})$ and $\mathbf{U}_2(\mathbf{x}, \mathbf{y})$, *i.e.*, $\mathbf{U}(\mathbf{x}, \mathbf{y}) = \mathbf{U}_1(\mathbf{x}, \mathbf{y}) + \mathbf{U}_2(\mathbf{x}, \mathbf{y})$.

To realize MPS between the two systems, the sub-controllers is designed as

$$\begin{cases} \mathbf{U}_1(\mathbf{x}, \mathbf{y}) = (\mathrm{D}^{\beta-\alpha} - \mathbf{I})\mathbf{g}(\mathbf{y}), \\ \mathbf{U}_2(\mathbf{x}, \mathbf{y}) = \mathrm{D}^{\beta-\alpha}[\boldsymbol{\gamma}(\mathbf{A} - \mathbf{B})\mathbf{x} - \mathbf{G}(\mathbf{y}) + \boldsymbol{\gamma}\mathbf{F}(\mathbf{x}) - \mathbf{K}\mathbf{e}], \end{cases} \tag{13}$$

where $\mathbf{I}$ is the identity operator, and $\mathbf{K} \in R^{n\times n}$ is a control matrix composed of constant parameters. Using eqs.(13) into eq.(12), one can obtain

$$\begin{aligned} D^{\alpha}\mathbf{y} &= D^{\alpha-\beta}[\mathrm{D}^{\beta}\,\mathbf{y}] \\ &= D^{\alpha-\beta}[\mathbf{g}(\mathbf{y}) + (\mathrm{D}^{\beta-\alpha} - \mathbf{I})\mathbf{g}(\mathbf{y}) + \mathbf{U}_2(\mathbf{x}, \mathbf{y})] \\ &= \mathbf{g}(\mathbf{y}) + \mathrm{D}^{\alpha-\beta}\,\mathbf{U}_2(\mathbf{x}, \mathbf{y}) \\ &= \mathbf{g}(\mathbf{y}) + \boldsymbol{\gamma}(\mathbf{A} - \mathbf{B})\mathbf{x} - \mathbf{G}(\mathbf{y}) + \boldsymbol{\gamma}\mathbf{F}(\mathbf{x}) - \mathbf{K}\mathbf{e} \\ &= \mathbf{B}\mathbf{y} + \boldsymbol{\gamma}(\mathbf{A} - \mathbf{B})\mathbf{x} + \boldsymbol{\gamma}\mathbf{F}(\mathbf{x}) - \mathbf{K}\mathbf{e}. \end{aligned} \tag{14}$$

The error dynamic system is obtained as

$$\begin{aligned} D^{\alpha}\mathbf{e} &= D^{\alpha}(\mathbf{y} - \boldsymbol{\gamma}\mathbf{x}) \\ &= \mathbf{B}\mathbf{y} + \boldsymbol{\gamma}(\mathbf{A} - \mathbf{B})\mathbf{x} + \boldsymbol{\gamma}\mathbf{F}(\mathbf{x}) - \mathbf{K}\mathbf{e} - \boldsymbol{\gamma}[\mathbf{A}\mathbf{x} + \mathbf{F}(\mathbf{x})] \\ &= (\mathbf{B} - \mathbf{K})\mathbf{e}. \end{aligned} \tag{15}$$

One can always choose a proper $\mathbf{K}$ to make all the eigenvalues $\lambda_i (i = 1, 2, \cdots, n)$ of $\mathbf{B} - \mathbf{K}$ satisfy inequality (2). Then system (15) is asymptotically stable and the error vector $\mathbf{e}$ converges to zero as $t \to \infty$. Hence MPS between the chaotic systems (11) and (12) is guaranteed.

4. MPS of Fractional-Order Dadras System and Cang System

In order to observe MPS behaviour of the two fractional-order systems, we select system (4) as the drive system and system (6) as the response system. α is chosen as 0.8 and β as 0.7. Therefore, the response one are described as

$$\begin{cases} d^{0.7}y_1 / dt^{0.7} = -50y_1 - 10y_4 + y_2y_3 + u_1, \\ d^{0.7}y_2 / \mathrm{dt}^{0.7} = y_1y_3 - 16y_2 + u_2, \\ d^{0.7}y_3 / dt^{0.7} = 10y_3 + 16y_4 - y_1y_2 + u_3, \\ d^{0.7}y_4 / dt^{0.7} = 0.2y_4 - 0.5y_3 + u_4. \end{cases} \tag{16}$$

Here u_1, u_2, u_3 and u_4 are four control functions to be designed so that MPS of the two systems can be realized. $\mathbf{U} = (u_1, u_2, u_3, u_4)^T$.

Let us define MPS errors as $e_1 = y_1 - \gamma_1 x_1$, $e_2 = y_2 - \gamma_2 x_2$, $e_3 = y_3 - \gamma_3 x_3$ and $e_4 = y_4 - \gamma_4 x_4$, where $\gamma_i(t) \in R(i = 1,2,3,4)$ are the scaling factors. $\mathbf{K}$ is chosen as $diag(k_1, k_2, k_3, k_4)$. According to the above proposed method, the controller function are adopted as follows:

$$\begin{cases} u_1 = (D^{-0.1} - 1)(-50y_1 - 10y_4) - y_2y_3 + D^{-0.1}[\gamma_1(58x_1 + 10x_4 - x_2x_3) - k_1e_1], \\ u_2 = (D^{-0.1} - 1)(-16y_2) - y_1y_3 + D^{-0.1}[\gamma_2(-24x_2 + x_1x_3) - k_2e_2], \\ u_3 = (D^{-0.1} - 1)(10y_3 + 16y_4) + y_1y_2 + D^{-0.1}[\gamma_3(-24x_3 - 16x_4 + 2x_1x_2 + 5x_1x_4) - k_3e_3], \\ u_4 = (D^{-0.1} - 1)(0.2y_4 - 0.5y_3) + D^{-0.1}[\gamma_4(-0.2x_2 + 0.5x_3 - 0.05x_4) - k_4e_4]. \end{cases} \tag{17}$$

If $\mathbf{K}$ is chosen as $\mathbf{K} = diag(5,5,15,5)$, the error system is calculated as

$$D^{0.8}\mathbf{e} = (\mathbf{B} - \mathbf{K})\mathbf{e} = \begin{bmatrix} -55 & 0 & 0 & -10 \\ 0 & -21 & 0 & 0 \\ 0 & 0 & -5 & 16 \\ 0 & 0 & -0.5 & -4.8 \end{bmatrix} \times \begin{bmatrix} e_1 \\ e_2 \\ e_3 \\ e_4 \end{bmatrix}. \tag{18}$$

The eigenvalues of the matrix $\mathbf{B} - \mathbf{K}$ are λ_1 = -55, λ_2 = -21 and $\lambda_{3,4}$ = -4.900$\pm$2.827i, which all satisfy $|\arg(\lambda_i)| > 0.8\pi / 2$. According to the stability condition of fractional-order systems, the error system is asymptotically stable and the synchronization is guaranteed.

To verify the effectiveness of the controller (17), we simulate the dynamic behaviours of the drive and response systems by MSFDTM. $\boldsymbol{\gamma}$ is chosen as *diag*(-1/2, -1/2, -1/2, -1/2) and the time step as 0.001. The initial conditions of the drive and response systems are (5, 5, 5, 0.5). The results are illustrated in fig. 3. It shows MPS between their state variables has been realized indeed. The splinters like tails, stretching from the linear parts, are brought out in the beginning of the synchronizing procedure, when the synchronization between the two systems aren't achieved totally.

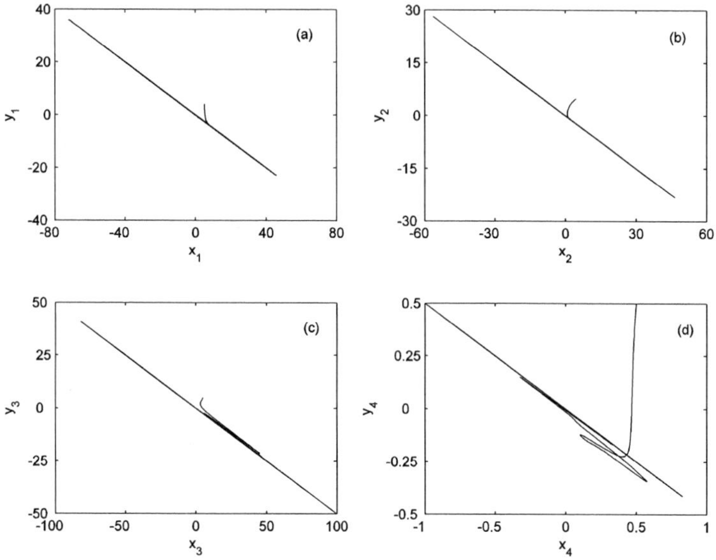

Fig. 3. MPS between of the driven system (4) and the response system (16) in the condition of the scaling matrix $\boldsymbol{\gamma} = \mathrm{diag}(-1/2,-1/2,-1/2,-1/2)$.

5. Conclusion

The behaviour and MPS of fractional-order Dadras system and Cang one were investigated according to the necessary condition of fractional-order systems and simulated by MSFDTM. It was found that the systems remain chaotic when their orders are in $(0, 1]$. MPS of the two systems with non-identical fractional orders were realized. All the theoretical analysis and simulation results showed the effectiveness of the proposed controller. The work in this paper may accelerate the application of fractional-order chaotic systems in practice.

References

1. Q. Jia, *Phys. Lett. A* **371**, 410(2007).
2. G.M. Mahmoud, E.E. Mahmoud and M.E. Ahmed, *Nonlinear Dyn.* **58**, 725(2009).
3. L. Wang, *Nonlinear Dyn.* **56**, 453(2009).
4. S. Dadras and H.R. Momeni, *Phys. Lett. A* **374**, 1368(2010).
5. S. Cang, G.Qi, and Z. Chen, *Nonlinear Dyn.* **59**, 515(2010).

6. R. Hilfer, *Applications of Fractional Calculus in Physics* (World Scientific, New Jersey, 2001).
7. X.J. Wu, H. Wang and H.T. Lu, *Nonlinear Anal. Real.* **13**, 1441(2012).
8. X.Y. Wang and Y.J. He, *Phys. Lett. A* **372**, 435(2008).
9. J. Bai, Y.G. Yu, S. Wang and Y. Song, *Commun. Nonlinear Sci. Numer. Simulat.* **17**, 1921(2012).
10. A. K. Alomari, *Comput. Math. Appl.* **61**, 2528(2011).
11. S.S. Ray and R.K. Bera, *Appl. Math. Comput.* **167**, 561(2005).
12. W.H. Deng and C.P. Li, *Phys. A* **353**, 61(2005).
13. J.K. Zhou, *Differential Transformation and Its Applications for Electrical Circuits* (Huazhong University Press, Wuhan, 1986, in Chinese).
14. A. Arikoglu and I. Ozkol, *Chaos Solitons Fract.* **34**, 1473(2007).
15. Z. Odibat, S. Momani and V.S. Erturk, *Appl. Math. Comput.* **197**, 467(2008).
16. Z.M. Odibat, C. Bertelle, M.A. Aziz-Alaoui and G.H.E. Duchamp, *Comput. Math. Appl.* **59**, 1462(2010).
17. M. Merdan, *Appl. Math. Model.* **37**, 6025(2013).
18. E. Ahmed, A.M.A. El-Sayed and H.A.A. El-Saka, *J. Math. Anal. Appl.* **325**, 542(2007).
19. D. Cafagna and G. Grassi, *Int. J. Bifur. Chaos* **13**, 2889(2003).
20. S. Dadras, H.R. Momeni, G.Y. Qi and Z.L. Wang, *Nonlinear Dyn.* **67**, 1161(2012).

Chapter 2

Computer Network and Information Security

Design and implementation of a RFID-based intelligent tools management system

Jia-Wen Xie*, Min-Li Wang and Jie Tan

Institute of Automation of, Chinese Academy of Sciences,
Haiding District, Beijing, China
**E-mail: xiejiawen2014@ia.ac.cn*

Safety is a priority for aviation companies. To ensure aviation safety, consistent maintenance is needed. The maintenance of airplanes requires good technology, and scientific management of maintenance tools help to improve maintenance efficiency and reliability. In this paper, we designed and implemented a Radio Frequency Identification (RFID) based intelligent tools management system for the scientific management of maintenance tools. The intelligent tool car that contains the tools needed for maintenance of airplanes has an enclosed metallic environment. Thus, its structure requires optimization and adjustments to be made before the RFID system can be embedded. RFID tags are then pasted on each tool. Due to the unique signals of each RFID tag, missing tools can be identified. Our experimental results show that the operation of the system is stable and reliable in the complex electromagnetic enclosed metallic environment.

Keywords: RFID Technology; Metallic Environment; Tool Car; Antenna Placement; Tool Management.

1. Introduction

After the occurrence of the MH370 aviation accident, the attention of people gradually focused on aviation safety. Maintenance is the important guarantee of aviation safety. Mismanagement of tools may result in maintenance errors, which greatly affect flight safety. For example, the lack of tools will seriously affect the efficiency of repair. If tools are accidentally forgot in certain parts of

---

* Corresponding author.

the aircraft, it may also pose a security risk. Therefore, it is meaningful to use a scientific tools management system in aviation service department.

RFID, an automatic identification technology[1], is widely deployed in security, assets tracking, supply chain management and many other fields from 1990. The advantages of RFID: first, RFID tag can pasted on different items, metallic or nonmetallic. Second, the system can use in different environments, liquid or metal. But the performance of RFID system can be strongly effected by environments. Many researches have been conducted to apply the RFID system to different environments. M.Periyasamy etc.[2] assessed and analyzed the performance of 13.56 MHz RFID in metal and liquid environment. C.A Diugwu etc.[3] researched field distributions and RFID reading in metallic roll cages. Xianming Qing etc.[4] studied the proximity effects of metallic environment and applied into the library shelf. On the maintenance tool management, the aircraft repair shops of China Southern Airlines Beijing Branch manage maintenance tools through the bar code management system[5]. ANA and NEC use UHF RFID technology to build the aircraft maintenance tools in and out management system[5]. Scholars have done many works and related applications in different environments. RFID technology also has been applied in relevant tools management. But less RFID applications are used for tools management in an enclosed metallic environment.

Tools are various, including metallic and nonmetallic tools. And most are stored in a complex closed metal environment. Therefore, with the RFID system embedded in, we design an intelligent tool car which has an enclosed metallic environment.

In this paper, a RFID-based intelligent tools management system is designed and implemented. Section 2 describes the configuration of tool car. Section 3 presents the placement of antennas based on the structure. Section 4 is software part of the system. Then experimental results are showed in section 5. Finally, section 6 draws some conclusions.

2. Set Up of Proposed System

The intelligent tools management system is composed of tool car and RFID system. For an ordinary tool car, RFID system cannot be embedded into it. The structure adjustment is needed. And the main consideration is the installation of antennas and the display panel of PDA.

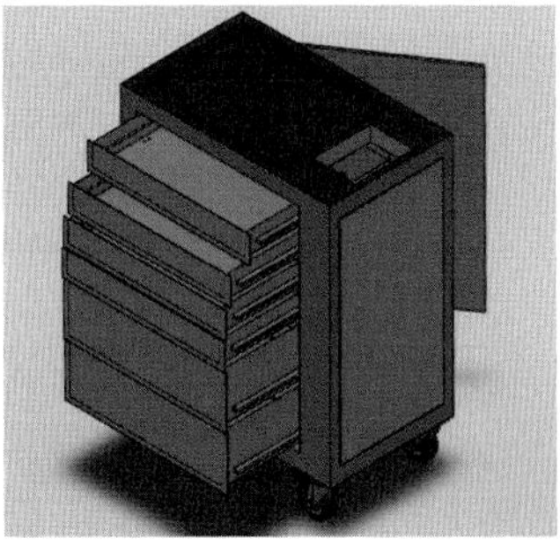

Fig. 1. The structure of tool car.

Figure 1 is the structure of tool car after adjustment. The display panel is installed on the surface of platform for it is very suited to operate and observation the data. However, the installation of antennas is affected by many factors. First, the space of tool car is small so that the installation location is limited. Then, inside of tool car has a complex electromagnetic environment. With these two situations considered, two installation methods are proposed, $antenna_{side}$ and $antenna_{up}$. $antenna_{side}$ represents the antennas mounted on the side of drawer. $Antenna_{up}$ represents the antennas mounted on the above of drawer. Section 3 will give a detailed description.

Once the structure of tool car sets up, we build the RFID system. The environment of the drawer is a narrow metallic environment. We select EastAmple E9238 high performance UHF directional slot antenna[6] which is suitable for a narrow signal area. The frequent ranges from 902MHz to 928MHz and the gain is 8dBi. Due to most tools are metal and wood, We choose Omni-ID Fit series tags, including Fit200, Fit210, Fit400, Fit400P, which have small size, high performance and especially suitable to pasted on the surface of tools. These series tags use Alien H3 IC which compliant with EPC Class 1 Gen2 protocol and the frequency range of EU and US. About the reader, we use FN-2184D four-channel UHF RFID reader to improve the reliability and read range. Finally, we build up the whole intelligent tool car. Figure 2 is the framework of the system.

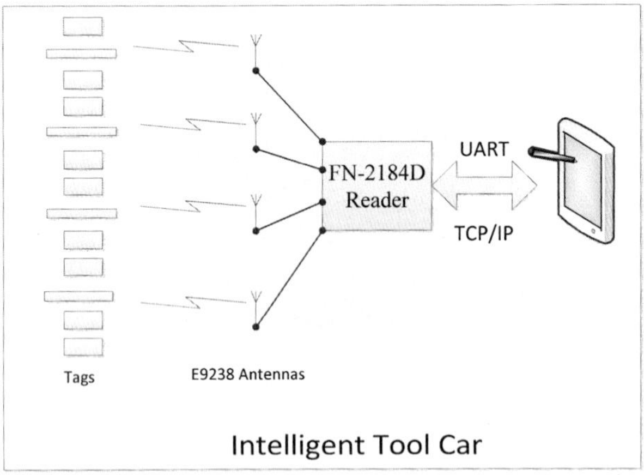

Fig. 2. The framework of the system.

3. Alternative of Antennas Placement

Intelligent tool car has a metal environment and metal surfaces are inside the drawers. There are large influence on the antennas which mainly include the absorption and reflection of electromagnetic signals[7]. The portion of High frequency electromagnetic signals which reach the metal surface are absorbed into heat, and another portion are reflected into all directions when they reach the face of a metal target. When the electromagnetic wave reflected by the metal and the reflected wave of tags meet by standing wave, the reflected wave of tags will be weakened, and it will greatly affect the performance of read. According to the radiation patterns of Fit 400 tags in [8], as shown in Figure 3. The read range will change along with the metal size. And when tag is pasted on the small metal sheet, some filed can not be read. So the mounting and placement of tags will also affect the read performance of system. Since we can not transform the tags,and tags can only pasted on the area which could not affect the use of tool. as well as the placement of tools is uncertain and the structure of tool car has been determined, so the installation position of the antenna will directly affect the read performance.

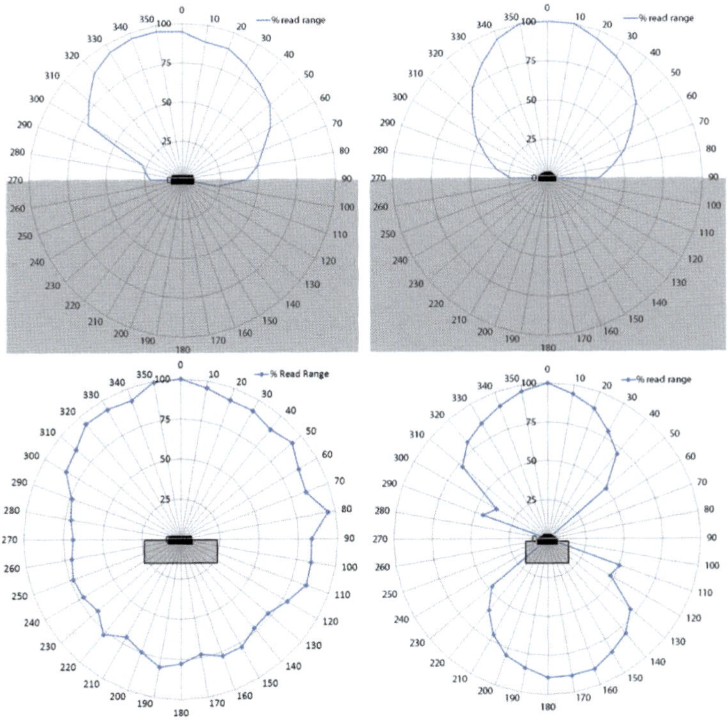

Fig. 3. The radiation patterns of fit 400 tags.

We have proposed two antenna installation methods in Section 2. One is mounted on the top of drawers, another is mounted on the side of drawers. In order to check the performance of these two installation, we designed the following experiment to determine the read range in the drawer.

Experiment 1.

Antenna is mounted on 15cm above the drawer which is enclosed. Evenly place 40 tags around the drawer. Continuously narrow the range of tags until all the tags can be read stably. This range is the read range which antenna is disposed above the drawer, as it is shown in Figure 4.

Experiment 2.

Antenna is mounted on the 3cm away from the side of the drawer where the metal plate is taken out from the drawer. Evenly place 40 tags around the drawer. Continuously narrow the range of tags until all the tags can be read stably. This range is the read range which antenna is installed on the side of drawer, as it is shown in Figure 4.

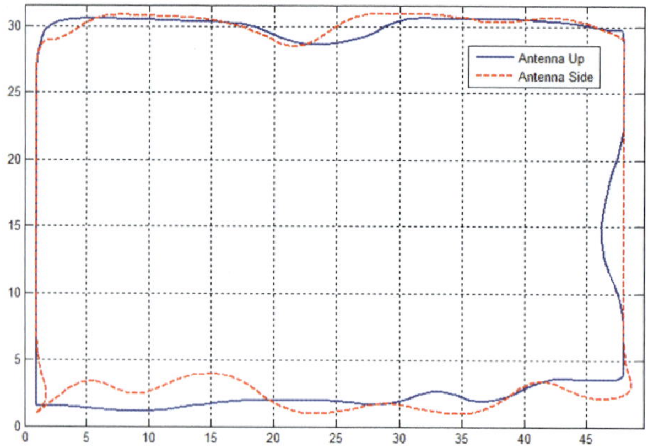

Fig. 4. The read range of two antenna installation methods (solid line is the read range of antenna up, dashed line is the read range of antenna side).

Figure 4 shows that the metal environment has influence on the read range. Not all the tags in the drawer can be read. When the antenna is installed on the top of the drawer, the read range of three edges of the drawer fluctuates, shown as the solid line in Fig. 4. When the antenna is installed on the side of drawer, the read range of edges of drawers which away from the antenna changes more, showed as the dashed line in Fig. 4. In the case of $antenna_{up}$, the strength of RF signal is relatively weak on the edge of the drawer and there are two reflection surfaces, so the fluctuation of read range produces. And in the case of $antenna_{side}$, the opposite side of antenna is metal surface. The electromagnetic wave reflection and absorption effect will be significantly stronger than other places, so the corresponding side has a greater fluctuation. Simulate the read range use matlab and then calculate the read area in the read range. From Table 1, two installation methods have similar read area, which means that the performance of the two installation methods are similar. Therefore, we can choose the right installation method according to the specific circumstance of the tool car. For example, when the intelligent tool car has a large number of drawers, we can choose the antenna mounted on the side. When with a small number of drawers, we can select the antenna installed on the top.

Table 1. The read area of two installation methods.

Antennas Installation Methods	Read Area
Antenna Up	130.14 cm^2
Antenna Side	131.33 cm^2

4. Software for Intelligent Tools Management System

Software system is an important part of the intelligent tool car. The software of the system is used to configure the RFID system, query the missing tools and provide a friendly user interface for the maintenance staffs. The software uses Visual Studio 2008 and C# to program and runs on WinCE platform to real-time monitor the tools in tool car and prevent the miss of tools. Three interfaces are designed in this paper; They are tool monitor, tool management and RFID configuration. RFID configuration interface is used to connect the reader and set the RF power and the work antennas of the reader. Tool management interface is used to add and delete the original tool data. Tool monitor interface is used to real-time monitor the deletion of tools.

The software provides two detection modes in tools detection. One is real-time mode,and another is fast switch mode. In real-time mode, after the program starts running, setup work *AntList* firstly according to the selected work antennas. Secondly, set antenna form *AntList* starting position. And then the reader begins to inventory and save the tag data. After the inventory, set the next work antenna. Until all the selected work antennas have worked, compare the queried tool data with the original tool data to obtain the missing tools. Finally, back to the starting position to repeat the query. The flowchart of real-time mode is shown in Figure 5. In fast switch mode, the differences with real-time mode is that do not need to select work antennas because all the antennas will work and the switching speed of antennas is faster to make the system read faster.

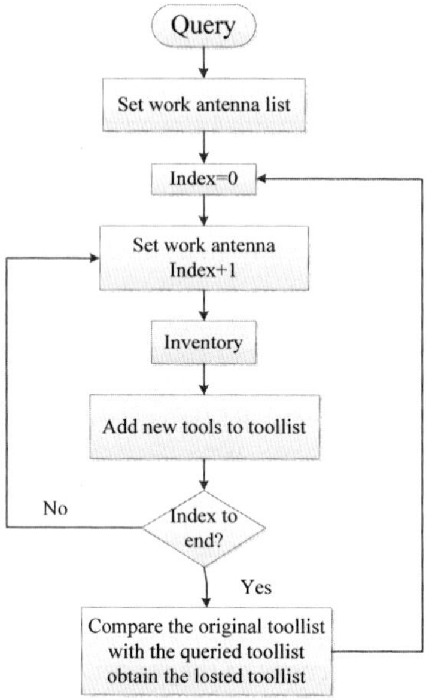

Fig. 5. The flowchart of real-time mode.

5. Experimental Results

A RFID-based intelligent tools management system is designed and complemented. First, in order to check the reading performance of the system to different types tools, choose Omni-ID series four different tags affixed to tools. Test and obtain its recognition effect as shown in Table 2. In the test , in order to ignore the effects of the rotation of the tools and tags, use different rotation combinations, such as 90°tag rotation with 135°tool rotation and 180°tag rotation with 180°tool rotation. Table shows that as the tags pasted to the applicable surface of items, the system recognition rate for individual items can reach more than 98%, even up to 100%. When tags pasted to non-applicable items, the recognition rate will greatly reduce.

Table 2. Recognition rate of tags affixed to defferent tools. (Legend: Fit200, Fit210, Fite400P, Fit400 are applicable to metal, metal and non-metal, non-metal, metal surface.)

Pasted Methods	Test Times	Recognize Times	Recognition rate
Fit200 pasted on metal	2160	2130	98.6%
Fit200 pasted on non-metal	2580	659	25.5%
Fit210 pasted on metal	2640	2640	100%
Fit400P pasted on non-metal	3320	3255	98.0%
Fit400 pasted on metal and Fit400P pasted on non-metal	300	295	98.3%

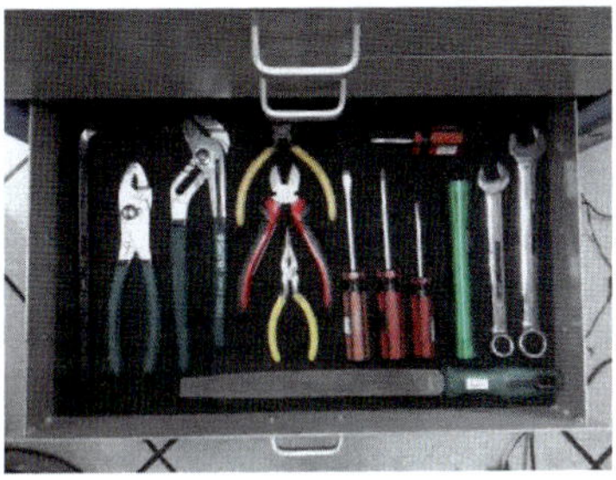

Fig. 6. The placement of tools

Then, test the system with the placement of tools shown in Figure 6 , Omni-ID Fit series tags pasted on the related tools, antennas mounted on the side of drawers. In the test, we have used 14 different tools, the reading rate is 100% in each of the drawers. When we remove the 19mm wrench, the query interface prompts the deletion of 19mm wrench and an alarm sounds, as shown in Figure 7. When freely remove the tools in the drawer, query interface prompts the deletion tool accurately and alerts, as shown in Figure 8.

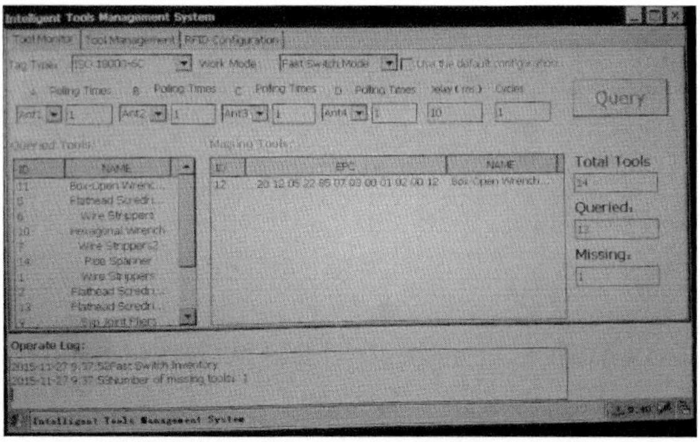

Fig. 7. Detection result of removing one tool.

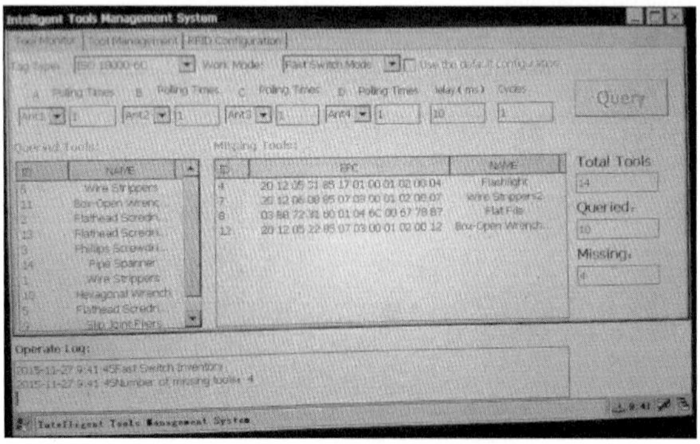

Fig. 8. Detection result of removing four tools.

In the complex electromagnetic closed metal environment, placement and posture of tools will affect the test results in a certain extent. In summery, the test result demonstrate the placement of tools in Figure 6 is a reliable scheme to ensure all the tools can be reliably detected. The test result in Figure 7 and Figure 8 further validate the reliability of the system.

6. Conclusion

Aviation security issues are getting more and more attention. Reliable management of maintenance tools is undoubtedly an important part to ensure aviation safety. In this paper, a RFID-based intelligent tools management system is designed and implemented to manage the maintenance tools effectively and reliably. Based on the feature of enclosed metallic environment, rebuild the inter structure of tool car. By using the RFID system mounted on the tool car, the status of tools can be detected real-time and the deletion tools can be get.The experimental results have confirmed the validity and the reliability of intelligent tools management system. The proposed system can be extended into the tools management of other fields in the future.

Acknowledgments

The research is supported by the National Natural Science Foundation of China under Grant U1201251 and National Science & Technology Support Plan of China under Grant 2014BAF07B04.

References

1. Landt. J, "The history of RFID", IEEE Potentials, Volume 24, Issue 4,Oct.-Nov. 2005,pp.8-11

2. M. Periyasamy, R. Dhanasekaran, "Assessment and Analysis of Performance of 13.56 MHz Passive RFID in Metal and Liquid Environment", International Conference on Communication and signal Processing, April 3-5, 2014

3. C.A. Diugwu, J.C. Batchelor and M. Fogg, "Field distributions and RFID reading within metallic rool cages", ELECTRONICS LETTERS, 9th November 2006, Vol. 42, No. 23

4. Xianming Qing and Zhi Ning Chen, "Proximity Effects of Metallic Environments on High Frequency RFID Reader Antenna: Study and Applications", IEEE transactions on antennas and propagation, Vol. 55, NO. 11, November 2007

5. Yong-Peng Teng, "Design and Implementation of Intelligent ToolboxSystem Based on RFID Technology", Beijing:Institite of Automation, Chinese Academy of Sciences, 2013:2-3

6. http://www.eastample.com

7. Ling-fei MO, Hong-jian ZHANG,Hong-liang ZHOU, “Analysis of dipole-like ultra high frequency RFID tags close to metallic surfaces”, Journal of Zhejiang University SCIENCE A ISSN 1673-565X:1217-1222

8. http://omni-id.com/pdfs/Omni-ID Fit 400 datasheet.pdf

sDOMO communication protocol for home robotic systems in the context of the internet of things

Marcel-Titus Marginean* and Chao Lu†

Computer and Information Sciences,
Towson University, Maryland, USA
**Email: mtm@mezonix.com, †clu@towson.edu*

sDOMO is a communication protocol developed for home automation and the building of robotic systems. Being optimized for small devices, sDOMO allows certain 8 bit microcontrollers to be a full featured, independent node in the domotic network, yet the protocol is powerful enough to provide soft-real-time communication for our computer-vision distributed system for domestic robots. Being designed with security and privacy concerns in mind, sDOMO has unique features to protect the residents.

Keywords: Home Automation; Domestic Robotics; Communication Protocol; Privacy.

1. Overview

Presented in our previous paper[1], sDOMO is a communication protocol developed originally to support our Distributed Processing Architecture for Domestic Robots [2] and during the development process, it was optimized for small devices transforming it into a general purpose soft-real-time protocol for domotics, and building of robotic systems. sDOMO uses low overhead binary data packing that can be implemented by very small, microcontroller based devices, and proposes a stand-alone integrated domotic system with the actual robots being just the mobile components of the smart house of the future. This distributed architecture, takes advantage of the already existing house network and provides by protocol design advanced security and privacy features intended to protect the residents.

sDOMO protocol from the beginning was designed to be implemented in a self-contained home automation and robotics system, and its design reflects this purpose. During development, it was observed that the protocol could be easily extended to any type of system integrating multiple devices (sensors and

* Corresponding author.

actuators) that need to behave plug-and-play; therefore, it is proposed also as a protocol for building robots and unmanned vehicles out of generic parts.

1.1. *sDOMO at a glance*

The typical architecture of a domotic network in the sDOMO model consists of a set of devices, a House Hub, Home Intelligence Unit, and an Internet Gateway. A device can be any functional node of the home automation network and is identified by a unique ID.

House Hub performs the function of a Rendezvous Server accepting and configuring devices in the network and routing sDOMO's messages between devices.

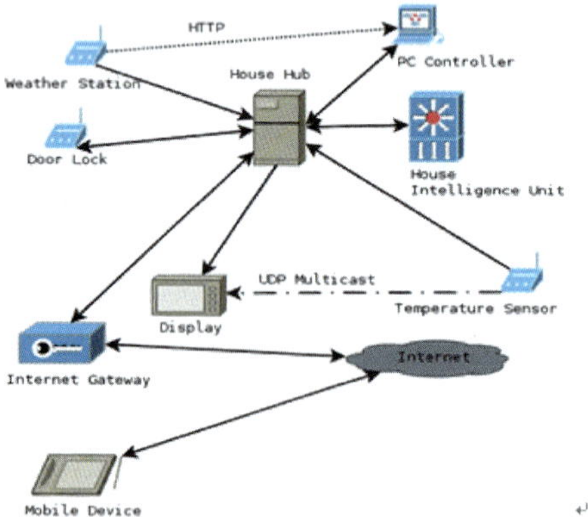

Fig. 1. sDOMO system architecture.

Devices communicate with the Hub via sDOMO Packets. They are packets for Discovery and Configuration, Session Control and Message Carrying Packets. Packets are carried inside the datagrams as provided by the underlying network technology. In the case of Ethernet and Wi-Fi, the UDP/IP datagrams are chosen to carry packets. From a software point of view, a device is composed from one to up to 255 software Objects. The Objects are the sender and target for Messages. The preferred method of data exchange in an sDOMO system is by passing Messages between devices routed by the House Hub. This allows small, microcontroller devices to be part of the network while benefiting from all the security and privacy features the House Hub can offer. Messages are binary chunks of data of up to 4GB in size subject to limitation of the device and Hub memory. Messages are carried between the Devices as one or more sDOMO Message Carrier Packets routed by the Hub.

There are 2 categories of messages exchanged by devices via the Hub: Direct Messages (one to one communication) and Notifications (one to many). A Direct Message is sent from an Object part of a device specifically toward a given object part of another device. Notifications are sent and usually triggered by external events, and are delivered by the Hub to all devices that have at least one object subscribing for them.

Besides communicating via sDOMO messages routed by the Hub, the model also allows devices to communicate directly using their own protocols, which is outside of sDOMO specifications, therefore allowing a transition path toward sDOMO. However, devices communicating that way must implement their own security and privacy mechanisms. A device using its own protocol can still use the House Hub as a simple Rendezvous Server in order to discover the other devices in the network. For the rest of this section, we will speak only about communication that takes place via sDOMO messages routed by the Hub.

As start-up devices advertise themselves and get invited by the House Hubs to join their network, they get configured and receive a Session Key used to sign all the packets using Hash based Message Authentication Code (HMAC). Encryption is optionally available for devices transmitting sensitive information. In the packet used to advertise the device, a Device Type String is provided from which the Hub is able to calculate a unique URL from where it can download an XML file called Device Specification File (Spec File).

Each device is described by an XML document called a Device Specification File, describing the capabilities of the device, configuration options, list of software objects exported by the device, defined requirements and limits, proposed Access Control List (ACL) entries, and links to a list of Interface Definition Files. Interface Definition Files are also XML files describing in a machine understandable way all the messages exchanged by the device and how those messages can be combined to implement Remote Method calls. Interfaces can inherit other interfaces, allowing a new device to extend the functionality of an old one therefore enabling backward compatibility. Interface files are the entry point for an interface compiler that generates code stubs for both the device itself and for the software communicating with the device. It also provides security information to help build an ACL Database. XML files are downloaded by the Hub and by Developer tools and there is no requirement for the devices to be able to parse or even provide them in order to accommodate small memory foot-print devices. However, envisioned enhancements to the protocol will add the option for devices to provide themselves the Spec Files (if hardware permits it) in order to allow them to operate in Internet disconnected environments.

The House Hub enforces strict access rights, by checking via Access Control Lists (ACL) each Direct Message sent and each Notification Subscription. Hub's ACL database is built automatically by the system from the Device Specification Files provided by the manufacturers and by analyzing Standardized Expert Advice Files provided by trusted third party experts.

The Session Key is generated by the Hub and uploaded to the device via a key exchange algorithm described in [1]. The key exchange algorithm is using the device unique Device ID and Device Unique Key.

In an sDOMO model, a device is not supposed to access the Internet directly, the model proposes allocating of a certain range of IP addresses for home automation devices and will have the firewall blocking those IPs from accessing the outside world. When a device needs to get updates or data from the Internet, it has to request the required information from an Internet Gateway which analyses the request, and performs a controlled access to the network to request the required resource. The type and purpose of the information required from the outside-world is to be described in the XML Spec file for the device and it may be subject to override by trusted expert advice.

House Intelligence Unit (HIU) is a piece of software running on the Base Station computer and having access privilege to the House Hub. The main role of HIU is to provide a higher level of automation by choreographing coordinated actions across multiple devices and external information.

1.2. *Highlights*

The compact binary packed protocol implemented by sDOMO and by the fact that the protocol outsources main routing and security decisions to the House Hub solves the problem of integration of cheap, small sensors in the domotic network, without the need to use protocol adapters. We demonstrated the scalability of sDOMO toward low end devices by implementing a minimalist but fully functional thermostat running on an Arduino Uno (8 Bit AVR Microcontroller, 2KB RAM, 32 KB Program Flash) speaking directly sDOMO over UDP/IP via an "Ethernet Shield".

Security is a big concern in any home automation network. A hacker sending fake temperature reports to an A/C unit can induce hypothermia in a diabetic patient unable to properly sense the temperature. If a thief can get control of the door-lock and the alarm system, the houses will become free-for-all self-serve depots. The advent of domestic robotics brings a whole new dimension to the security and privacy problem since a robot is capable of directly inflicting bodily harm or even killing the unsuspecting residents.

The “Internet of Things Research Study, 2014 Report” [14] conducted by Hewlett-Packard examined the security of some commercial devices directly connected via Internet to their manufacturers (Cloud based IoT model) and the findings were alarming. For example 70% of the devices used unencrypted network services, 80% used weak passwords schema, while 90% of the devices and their services collected personal information about the users. To help alleviate these security and privacy problems, these issues have to be considered from the design phase of the communication protocol and home automation model. To gain trust, the communication protocols can be fully documented and the specifications available for public scrutiny by other experts. We are doing this with sDOMO and in our work we tried to incorporate from the protocol design layers of protective measures for the security and privacy of the users.

The first line of defense is the fact that an sDOMO home automation system is designed to operate as a self-contained network, independent of Internet connectivity or protected behind a firewall. The data traffic between devices is taking place on the local house-network reducing the opportunity for snooping or direct attack.

To prevent attacks, in cases when an intruder got a foothold inside the local network, sDOMO requires each packet to be signed and verified at its destination, it also implements a schema for enforcing the uniqueness of each packet preventing even reply attacks. This works by attaching a unique id to each packet and forcing a generation of a new Session Key when the packet counter approaches the upper limit.

We also implemented a unique feature that discourages a Trojan horse from even attempting to hijack a device by the threat of automatic disclosure of the attackers and by allowing HIU to run scripts that retaliates against the detected intruder as presented in details in [1].

Privacy is a huge concern with a home automation network, especially since we live in a time when an increasingly large number of companies purposely violate the expectation of privacy of people and collect whatever data they can gather to make money by selling it to marketers. Inside our homes most people have a total expectation of privacy, but with the model of “Cloud Based Internet of Things” this is nothing but a false hope as presented above.

By proposing a self-sufficient domotic network that can operate locally independent of “The Cloud” and having all the Internet requests filtered by Internet Gateway, we are taking the first step toward protecting the privacy of the users by avoiding the reliance on services of various companies that prey on user's data. Analyzing plug-ins running on the Gateway can help monitor and raise alarm if a device is trying to upload data to the cloud by disguising it as a request for updates, or other "creative ideas".

The enforcement by ACL of which a device can access data will also reduce the risk that companies will design Trojan devices that collect information from other sensors into the house to pack and upload it to the manufacturer site.

Another layer of enforcement is provided by an alternative use of HIU to increase the privacy by coordinating actions between devices that we do not trust to have access to each other. For example, a thermostat may need access to motion information from the cameras to determine if the house is occupied or not. But if the parent company of the thermostat maker is in the business of selling data to marketers, we may be concerned to let it access the video stream from cameras. However, HIU can subscribe to cameras and spoof toward the thermostat motion information without any identifiable feature. That way, commercial off the shelf devices from vendors that we don't trust can be safely integrated in our homes.

Finally, since device Specification Files must be public documents, they can be audited by third party experts, who can, not only raise red-flags but can override ACL proposed by manufacturer to solve discovered security and privacy issues. The core idea is that each Hub will have a list of trusted experts and their public keys and periodically scan their sites for XML files that provide machine understandable expert advice files. That way, third party experts can prevent dishonest manufacturers from spying on their customers.

2. sDOMO and IoT

The concept of an Internet of Things is a very broad concept covering various aspects from embedded RFID tags into clothing, to fully integrated city wide utility information systems [9]. Therefore, the scope of communication protocols used in IoT is equally wide. In [13] we see a few critical considerations about the design of smart homes.

A survey [12] conducted at the university of Essex highlighted the major concerns of people in regard with intelligent homes, among which some are important to note: the feeling of being in control, privacy and cognitive workload. As we are headed toward pervasive intelligent environments, the human factor must also be considered and we believe that more studies of this kind are necessary, since the design engineers need to understand the response of the users living inside smart environments.

In the following literature review, we will look only at the protocol with a somehow similar scope with sDOMO highlighting the differences and similarities. Extensible Messaging and Presence Protocol (XMPP) [3] developed originally for Instant Message - "Jabber" provides near-real-time data exchange via XML based messages. The protocol is implemented as an open standard, using a client-server architecture. System speaking, the protocol can be isolated from the internet by implementing behind a firewall a private XMPP server. However, the verbosity of XML and the lack of native binary data transfer (which must be encoded base64 inside XML) makes it difficult to be used for small microcontroller based devices. Nest Service Data Model is tied to direct Internet access. All Nest devices connect directly to Nest Service which is hosted online by company servers and are used to access the devices via web-site or applications. The system relies on JSON data packing using REST interface over HTTPS excluding small microcontroller devices from being able to speak to the Nest API directly. Hosting the data services on company websites also raises questions of privacy and security of personal information.

Because of the scope of the protocol, sDOMO is designed to be able to operate on local networks, isolated from the Internet by a firewall. The boundaries of the system are defined by a local network (domestic or robot based) and not the provider of the devices. SDOMO is more compact and able to work on devices that are out of reach for XMPP or Nest JSON.

The Constrained Application Protocol (CoAP) [4] is a very interesting recent development in IoT. It appears to be even more compact than sDOMO fitting inside 8-bit microcontrollers. The minimally binary packed header allows very small CPU and bandwidth overhead but it lacks security and privacy features. The message size in CoAP is also limited to the datagram size (about 1.1KB over UDP) while sDOMO is using the Message Carrier Packs to allow messages with a theoretical limit of 4GB. CoAP has been designed to have their message easily translated to HTTP by an adapter but does not in itself define a network model beyond that of device to device communication. sDOMO in contrast defines the network model and services like HIU and Hub's virtual Devices.

In [7] there has been presented a framework for connecting devices to "the cloud" via REST Web Services. In the proposed architecture sensors speaking their own native protocol are connected to the Internet via an adapter connected to an open source "Internet of Things" website that allows publishing sensor data using HTTP. The security implications of this approach are immediately observed. Publishing data on the Internet sometimes even using unsecured HTTP protocols is incompatible with requirements for a home automation system where security and privacy must be taken very seriously. By contrast sDOMO is addressing these issues from protocol design.

Originally, accepted with enthusiasm without too much attention to security and privacy, in recent years IoT is finally receiving the needed scrutiny. In [8] the authors analyze the current state of cloud-based IoT and discuss a number of consideration about it. And in [10, 11] we can find another quick overview of the issues.

3. Usage Example

The protocol has been originally developed to support our Distributed Processing Architecture for Domestic Robots [2]. The modules of this system are currently fully functional communicating using sDOMO to provide soft-real-time data exchange, making the system a good illustration of how sDOMO is intended to operate.

From the software point of view the system is composed of Camera Modules (CM) collecting images from fixed cameras and processing them with the object tracking algorithm MPTracker [6]. As a result to tracking objects CMs emit a notification to which subscribes Situation Awareness Module (SAM) and an Engineering Console.

SAM and the Console can request via Direct Messages images or extended information from the tracker. The request is answered with another Direct Message by CM, the request/response being in fact an sDOMO remote method call.

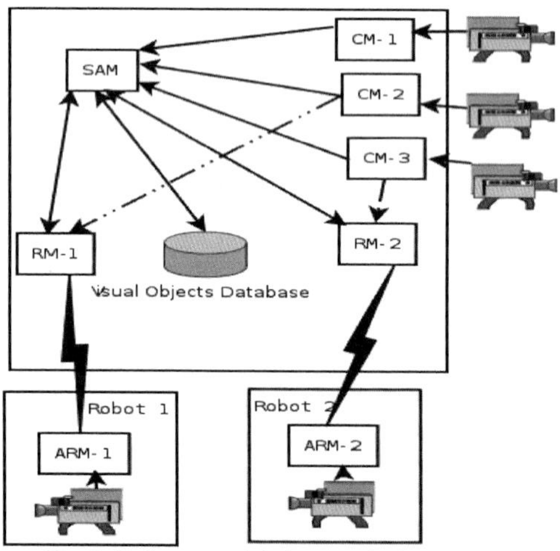

Fig. 2. Software module in our robotic system.

SAM is using homography to build from images a 3D model of the room and then translates the tracking information from CMs into the model coordinates. The translated tracking coordinates are also broadcasted as an sDOMO Notification.

Robot Module (RM) running on the Base Station subscribes to these Notifications and uses them to track its own movement. RM can also request as needed images from CM and ARM to perform epipolar geometry matching. RM sends high level movement commands to ARM. Autonomous Robot Modules (ARM) runs on the embedded board on the robot itself. It receives Direct Messages from RM with requested movement vectors and responds to requests. An optical flow Notification is sent by ARM from locally processed images acquired by the robot camera.

4. Conclusions and Future Work

sDOMO is a communication protocol optimized for Home Automation and Robotic Systems, small enough to accommodate microcontroller based devices as peers in the network but powerful enough to handle a house wide computer vision system for controlling domestic robots.

With a CPU/Bandwidth overhead a little over that of CoAP, sDOMO is able to provide security and privacy features some of them unmatched [1] by more heavy-weight protocols, like the security features that discourages a Trojan Horse from even attempting to hijack a device or the ability of trusted third party experts to publish XML documents that fixes on the flight privacy features discovered with devices already deployed.

The protocol is in active development under a liberal open source license. We hope to be able to provide a practically-functional release by the end of 2016 or so. A companion protocol to sDOMO is being envisioned as an Inter-Hub protocol. This will allow multiple Hubs to coexist on the same domain increasing reliability and throughput. An extension to the protocol to allow devices to provide themselves the Spec File upon request, combined with network topology detection, will enable sDOMO devices to be used as building blocks for robotic systems, where a "hand" or a "leg" can be plugged into a body, detect it position and be configured on the fly.

The Public Key Infrastructure and Trust Management rules required to support the Expert Advice Files is still pending to be defined.

References

1. Marcel-Titus Marginean and Chao Lu, "sDOMO – A Simple Communication Protocol for Home Automation and Robotic Systems", IEEE International Conference on Technologies for Practical Robot Applications; May 11 - 12 2015.

2. Marcel-Titus Marginean and Chao Lu, "A Distributed Processing Architecture for Vision Based Domestic Robot Navigation", International Conference on Computers, Communications and Systems; Nov 1 2013.
3. Internet Engineering Task Force, RFC-6120, RFC-6121. P. Saint-Andre, March 2011. ISSN: 2070-1721.
4. Internet Engineering Task Force, RFC-7252, Z. Shelby et al, June 2014. ISSN: 2070-1721.
5. Castro M, et al. "Enabling end-to-end CoAP-based communications for the Web of Things". Journal of Network and Computer Applications (2014), http://dx.doi.org/10.1016/j.jnca.2014.09.019i.
6. Marcel-Titus Marginean and Chao Lu, "A Multi-Paradigm Object Tracker for Robot Navigation Assisted by External Computer Vision". ACM RACS 2014.
7. Rajeev Piyare and Seong Ro Lee, "Towards Internet of Things (IoTs): Integration of Wireless Sensors Network to Cloud Services for Data Collection and Sharing".International Journal of Computer Networks & Communications (IJCNC) Vol.5, No.5, September 2013.
8. Singh, J.; Pasquier, T.; Bacon, J.; Ko, H.; Eyers, D., "Twenty Cloud Security Considerations for Supporting the Internet of Things," in Internet of Things Journal, IEEE , vol.PP, no.99, pp.1-1.
9. Ángel Asensio et al., "Managing Emergency Situations in the Smart City: The Smart Signal".Sensors 2015, 15, 14370-14396; doi: 10.3390/s150614370.
10. M.U. Farooq et al. "A Critical Analysis on the Security Concerns of Internet of Things (IoT)". International Journal of Computer Applications (0975 8887) Volume 111 - No. 7, February 2015.
11. Rodrigo Roman, Pablo Najera, and Javier Lopez, "Securing the Internet of Things". IEEE Computer, vol. 44, no. 9, pp. 51-58, September 2011.
12. Matthew Ball and Vic Callaghan, "Who is in Control of Intelligent Environments? A Question of Autonomy". 10th International Conference on Pervasive Computing, June 18, 2012.
13. Sam Solaimani et al. "Critical design issues for the development of Smart Home technologies". J. Design Research, Vol. 11, No. 1, 2013.
14. Hewlett-Packard, "Internet of Things Research Study, 2014 Report". Online: http://h20195.www2.hp.com/V2/GetDocument.aspx?docname=4AA5-4759ENW.

A research on chinese selectional preference based on online materials

Zhang Qing*
School of Computer & information science, Huanggang Normal University, Huanggang, Hubei, China
**E-mail: 823963996@qq.com*
www.hgnc.net

Selectional preference is an important concept widely used for syntactic and semantic analysis of natural language. However, works published so far on the automatic acquisition of Chinese selectional preference are somewhat lacking in terms of accuracy. In this paper, a Chinese selectional preference model based on online resource *Balanced Corpus of Modern Chinese* developed by *State Language Work Committee* is introduced. Test results have indicated that the proposed model is effective, and demonstrate its fault tolerance ability in handling online materials with less accurate annotation.

Keywords: Selectional Preference; Probability Distribution; Knowledge Acquisition.

1. Introduction

Many words give semantic restrictions to the words that co-occur in a syntactic relationship. For instance, verbs have semantic restrictions on nouns as their objective. This kind of restrictions is known as selectional restrictions or selectional preference. Selectional preference has long been investigated and used in syntactic and semantic analysis of natural language.[1-5]

For Chinese selectional preference, there are also various studies concerning its characteristics or combine it with some natural language processing tasks.[6-10] But few has been done about automatic acquisition of Chinese selectional preference knowledge.[11-14] These automatic acquisition studies use precisely annotated corpora which are not abundant in Chinese. An online material based model is built in this paper. It is convenient for construction and use. Experiment indicates that this kind of model will be efficient for learning Chinese selectional preference knowledge.

* Corresponding author.

2. Related Work

2.1. *Researches based on semantic taxonomy knowledge*

Researches on Chinese selectional preferences based on semantic taxonomy knowledge began earlier. Using the noun taxonomy of *HowNet*, Wu Yufang et al investigate characteristics of selectional restrictions on objects by verbs.[6] They found various distributions of nouns as each verb's objective over semantic classes. They illuminate that selectional restrictions can be applied in Chinese language processing, but have some limitations. There are applications concerning Chinese selectional preferences based on different semantic taxonomy systems as well. J.Gao use concept system of *HowNet* for verb-object collocation identification. Based on taxonomic system of *Tongyici Cilin*, a thesaurus of Chinese, developed by *Information Retrieval Laboratory, Harbin Institute of Technology*,[15] Y.Cheng et al investigated verb-objective collocation, using probability distribution of noun over semantic classes.[9] Concept-based selectional preferences is also used for other natural language processing tasks such as automatic construction for Chinese verb frame repository,[10] Chinese verb metaphor recognition[11] etc.

Based on semantic taxonomy knowledge, results are easily understandable for human being, but for computer programming, they are not evaluated as good as that achieved from researches based on probability distributions.

2.2. *Researches based on probability distributions*

Latent variable model is the most commonly used model of this kind. Latent Semantic Clustering (LSC) is a probabilistic latent-variable model introduced by Rooth et al for selectional preferences.[16] D.Li et al used LSC to obtain Chinese selectional preferences knowledge.[12,13] Chose three word sets of different size from the *Corpus of People's Daily*, annual collection 1998 as training corpus. The largest vocabulary is 49 verbs and 97 nouns. The number of classes is set to 25. Results of their experiment indicate that LSC provide a way for Chinese selectional preferences acquisition.

The disadvantage of LSC is that the number of clusters has an influence on the quality of the semantic classification, and results might be influenced by the size of training corpus. No large scale result of LSC research has been reported for Chinese selectional preferences.

Another latent variable model has been used for Chinese selectional preference is Latent Dirichlet allocation (LDA).

LDA has been a successful topic model in modeling document-word co-occurrences. In LDA topic model, documents are represented as random mixtures of words that distributed over latent topics. D.O.Séaghdha adapted LDA to the task of selectional preference learning.[17] In LDA selectional preference model, regard predicates as documents, arguments as words, and semantic classes of arguments as topics. Associate each predicate type with a single relation r, the resulting parameters Θv denote predicate-specific distribution over semantic classes $P(z|v,r)$ and Φz denote distribution of arguments over those classes $P(n|z)$, the plausibility of a predicate-argument pair can be figured out as:

$$P(n|v,r) = \sum_z P(n|z)P(z|v,r) \tag{2}$$

As LDA models derive their predictions by modeling predicate-argument plausibility through the intermediary of latent variables, they are especially effective for estimating plausibility of infrequent items.

Y.Jia et al used this model for Chinese selectional preference acquisition.[14] Define $spLDA(n,v,r)=P(n|v,r)$. Choose certain syntactic relationship such as verb-object, the training corpus consists of numbers of observations of verb-object pairs. To evaluate the model, 93535 verb – noun object pairs from *Corpus of People's Daily*, annual collection 2000 are used as training corpus. 1592 different verb – noun object pairs from the *Corpus of People's Daily*, monthly collection of January 1998 are used as testing corpus. LDA-SP proved better at both coverage and correctness than results from concept-based selectional preferences based on *HowNet*.

Researches based on probability distributions do not need Comprehensive language knowledge base. Instead, statistical data of pairs of verbs and nouns belong to certain grammatical relations from a training corpus is needed. D.Li et al and Y.Jia et al use the *Corpus of People's Daily*, annual collection of 1998, 2000 respectively. The corpus has precise Chinese word segmentation and part of speech tagging. The size of accurately annotated corpus of Chinese like this is limited. They also has the disadvantage of restriction on the use. In next sections, training corpus collected from online materials is recommended for Chinese selectional preferences research.

3. Get Training Corpus from *Balanced Corpus of Modern Chinese* Developed by *State Language Work Committee*

Balanced Corpus of Modern Chinese (*BCMC*) is an online resource built by *State Language Work Committee*.[18] It consists of primordial language materials comprising 100 million Chinese Characters, and an annotated corpus comprising

50 million Chinese Characters. It is still supplemented with about 3 million Chinese Characters annually. The annotated corpus is built by Segmentation software and part of speech Tagging software. It has been manually revised three times and has an precision over 98%. The scope of the corpus covered humanities, natural science, and miscellaneous other aspects. Time span of the material is from 1919 till now. Sampling for this corpus has been well planned to guarantee the balance. Contrast with the *Corpus of People's Daily*, this corpus is less accurate in annotation, but much more general and balanced.

Corpus Online provides services of inquiry, download, and statistics etc for *BCMC*.[18] Needed material can be collected from inquiry results. Some mistakes like repetition in the collection may be corrected by simple computer programs.

BCMC has not provided grammatical annotation such as dependency parsing sequence that might be needed in selectional preferences studies. Some online software like that available on *Language Technology Platform* (*LTP*) of *HIT IR-Lab* might be used to address this shortcoming. The advantage of *LTP* software is that its symbolic set for part of speech tagging is almost the same as that used in *BCMC*. Few revision is needed.

4. Experimental Results

For convenience to compare with reference[14], choose verb-object relation. The training corpus was built based on the *Balanced Corpus of Modern Chinese*. The free software *GibbsLDA++* was used to build LDA models.[19] The parameter setting was: number of topic (semantic classes) K=200, alpha=50/K, beta=0.1, sampling iterations=2000.

Models are evaluated by pseudo-words evaluation which is now commonly used to evaluate selectional preferences models.[20] Different ways of choosing a partner word for a word that original in a document may produce quite different results. The tactic of reference[14] is to select the nearest neighbor. A test corpus similar to that used in reference[14] was built. Original verb - nominal-object pairs are chosen from the *Corpus of People's Daily*, monthly collection of January 1998. LDA gives Probabilistic results. In order to investigate the stability of the model, five models were trained and evaluated.

Table 1 shows the results, and Table 2 shows the results from reference[14]. Table 1 listed only accuracy. Since the precision of *GibbsLDA++* processing is up to 15 digits, actually there is no tie in evaluation; and as long as verbs and nouns of the testing set are within those of training set, which is easily guaranteed by a large training set, the coverage of LDA models is 100%. The average accuracy of our model is 9.95% higher than *HowNet*-SP. It illuminate again that LDA models easily work better than models based on semantic taxonomy knowledge. Compare to LDA-SP, The average accuracy of our model is 2.4% lower. Take account of the original similarity between the test corpus and training corpus used in reference[14], this is not a determinant difference. It illustrates that even though precision of Chinese word segmentation and part of speech tagging in *BCMC* developed by *State Language Work Committee* is lower than the *Corpus of People's Daily*, it is qualified to be a resource for Chinese selectional preferences research.

Table 1. Experimental results of five models.

Model	Correct	Accuracy(%)
model-1	1546	79.61
model-2	1548	79.71
model-3	1522	78.37
model-4	1541	79.35
model-5	1545	79.56
Average	1540	79.32

Test corpus consists of 1942 collocation pairs.

Table 2. Experimental results in reference [14].

Model	covered	correct	tie	coverage(%)	accuracy(%)
HowNet-SP	831	472	209	62.53	69.37
LDA-SP	1329	1068	0	100	81.72

Test corpus consists of 1329 collocation pairs.

5. Conclusion

Selectional preferences knowledge has been widely used in various natural language processing tasks. Few works has been done for learning Chinese selectional preferences knowledge. The existing reports are based on high precision corpora, which are not abundant in Chinese. A Chinese selectional preferences model based on *BCMC* developed by *State Language Work Committee* was built. Testing result illustrates that even though precision of online materials are not as precise as accurately built corpora, they delivers a good substitute.

Acknowledgements

This work is supported by Research project of the Natural Science Foundation of Hubei, China (Grant No.2014CFC1100), Research project of Hubei

Provincial Education Department (Grant No.D20152903, 15Y159), Foundation of Huanggang normal university (No. xfg2015A11, 2014015103).

References

1. D. McCarthy, S.Venkatapathy., A.K. Joshi. Detecting compositionality of verb-object combinations using selectional preferences. *EMNLP-CoNLL*. 2007: 369-379.
2. D. McCarthy, J.Carroll. Disambiguating nouns, verbs, and adjectives using automatically acquired selectional preferences. *Computational Linguistics*, 2003, 29(4): 639-654.
3. S.Bergsma, D.Lin, R.Goebel. Discriminative learning of selectional preference from unlabeled text. *Proceedings of the Conference on Empirical Methods in Natural Language Processing. ACL*, 2008: 59-68.
4. Z.J.Mason. CorMet: a computational, corpus-based conventional metaphor extraction system. *Computational Linguistics*, 2004, 30(1): 23-44.
5. W.Roberts, M.Egg. A Comparison of Selectional Preference Models for Automatic Verb Classification. *Proceeding of EMNLP*. 2014.
6. Y.Wu, H.Duan, S.Yu. A Study of Verb Selectional Restrictions on Objects. *Applied Linguistics*,2005,02:121-128.
7. Y.Shao, Z.Sui,Y.Wu. Chinese Semantic Role Labeling Based on Lexical Semantic Features. *Journal of Chinese Information Processing*,2009,06.
8. J.Gao. Automatic Identification of Verb-0bject Phrase for Chinese Language[D].*Beijing Language and Culture University*,2000.
9. Y.Cheng, X.Chen, B.Li. Verb-object collocation investigation and experiment based on Thesaurus-based Classification. *Proceedings of the 7th International Conference on Chinese Information Processing*, 2007:5.
10. F.Zeng, N.YE, G.Zhang. The research on automatic construction technology of Chinese verb frame repository. *Journal of Shenyang Aerospace University*,2011,05:63-66.
11. Y.Jia , S.Yu. Automatic acquisition of selectional preference and its application to metaphor processing. *Proceedings of the 4th SWCL*, 2008:7.
12. D.Li, L.Zhang, W.Zhao, J.Shi. Semantics preference method based 0n MDL and LSC. *Computer Engineering*,2011,17:15-18.
13. D.Li, L.Zhang, W.Zhao, J.Shi. Unsupervised learning of semantic selectional preferences. *Computer Applications and Software*, 2012, 01:155-158.

14. Y.Jia, H.Wang, H.Zan, S.Yu, Z.Wang. Research on Chinese Selectional Preferences Acquisition. *Journal of Chinese Information Processing*, 2014,5.
15. Language Technology Platform, http://www.ltp-cloud.com
16. M.Rooth, S.Riezler, D.Prescher, et al. Inducing a semantically annotated lexicon via EM-based clustering. *Proceedings of the 37th annual meeting of the Association for Computational Linguistics,ACL*, 1999: 104-111.
17. D.O.Séaghdha. Latent variable models of selectional preference. Proceedings of the 48th Annual Meeting of the Association for Computational Linguistics, ACL,2010: 435-444.
18. *Corpus Online*, http://www.cncorpus.org
19. *GibbsLDA++*, http://gibbslda.sourceforge.net
20. N.Chambers, D.Jurafsky. Improving the use of pseudo-words for evaluating selectional preferences. P*roceedings of the 48th Annual Meeting of the Association for Computational Linguistics, ACL*,2010: 445-453.

Multi-route avoidance routing without location information of malicious wireless nodes in MANET

T. Kanachi* and H. Higaki†
Department of Robotics and Mechatronics, Tokyo Denki University, Tokyo, Japan
*E-mail: *kanachi@higlab.com, †hig@higlab.com*

In Mobile Ad-Hoc Networks (MANETS), data messages are transmitted along a wireless multihop transmission route consisting of multiple intermediate wireless nodes. Hence, intermediate wireless node and their neighbor nodes can receive the data messages. Due to the high performance of computers and advancement of various secure communication technologies, encrypted data messages can be decrypted by gathering a number of data messages. Works have been published on multi-route avoidance routing, which avoids sending data messages to malicious wireless nodes. Nevertheless, conventional avoidance routing operates under the assumption that the locations of malicious wireless nodes are already known by source wireless nodes. In order to solve this problem, this paper proposes a novel neighbor-disjoint multi-route ad-hoc routing, which could detect multiple wireless multihop transmission routes from the source node to the destination node where no other wireless node receives all the data messages. The simulation results show that with enough coverage of wireless networks, the wireless multihop transmission routes can be detected even at low ratio of the critical area. This paves the way for future studies of the route detection ratio and the size of the secure zones in critical areas caused by various disjoint route.

Keywords: Ad-Hoc Network Security; Routing Protocol; Secret Sharing Communication; Neighbor-Disjoint Multiple Routes.

1. Introduction

In wireless multihop networks such as mobile adhoc networks (MANETs), mesh networks and wireless sensor networks, data messages are transmitted by cooperation of multiple intermediate wireless nodes which forward the data messages from their previous-hop nodes to their next-hop nodes. That is, a wireless multihop transmission route is a sequence of neighboring wireless nodes from a source wireless node to a destination one. Due to their mobility, low-reliability and vulnerability, dedicated routing protocols based on the properties of mobile wireless networks different from those of wired networks

† Corresponding author.

such as widely available Internet routing protocols are required to be desined [1].

One of the most important different properties of wireless networks from wired ones is the broadcast based data message transmission. Even when a wireless node transmits a data message to one of its neighbor nodes, i.e. even in a unicast data message transmission, the data message is broadcasted to all its neighbor nodes which are in its wireless transmission range. That is, all its neighbor nodes *overhear* the data message. Here, the network security becomes critical since not only the intermediate wireless nodes in a wireless multihop transmission route forward the data messages but also their neighbor nodes overhear them. One of the well-known methods for achieving secure networks is the secret sharing communication [3], [4]. In the secret sharing communication, the secret data is divided into multiple data messages and these m data messages are transmitted along n different transmission routes. It is impossible to avoid receptions of m/n data messages in case a malicious wireless node poses a proper intermediate one. However, in order to make difficult for the malicious node to estimate the secret data, it is required to avoid that it receives more data messages. Though various multipath routing protocols for wireless multihop networks have been proposed, there are no appropriate protocols for the secret-sharing communication. This paper proposes neighbor disjoint wireless multihop transmission routes for avoidance of secret estimation by malicious wireless nodes and design a routing protocol to detect the neighbor disjoint routes.

2. Related Works

In order to avoid an illegal achievement of secret data, the secret sharing [2], [14] has been proposed. Here, the secret data and its encryption key data are divided into multiple data chunks and each of them are separately stored into storages or transmitted through networks [6], [16]. In order to achieve the secret data, achievement of required part of the data chunks more than the predetermined threshold is mandatory. In addition, more part of the data chunks are required to estimate the secret data more precisely. Thus, it is critical for keeping secret data secure to prevent that more data chunks are achieved illegally.

For the secret sharing communication where each data chunk is stored into a data message which is transmitted from a source node to a destination node through a network, the multiple data messages are transmitted along different routes in order for a malicious node not to achieve more data messages. That is, the secret data is divided into *m* data messages and these messages are transmitted along *n* different transmission routes to make difficult for the malicious node to achieve and/or estimate the secret data.

In [3], [4], methods for the secret sharing communication in mobile ad-hoc networks have been proposed. Here, the data messages carrying a part of the secret data are transmitted from one of the detected wireless multihop transmission routes to make estimation of the secret data difficult. In [3], one of the well-known multipath routing protocol for mobile ad-hoc networks AOMDV is extended for the secure communication especially in order to tolerate the unstable network topology due to the mobility of wireless nodes. In [4], a novel coding method has been proposed and the data message is transmitted along one of the wireless multihop transmission routes detected a conventional multipath routing protocol for mobile ad-hoc networks.

Until now, various multipath routing protocols for wireless multihop networks have been proposed [11]. Each routing protocol has been designed for its dedicated objective such as for tolerance to mobility, vulnerability of wireless nodes, unreliability of wireless links, mobility of obstacles and noisy wireless environment and for achievement of high performance such as shorter delay, lower loss ratio and higher throughput. Here, an important concept is *it independence among routes*.

The *it link-disjoint* multiple wireless transmission routes tolerate link-breakages due to mobility of intermediate wireless nodes [7]. If multiple routes share a wireless link and it is broken due to mobility of its end nodes, that is, the distance between the nodes becomes longer than their wireless transmission range, all the wireless transmission routes are simultaneously broken. In order to avoid such cases, no wireless links are shared among the detected wireless transmission routes.

The *node-disjoint* multiple wireless transmission routes tolerate node failures due to faults and battery-limitation of intermediate wireless nodes [8].

If an intermediate wireless node commonly included in multiple wireless transmission routes fails, all the wireless transmission routes are simultaneously broken. In order to avoid such cases, no intermediate wireless nodes are shared among the detected wireless transmission routes.

Performance of wireless multihop transmissions of data messages depends on the occurrence of collisions and contentions among neighbor wireless nodes. Due to the property of broadcast based communication in wireless networks, the exposed and the hidden terminal problems should be avoided. Various routing protocols realizes the detection of contention- and collision-free routes among them. [9], [15], [5] propose the methods to detect *radio-disjoint* multiple wireless transmission routes. In case that two wireless multihop transmission routes $\mathcal{R} := ||N_0(= N^s) \dots N_n(= N^d)$ and $\mathcal{R}' := ||N'_0(= N^s) \dots N'_{n'}(= N^d) \gg$ from a source wireless node N^s to a destination one N^d are detected for data message transmissions, no intermediate nodes $N_i \in \mathcal{R}$ and $N'_{i'} \in \mathcal{R}'$ are included in wireless transmission ranges of any intermediate wireless nodes $N'_{j'} \in \mathcal{R}'$ and $N_j \in \mathcal{R}$, respectively. That is, where $R(N)$ represents a wireless transmission range of a wireless node N, $N_i \notin N'_{j'}$ and $N'_{i'} \notin N_j$ as shown in Figure 1.

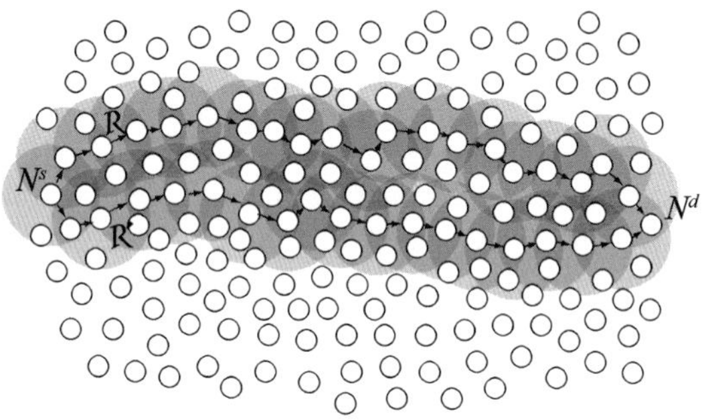

Fig. 1. Radio-disjoint wireless multihop transmission routes.

These three disjoint wireless multihop transmission routes are classified by the degree of independence among the detected multiple wireless transmission routes. The radio-disjoint routes are the most independent ones among these three. However, it is still insufficient for the secret sharing communication.

This is because there might be common areas among the wireless transmission ranges of the intermediate wireless nodes in multiple wireless transmission routes .That is, $R(N_i) \cap R(N'_{i'}) \neq \emptyset$ and a mobile wireless node N where $N \in R(N_i) \cap R(N'_{i'})$ overhears both the data messages transmitted along $\mathcal{R}$ and $\mathcal{R}'$. Thus, if N is malicious, N can estimate the secret data more precisely than our predetermined threshold.

3. Proposal

3.1. *Neighbor disjoint routes*

In the secret sharing communication, the more data messages a wireless node achieves especially it is a malicious wireless node, the more precisely it estimates the secret data. Thus, in wireless multihop transmission networks, it is required for the data messages to be transmitted along multiple wireless transmission routes where any wireless node achieves at most the data messages transmitted along one of the multiple wireless transmission routes except for the destination wireless node. By applying the radio-disjoint route detection protocol, all the intermediate nodes receives only the data messages transmitted along the wireless transmission route which contains the intermediate nodes. However, a wireless node might be a neighbor node of intermediate nodes included in different wireless multihop transmission routes. In this case, the node overhears all the data messages transmitted along multiple wireless transmission routes and in the consequence it might be possible to estimate the secret data more precisely.

In order to solve this problem, this paper proposes a novel independence among wireless multihop transmission route between the same pair of a source and a destination wireless nodes. This paper proposes the *strict neighbor-disjoint* wireless multiple multihop transmission routes where there are no neighbor wireless nodes among multiple wireless transmission routes. The definition of the strict neighbor-disjoint wireless multiple multihop transmission routes $\mathcal{R}_i := ||N_{0(i)}(= N^s) \ldots N_{Ni(i)(=N^d)\gg}$ from a source wireless node N^s to a destination one N^d is that $\forall N \in R(N_{j(i)})(0 < j < n_i)$, $N \notin R(N_{j'}(i'))(i \neq i', 0 < j' < n_{i'})$ is satisfied where $R(N)$ represents a wireless transmission range of a wireless node N. This means that no overlap areas of wireless transmission ranges of intermediate nodes in different wireless multihop transmission routes as shown in Figure 2. Thus, no intermediate nodes and their neighbor nodes of a wireless multihop transmission route receives the data messages transmitted along a different wireless multihop transmission route.

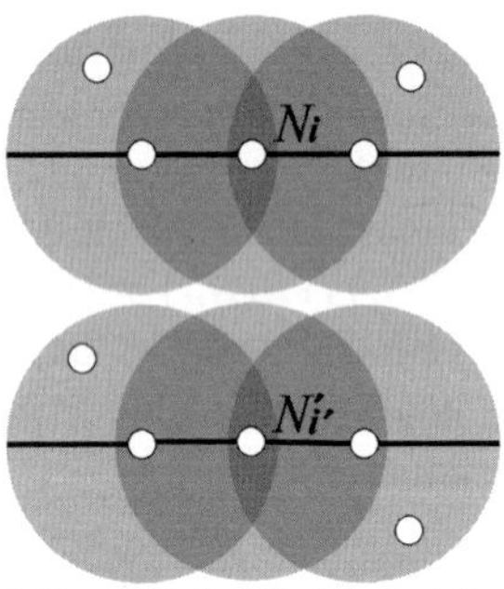

Fig. 2. Strict neighbor-disjoint wireless multihop transmission routes.

The strict neighbor-disjoint wireless multihop transmission routes are ideal for the secret sharing communication. However, since the wireless transmission ranges of the intermediate nodes are not always the same due to the wireless noises and various obstacles in the environments, it is difficult to verify the satisfaction of the conditions of the strict neighbor-disjoint routes. In addition, due to the requirements for the geometrical distribution, it is difficult to detect many disjoint wireless multihop transmission routes. Even if there is an overlap area of two intermediate wireless nodes in different wireless multihop transmission routes, there are no security flaws if there are no wireless nodes in the area as shown in Figure 3.

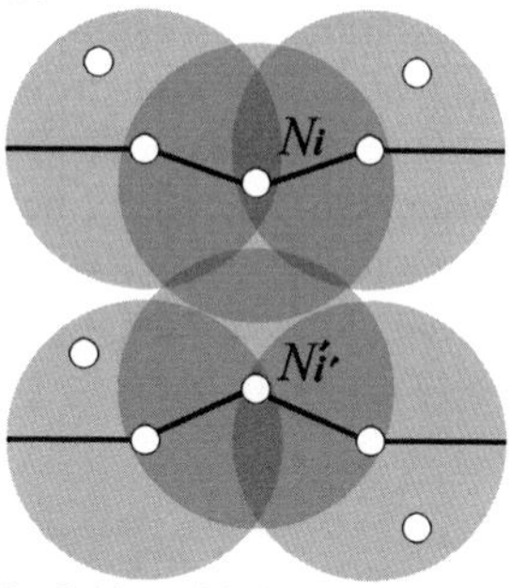

Fig. 3. Observation based neighbor-disjoint wireless multihop transmission routes.

Now, we define the *it observation based neighbor-disjoint* wireless multihop transmission routes where no wireless nodes are observed in the common area of wireless transmission ranges of intermediate wireless nodes in different wireless multihop transmission routes (Figure 4). This is reasonable for the secret sharing communication in wireless multihop transmission networks under an assumption that most of wireless nodes are not malicious and are engaged in the route detection protocol discussed in the next subsection.

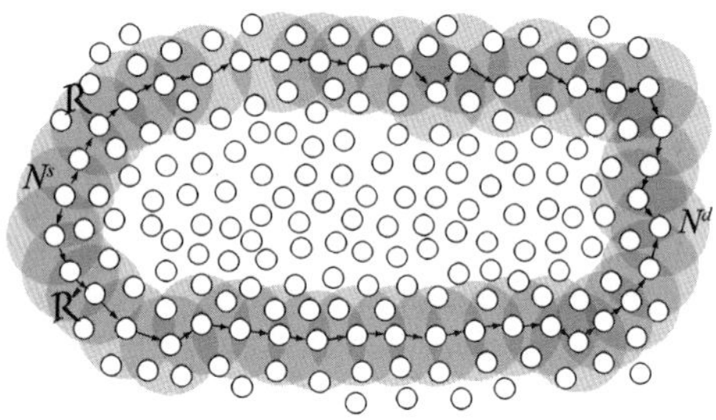

Fig. 4. Observation based neighbor-disjoint wireless multihop transmission routes.

The wireless transmission ranges of all the next-hop wireless nodes $N_{1(i)}$ of

N^s along $\mathcal{R}_{(i)}$ overlap since they surely include N_s. On the other hand, the wireless transmission ranges of all the previous-hop wireless nodes $N_{n_i-1(i)}$ of N^d along $\mathcal{R}_{(i)}$ overlap since they surely include N^d. Thus, it is impossible to satisfy the restriction of the strict neighbor-disjoint wireless multihop transmission routes and it is also difficult to satisfy the restriction of the observation based neighbor-disjoint wireless multihop transmission routes. Here, we consider the situation where the secret sharing communication is applied. It is possible that both N^s and N^d are in secure zones SZ^s and SZ^d, respectively, such as office buildings, laboratories, campuses and so on in which there are no malicious wireless nodes to estimate the secret data. Thus, we assume the existence of SZ^s and SZ^d and it is allowed for the wireless transmission rages of intermediate wireless nodes in different wireless multihop transmission route to be overlapped. Therefore, newly defined the *observation based neighbor-disjoint* wireless multiple multihop transmission routes with the secure zones $\mathcal{R}_{(i)}$ is that $\forall N \in R(N_{j(i)}) SZ^s \cup SZ^d (0 < j < n_i)$, $N \notin R(N_{i'}(i'))(i \neq i', 0 < j' < n_{i'})$ is satisfied.

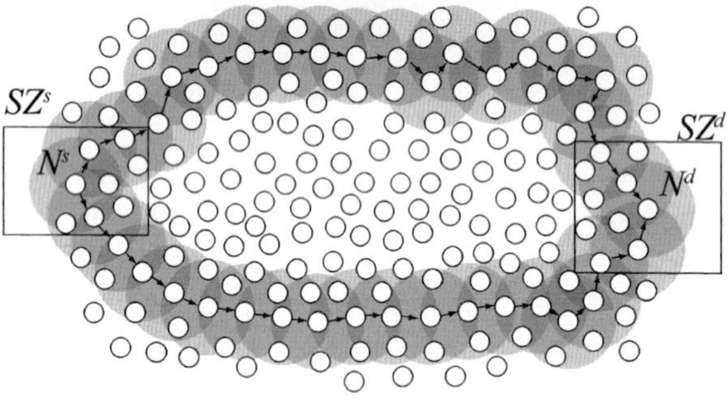

Fig. 5. Observation based neighbor-disjoint wireless multihop newline transmission routes with secure zones.

3.2. *Simulation experiments*

The proposed ad-hoc routing protocol in the previous subsection for detection of the observation based neighbor-disjoint multiple multihop wireless transmission routes avoids overlap areas of wireless transmission ranges of intermediate wireless nodes in different wireless transmission routes if all the wireless nodes properly works in accordance with the routing protocol. However, a malicious wireless node transmitting no control messages is not detected by the other wireless nodes and might achieve more data messages by locating in an overlap area where no other wireless nodes are included. That is, for two wireless multihop transmission routes $\mathcal{R} :=\| N_0(= N^s) \cdots N_n(= N^d) \gg$ and $\mathcal{R}' :=\| N'_0(= N^s) \cdots N'_{n'}(= N^d) \gg$ and intermediate wireless nodes $N_i \in \mathcal{R}$ and $N'_i \in \mathcal{R}'$, there are no proper wireless nodes N in the common area of the wireless transmission ranges $R(N_i)$ and $R(N'_i)$ of N_i and N'_i, respectively, and a malicious wireless node N^m is in $R(N_i) \cap R(N'_i)$. Here, N^m achieves the data messages transmitted through both $\mathcal{R}$ and $\mathcal{R}'$ and it might estimate the secret data more precisely. Thus, in this subsection, we evaluate the ratio of such the overlap areas by simulation experiments.

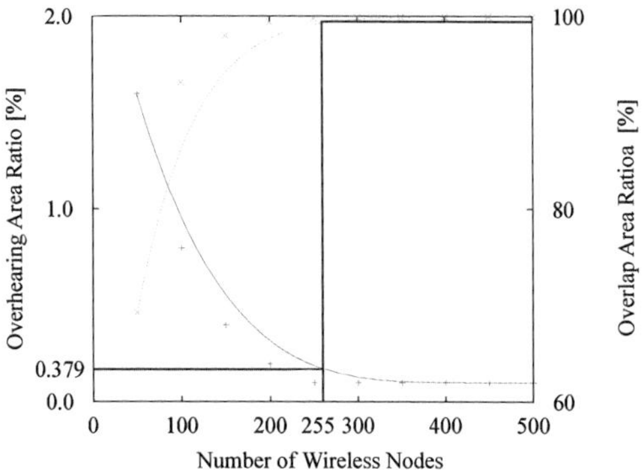

Fig. 6. Overlap area for malicious wireless node to achieve more data messages.

A simulation field is a 1,000m×1,000m square and 50-500 wireless node with 100m wireless transmission range are randomly distributed according to the unique distribution randomness. The ratio of the overlap areas called the critical areas in this subsection of two wireless nodes without other wireless nodes is evaluated. At the same time, we also evaluate the ratio of the overlap areas of two wireless nodes is evaluated. If the ratio is too low, each wireless node has less neighbor wireless nodes and it is difficult to detect wireless multihop transmission routes. Figure 6 shows the result. The ratio of the critical areas is between 0.217% and 1.72% which is monotonically decreased according to the increase of the number of the wireless nodes. On the other hand, the ratio of the overlap areas increases according to the number of wireless nodes and it reaches 99.0% with 200 wireless nodes. In [13], for detection of wireless multihop transmission routes with high probability, avaragely more than 8 neighbor wireless nodes are required and in this simulation experiments 255 wireless nodes are needed. In this case, the ratio of the critical areas is less than 0.379% and it is enough difficult for a malicious node to locate itself in such areas.

4. Conclusion

This paper proposes the neighbor-disjoint multiple multihop wireless transmission routes for the secret sharing communication in wireless multihop networks. Here, a malicious wireless node achieves only part of the data messages transmitted along only one wireless multihop transmission route and is impossible to estimate the secret data precisely. We design an applicable routing protocol where limited overlap areas of wireless transmission ranges of intermediate wireless nodes in different wireless multihop transmission routes are allowed and the source and the destination wireless nodes are in secure zones without malicious wireless nodes. The simulation results show that with enough dense wireless networks where wireless multihop transmission routes are detected there are enough low ratio of the critical area is achieved. In future work, the ratio of the critical areas caused by various disjoint route detection protocols are compared and the relation between the route detection ratio and the size of the secure zones is evaluated.

References

1. Boukerche, A., Turgut, B., Aydin, N., Ahmad, M.Z., Boloni, L. and Tugut, D., Routing Protocols in Ad Hoc Networks : A Survey, Computer Networks, Elsevier, vol. 55, no. 13, pp. 3032--3080 (2011).
2. Blakley, G.R., Safeguarding Cryptographic Keys, Proceedings of the National Computer Conference, vol. 48, pp. 313--317 (1979).
3. Chandrasekar, C. and Baboo, S.S., Proactive Bai's Secret Sharing for AOMDV Routing Protocol for Secured Communication in MANET, International Journal of Engineering Research and Applications, vol. 2, no. 2, pp. 655--659 (2012).
4. Imai, K., Fujii, M. and Watanabe, Y., A Study on Secure Multi-Path Wireless Networks, IEICE Technical Report, vol. 108, no. 205, pp. 25--30 (2008).
5. Komatsu, T. and Shiokawa, S., A Multi-Path Routing Protocol Considering Inter-Path Interference in MANET, Proceedings of the 2011 IEICE Communications Society Conference, p. 477 (2011).
6. Kurihara, M. and Kuwakado, H., Secure Regenerating Codes based on MSR Codes for Distributed Storage Systems, IEICE Transactions on Fudamentals of Electronics, Communications and Computer Science, vol. J96-A, no. 4, pp. 166--174 (2013).
7. Le, P.H. and Pujolle, G., A Link-Disjoint Interference-Aware Multi-Path Routing Protocol for Mobile Ad Hoc Network, Digital Information and Communication Technology and Its Applications, Springer, pp. 649--661 (2011).
8. Lee, S.J. and Gerla, M., Split Multipath Routing with Maximally Disjoint Path in Ad Hoc Networks, Proceedings of the IEEE International Conference on Communications, pp. 3201--3205 (2001).
9. Maimour, M., Maximally Radio-Disjoint Multipath Routing for Wireless Multimedia Sensor Networks, Proceedings of the 4th ACM Workshop on

Wireless Multimedia Networking and Performance Modeling, pp. 26--31 (2008).
10. Perkins, C.E., Ad Hoc On-Demand Distance Vector (AODV) Routing, RFC 3561 (2003).
11. Radi, M., Dezfouli, B., Bakar, K.A. and Lee, M., Multipath Routing in Wireless Sensor Networks: Survey and Research Challenges, Sensors, MDPI, vol. 12, no. 1, pp. 650--685 (2012).
12. Ravindra, E., Kohir, V.D. and Mytri, V.D., A Local Route Repair Algorithm based on Link Failure Prediction in Mobile Ad Hoc Network, World Journal of Science and Technology, vol. 1, no. 8, pp. 64--67 (2011).
13. Seyama, T. and Higaki, H., G-AODV+PCMTAG: Routing in MANET with Low Overhead Flooding and Route-Shortening, Proceedings of the 7th International Conference on Parallel and Distributed Computing and Networks, pp. 103--110 (2008).
14. Shamir, A., How to Generate and Exchange Secret, Communications of the ACM, vol. 22, no. 11, pp. 612--613 (1979).
15. Waharte, S. and Boutaba, R., Totally Disjoint Multipath Routing in Multihop Wireless Networks, Proceedings of the IEEE International Conference on Communications, pp. 5576--5581 (2006).
16. Yamanaka, M., Miyamoto, S. and Sampei, S., A Secure Wireless Link Creation based on Distributed Transmission Using a Secret Sharing Method, IPSJ Journal, vol. 54, no. 12, pp. 2440--2450 (2013).

Computerized flat knitting machine network control research

Cheng-Jun Zhang and Xiao-Yan Zuo*

School of Mechanical Engineering and Automation, Wuhan Textile University, Wuhan, 430073, China

**E-mail: zchengj_wuse@163.com*

www.wtu.edu.cn

By analyzing the computerized flat knitter's control system, we realized that data processing rate and storage speed are the main linchpins in improving the working speed of the machine. Since PCs have large capacity, high-frequency and good transmission characters, we propose a computerized flat knitting machine with network control suggestion which connects the PC and the flat knitter. To test the effectiveness of the proposed control system, a network control system model is created and the Truetime toolbox is used to simulate the sequence of control data timing. The results show that the flat knitters' network control system transmission format is able to meet the industry's requirements as well as increase the working speed of the flat knitter.

Keywords: Computerized Flat Knitter; Network Control System; CAN Bus; Truetime Toolbox.

1. Networked Computer Control System Design Scheme

1.1. *The traditional computer flat knitting machine control system design scheme*

The action of computer flat knitting machine is very complicated, in order to achieve the complex process, the design of computer flat knitting machine actuators adopt servo motor, motor, raw, motor, tableting electromagnet and needle selection and many other components [1].

ARM9 controller is the main control part of computer flat knitting machine, which responsible for human-computer interaction interface, servo motor driver and alarm function modules such as the outside module. However, the traditional machines' pattern design was conducted by PC, then PAT, CNT and FKW format [2] control data transfer by U disk and story in the FLASH memory of ARM9. When working, the data is opened, calculated and transmitted to upper machine by CAN bus.

---

* Corresponding author.

The frequency of ARM9 limits the data processing speed, so the current computer flat knitting machine working speed is generally within 2 m/s. In order to improve the working speed of flat knitter, the master control CPU bus transmission and data processing speed needs to be further accelerated to satisfy the requirement of high speed working. Therefore, this paper presents a design of direct using PC to replace the ARM9.

1.2. *Networked flat knitting machine control system design scheme*

Figure.1 shows the PC directly replace ARM9 to form the networked control system. The computational speed of PAT, CNT and FKW data generated by a computer flat knitting machine pattern design software is much higher than ARM9 processor [2]. Direct extending PCI-CAN the board on computer PCI slots, the data can be directly converted to CAN bus through the PCI bus data, which can directly communicate with multiple under computer flat knitting machine, constitute a new networked computer flat knitting machine control system.

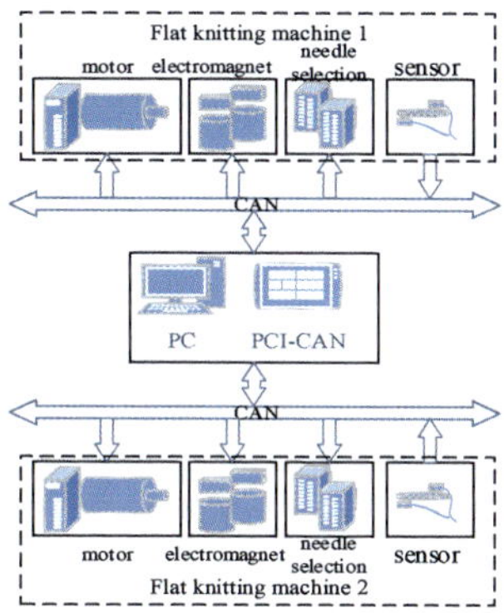

Fig. 1. Structure of flat knitting machine network control system.

There are a lot of advantages for network control system of automatic computerized knitting machine using CAN bus, such as long transmission distance, strong anti-interference ability, strong real-time, more node numbers and convenient maintenance. But a large number of information sources will appear in the whole network flat knitting machine control system. Information transmission process will take up the bandwidth of the network communication lines, cause the phenomenon of packet loss and delay in the process of information transmission inevitably. In order to solve this problem, the data factor influencing network control system is analyzed through modeling and simulation, which can make the distributed network control system to obtain ideal control effect.

2. Networked Computer Flat Knitting Machine Driven Modeling Based On Truetime Toolbox

2.1. *Truetime toolbox description*

Truetime is a toolbox developed by DanHenriksson and Anton Cervin of Sweden Lund university based on Matlab in 1999 [3]. The toolbox mainly contains Truetime Kernel, Truetime Network, Truetime, and Truetime Wireless Network module.

Truetime Kernel module can structure each node in the network, such as servo motor driver module of computer flat knitting machine, sensor module and controller module, etc. In additionally, Truetime Network is used to construct the cable Network of network control system, such as the CAN Network, Ethernet, switched Ethernet and token ring Network.

2.2. *Network flat knitting machine drive system modeling based on Truetime toolbox*

2.2.1. *Network flat knitting machine drive system model as a whole*

The main movement of computer flat knitting machine is driven by the servo motor. If a computer host controls two computer flat knitting machine at the same time, a computer is needed to control two servo motors simultaneously. The model built by Truetime is shown in Figure 2.

The Network model is composed of 5 nodes (Node1 ~ Node5) in Fig. 2. Among these nodes, Node1 and Node4 are servo motor drive module, responsible for the receiving the control information of the controller Node2 from Network, and output the speed of analog servo motor. Node3 and Node5 work as the sensor unit, collecting the amount of the module of the servo motor, and then converting the amount into digital form. Another Node2 is the computer controller, according to the servo motor data from the sensor nodes and the inputting reference data, Node2 implements PID compensation for the control data of each servo drive unit, so as to realize a computer controller to control both computer flat knitting machine drive system at the same time.

2.2.2. *Network flat knitting machine computer controller model*

The computer controller (Node2) is set up with Truetime Kernel module of Truetime shown in Figure 3. In this paper, when a main controller control two computer flat knitting machine at the same time, there is existing two analog input (r1 and r2) on driver input, they are used to control two servo motors respectively. In additionally, the computer controller can not only receive the data from the Network, but can give out the data to the Network control. Therefore, there is a data input interface (Rcv interface) and a data output interface (Snd interface) on computer controller.

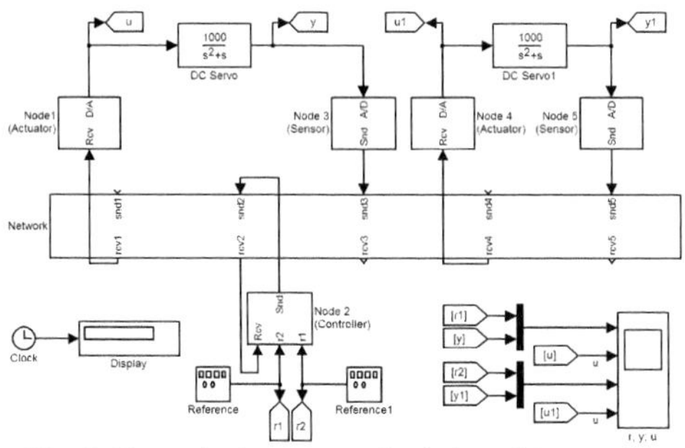

Fig. 2. Networked computer flat knitter drive model.

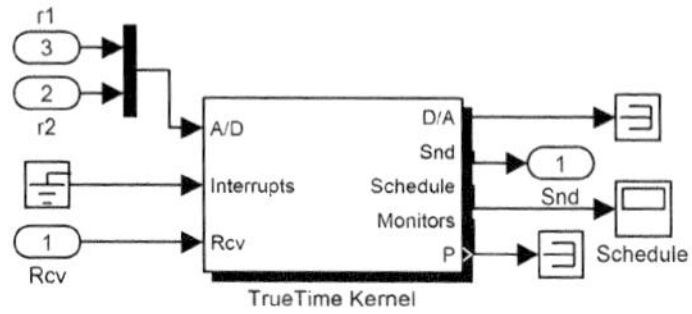

Fig. 3. Computer controller model.

2.2.3. *PID algorithm of network flat knitting machine controller*

In order to keep the servo drive controller with send data, PID algorithm is used.

Suppose that the transfer function of servo motor [4] is

$$G(s) = \frac{1000}{s^2 + s} \tag{1}$$

The discrete PID algorithm is as follows:

$$\begin{cases} P(k) = K(r(k) - y(k)) \\ I(k+1) = I(k) + \dfrac{Kh}{T_i}(r(k) - y(k)) \\ D(k) = \alpha_d D(k-1) + b_d(y(k) - y(k-1)) \\ u(k) = P(k) + I(k) + D(k) \end{cases} \tag{2}$$

Where

$$\begin{cases} \alpha_d = \dfrac{T_d}{Nh + T_d} \\ b_d = \dfrac{NKT_d}{Nh + T_d} \end{cases} \tag{3}$$

In Eq. (2) and Eq. (3), K is the proportion coefficient, Ti is the integral coefficient, Td is the differential coefficient, and h is the sampling period.

3. Network Flat Knitting Machine Drive System Simulation Based On Truetime Toolbox

3.1. *The simulation parameters settings*

Based on the simulation of Truetime toolkit, the first is to set the network parameters of the Network. This paper transports the data through the CAN bus. Therefore, when setting Truetime Network module parameters, CSMP/AMP (CAN) is selected as the Network type, which includes five nodes. In order to keep industrial site speed of the CAN bus transmission, remote node data can obtain stable data transmission. The data transmission speed is 250 KBPS, the minimum data frame is 40 bytes, and the data packet loss rate of 0-10% in this paper.

Computer controller uses event triggers scheduler to perform. When the controller receives the CAN network data, the task of start begins to execute PID algorithm. At this time, according to the empirical formula, proportion coefficient K value is 1.5, Td integral coefficient is 0.035, and sampling period h is 0.01s.

3.2. *Network flat knitting machine drive system simulation based on Truetime toolkit*

According to the above simulation parameters, combining with Truetime toolkit and matlab software, network computer flat knitting machine drive model shown in Fig.2 is simulated to observe the response of the servo motor in the network. In the ideal network, namely, under the condition of no packet loss in the data transmission process, networked motor response in 1 seconds is shown in Fig.4. In Fig.4, r is analog input reference value of the servo motor, y is the actual response value of servo motor, u is the output value of the servo driver simulation. r1, y1 and u1 are the values another motor within the network, the same as aforesaid. Through observing the ideal model of no packet loss, it can be found that the servo motor is better response signal input using computer controller PID algorithm, which proves the networked computer flat knitting machine drive system feasible. If network exists packet loss in data transmission process, assuming that the probability of packet loss is 10%, the response of the servo motor is shown in Fig.5. From Fig. 5, you can see that the response data will produce mutation in addition to the analog input of high frequency, the rest of the motor response is in good condition. It also proves that networked computer flat knitting machine drive system is basically a normal response under the condition of low probability of the network data packet loss. If increasing the data packet loss probability further, the servo motor will be unable to response the situation, it cannot meet the requirement of the computer flat knitting machine drive.

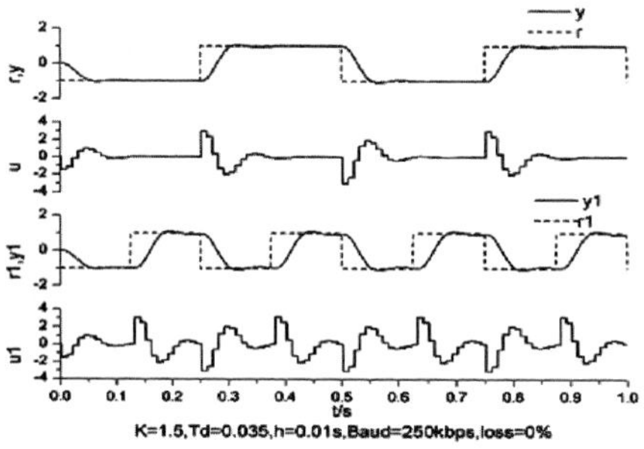

Fig. 4. The network drive motor response curves without a packet loss.

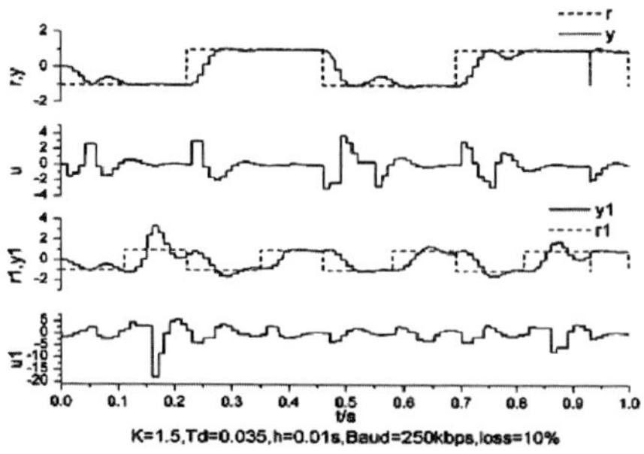

Fig. 5. The network drive motor response curves with a packet loss.

4. Conclusion

Based on the storage and transmission speed of PC, this paper designs a networked control system. Meanwhile, a networked computer controller driver model is established adopting Truetime toolkit. Through simulating and analyzing the model, it can be concluded that the networked control system can meet the requirements of servo drive under the circumstances of the low probable packet loss.

Acknowledgments

This work is supported by Innovation Science Foundation of Wuhan Textile University and partially supported by grant 51305309 of National Natural Science Foundation of China.

References

1. Jian-fei L, FU J, LIU D. Development of the control system of automatic flat knitting machine based on μC/OS-II embedded technology. Journal of Textile Research, 2006, 1: p.8-11.

2. Zhang Chi, Zhang Chengjun, Wu Xiaoguang. Control system data preparation of computer flat knitting machine. Journal of Textile Research, 2012. 33(8): p. 124-129.
3. Cervin, a., d. Henriksson, b. Lincoln, et al. How does control timing affect performance analysis and simulation of timing using jitterbug and Truetime. Control systems, IEEE, 2003. 23(3): p. 16 - 30.
4. Henriksson D, Cervin A, Arzén K E. Truetime: Simulation of control loops under shared computer resources. Proceedings of the 15th IFAC World Congress on Automatic Control. Barcelona, Spain. 2002.

A novel dependability optimization method in network layer of IOT

Rui-Juan Zheng, Tong Zhang*, Ming-Chuan Zhang, Qing-Tao Wu and Wang-Yang Wei
College of Information Engineering, Henan University of Science and Technology, Luoyang, Henan, China
**E-mail: rjwo@163.com*

Based on the multi-layer characteristics of IOT security elements, this paper attempts to fine-tune dependability elements in network layer from the "internal cause" of the dependability situation, trying to make the optimization of system dependability indicators. Firstly, linear programming theory is used to analyze the change rule between the dependability elements and dependability indicators in the network layer, meanwhile, combined with the features and dependability properties, the change regulation function of the dependability is established from all details. At last, the optimization effect of network layer is simulated, and the results show that, the optimization method proposed in this paper can optimize the dependability of network layer in IOT to a large degree.

Keywords: Dependability Optimization; Network Layer; Network Layer; Application Layer.

1. Introduction

With the advent of the era of IOT, many industry operation modes and the ways of human life have earth-shaking changes. In the near future, as networking technology continues to mature and the development of the applications, IOT plays a significant role in the social life of the specific application. At the same time, the social activities in IOT and strategic resources infrastructure built on the interconnected network over the entire structure, all of the activities and facilities have high transparency, once attacked, the security and privacy will face enormous threat.

As we all know, the IOT has the characteristics of multi-level, which could be divided into perception layer, network layer and application layer with the bottom-up. In contrast, the network layer has the direct relation between all kinds of specific applications and actual users.

According to the actual demand in the process of IOT specific application, and based on multi-layer features of IOT dependability elements, this paper will

* Corresponding author.

fine-tune elements among each layer from the "internal cause" of IOT dependable status, trying to find the system optimization dependable indicators. The main contents of this paper are organized as follows: The second part will introduce and analyze the related work; The third part will propose the method of dependable optimization by calculating the dependable indexes; The fourth part will simulate the optimized results and analyze the simulation results.

2. Related Works

As all kinds of systems have the similar problems in maintaining or enhancing the dependability, therefore, in recent years, the researches of dependability and reliability are also increasing. The configuration error is one of the most important causes [1]. Autonomic configuration and management of complex and dynamic networks is a challenging problem for network researchers and designers. Such online optimization of systems can be performed in two ways: (i) using a separate model of the system for experimenting new configurations, (ii) using the system itself for experimentation without a separate system model [2]. At the same time, Self-organizing network, or SON, technology, which is able to minimize human intervention in networking processes, was proposed to reduce the operational costs for service providers in future wireless systems [3]. In the aspect of autonomic configuration algorithm, [4] demonstrated the technique as a solution to spatial reuse time-division multiple access (STDMA) scheduling with reconfigurable antennas. Measurement-Based Admission Control for Flow-Aware Networks is presented in [5] and extended in [6], and according to the static limitation mechanism, a dynamic limitation mechanism shows that it is possible and more efficient to provide dynamic limitations, where the admission limit is not static, but dynamically calculated [7]. [8] proposed a new auto-configuration algorithm called Pivot-based Fast Automatic Configuration (PFAC). Leveraging properties of the given data center network topology, PFAC can assign addresses to physical devices in the data center automatically and efficiently. [9] proposed an approach for improving the enterprise fem to cell network's performance by automated optimizing the fem to cell base station's (FBS's) pilot power as well as antenna pattern, and the recently proposed multi-element antenna which is appropriate for fem to cell is also introduced.

3. Optimization Approach

3.1. *Dependability requirement*

Among Dependability Element of Network Layer (E_N), the route coverage (Rc), response time (Rt), packet loss rate (Pr) and anti directional analysis (Ad) will be the key elements to affect the survivability of network(SV_N), safety of network (SF_N), completion of network (CP_N) and availability of network (AB_N).

The survivability of network SV_N is decided by response time Rt, packet loss rate Pr, anti directional analysis Ad and the route coverage Rc, and the calculation formula is as Eq. (1).

$$SV_N = \sum_i \varpi_{1i} E_{N1i} = \varpi_{11} \cdot Rt + \varpi_{12} \cdot Pr + \varpi_{13} \cdot Ad + \varpi_{14} \cdot Rc, \sum_i \varpi_{1i} = 1 \cdot \quad (1)$$

Safety of network(SF_N) is decided by anti directional analysis Ad, packet loss rate Pr, the route coverage Rc and response time Rt, and the calculation formula is as Eq. (2).

$$SF_N = \sum_i \varpi_{2i} E_{N2i} = \varpi_{21} \cdot Ad + \varpi_{22} \cdot Pr + \varpi_{23} \cdot Rc + \varpi_{24} \cdot Rt, \sum_i \varpi_{2i} = 1 \cdot \quad (2)$$

Completion of network CP_N is decided by response time Rt and packet loss rate Pr, and the calculation formula is as Eq. (3).

$$CP_N = \sum_i \varpi_{3i} E_{N3i} = \varpi_{31} \cdot Rt + \varpi_{32} \cdot Pr, \sum_i \varpi_{3i} = 1 \cdot \quad (3)$$

Availability of network AB_N is decided by packet loss rate Pr and response time Rt, and the calculation formula is as Eq. (4).

$$AB_N = \sum_i \varpi_{4i} E_{N4i} = \varpi_{41} \cdot Pr + \varpi_{42} \cdot Rt, \sum_i \varpi_{4i} = 1 \cdot \quad (4)$$

3.2. *Optimization flow*

On the premise of given the initial values to dependability factors, the local optimization in layers, values of dependability factors for dependability maximization will change, and the deviation of this change is called the cost optimization. For example, the dependability factor set in business layer is $E_B = \{Ft, Ai, Nc, Ir\}$, if the initial values of dependability factors separately are Ft_0, Ai_0, Nc_0, Ir_0, and the optimal configuration values are $Ft_{OC}, Ai_{OC}, Nc_{OC}, Ir_{OC}$, then the optimization cost of this local optimization can be expressed as $OC = \sum(| Ft_{OC} - Ft_0 |, | Ai_{OC} - Ai_0 |, | Nc_{OC} - Nc_0 |, | Ir_{OC} - Ir_0 |)$.

4. Optimization Example

We derive the calculation formula of the optimization objective in network layer as Eq. (5).

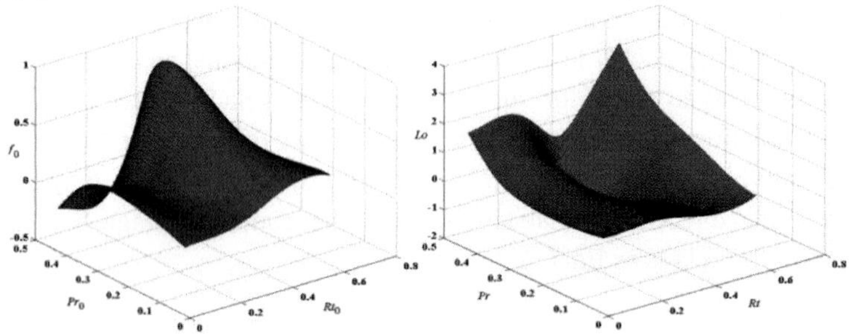

(a) Comparing of optimization effect betweenRt_0,Pr_0,f_0 and Rt, Pr, Lo

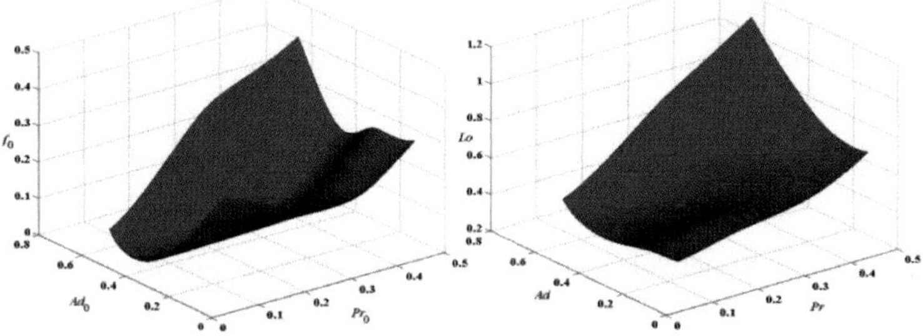

(b) Comparing of optimization effect betweenPr_0,Ad_0,f_0 and Rt, Pr, Ad, Lo

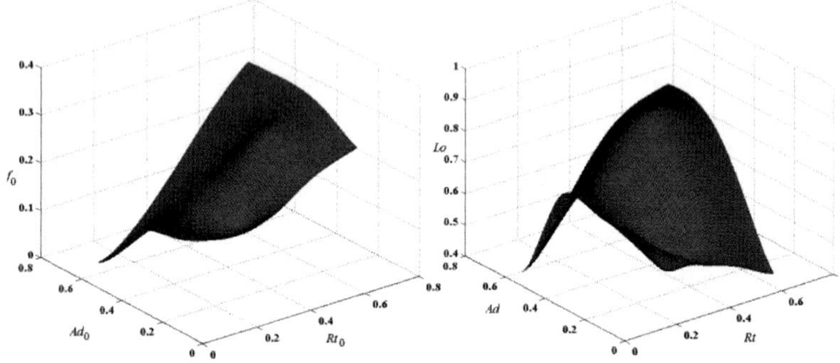

(c) Comparing of optimization effect betweenRt_0,Ad_0, f_0 and Rt, Ad, Lo

Fig. 1. Comparing of optimization effect in network layer.

$$Lo_{en}=\begin{cases}\sum_{i} en_{i_0},\forall en_i\in En,\text{satisfies } en_{i_0}\le en_{i_{\alpha}};\\ \sum_{j,k}(|2en_{j_{\alpha}}-en_{j_0}|,en_{k_0}),\forall en_j\in En,\text{satisfies } en_{j_0}>en_{j_{\alpha}},\forall en_k\in En,\text{satisfies } en_{k_0}\le en_{k_{\alpha}};\\ \sum_{i}|2en_{i_{\alpha}}-en_{i_0}|,\forall en_i\in En,\text{satisfies } en_{i_0}>en_{i_{\alpha}}.\end{cases} \tag{5}$$

Accordingly, the comparing of dependability between the condition of Rt_0, Pr_0, Ad_0 and optimized elements is shown in Figure.1.

As shown in Figure.1, (a), (b), (c) separately express the comparing of optimization effect between Rt_0, Pr_0, f_0 and Rt, Pr, Lo, the comparing of optimization effect between Pr_0, Ad_0,f_0 and Rt, Pr, Ad, Lo, the comparing of optimization effect between Rt_0, Ad_0, f_0 and Rt, Pr, Ad, Lo. As we can see, in this example, the minimum optimization range of dependability is 49%, and the highest optimization range is 3.8 times. The average optimization range is 2.7 times, so the optimization effect is remarkable.

5. Conclusion

The rapid development and the popularization of IOT put forward higher requirements to the dependability problems in the process of its application, and the dependability of network layer in IOT is vital. Based on layered thought in IOT, this paper researches and proposes a dependability optimization approach to network layer in IOT system. With the change of dependability elements directly affecting the dependability of network layer as the internal cause, the complex multi-source dependability parameters variables, configuration requirements of network layer in IOT and dependability environment variables are analyzed. Deriving the mapping relationship between the dependability elements and the dependability in each layer, the optimization approach of network layer in IOT is obtained. The last simulation results in network layer and application layer show that, in the condition of random assignment to key elements, the optimization effect of proposed approach is obvious. Each layer can improve the dependability of this layer in different degrees, and the optimization effect is significant. In the future work, we will focus on the global optimization in IOT.

Acknowledgments

This work has been supported by the Project of Natural Science Foundation of China (No. U1204614, 61370221,U1404611), in part by Program for Innovative Research Team (in Science and Technology) in University of Henan Province

(No. 14IRTSTHN021), in part by Key Project of Science and Technology Department of Henan Province (No.112102210187).

References

1. F Li, J Yang, J Wu. C An,N Jiang,*Research on automatic configuration of Internet,J.Journal of Software.A*25,118 (2014).
2. B Gonen, M Yuksel,Network configuration and management via two-phase online optimization, in *Proc. IEEE Int.Global Telecommunications Conference (GLOBECOM 2011)*, (Houston, TX, United states, 2011).
3. M Peng, D Liang,Y Wei, J Li,*Self-configuration and self-optimization in LTE-advanced heterogeneous networks, J. Communications Magazine. A***51**, 36 (2013).
4. E Anderson, C Phillips, D Sicker, D Grunwald,*Optimization decomposition for scheduling and system configuration in wireless networks, J.IEEE/ACM Transactions on Networking (TON).A***22**,271 (2014).
5. R Wójcik, J Domzal, A Jajszczyk. Fair Rate Degradation in flow-aware networks, in *Proc. IEEE Int. International Conference on Communications (ICC)*, (Cape Town, South Africa,2010).
6. R Wójcik, J Domzal, A Jajszczyk. Enhanced measurement-based admission control for flow-aware networks, in *Proc. IEEE Int. International Conference onComputing, Networking and Communications (ICNC)*,(San Diego, CA, United states, 2013).
7. R Wójcik, D Garbacz, A Jajszczyk. A dynamic limitation mechanism for flow-aware networks. in *Proc. IEEE Int. International Conference on Communications (ICC)*,(Budapest, Hungary,2013).
8. C Lu, Z Jiang, Y Cai, G Zhang, J Yu, W Liang, J Bi. Pivot-based fast automatic address configuration for data center network.in *Proc. IEEE Int.International Conference on Cloud Networking (CLOUDNET)*, (Paris, France,2012).
9. Y Li, Z Feng, D Xu, Q Zhang, H Tian. Automated optimal configuring of femtocell base stations' parameters in enterprise femtocell network. in *Proc. IEEE Int.Global Telecommunications Conference (GLOBECOM 2011)*, (New York, United states,2011).

The effectiveness of applying moodle E-learning platform on resource classroom students

Hua-Yi Lin*, Jiann-Gwo Doong and Xiao-Wei Lin
Department of Information Management, China University of Technology, Taiwan
**E-mail: calvan.linmsa@gmail.com*

This study was designed to investigate hybrid teaching, which is a combination of traditional teaching with Moodle e-learning platform for classroom curriculum. Here, we will discuss the learning effectiveness and the learning attitude on special education students. This study analyzes the learning efficiency, motivation and satisfaction of 20 high school students before and after using Moodle online teaching modules. We evaluate the learning effectiveness on the hybrid teaching manner using a paper exam and a questionnaire to determine the learning attitudes of students. Our results show that the hybrid teaching model improves the learning effectiveness, and the students hold positive attitude towards learning.

Keywords: Moodle; Hybrid Teaching; Special Education Students.

1. Introduction

In response to the rapidly changing of the global society and trend in curriculum and teaching reform. The world highly uses information and network on education and training in order to enhance the effectiveness of teaching. In this trend, school curriculum developments are toward to digitization and interaction. This study expects to provide teachers and students with a common platform to discuss curriculums and share their study achievement. Thus, we can highlight the importance of using Internet resources to learn and study. Teachers can abandon the traditional input-teaching skills and create diverse environments for fitting the rapidly changing of social characteristics. Thus, the concept of lifelong learning can be implemented in each student and create an e-generation learning type. Nowadays, Ministry of Education in Taiwan builds up a free curriculum platform named Moodle. It is an open environment, and every school can apply for its curriculums and implement a digital e-learning platform like the business of e-learning software platform. Therefore, recently the Moodle

* Corresponding author.

platform is gradually be taken seriously. Some schools construct the Moodle e-learning platform for teaching, and many scholars start discussing the achievement of after adopting Moodle platform on curriculums.

2. Explore the Digital Learning Platform

Moodle (modular object oriented dynamic learning environment) is a curriculum management system via Internet which is to be designed to offer e-learning education and training courses [1,2]. The design concept of Moodle is different from the traditional teaching manner relying on teachers in classroom. Moodle gets rid of the hierarchical learning method, and gives a learner-centered platform. By sharing and interactive platforms, students can control his learning progress, plan his learning schedule, and cultivate active learning habits. Through interactive forums or bulletins, students form a learning social network [3]. Moodle uses the modular design and supports many kinds of databases. The following presentations describe the main function of Moodle:

(1). Course description: This interface gives the beginner an instinct and a quick architecture to realize how to use this platform.

(2). Administration interface: This interface provides number of modules, and thus administrators can systematically plan and manage courses.

(3). Interface of a course design: this interface provides a course administration, instructions and class rules.

(4). Interface of user administration: It provides an interface for users and teachers to administrate the students who participates in this course.

(5). Learning platform framework: The main architecture is divided into three parts, and there are system administration, course management, and student management. Each part provides diverse modules to use.

Here, we discuss three teaching manners as follows. (1)Traditional teaching method is a collective teaching [4]. It is a plain and easy way to instruct students in a specific place. Teachers introduce and demonstrate teaching materials to students according to the designated teaching material, target, and progress. This way is the most popular manner for teachers to proceed teaching. The transmission of information is one-way type, and thus it less likely arouses students' interest and active learning. (2) E-learning becomes popular and it consists of three elements of learning culture, values and instructional design and technology, and is to be produced by their intersection [5]. Learners want to use e-learning with student-centered, active learning of students, and guiding learning of teachers. E-learning exploits a group or a community group to study, so that students can freely learn without constraints of time. Digitized curriculum usually contains information contents of digital teaching materials, teaching methods, and teaching activities, and they are to be presented in the way of texts, graphics, animations, videos, etc. (3)Hybrid learning refers to the integration of the traditional face to face classroom teaching and self-learning on network. Now, it is applied to education and training and is a new educational trend of e-learning. Hybrid learning effectively reduces the dropout rate of learning. Hybrid learning is a student-centered teaching mode. In hybrid learning, there are usually some part of the implementation of traditional teaching. Teachers guide students to learn, after the students have enough knowledge, and then teachers combine with the characteristics of online education to meet students' active construction of knowledge. Thus hybrid learning can enhance the motivation of the cooperative discussion with others.

3. Research Methods

The object of this study is in a public junior resource class with special education students in Taoyuan City. In this school, there are 16 classes, and special resource students randomly join the regular classes and are also arranged into special resource classes to be educated with special education curriculums. We arrange six groups of students, and there are 20 students. The Chinese curriculums have two classes per week and total 90 minutes. Participants are special education students of the resource class. These students are learning disabilities or mild mental retardations. This study adopts the Moodle web address: http://learning.edu.tw/moodle/tyc/zmjhs/. We edit specific materials and examinations for those special education students, including two categories of vocabulary test and comprehension test. The policy of choosing subjects are as follows: (1). The same number of questions of each topic; (2). Each question has the same level of difficulty. Eventually, this study evaluates what kind of teaching manners is better to them. Then, according to the senior resource education teachers' opinion to modify the teaching materials.

We prepare three categories of questionnaires to discuss special education students who are educated by traditional teaching manners versus hybrid teaching manners. We would like to realize whether students like the hybrid teaching manners.

Learning motivation: It focuses on learning interest and learning willingness for special education students via traditional manners versus hybrid (Moodle with traditional education) manners in Chinese fields.

Learning assistance: It focuses on the feelings of learning assistance for special education students via traditional manners versus hybrid (Moodle with traditional education) manners in Chinese fields.

Learning satisfaction: It focuses on the satisfaction degree of learning for special education students via traditional manners versus hybrid (Moodle with traditional education) manners in Chinese fields.

The questionnaire of learning attitude has four subjects in each category. This study adopts Likert Scale five-point scale. 1 presents strongly disagree; 2 presents disagree; 3 presents ordinary; 4 presents agree; 5 presents strongly agree. The scoring mode is as follows. Answering 1 gets 1 point; Answering 2 gets 2 point; Answering 3 gets 3 point, and so on. This study considers the regular examinations and the basic curriculum of the experimental school. Our study is implemented in two phases. The first phase is executed from Sep. 1, 2014 to Oct. 19, 2014. During this phase, we perform the traditional teaching. The second phase is executed from Oct. 19, 2014 to Dec. 7, 2014. During this phase, we perform the hybrid teaching. The whole phases pass through fourteen weeks.

This study designs hybrid teaching manners which combines traditional education and online e-learning educations via Moodle. Then we investigate the difference of learning effectiveness and learning attitude between traditional educations and hybrid educations. Our research adopts pre-test and post-test per single group. In order to realize the learning effectiveness, we take *t* verification to analyze the result of before and after teaching. Subsequently, this study would like to realize the learning attitude and opinion on hybrid teaching for special education students. Here, this study adopts a questionnaire to analyze the learning attitude. Finally, according to the collected information, this study synthesizes and sorts out the results.

4. Analysis of Effectiveness

Special education students after accept the hybrid teaching in Chinese curriculums. This study would like to realize the learning performance, and we evaluate the first (after traditional teaching) and the second (after hybrid teaching) experimental teachings on resource classes. This study takes SPSS 22.0 statistic software to verify the difference on learning score in different teaching modes. The analysis results are as follows:

After the traditional teaching and the hybrid teaching, the participators accept a paper test before and after the two mode teaching manners. Their learning effectiveness are presented in table 1.

Table 1. Learning effectiveness analysis.

	First Score (Traditional Teaching)		**Second Score (Hybrid Teaching)**	
Variable	Mean(M)	Standard Deviation (SD)	Mean(M)	Standard Deviation (SD)
All students	65.65	8.911	74.70	7.079

Table 2. Test table of the learning effectiveness.

	Pair Deviation							
Mode	Mean	SD	SD of the mean	95% Confidence interval of variance		T	DF	Significant Difference
				Lower limit	Upper limit			
Trad. vs Hybrid	-9.05	10.07	2.253	-13.76	-4.334	-4.01	19	.001

From table 1, this study observes that the hybrid teaching has a better mean value than the traditional teaching. Besides, in the traditional teaching, *SD* value is higher than the hybrid teaching that means the traditional teaching has higher dispersion degree than the hybrid teaching. In order to realize the difference, the participators accept a paper test before and after the two mode teaching manners. Table 2 describes the *T* test table of the learning performance.

From table 2, the statistic reveals that paired-sample *T* test shows T=-4.01, significant difference (P) is equal to 0.001, $P < \alpha$ (=0.05). This result indicates that the hybrid teaching reaches a significant level. In other words, the learning performance on hybrid teaching outperforms the traditional teaching. The statistic results reveals that the effectiveness of the hybrid teaching causes significant differences. Additionally, the learning attitudes have three faces, and the scoring mode is as follows. Answering 1 gets 1 point; Answering 2 gets 2 point; Answering 3 gets 3 point, and so on. The higher score indicates that the learning attitude tends to aggressive. On the contrary, the lower score indicates that the learning attitude tends to passive negative. The following figures depict that the analysis of leaning motivation is as shown in figures 1-1 and 1-2. Learning assistance is as shown in figures 2-1 and 2-2. Learning satisfaction is as shown in figures 3-1 and 3-2.

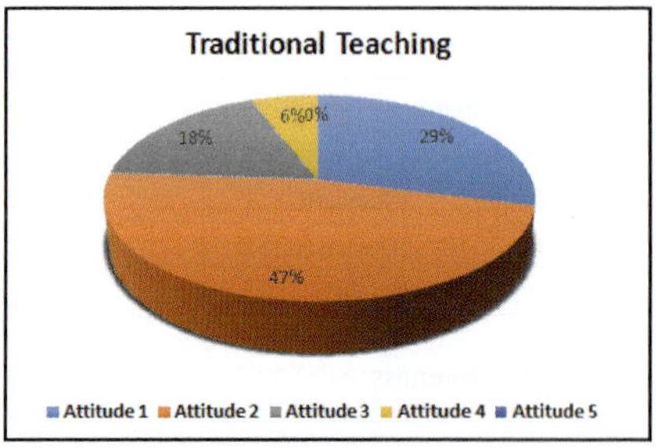

Fig.1-1. Pie chart of leaning motivation.

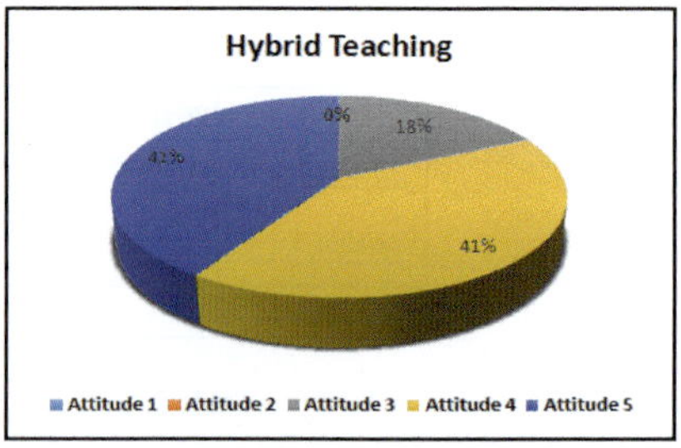

Fig.1-2. Pie chart of leaning motivation.

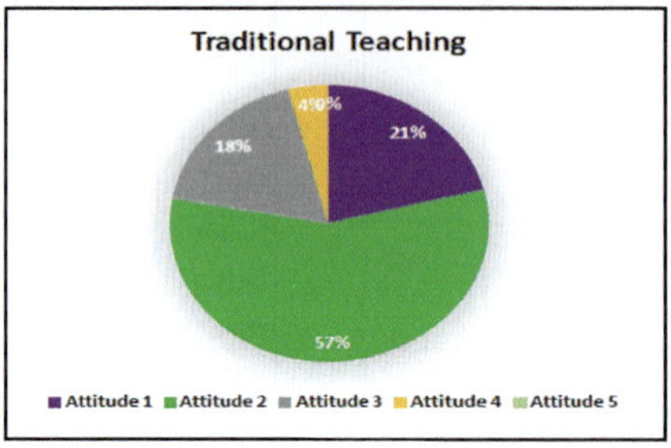

Fig.2-1. Pie chart of leaning assistance.

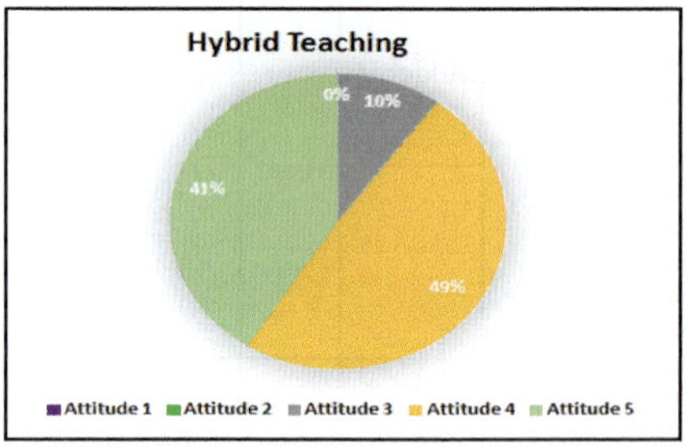

Fig.2-2. Pie chart of leaning assistance.

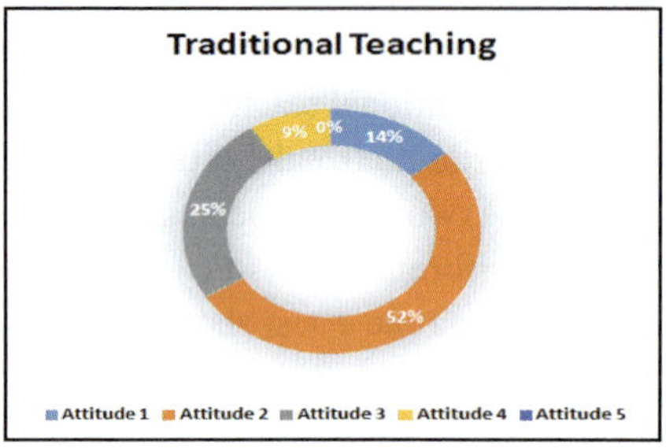

Fig.3-1. Pie chart of leaning satisfaction.

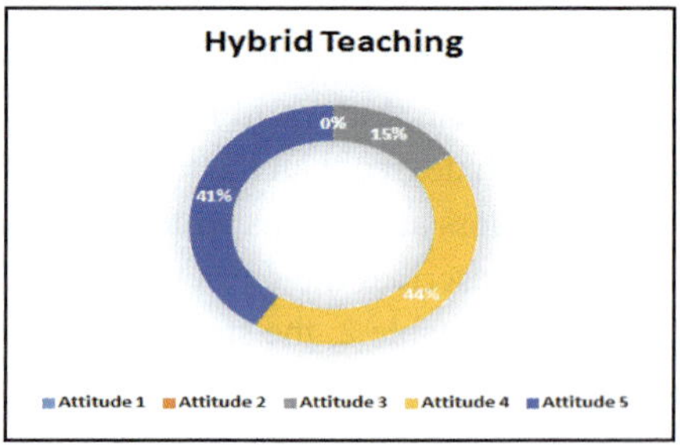

Fig.3-2. Pie chart of leaning satisfaction.

Figure 1 indicates that the traditional teaching manner is 76.25% which tends towards negative (attitude values are 1 and 2), and the positive learning attitude (attitude values are 4 and 5) only takes 6.25%. From the above value, over half participants think that the traditional teaching manner lacks learning motivation. However, the attitude of hybrid teaching is 82.5% which tends towards positive (attitude values are 4 and 5). As a result, over half participants think that the hybrid teaching manner increases learning motivation. Figure 2 indicates that the traditional teaching manner is 78.75% which tends towards negative (attitude values are 1 and 2), and the positive learning attitude (attitude values are 4 and 5) only takes 3.75%. In other words, over half participants think the traditional teaching manner cannot improve learning assistance. However, the attitude of hybrid teaching is 90% which tends towards positive (attitude values are 4 and 5). As a result, over half participants think the hybrid teaching manner improves learning assistance. Figure 3 indicates that the traditional teaching manner is 66.25% which tends towards negative (attitude values are 1 and 2), and the positive learning attitude (attitude values are 4 and 5) only takes 8.75%. In other words, over half participants think the traditional teaching manner cannot meet learning satisfaction. However, the attitude of hybrid teaching is 85% which tends towards positive (attitude values are 4 and 5). As a result, over half participants think the hybrid teaching manner improves learning satisfaction.

5. Conclusions

traditional teaching combines with Moodle online learning named hybrid teaching that has a better learning efficiency than the pure traditional teaching mode. Scholars think that Moodle platform gives vivid and abundant materials, diverse multimedia, such as texts, photos, motion pictures and videos, to present the curriculums and to do a lot of practice opportunities. In this study, the special education students have positive learning attitude on the hybrid teaching manner, which gives many interactive communications between teachers and students. Additionally, students can co-work and exchange learning experience with each other. Thus, the hybrid teaching manner improves the learning willingness, and makes learning motivation of students become positive, and also encourages the positive learning attitude of students.

References

1. Marton, D., Bednar, S., & Modrak, J., "Enhancing Learning Management System Through Web Academic Library Information Systems", Polish Journal of Management Studies, Vol. 10(1), pp.63-73, 2014.
2. Oyler, C., "Teacher preparation for inclusive and critical (special) education. Teacher Education and Special Education, vol. 34, pp.201-218, 2011.
3. Seo, Y. J., &; Bryant, D., "Multimedia CAI program for students with mathematics difficulties". Remedial and Special Education, 2010.

4. H. Bär, G. Häussge, G. Rößling, "An integrated system for interaction support in lectures", Proceedings of the 12th annual SIGCSE conference on Innovation and technology in computer science education, 2007.
5. Z. Putnik, M. Ivanović, Z. Budimac, K. Bothe, M. Ivanović, Z. Budimac, K. Bothe, "Analysis of students' behaviour based on participation and results achieved in Wiki-based team assignments", Proceedings of the 6th Balkan Conference in Informatics, 2013.

A general threshold scheme based on elliptic curve cryptosystem

Yu-Lian Shang[1], Wu-Li Song[1], Wu-Yuan Jia[2] and Zhao-Chen Zhang[1*]

[1]*College of Information Engineering, Taishan Medical University,*
[2]*College of pharmacy, Taishan Medical University,*
Taian 271016, China
[*]*E-mail: zczhang@tsmc.edu.cn*

Based on Elliptic Curve Cryptosystem, a threshold authenticated encryption and signature scheme characterized by (t, n) shared verification for (k, l) signature was put forward. Information m is transmitted to a particular verifier company, and the signature is then verified through the cooperation of t ones from the verifier company with n members after being signed by a signer company employing (k, l) threshold signature scheme. Similarly, by integrating (k, l) threshold signature scheme with the message recovery technique and after being encrypted singed by any k ones from a group with l members, information m was transmitted to a particular verifier group with n members, and then recovered through the cooperation of any t ones from the verifier group. Application examples of the company encryption communication system and the generation of the polynomial of the company private key and public key were given. The security of this scheme is based on Shamir threshold scheme and the safety of Elliptic Curve Cryptosystem. The scheme realized a directional transmission between different companies, and due to the advantages of Elliptic Curve, it enjoys wider applications in practice.

Keywords: Authentication Encryption; Elliptic Curve; Threshold Signature; Interpolation Polynomial; Company-oriented.

1. Introduction

The research presented in this paper relies on the following facts in the digital contract:

A contract for signature and authentication between two companies cannot be exposed to any outsider; otherwise it will have forged and altered risk. In other words, the signer and verifier company must be regulated under a pre-stipulated security environment.

For the example above, suppose that several managers on behalf of the company signed a digital contract with the other one; they have not only to

---

[*] Corresponding author.

verify the authenticity of the signing of the contract, but also to ensure that only they can see the contents of the contract and the other people cannot see it.

In view of the above facts, based on Elliptic Curve Cryptosystem, a threshold signature scheme characterized by (t,n) shared verification for (k,l) signature is proposed to solve the first problem. There are l members in a signer company and it's stipulated that only through the cooperation of at least k ones can some information be effectively signed; On the other hand, only through the cooperation of at least any t ones from a verifier company with n members can the signature be verified. A general authenticated encryption scheme characterized by (t,n) shared verification was proposed to solve the second problem through integrating (k,l) threshold signature scheme and message recovery technique. It's stipulated that there are l members in a signer group, and only through the cooperation of at least k ones can some information be effectively encrypted signed; on the other hand, only through the cooperation of at least t ones from a verifier company with n members can the signature be recovered.

Thus, the previously proposed scheme schemes [1-6] before can not meet practical needs, thinking that in the signature and verification stage anyone who receives the document can practice verification. Therefore, there shall be a general schemes [7,8] in which the signer company and verifier company (or recovery company) can be regulated under a pre-stipulated security environment. The security of the scheme is based on the intractability of discrete logarithm calculation on Elliptic Curve [9] and Shamir threshold scheme [10], and enjoys the characteristics of reduced communication load, lowered calculation complexity and wider application in practice, etc.

As a caveat: what we described in this paper is based on [11] and some points have published in Advanced Materials Research with title of "A General Threshold Signature Scheme Based on Elliptic Curve", on Vols. 756-2759(2013),pp:1339-1343, at Qingdao.

2. Description of the Scheme

2.1. *Group-oriented general threshold signature scheme*

The scheme is consists of three parts, i.e., sharing distribution center (SDC), Signer Company and Verifier Company. The scheme is consists of three stages, i.e., generation stage of parameters, generation and verification stage of private signatures, and generation and verification stage of company signature.

2.1.1. *Generation stage of parameters*

In this stage, SDC defines the parameters of system member and he firstly select the Elliptic Curve $E: y^2 = x^3 + ax + b(\text{mod}\, p)$ and big prime numbers P and n, where, $a, b \in Z_p, 4a^3 + 27b^2 \neq 0(\text{mod}\ p)$, $GF(p) = \{0, \cdots, \ p-1\}$, and n is the order of primitive root. The value of $\# E(GF(p))$ is ranged between $p+1 \pm 2\sqrt{p}$. G is base point, and represents a primitive root of the Elliptic Curve, where, $G \in E(GF(p))$ and order is n. p, n, G are used and made public by the company. Next, SDC accepts the registration of a group of signer or verifier, of whom parameter properties are marked by subscript s or v. $G_s = \{u_{s1}, u_{s2}, \cdots, u_{sl}\}$ and $G_v = \{u_{v1}, u_{v2}, \cdots, u_{vn}\}$ are respectively the companies combined by l singer and n verifier. Finally, SDC randomly select interpolation polynomials of the signer company G_s and the verifier company G_v :

$$f_s(x) = a_{k-1}x^{k-1} + a_{k-2}x^{k-2} + \cdots + a_1 x + a_0 \bmod n$$

$$f_v(x) = c_{t-1}x^{t-1} + c_{t-2}x^{t-2} + \cdots + c_1 x + c_0 \bmod n$$

Where, $a_i, c_j \in [1, n-1], 1 \leq i \leq k-1, 1 \leq j \leq t-1$. The private key of the signer company is defined as $f_s(0) = a_0$, with which the public key associated is $Y(s) = f_s(0) G \bmod p$ · $x_{si} \in [1, n-1]$ are the randomly-selected integral numbers for every member of the signer company of SDC, whose private shadow $f_s(x_{si}), i = 1, 2, \cdots, l$ are calculated and the public key associated with are determined by $y_{si} = f_s(x_{si}) G \bmod p, \ \ i = 1, 2, \cdots, l$. Similarly, the private and public key of the verifier, the private and public key of the verifier group can also be got. On the other hand, in order to generate group signature, it is required for SDC to select a secret polynomials for the signer company:

$$f_b(x) = b_{k-1}x^{k-1} + b_{k-2}x^{k-2} + \cdots + b_1 x + b_0 \bmod n$$

Where, $f_b(x_{si})$ are the session keys of the signer member, with which the public values associated are $y_{bi} = f_b(x_{si}) G \bmod p$ .The company value y_{si} = $Y(b) = f_b(0) G \bmod p$ is made public. For safety purposes, SDC select a one-way hash function $h()$ and make it public. To sum up, all the parameters of the system are shown as follows:

1. Shared information in SDC: $p, n, G, h()$
2. Secret information in SDC: $f_s(x), f_v(x), f_b(x)$
3. Shared information of members: xsi, xvj, ysi, yvj and ybi
4. Secret information of members: $f_s(x_{xi}), f_v(x_{vi}), f_b(x_{si})$
5. Shared information of companies: *Ys* (or *Yv*), Y_b
6. Secret information of companies: *fs* (0), *fv* (0)

2.1.2. *Generation and verification stage of private signatures*

According to the security strategy, the scheme allows any k members to sign a piece of information on behalf of the company and the members can simultaneously and independently sign it. Without loss of generality, the k members for signing were expressed as $u_{s1}, u_{s2}, \cdots$ and u_{sk}.

Every member $u_{si}, 1 \le i \le k$, with its own private key $f_s(x_{xi})$ and the public key of the verifier company Y_v, carry out the following calculation:

$$r_{si} = (Y_v) \cdot f_s(x_{si}) \cdot \prod_{j=1, j \neq i}^{k} \frac{0 - x_{sj}}{x_{si} - x_{sj}} \bmod P \tag{1}$$

The result $\{r_{si}\}$ is sent to all participating members in signing through a secure channel, and each member calculates r as follows:

$$r = \prod_{i=1}^{k} r_{si} \bmod P \tag{2}$$

Then, through Elliptic Curve signature scheme, each member u_{si} signs the information m by using his private key $f_s(x_{xi})$ and the session key $f_b(x_{si})$. i.e., u_{si} calculates his private signature with the following formula:

$$s_i = h(m) f_s(x_{si}) \prod_{j=1, j \neq i}^{k} \frac{0 - x_{si}}{x_{si} - x_{sj}} - r f_b(x_{si} \prod_{j=1, j \neq i}^{k} \frac{0 - x_{sj}}{x_{si} - x_{sj}} \bmod n \tag{3}$$

In order to verify the validity of private signature s_i, one member is randomly selected as a full-time secretary to be responsible verifying private signatures and calculating company signatures. The full-time secretary verifies the private signatures according to the following formula:

$$h(m) y_{si} \cdot \prod_{j=1, j \neq i}^{k} \frac{0 - x_{sj}}{x_{si} - x_{sj}} \stackrel{?}{=} s_i G + r y_{bi} \cdot \prod_{j=1, j \neq i}^{k} \frac{0 - x_{sj}}{x_{si} - x_{sj}} \bmod p \tag{4}$$

If the equation is right, the private signatures of the information are valid. It's testified as follows:

$$h(m) y_{si} \cdot \prod_{j=1, j \neq i}^{k} \frac{0 - x_{sj}}{x_{si} - x_{sj}} = h(m) \cdot f_s(x_{si}) \cdot G \cdot \prod_{j=1, j \neq i}^{k} \frac{0 - x_{sj}}{x_{si} - x_{sj}} \tag{5}$$

It's known from (3) that:

$h(m) f_s(x_{si}) \prod_{j=1, j \neq i}^{k} \frac{0 - x_{si}}{x_{si} - x_{sj}} = s_i + r f_b(x_{si}) \prod_{j=1, j \neq i}^{k} \frac{0 - x_{sj}}{x_{si} - x_{sj}}$; Substituted to (5):

$$h(m)\cdot y_{si}\cdot\prod_{j=1,j\neq i}^{k}\frac{0-x_{si}}{x_{si}-x_{sj}}=s_iG+rf_b(x_{si})\cdot G\cdot\prod_{j=1,j\neq i}^{k}\frac{0-x_{sj}}{x_{si}-x_{sj}}$$
$$=s_i\cdot G+r\cdot y_{bi}\cdot\prod_{j=1,j\neq i}^{k}\frac{0-x_{sj}}{x_{si}-x_{sj}}\bmod p$$

2.1.3. *Generation and verification stage of company signature*

After k private signatures are verified, the full-time secretary calculates the company signatures.

$$s=\sum_{i=1}^{k}s_i\ \bmod q \tag{6}$$

The company signatures of information m are sent to the verifier company G_v, and any t ones of the n members in company G_v can cooperate to verify the validity of the signatures. With its own private key $f_v(x_{vi})$ and the public key of the signer company Y_s, every member $u_{vi}, 1\le i\le t$ carries out the following calculation:

$$r_{vi}=f_v(x_{vi})\cdot(Y_s)\cdot\prod_{j=1,j\neq i}^{t}\frac{0-x_{sj}}{x_{si}-x_{sj}}\bmod P \tag{7}$$

Similarly, a full-time secretary is randomly selected from the verifier company and he calculates the following forma:

$$r=\prod_{i=1}^{t}r_{vi}\bmod P \tag{8}$$

Therefore, by formula (9), the company signature can be verified:

$$h(m)\cdot Y_s\overset{?}{=}s\cdot G+r\cdot Y_b\bmod P \tag{9}$$

Theorem 2.1 if (9) is right, information m is verified.

It's testified as follows:

Substituted formula (1) to formula (2), it's got that:

$$r=\sum_{i=1}^{k}f_x(x_{si})\prod_{j=1,j\neq i}^{k}\frac{0-x_{sj}}{x_{si}-x_{sj}}\bmod P\ =f_s(0)\cdot f_v(0)\cdot G\bmod P \tag{10}$$

For i=1, 2, 3, …, k, multiplied by the left and right sides respectively, it's known from (4) that:

$$h(m)\cdot\sum_{i=1}^{k}y_{si}\prod_{j=1,j\neq i}^{k}\frac{0-x_{sj}}{x_{si}-x_{sj}}=\sum_{i=1}^{k}s_i\cdot G+r\cdot\sum_{j=1,j\approx i}^{k}y_{bi}\prod_{j=1,j\neq i}^{k}\frac{0-x_{sj}}{x_{si}-x_{sj}}\bmod p \tag{11}$$

It can determine a $k-1$ degree polynomial utterly by Lagrange interpolation polynomials:

$$f(x)=\sum_{i=1}^{k} y_i \cdot \prod_{j=1, j\neq i}^{k} \frac{x-x_j}{x_i-x_j}$$

The left of (11) can rewrite by:

$$h(m)\cdot \sum_{i=1}^{k} y_{si} \prod_{j=1, j\neq i}^{k} \frac{0-x_{sj}}{x_{si}-x_{sj}} \cdot G \bmod p$$

$$= h(m)\cdot f_s(0)\cdot G \bmod p = h(m)\cdot (Y_s) \bmod p \quad (12)$$

For the member of verifier company u_{vi} can calculate r_{vi} from (6), multiplied by both sides for $i=1,2,\cdots,k$, r' can be calculated from the formula (8):

$$r' = \sum_{i=1}^{t} f_v(x_{si}) \prod_{j=1, j\neq n}^{t} \frac{0-x_{sj}}{x_{si}-x_{sj}} \cdot (Y_s) \bmod P = f_v(0)\cdot f_s(0)\cdot G \bmod P \quad (13)$$

The value of r' is equal to the r from (2). The company signature can be calculated from (5). On the other hand, the right half part of (11) can be rewritten by:

$$\sum_{i=1}^{k} s_i \cdot G + r\cdot \sum_{j=1, j\approx i}^{k} y_{bi} \prod_{j=1, j\neq i}^{k} \frac{0-x_{sj}}{x_{si}-x_{sj}} \bmod p$$

$$= s\cdot G + r\cdot f_b(0)\cdot G \bmod P = s\cdot G + r\cdot y_b \bmod P \quad (14)$$

2.2. *Company-oriented general authenticated encryption threshold scheme*

There are also three parts in the whole scheme, i.e., sharing distribution center (SDC), Signer Company and Verifier Company. The scheme is consists of three stages, i.e., generation stage of parameters, encryption and signature stage, and message recovery stage.

2.2.1. *Generation stage of parameters*

In this stage, public information p, n and G are similarly to 2.1.1. Next, SDC accepts the registration of a group of signer or verifier, of whom parameter properties are marked by subscript s or v . $G_s=\{u_{s1},u_{s2},\cdots,u_{sl}\}$ and $G_v=\{u_{v1},u_{v2},\cdots,u_{vn}\}$ are respectively the companies combined by l singer and n verifier. Most of the private key, public key of the members (and companies) are also similarly to 2.1.1.To sum up, all the parameters of the system are shown as follows:

1. Shared information in SDC: p, n, G
2. Secret information in SDC: $f_s(x), f_v(x), f_b(x)$
3. Shared information of members: xsi, xvj, ysi, yvj and ybi
4. Secret information of members: $f_s(x_{xi}), f_v(x_{vi}), f_b(x_{si})$
5. Shared information of companies: *Ys* (or *Yv*)
6. Secret information of companies: *fs* (0), *fv* (0)

2.2.2. *Encryption and signature stage*

Assuming that the signer group G_s sends the secret information m to the verifier group G_v. After the information m is recovered, there is enough redundancy for validation in it. According to the threshold signature scheme, any k member can represent the group to sign the message. Without loss of generality, the k members for signing are expressed as $u_{s1}, u_{s2}, \cdots, u_{sk}$. Each member u_{si}, $1 \le i \le k$ uses his (or her) own key $f_s(x_{si})$ and public key Y_v of the signer group to calculate:

$$z_{si} = f(x_{si}) \cdot \prod_{j=1, j \neq i}^{k} \frac{0 - x_{sj}}{x_{si} - x_{sj}} \cdot (Y_v) \bmod p \tag{15}$$

and

$$r_i = f_b(x_{si}) \cdot \prod_{j=1, j \neq i}^{k} \frac{0 - x_{sj}}{x_{si} - x_{sj}} \cdot G \bmod p \tag{16}$$

After the result $\{ z_{si}, r_i \}$ be sent to all signers by security channels, each me mber calculates the public session key z :

$$z = (\sum_{t=1}^{k} z_{si} \bmod p) \bmod n \tag{17}$$

Then u_{si} uses his(or her) private key $f_s(x_{si})$ and session key $f_b(x_{si})$ to sign the information m according to the modified Elliptic Curve signature scheme. Member u_{si} calculates the stipulated value r and the private signature s_i:

$$r = m + z \cdot \sum_{t=1}^{k} r_i + z \cdot G \bmod p \tag{18}$$

and

$$s_i = z f_b(x_{si}) \prod_{j=1, j \neq i}^{k} \frac{0 - x_{sj}}{x_{si} - x_{sj}} - r f_s(x_{si}) \prod_{j=1, j \neq i}^{k} \frac{0 - x_{sj}}{x_{si} - x_{sj}} \bmod n \tag{19}$$

In order to verify the validity of the private signatures, a member is randomly selected as a full-time secretary with the responsibility for the validation of private signatures and the calculation of group signature. After received the private signatures, the secretary validates the private signatures using the following formula:

$$r_i \cdot z \overset{?}{=} s_i \cdot G + r \cdot \prod_{j=1, j\neq i}^{k} \frac{0 - x_{sj}}{x_{si} - x_{sj}} \cdot y_{si} \bmod p \tag{20}$$

If the equation is right, the private signatures s_i of the information m are valid. Then the full-time secretary calculates the group signature as follows:

$$s = \sum_{t=1}^{k} s_i \bmod n \tag{21}$$

2.2.3. *Message recovery stage*

Without loss of generality, we express t validation members as $u_{v1}, u_{v2}, \cdots, u_{vt}$. Firstly, each member u_{vj}, $1 \leq j \leq t$ uses signature (r, s) to calculate $z \cdot f_b(0) \cdot G$:

$$z \cdot f_b(0) \cdot G = s \cdot G + r \cdot Y_s \bmod p \tag{22}$$

Then each verifier uses his private key $f_v(x_{vj})$ and public key Y_s of the signer company to calculate the value of z_{vi}:

$$z_{vi} = f_v(x_{vi}) \cdot \prod_{j=1, j\neq i}^{t} \frac{0 - x_{vj}}{x_{xi} - x_{vj}} \cdot (Y_s) \bmod p \tag{23}$$

The public session key z can be determined by the verifier group through combining t $z_{vi}, i = 1,2,\cdots,t$:

$$z = (\sum_{i=1}^{t} z_{vi} \bmod p) \bmod n \tag{24}$$

Therefore, information m can be recovered from $z \cdot f_b(0) \cdot G$ and z:

$$m = r - z \cdot f_b(0) \cdot G - z \cdot G \bmod p \tag{25}$$

Finally, the recovered information m can be authenticated through checking the validity of its interior redundancy.

Theorem 2.2: Information m can be decrypted precisely during the stage of message recovery if (r, s) is the signature pair generated during the stage of signature encryption.

It's testified as follows:

During the stage of signature encryption, each user $u_{si}, 1 \le i \le k$ calculates z_{si}, then the signer group calculate public session key z :

$$z = (\sum_{i=1}^{k} f_s(x_{si}) \prod_{j=1, j\neq i}^{k} \frac{0 - x_{sj}}{x_{si} - x_{sj}} \cdot (Y_v) \bmod p) \bmod n = (f_s(0) \cdot f_s(v) \cdot G \bmod p) \bmod n$$

The (r, s) signature can be got from formula (18) and formula (21), and then be sent to the verifier group. On the other hand, member $u_{vj}, 1 \le j \le t$ of the verifier group calculates $z \cdot f_b(0) \cdot G$ with formula (22) and z_{vi} with formula (23).Then the t members can cooperate to get the public session key using formula (24).

$$z' = (\sum_{i=1}^{t} f_v(x_{vi}) \prod_{j=1, j\neq i}^{t} \frac{0 - x_{vj}}{x_{xi} - x_{vj}} \cdot (Y_s) \bmod p) \bmod n = (f_s(0) \cdot f_s(v) \cdot G \bmod p) \bmod n$$

Obviously the value of z' is equal to the value z in formula (16). So information m can be recovered by formula (25):

$$r - z \cdot f_b(0) \cdot G - z \cdot G \bmod p$$

$$= m + z \cdot f_b(0) \cdot G - z \cdot f_b(0) \cdot G + z \cdot G - z \cdot G \bmod p = m \bmod p$$

3. Application of the Scheme

Here we only give the application examples of company encryption communication system and the generating polynomial of company private key and public key. For the examples of the threshold signature scheme characterized by (t,n) joint verification for (k,l) signature, it is also very easy to be given and we omit it.

Suppose that company S and company V have three members $S = \{s_1, s_2, s_3\}$, $V = \{v_1, v_2, v_3\}$, respectively; S is the encryption of plaintext P_m ,the sender of cipher text C_m ; V is the receiver and decryption of C_m ; $f_s(x_{s1}) = 2$, $f_s(x_{s2}) = 3, f_s(x_{s3}) = 5$ are respectively the private key of s_1, s_2, s_3 and $f_v(x_{v1}) = 11, f_v(x_{v2}) = 13, f_v(x_{v3}) = 17$ of v_1, v_2, v_3 ; $f_s(0)$ is the private key of company S, and Y_s is the public key of company S. $E_{23}(1,1): y^2 = x^3 + x + 1 \bmod 23$ is the elliptic curve that be common selected by company S and V. $a = 2, b = 7$ are respectively the primitive roots selected by company S and V, and $p = 11$ is the mode selected by company S. Give modular arithmetic to $E_{23}(1,1)$, we get the point of $E_{23}(1,1)$, then:

$$y^2 = x^3 + x + 1 \bmod 23$$

Let $x = 0$, it's known that:

$$y^2 = 1 \bmod 23 \qquad y = \begin{cases} 1 \bmod 23 = 1 \bmod 23 \\ -1 \bmod 23 = 22 \bmod 23 \end{cases}$$

Then there are two points: $(0,1),(0,22)$; and so on, the points met with $E_{23}(1,1)$ are $(0,1),(0,22)$, $(1,7),(1,16)$, $(3,10)$, $(3,13),(4,0),(5,4),(5,19),(6,4)$, $(6,19),(7,11)$, $(7,12)$, $(9,7)$, $(17,3),(17,20)$, $(18,3)$, $(9,16),(11,3),(11.20)$, $(18,20)$, $(19,5)$, $(19,18)$, the Sequence of who are $n = 7$, $(n+1) = (27+1)$ $= 28G = O$, O is the infinity point of Elliptic Curve $E_{23}(1,1)$.

The company S and V select the base point $G = (5,4)$ and plaintext $P_m = (9,7) \in E_{23}(1,1)$, where, the true plaintext of P_m has been compressed to a point of $E_m(a,b)$ using by technology of data compression.

The private key $f_S(0)$ and the public key Y_S of company S are, respectively:

$$f_s(0) = 9$$

$$Y_s = f_s(0)G = 72G = 56G + 16G = 16G = (11,3)$$

The private key $f_V(0)$ and the public key Y_V of company V are, respectively:

$$f_v(0) = 10$$

$$Y_v = f_v(0)G = 24G = (18,3)$$

$\{s_1, s_2, s_3\}$ decrypt P_m into C_m and send C_m to company V

1. $\{s_1, s_2, s_3\}$ take the positive integer $k = 3$, base point $G = (5,4)$, plaintext $P_m = (9,7)$, the public key $Y_v = 24G = (18,3)$ of $\{v_1, v_2, v_3\}$ and give:

$$C_m = \{kG, P_m + kY_v\} = \{3G, 14G + 3 \times 24G\}$$

$$= \{3 \times 8G, 14G + 72G\} = \{24G, 86G\} = \{24G, 2G\} = \{(18,3),(0,22)\}$$

2. $\{s_1, s_2, s_3\}$ send $C_m = \{(18,3),(0,22)\}$ to $\{v_1, v_2, v_3\}$

$\{v_1, v_2, v_3\}$ receive C_m, and decrypt C_m into P_m

$\{v_1, v_2, v_3\}$ take the second term $C_{m,2} = P_m + kY_v$ of C_m, take the product of the first term $C_{m,1} = kG$ of C_m and $f_v(0)$, and give:

$$C_{m,2} - C_{m,1} = P_m + kY_v - f_v(0)(kG)$$

$$= 2G - 10(24G) = 2G - 240G = 2G - 16G$$

$$= 2G + (-16G) = 2G + 12G = 14G = (9,7) = P_m$$

Acknowledgment

This work was financially supported in part by Institutes of Higher Education Science and Technique Foundation of Shandong Province (Grant no. J14LN22). The authors thank the College of Information and Engineering Taishan Medical University colleagues for manuscript comments. The authors thank the anonymous reviewers for their valuable comments and suggestions.

References

1. Harn L.: IEEE Proceedings of Computers and Digital Technique Vol. 141 (1994), pp. 307-313.
2. Bin Wang and Jianhua Li: Chinese Journal of Computers Vol. 26 (2003), pp. 158l-1583.
3. Y. Ming and Y. Wang: in Proceedings of the 2009 Fifth International Conference on Information Assurance and Security. Washington: IEEE Computer Society, 2009, pp. 87-90.
4. R. Tso, C. Gu and T. Okamoto: Cryptology and Network Security 2007(CANS 2007), LNCS 4857. Berlin: Springer-Verlag, 2007, pp. 47-59.
5. B. Kang, J. Park, and S. Hahn: Topics in Cryptology - CT-RSA 2004, LNCS 2964. Berlin: Springer-Verlag, 2004, pp. 99-111.
6. J. Zhang: Information Security Practice and Experience 2009, LNCS 5451. Berlin: Springer-Verlag, 2009, pp. 47-58.
7. GENNAROR: Massachusetts Institute of Technology, Cambridge, 1996.
8. SCHNEIERB: John Wiley & Sons, Inc., 1994.
9. Jurisic A, Menezes. Dr Dobb's Journal. A pril 1997.22, pp. 26-37.
10. A. Shamir: Communication of the ACM Vol. 22 (1979), pp. 612-613.
11. Yulian Shang.: Advanced Materials Research. Vols.756-759(2013), pp.1339-1343.
12. B.W. Kernighan, *PIC—A Graphics Language for Typesetting*, Computing Science Technical Report 116, AT&T Bell Laboratories (Murray Hill, NewJersey, 1984).
13. H.C. Heard, I.V. Borg, N.L. Carter and C.B. Raleigh, VoQS: Voice QualitySymbols, Revised to 1994, (1994).
14. M.E. Brown, An interactive environment for literate programming, PhD thesis,Texas A&M University, (TX, USA, 1988), pp. ix + 102.
15. G.S. Lodha, Quantitative interpretation of ariborne electromagnetic responsefor a spherical model, Master's thesis, University of Toronto (1974).
16. D.Jones, The term 'phoneme', in *Phonetics in Linguistics: A Book of Reading*, eds. W.E. Jones and J.Laver (Longman, London, 1973) pp.187–204.
17. B.Davidsen,Netpbm(1993),ftp://ftp.wustl.edu/graphics/graphics/packages/NetPBM.

New short interleaver design for turbo codes

Li-Fang Ma[1,*], Shi-Zhan Lan[2,†], Zhao-Jing Zhang[3,a] and Hua-Hua Xie[3]

[1]*Guangxi Polytechnic of construction,*
Nanning, China
[*]E-mail: mlf06@163.com
www.gxjsxy.cn

[2]*China Mobile Communications Group Co., Ltd.*
Guangxi Nanning, China
E-mail: [†]lanshizhan@gx.chinamobile.com

[3]*Key Laboratory of Universal Wireless Communication, Ministry of Education*
Beijing University of Posts and Telecommunications,
Beijing, 100876, China,
Email: [a]buptzzj@163.com

Since structured interleavers perform better than random interleavers for short block, a structured interleaver used in turbo codes for services requiring less latency time is preferred. Thus, we propose a new method to design interleavers for turbo codes[1] with short block. The designed interleaver is generated through ascertaining adjacent information bits distances and position of the first bit, which guarantees large minimal self interleaving distance and minimal correlative interleaving distance. Compared with other interleavers, the interleaver is able to minimize the negative impact of noise bursts and is easy to implement.

Keywords: Turbo Codes; Interleaver; Self Interleaving Distance; Correlative Interleaving Distance.

1. Introduction

Interleavers used in turbo codes play an important role, which influence the performance of turbo codes to a great extent. At present, there are mainly two kinds of interleavers: random interleavers and structured interleavers. For large block, random interleavers perform well; On the other hand, structured interleavers perform better than random interleavers for short block. Thus it is preferable to have a structured interleaver used in turbo codes for some services requiring less latency time, and the question how to rationally design structured interleavers for Turbo codes becomes more meaningful.

---

[*] Corresponding author.

An outline of the remainder of the paper is as follows. Section 1 outlines the structure of turbo encoders and principles of designing interleavers. Section 2 describes linearity iterative interleavers[2] which perform well in the noise bursts channel, and puts forward a new kind of short interleaver, and proves their distance properties. Section 3 compares the performance of the new interleaver with that of other interleavers. Section 4 offers some conclusions.

2. Principle of Designing Interleaver

A typical turbo encoder is formed by parallel concatenation of two identical recursive systematic convolutional (RSC) coders separated by an interleaver (see figure 1). In this paper, we employ recursive systematic convolutional (RSC) coders with memory m=3 and with feedback and feedforward generator polynomial and (see figure 2).

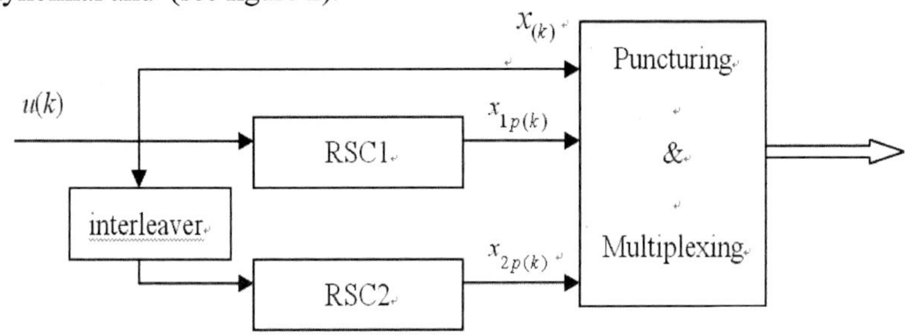

Fig. 1. Structure of a turbo encoder.

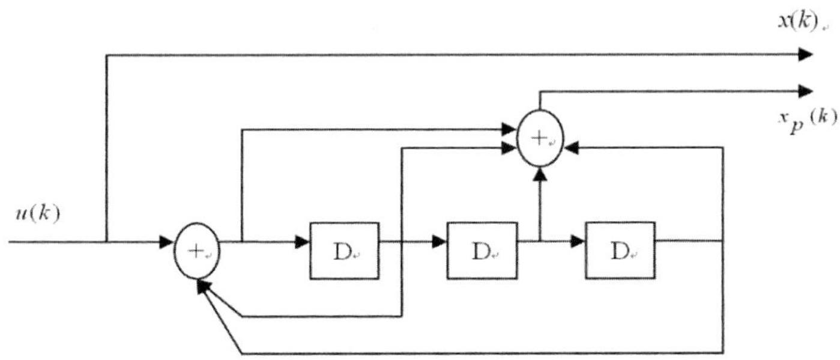

Fig. 2. Structure of a RCS coder.

Interleavers in turbo codes is to scramble data sequences as random as possible, which are helpful to boost up error correcting performance of turbo codes by breaking low weigh input sequence[3]. Two criteria can become performance evaluation of interleaver. One is minimal distance between the position of input bit $u_{(k)}$ pre-interleaving and the position of $u_{(k)}$ interleaved, where $k \in \{0,1,2\cdots N-1\}$, which is defined as minimal self interleaving distance s_1

$$s_1 = \min|x_i - y_i| \tag{1}$$

Where x_i denotes the position of i_{th} bit pre-interleaving, namely $x_i = i$, and y_i is the position of i_{th} bit interleaved.

The other is minimal adjacent bits distance after interleaving, which is defined to be minimal correlative interleaving distance s_2

$$s_2 = \min|y_i - y_{i-1}| \tag{2}$$

In figure 1, output bits are $\{x_{(k)}\}$, $\{x_{1p(k)}\}$ and $\{x_{2p(k)}\}$, where $\{x_{(k)}\}$ sequence is the same as input sequence $\{u_{(k)}\}$. For two identical RSC coders are employed with constraint length m+1=4 (see figure 2), input bit $u(k)$ bits is directly relative to m+1 output bits from $x_p(k)$ to $x_p(k+m)$. Above all, we suppose that code rate is 1/3, $\{x_{(k)}, x_{1p(k)}, x_{2p(k)}\}$ comprises output bits after input bit $u(k)$ is encoded, $x_k = k$ stands for position of input bit $u(k)$ and y_k denotes the position of bit $u(k)$ interleaved. According to analysis above, $u(k)$ is directly relative to m+1 codes from $\{x_{(k)}, x_{1p(k)}, x_{2p(k)}\}$ to $\{x_{(k+m)}, x_{1p(k+m)}, x_{2p(k+m)}\}$; Because the position of $u(k)$ interleaved becomes y_k, $u(k)$ is also directly relative to m+1 codes from $\{x_{(y_k)}, x_{1p(y_k)}, x_{2p(y_k)}\}$ to $\{x_{(y_k+m)}, x_{1p(y_k+m)}, x_{2p(y_k+m)}\}$, If $|y_k - x_k|$ isn't large enough, m+1 code from $\{x_{(k)}, x_{1p(k)}, x_{2p(k)}\}$ to $\{x_{(k+m)}, x_{1p(k+m)}, x_{2p(k+m)}\}$ is near codes $\{x_{(y_k)}, x_{1p(y_k)}, x_{2p(y_k)}\}$ ~ $\{x_{(y_k+m)}, x_{1p(y_k+m)}, x_{2p(y_k+m)}\}$.

Once noise bursts occur in channels, there is a great probability to lose most of 2(m+1) codes which directly connect with bit $u(k)$, and result in judging $u(k)$ wrong in turbo decoders. Thereby the minimal self interleaving distance $s_1 = \min|x_i - y_i|$ affects performance of interleavers to some extant.

It is well known that two decoders exchange extrinsic information[4] in the process of decoding; the less extrinsic information correlation is, the better error correcting performance of decoders in turbo codes is; Extrinsic information correlation depends on the correlation between input bits pre-interleaving and the sequence interleaved. We know $u(k)$ connects with $x_p(k) \sim x_p(k+m)$ and $u(k+1)$ connects with $x_p(k+1) \sim x_p(k+m+1)$, so we can deduce that m bits $x_p(k+1) \sim x_p(k+m)$ is relative to $u(k)$ and $u(k+1)$.

The worst situation is enumerated here, if the positions of $u(k)$ and $u(k+1)$ still are adjacent after bits $u(k)$ and $u(k+1)$ are interleaved, we set their positions as y_k and y_{k+1}, output codes $\{x_{(y_k+1)}, x_{1p(y_k+1)}, x_{2p(y_k+1)}\} \sim \{x_{(y_k+m)}, x_{1p(y_k+m)}, x_{2p(y_k+m)}\}$ are also relative to $u(k)$ and $u(k+1)$, which are connected again by encoding to result in making the two sequences correlation more and alleviating the error correcting performance. So it is also important to keep minimal correlative interleaving distance $s_2 = \min|y_i - y_{i-1}|$ large enough.

Certainly, a good interleaver should have two large distances above at the same time.

3. Designing of Interleavers

3.1. *Linearity iterative interleaver*

Linearity iterative interleaver expression is given by

$$y_n = (a * y_{n-1} + b) \bmod N, 0 \le n \le N-2 \tag{3}$$

To insure the interleaver one-onea attribute, some conditions have to be fulfilled as follows

l_1.$a < N, b < N$

l_2. a and b both are coprime with N

l_3. If N is divisible by 4, $a-1$ is multiples of 4

l_4. $a-1$ exactly divides any prime numbers which N is multiples to

Linearity iterative interleaver is able to guarantee s_1 and s_2 large thought adjusting a and b according to different block size. Thus its bit error performance is no bad.

3.2. *A new short interleaver is proposed and its arithmetic analysis*

A new short interleaver is proposed in the section based on 2 parameters mentioned above s_1 and s_2, which is given by $y_0 = a$, $0 \le a \le N-1$ (4)

$$\begin{cases} y_{n+1} = (y_n + \mathrm{mod}(i_{(k)}, 2) * (N/2) + b) \bmod N \\ y_{n+2} = (y_n + \mathrm{mod}(i_{(k)} + 1, 2) * (N/2) + b) \bmod N \end{cases}$$

$$1 \le n \le N-4, \quad n \text{ is even} \tag{5}$$

$$y_{N-1} = N/2 + a \tag{6}$$

Where $b < N/4$, and $i_{(k)}$ ($0 \le k \le N/2-1$) is pseudorandom sequence, which makes the sequence more random in the process of interleaving. In this paper, $i_{(k)}$ are generated by M-sequence ($f(x) = a_0 + a_1 x + \cdots + a_n x^n$) (see fig.3), which generates a long periodic sequence.

c_n stands for the corresponding register, $a_i = 1$ denotes the corresponding line swiching on, $a_i = 0$ means the line off.

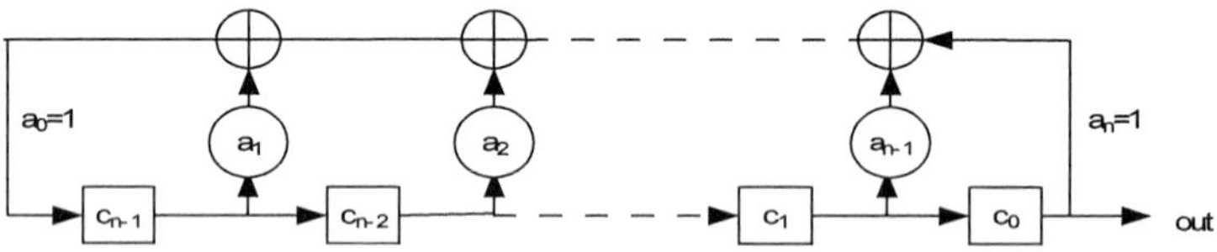

Fig. 3. Structure of pseudorandom sequence generator.

To insure the interleaver one-onea attribute, N and b in the four equations above have to meet two conditions as follows

(1) N is even

(2) b is coprime with N

Proof. If we ignore above four equations "mod N", they is shown to be

$$y_0' = a \tag{7}$$

$$\begin{cases} y_n' = \frac{n+1}{2} * b + \mathrm{mod}(i_{(\frac{n+1}{2})}, 2) * \frac{N}{2} + a \\ y_{n+1}' = \frac{n+1}{2} * b + \mathrm{mod}(i_{(\frac{n+1}{2})} + 1, 2) * \frac{N}{2} + a \end{cases} \quad 1 \le n \le N-1, n \text{ is odd} \tag{8}$$

$$y'_{N-1} = N/2 + a \quad (9)$$

It is easy to prove the equation $y_n = y'_n \bmod N$ comes into existence.

Since $y_n < N, 0 \le n \le N-1$, if we can prove that $y_0, y_1 \cdots y_{n-1}$ aren't equal to one another, the interleaver one-onea attribute also is guaranteed, which is equivalent to the event.

$$\left| y'_m - y'_n \right| \bmod N \ne 0, \quad \forall m, \forall n, m, n \in \{0, 2, \cdots N-1\}, m \ne n \quad (10)$$

Above all, we prove the inequation

$$\left| y'_m - y'_n \right| \bmod N \ne 0, \quad \forall m, \forall n, m, n \in \{1, 2, \cdots N-2\}, m \ne n \quad (11)$$

Utilizing Eq.(8), we can get

$\left| y'_m - y'_n \right| \bmod N = \left| \frac{m'+1}{2} - \frac{n'+1}{2} \right| b + rN/2$, where

$r = \left| \bmod(i_{m_1}, 2) * \frac{N}{2} - \bmod(i_{n_1}, 2) * \frac{N}{2} \right|$, and both are integer

$\because r = \left| \bmod(i_{m_1}, 2) * \frac{N}{2} - \bmod(i_{n_1}, 2) * \frac{N}{2} \right| < 2$, r is integer. $\therefore r$ =0 or 1

$\therefore \left| y'_m - y'_n \right| \bmod N$ has two potential values.

Since N is coprime with b, eq.2.are impossible to be equal to 0.

$\because 1 \le m' < N-1, 1 \le n' < N-1 \quad \therefore \left| \frac{m'+1}{2} - \frac{n'+1}{2} \right| < N/2$

Since N is coprime with b, the equation $\left(\left| \frac{m'+1}{2} - \frac{n'+1}{2} \right| b + N/2 \right) \bmod N = 0$ comes into existence only if $\left| \frac{m'+1}{2} - \frac{n'+1}{2} \right|$ is divisible by $N/2$, but $\left| \frac{m'+1}{2} - \frac{n'+1}{2} \right| < N/2$

$\left| y'_m - y'_n \right| \bmod N \ne 0, \forall m, \forall n, m, n \in \{1, 2, \cdots N-2\}, m \ne n$ In this way, two inequations $\left| y'_k - y'_0 \right| \bmod N \ne 0$, $\left| y'_k - y'_{N-1} \right| \bmod N \ne 0$ also are true, $k \in \{1, 2 \cdots N-2\}$. So the interleaver is able to fulfill one-onea attribute.

The new interleaver can keep $s_2 = \min|y_i - y_{i-1}| = b$ and regulate $s_1 = \min|x_i - y_i|$ by adjusting a in the equation $y_0 = a$. □

Proof.

According to eq.(5) $\begin{cases} y_{n+1}=(y_n+\text{mod}(i_{(k)},2)*(N/2)+b)\bmod N \\ y_{n+2}=(y_n+\text{mod}(i_{(k)}+1,2)*(N/2)+b)\bmod N \end{cases}$ when n is even,

$$|y_{n+2} - y_{n+1}| = (y_n + \text{mod}(i_{(k)},2)*(N/2)+b)\bmod N - (y_n + \text{mod}(i_{(k)}+1,2)*(N/2)+b)\bmod N$$

$$\Rightarrow |y_{n+2} - y_{n+1}| = (y_n + \text{mod}(i_{(k)},2)*(N/2)+b - y_n + \text{mod}(i_{(k)}+1,2)*(N/2)+b))\bmod \text{N}$$

$$\Rightarrow |y_{n+2} - y_{n+1}| = (N/2)\text{mod}N$$

$$\Rightarrow |y_{n+2} - y_{n+1}| = N/2$$

$|y_{n+1} - y_n|$ is divided into two situations:

When $y_n + \text{mod}(i_{(k)}, 2) * (N / 2) + b < N$, Since $b < N/4$,

$|y_{n+1} - y_n| = \text{mod}(i_{(k)},2)*(N/2)+b$, $|y_{n+1} - y_n| = b$ or $N/2+b$

When $y_n + \text{mod}(i_{(k)}, 2) * (N / 2) + b > N$, For

$|y_{n+1} - y_n| = |\text{mod}(i_{(k)},2)*(N/2)+b-N|$, $|y_{n+1} - y_n| = N-b$ or $N/2-b$

Through analysis above, five distances of adjacent bits can be given by

(1) $N/2$ (2) b (3) $N/2+b$ (4) $N-b$ (5) $N/2-b$

Since $b < N/4$, b is the minimal number among them, in this way, $\min|y_{n+2} - y_n|$ is also proved that is equal to .

Once sequence distances of every two adjacent bits have been ascertained and $y_0 = a$ is fixed, the whole sequence is sole. So it is flexible to regulate $s_1 = \min|x_i - y_i|$ by adjusting a in the equation $y_0 = a$. Summary of above points, it is convenient to design an ideal interleaver produce large minimal self interleaving distance and minimal correlative interleaving distance through adjusting the constant a and b in eq. (4), Eq(5), Eq.(6).

4. Simulating Result and Comparing

In this paper, the encoder employs two identical recursive systematic convolutional (RSC) coders with memory m=3 and with generating matrix $(1, 1+D+D^2+D^3/1+D+D^3)$ (see figure 2). The decoder employs soft input soft output MAX-LOG decoding algorithm with iterative number 5, code rate $1/3$.

In the process of simulation, noise bursts channel is chosen with probability of noise bursts 0.01(see figure 3), eq.(12) is its expression

$$r_k = a_k s_k + n_k \tag{12}$$

Where a_k is information bits reversal, s_k is information bits, n_k BPSK modulating gauss noise, and $eps = 0.01$ is information bits reversal probability.

$$a_k = \begin{cases} 1, p(a_k = 1) = 1 - eps \\ -1, p(a_k = -1) = eps \end{cases}$$

We simulate performance of four kinds of interleavers, block Interleaver, random interleaver,linearity iterative interleaver above, and new interleaver. Block sizes 280 and 144 are chosen.

When block sizes is 144, block interleaver is (12*12), linearity iterative interleavers (a =73, b =85, $y_0 = 0$). In the new interleaver, b =13, $a = 1$; Pseudorandom sequence generator $f(x) = x^8 + x^4 + x^3 + x^2 + 1$.

While block size is 280, block Interleaver(20*14), linearity iterative interleavers(a =141, b =153, $y_0 = 0$), the new interleaver(b =15, $a = 1$, $f(x) = x^8 + x^4 + x^3 + x^2 + 1$).

The simulation result with block size 140 is shown in figure 4, the new interleaver gets 0.3dB coding gain referring to the block when error rate 3×10^{-4}, about 0.1dB comparing with iterative interleaver when error rate 2×10^{-4}.

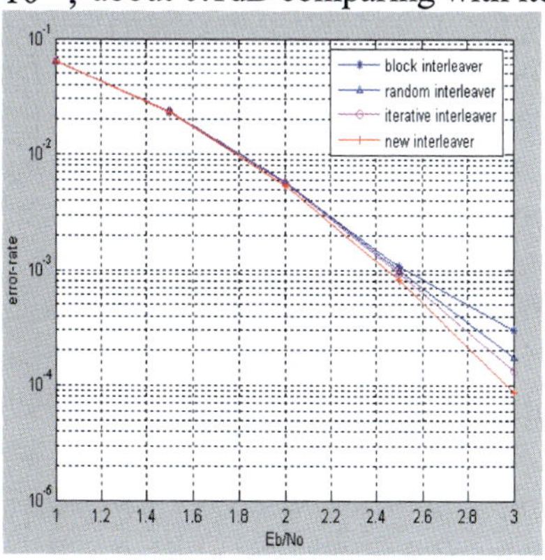

Fig.4. Performance of interleavers with block size 144 in noise bursts channel

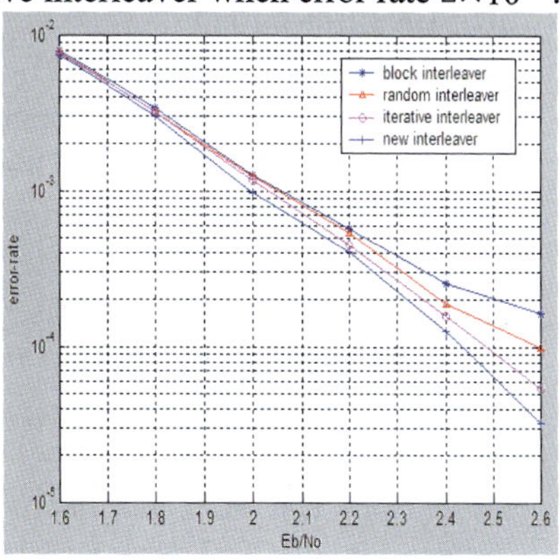

Fig.5. Performance of interleavers with block size 280 in noise bursts channel

With block size 280, the new interleaver also express better performance, which gets 0.2dB coding gain comparing with the random when error rate 10^{-4}, about 0.1dB referring to iterative interleaver when error rate 6×10^{-5}.

5. Conclusions

Interleavers based on minimal self interleaving distance and minimal correlative interleaving distance effectively restrain negative affection noise bursts bring. The new interleaver proposed in this paper achieves better bit error performance than that of traditional structured interleavers, including iterative interleaver. In addition, the kind of interleaver can be flexibly designed according to different block size through adjusting parameters to get the best interleave. Thus it is very valuable for the new interleave to apply to some services requiring little time latency.

Acknowledgments

This work is supported by The New Century Higher Educational Reform Projects of Guangxi 2011. Project Name: Research on the sharing of Muti-campus Network teaching resources (2011JGA236)

References

1. Berrou C. Glavieux A. Thtimajshima P. *Near Shannon Limit Error Correcting Coding and Decoding: Turbo Codes* [A]. Proceedings of ICC [C]. 1993. 1064-1070.
2. L.Wenming, Z. Guangxi, and H. Yejun, *New Short Interleaver Design for Turbo Codes* [J], Journal on Communications, November, 2005. 62-67.
3. S. Dolinar and D. Divsalar, *Weight distribution for turbo codes using random and nonrandom permutations*, JPL Progress report 42-122, Aug,15,1995
4. J.Hokfelt,O.Edfors, and T.maseng, *Turbo codes: Correlated Extrinsic Information and Its Impact on Iterative Decoding Performance*, in Proc. IEEE 49th Vehicular Technology Conf., vol. 3, Houston, TX, May 1999. 1871-1875.
5. H. Ogiwara and M. Yano, *Improvement of turbo trellis-coded modulation system*, in IEICE Trans. Fun-damentals., Oct. 1998, vol. E81-A. 2040-2046.

6. J. Han and O. Y. Takeshita, *On the Decoding Structure of Multiple Turbo Codes*, in Proc. IEEE International Symposium on Information Theory, Washington D.C., June 2001, p. 98.
7. D. Divsalar and F. Pollara, *Multiple turbo codes*, in Proc. IEEE Military Communications Conference, San Diego, CA, Nov. 1995, pp. 279-285.
8. Andrews K S, Heegard C , Kozen D. *Interleaver Design Methods for Turbo Codes*. Proceedings of the IEEE International Conference on Information theory, 1998. 16-21 ,420.
9. Eroz M, Hammons A R. *On the Design of Prunable Interleavers for Turbo Codes*. Proc IEEE VTC'99, Houston, May 1999.1669-1673.

Trusted network management model based on clustering analysis

Hong-An Xie*
Electronics department Engineering University of Chinese Armed Police Forces,
Key laboratory of CAPF for information security,
Xi 'an Shaanxi China
**E-mail: 155293322695@163.com*

Dong Li
Information engineering department,
Engineering University of Chinese Armed Police Forces,
Xi 'an Shaanxi china
E-mail:563783912@qq.com

In order to solve the problems of accuracy and availability in traditional network trust evaluation, a new trusted network management model based on clustering analysis was built. The model evaluates the trust value in users' behavior based on behavior expectation. Through clustering analysis of users' past behavior data, the network manager can determine the user behavior expectation, which is then used to evaluate the trust in users' behavior. Our model can manage the network based on users' trust value. Experimental results have shown that our model can rapidly detect and isolate malicious users and accurately differentiate trusted users from malicious ones. In conclusion, our model has achieved the purpose of improving network security.

Keywords: Trusted Network; Clustering Analysis; Trust Model.

1. Introduction

TCG[1,2](Trusted Computing Group)has proposed a trust network connect (TNC) to build a trusted network. But it is not a good solution to solve the problem. So there are many scholars researches on the trust model under different network environment[3] Typically there is: Trust model based on Bayesian theory [4]; The trust model based on DS evidence theory [5,6]; The trust model based on behavior state correlation method [7] etc. But the above model applicable field is limited and still has certain limitations at the availability and the accuracy of trust evaluation.

So in order to solve the above problems, this paper proposes a new reliable network management model based on clustering analysis.

2. The *N*etwork *M*anagement *M*odel *B*ased on *C*lustering *A*nalysis

2.1. *Build behavior expectations*

In this paper, the definition of trusted reference TCG[2] organization.

Definition 1 A state of computing platform is trusted when its behavior is in line with the trust strategy.

Model is the main process includes: build behavior expectations, trust evaluation, update and decision making. The model can be described in detail below:

The first step of model is build user behavior expectation. and we choose the clustering methods to analysis the user behavior.

Cluster is a method of Knowledge discovery in Database (KDD). Through this technology [9] we can get certain knowledge by mining existing data of user behavior. So in order to build behavior expectations, adopted the cluster analysis method for collection of user behavior. DBSCAN[9] (density-based spatial clustering of applications with noise) is a density based clustering algorithm, Based on density clustering can be according to the different density of separation in the sample data, to build user behavior expectations. In this article user behavior are defined as follows:

Definition 2 user behavior

User behavior $B = \{b_i(O, S, A, R) \mid b_1, b_2, b_3, b_4 ... b_n\}$ contains:

(1)Set of entity $O, o_1, o_2, o_3, o_4 ... o_n$ denote elements.

(2)Set of entity state, S , $s_1, s_2, s_3, s_4 ... s_n$ denote elements.

(3)Set of entity action, A , $a_1, a_2, a_3, a_4 ... a_n$ denote elements.

(4)Trust entity behavior state entropy $R_{ij} = H(S_i, A_j)$, the entropy represent the possibility of the action.

Definition 3 algorithm parameters $(\varepsilon, MinPts, MinDs)$

Radius of cluster ε ; Density threshold $MinPts$; The minimum distance $MinDs$

Algorithm process is as follows:

Algorithm 1 DBSCAN(B ; ε ;*MinPts; MinDs*)

```
begin
 Init_Cluster (C);              //The cluster is initialized
 for each unvisited point p in B
  Mark p as visited ;      /On a visit to an arbitrary point and tags
  N=get_Neighbours(p, ε);      // Access to all the neighboring points
  K=get_Neighbours(p,MinDs);
  if K!=0 then
    mark p as P;               // Find out the point P
  end if
  else
  C=nextcluster;              // Find a cluster, and record
  Expand_Cluster (p,N,C,ε,MinPts);  // Cluster expansion algorithm
  end if
  end for
  end                       // End of the algorithm
```

Algorithm 2 *Expand_Cluster*(p, *N, C,* ε , *MinPts*)(Cluster expansion algorithm)

```
 Add p to cluster C;                  // Add point P to cluster C
 for each point p' in N          // Since any adjacent points expand clusters
  Mark p' as visited
   N' =get_Neighbours(p',ε);
   if size of(N')>= MinPts
    then N= N+ N'           // To meet the requirements expand clusters
   end if
   if p' is not member of any cluster
      add p' to cluster C; // If the point P can't expand, continued to algorithm
   end if
  end for
  end                          // End of the algorithm
```

Algorithm for cluster based on two parameters ε and *MinPts* : For any Point P, all point with in the distance of ε are all ε nearby. If the number of ε nearby P points is not less than *MinPts* , so these points (including point *P*) are part of the data set C, belong to the same cluster, algorithm can extract on the high-density portion of collection. Thus by using the above algorithm to cluster of user behaviors evidence collection to select high density behavior clusters, can build behavior expectations. Algorithm according to the density difference between selected reliable behavior. Moreover the algorithm can accord the behavior of the deviation to isolate outliers. Outliers are "with the data set is greatly different from other data. Separation of outliers can be effectively separated with the user common behavior the behavior of the gap is giant enough, which can be used as the basis for reduce user's trust.

So by cluster the data set of the user history action, we can get the *(C,P)* as the set of user behavior expectation, and use this set to evaluate the trust of the network user.

2.2. *Trust evaluation*

Trust evaluation is to verify the proximity between user actual behavior and behavior expectation, the closer between the expectation and behavior, the higher of user's trust level. According to the result of quantitative classification, network manager assign different levels of permissions to the users for different levels of user's trust .

So we can classify the Trust hierarchy below :

Definition 4 Trust hierarchy

In order to realize network management based on trust, to quantify the entity trust and hierarchical management, the higher the levels of trust entity are more credible. And the Trust level are described in Table 1.

Table 1. Trust lever.

Trust level Ti	Value	Semantics	Classify
T1	[0.0, 0.2)	Top Untrusted	Malicious user
T2	[0.2, 0.4)	Untrusted	
T3	[0.4, 0.6)	Critical credible	Trusted User
T4	[0.6,1.0)	Trusted	

By definition 1, trust computation relay on the *Jaccard* similarity of user behavior consistent with behavior expectation. Intercept users for a period of time the behavior of the data set ***N*** as the current sample. ***N*** is the window of a user behavior. And we can get the set of user behavior expectation *(C, P)*.Consider the similarities between the two set. The two set of set A and set B overlap between elements in A and B two collections and focus the proportion, refer to the Trust value. $J(A,B)$ denotes element. Based on the expectation *(C, P)*, and Set ***N*** the user's trust value can be written as:

$$T_i = J_\theta(N,C) - J_\theta(N,P) \tag{1}$$

The more close to the user behavior expectation, the higher trust of the behavior. But problem is that the abnormal behavior cannot rapidly decrease the trust value of the user. So in order to solve the problem by comparing user behavior expectations and user behavior records set over a period of time to assess the user's trust. Therefore trust evaluation must be taken in punish mechanism; the model constructs a short-term behavior of the small window of *m*, and use it to assess the user behavior credibility. The recent untrustworthy behavior will quickly affect the overall trust. So this paper improves the formula 1 and gets the Eq.(2).

$$T_i = J_\theta(N,C) - J_\theta(N,P) + J_\theta(m,C) - 1 \tag{2}$$

The recent credible behavior will rapidly reduce the trust of users, at the same time, improve the trust can only rely on long-term exchanges between trust network user, model achieved a combination of slow rise and fast drop of trust evaluation[8], evaluation of trust is more reasonable.

3. Model and Experimental Analysis

3.1. *Functional test*

Functional test verify the function of the model. In the experimental conditions, acquisition and abstract terminal 1 behavior data inputs to the model, At the same time, the terminal 2 infection malicious code and network interaction, acquisition and the abstraction of the behavioral data input model, 50 consecutive experiments, continuous experiment 50 times. The final results are shown in Figure1.

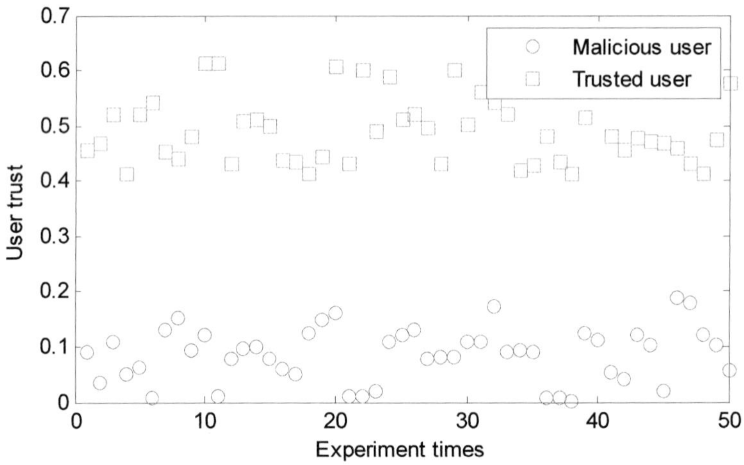

Fig. 1. The results of user distinguish experimental.

In Fig1. the abscissa is the time of experiments, the ordinate is the user trust evaluation value *T*. Due to infection of the malicious code of the user terminal network behavior will be different from the normal user; the model can be used to assess the trust degree to achieve recognition of malicious users, and malicious terminal isolation purposes.

3.2. Performance test

Performance testing will compare the model with SHTrust [8] model and DWTrust [10] model, Calculate the trust value of the same entity, examine the accuracy of the trust evaluation.

Experiments using three kinds of models and the experimental terminal 1 trust evaluation, mainly of trust evaluation accuracy of analysis, aimed at different sample states of the same entity of trust evaluation to calculate trust evaluation value of the standard deviation of experimental analysis. Standard deviation and accuracy is inversely proportional Each experiment 10 times to get the trust evaluation results, the calculation results between the standard deviation of the results of experiments is shown in Figure 2:

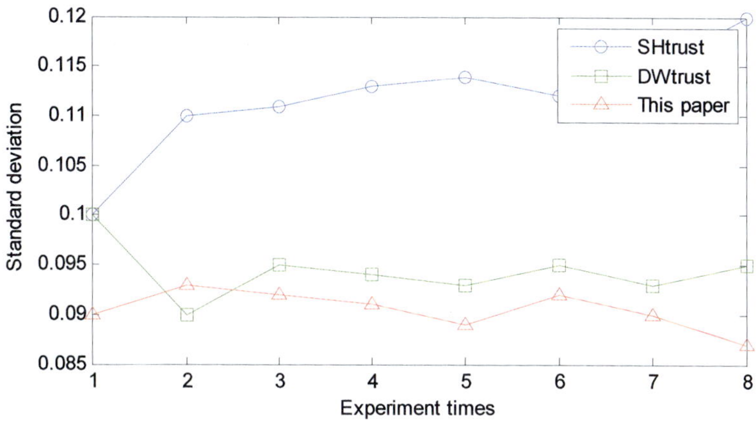

Fig. 2. Accuracy of the model.

Experimental results show that the three kinds of models of the same entity trust evaluation with an accuracy approaching that of three kinds of models for the legal entity of trust evaluation has a certain degree of accuracy.

4. Conclusions

This paper evaluates the user's trust by build user behavior expectation, and a new network management model based on clustering analysis was proposed. The model build the user behavior expectation by clustering analysis the historical evidence of network user and use the behavior expectation to calculate the actual trust value of user behavior. According to the experimental results, this model can generate trust evaluation results firmly, detect the malicious users rapidly and isolate it, it was more accurate and available than tradition model, basically achieved the purpose of improving network reliability.

References

1. YAN X X. *The trusted network framework research and design[D]. Beijing: Beijing University of Technology, 2014.*
2. FENG D G, QIN Y, WANG D, *et al. Research on trusted computing technology[J]. Journal of Computer Research and Development, 2011, 48(8): 1322-1349.*
3. LIN C, TIAN L Q, WANG Y Z. *Research on user behavior Trust in trustworthy network[J]. Journal of Computer Research and Development, 2008, 45(12): 2033-2043.*
4. LIANG H Q, WU W. *Research of trust evaluation model based on dynamic bayesian network[J]. Journal on Communications, 2013, 34(9): 68-76.*
5. Feng R J, Xu X F, Zhou X. *A trust evaluation algorithm for wireless sensor networks based on node behaviors and D-S evidence theory[J]. Sensors, 2011(11): 1345-1360.*
6. JIANG L M, ZHANG K, XU J, *et al. A new evidential trust model based on graph theory for open computing System [J]. Journal of Computer Research and Development, 2013, 50(5): 921-931.*
7. LI D F, YANG Y X, GU L Z, *et al. Study on dynamic trust metric of trusted network based on state and behavior associated[J]. Journal on Communications, 2010, 31(12): 12-19.*
8. WU X. *Enhanced stable group model-based trust evaluation scheme for model P2P Network[J]. Chinese Journal of Computers, 2014, 37(10): 2118-2127.*
9. J H, M K, J P. *Data mining : concepts and techniques[Z]. Morgan Kaufmann, 2011.*
10. HUANG C L. *Research dynamic trust relationship modeling and management technology[D]. Changsha: National University of Defense Technology, 2005.*

Multi-level sleep scheduling for heterogeneous wireless sensor networks

Shahzad Hassan*, Muhammad S.Nisar and Hong-Bo Jiang

School of Electronic Information and Communications,
Huazhong University of Science and Technology,
Wuhan, Hubei, China
**shahzad@hust.edu.cn*

For uniform consumption of network energy and enhancement of network lifetime, clustering has gained much attention in wireless sensor networks. Clustering can potentially decrease energy usage and improve the network life time by power control and data aggregation. In order to make clustering more energy efficient, various researchers have proposed putting idle nodes in sleep mode, allowing significant amounts of energy to be conserved. In this paper, we proposed a multi-level sleep scheduling technique which is based on exponentially smoothed average in clustered heterogeneous wireless sensor networks. The duration of sleeping is calculated by the monitoring the inter arrival time between the two packets transmitted by node to cluster head. The experimental simulation has shown that the proposed method significantly enhances the network life time by conserving the energy of nodes otherwise wasted in idle mode.

Keywords: Clustering; Sleep Scheduling; Idle nodes; Exponential; Inter-Arrival Time; HWSN.

1. Introduction

Energy management is one of the critical parameters in large scale wireless sensor networks (WSNs). If individual node in the network communicates and forwards its data directly to base station (BS), it increases the probability of network congestion and data collisions. As the sensor nodes have limited battery power, it will result in faster energy depletion, degrading the overall network life time. To optimize the network life time, scalability, coverage and network simplicity; it is essential to purpose an efficient and scalable network layer protocol.

To address the above mentioned issues, clustering [1,2] has been proposed by various researchers. Clustering protocols provide the solution to utilize the network energy uniformly which enhances network life time, maximizes the

* Corresponding author.

packet delivery ratio and throughput. Clustering of nodes does also avoid long distance communication between nodes and BS. In clustering, the nodes are divided into clusters and generally one cluster head (CH) is selected from each cluster and rest of the cluster members communicate and forwards their sensed data to the CH. The CH performs data aggregation and forwards the aggregated data to the BS for end user processing.

To make clustering more energy efficient various techniques have been proposed by putting the idle nodes in sleep mode through which significant amount of energy can be preserved [3].Various research works have identified idle listening as a major source of energy consumption [4,5] which utilizes the same amount of energy consumed in receiving [6]. Thus by introducing sleep scheduling in clustering can greatly reduce the energy consumption in the idle state and ultimately the network life time can be prolonged

2. Related Work

Clustering protocols like DTRE-SEP, DEEC and DDEEC for heterogeneous wireless sensor networks (HWSN) have a reasonably enhanced network life time, but however in these protocols the energy wasted in idle listening is not preserved. DEEC [7] is a distributed multilevel clustering algorithm for HWSN. DEEC selects the CHs with the help of probability based on the ratio between residual energy of each node and the average energy of the network.

DDEEC [8] is improved version of DEEC protocol, in which each network node calculates the initial and residual energy level for CH election. Each node in network requires the global knowledge of the network to compute the reference energy which will be expended by each node in a given round. . DTRE-SEP [1] engages both direct transmission and residual energy of nodes in order to form a new routing protocol on familiar foundations. Each node computes its distance with the associated CH as well as the BS. If the compared distance between the node and the BS is less than that of CH, instead of transmitting data to CH the node transmits directly to BS. The probability of CH election is both weight and residual energy based. If the energy of advance nodes becomes less than any on the normal nodes, both normal and advance node will have equal probability to become CH on basis of residual energy. In EESA [9] the author presented the idea of pairing between the nodes of same application which are deployed closed to each other will form a pair for sensing and data communication. Only one member of active pair nodes take part in communication while other member enters into sleep mode in order to preserve energy. The drawback of this scheme is that some of the nodes remains unpaired and stay active throughout the life time of the network and their energy is depleted at a higher rate which minimizes the network stability, while the pairing nodes alternatively on/off their radio.

3. Proposed Sleep Scheduling

3.1. *Assumptions and system model*

- An adequate number of heterogeneous nodes are randomly deployed over a region M*M.
- Static Clustering has been adopted [10].
- Node membership remains static i.e. the association of nodes with their respective CH remains unchanged throughout their life time.
- An Exponential Smooth Average (ESA) scheme is employed for predicting the next arrival rate of packet.
- The weighting factor of old data reduces exponentially giving more importance to recent data
- Only uplink traffic is considered i.e. traffic from nodes to CHs.

3.2. *Energy consumption model*

To transmit a P-bit Packet across a distance d_{TX} [2], the energy expended by the system is given below,

$$E_{Tx}(P, d_{TX}) = \begin{cases} P.\,E_{Ckt} + P.\,\varepsilon_{fs}\,.\,d_{TX}^2 \text{ if } d_{TX} < d_{NS} \\ P.\,E_{Ckt} + P.\,\varepsilon_{mp}\,.\,d_{TX}^4 \text{if } d_{TX} \geq d_{NS} \end{cases} \tag{1}$$

Where ECk is the energy dissipated per bit to run the circuit, εfs is free space and εmp is multipath (εmp) power loss, dTX is the distance between the Tx (transmitter) and Rx (receiver) while the distance between the network nodes and the BS is at all times less than or equal to d_{NS}. To receive P–bit message the systems utilizes energy equal to "$E_{RX=}P.E_{Ck}$

3.3. *Multi-level sleep scheduling (MLSS)*

In HWSN due to node heterogeneity, arrival time of the network nodes may vary and it is not feasible to implement a static node sleep scheduling. To overcome this problem we have proposed a new sleep scheduling algorithm on the basis of inter arrival time between the two packets based on three level .i.e. (i) Short Sleep (SS) (ii) Moderate Sleep (MS) (iii) Extended Sleep (ES).

In MLSS the sleep scheduling is implemented on the basis of predicted inter-arrival time calculated by ESA. The value of ESA of is calculated for dual tasks (i) prediction of next inter arrival time and (ii) to activate the multi- level sleep mode. If the value of ESA is below a certain threshold, the nodes will enter in sleep mode for a specified number of rounds.

For example, there are three types of nodes transmitting on different time intervals and the above mentioned method is applied. The node having minimum time interval between two packets will enter in a sleep mode for short duration(SS) , in case of second node which is transmitting less frequently will enter in moderate sleep mode (MS). While node having maximum time interval between the two packets will have longer sleep mode (ES).

3.4. *Prediction of idle periods*

The key mechanism is based on the monitoring of inter arrival time between two packets. Firstly we compute the mean value of inter arrival time (IAT) for current n number of packets by means of arithmetic average.

$$Avg(m) = \frac{\sum_{j=0}^{n-1} IAT(m-j)}{n}, m \geq n-1 \quad (2)$$

Where m is used to count the arrival events of packets and IAT is the inter arrival time between the two packets. Secondly the real estimation of following inter arrival time between the two packets is computed by exponentially smoothed average (ESA) which is given as

$$ESA(t) = \alpha.X(t-1) + (1-\alpha).ESA(t-1), t > 0 \quad (3)$$

Else

$$ESA(0) = 0 \quad (4)$$

where t is discrete variable representing discrete time values, α is the coefficient of weighting decrease, and X (t-1) is the monitoring time. The ESA average decreases exponentially and allocates different weight factors to the distribution, thus it allocates more weight to recent data by changing the value of α.

3.5. *Algorithm*

If ((Node== Idle) && (IAT == Minimal))
Sleep Duration =Shorter
Else if ((Node== Idle) && (IAT == Medium))
Sleep Duration = Moderate

Else if ((Node== Idle) && (IAT == Maximum))
Sleep Duration = Extended

Table 1. Parameters.

Parameter	Value
Area	150*150
N	100
E_{AD}	5nJ/bit/message
E_0	0.5J
Packet Size	4000
P_{opt}	0.1
ε_{fs}	10pJ/bit/m^2
ε_{mp}	0.0013pJ/bit/m^4

4. Simulations And Discussions

This section demonstrates the results of power preservation and network life time enhancement through simulations of the proposed multi-level sleep scheduling protocol using MATLAB. Our deployment scenario comprised of 100 nodes randomly deployed in 150m * 150m area. The values of simulation parameters are given in table 1.

Figure 1 presents the total amount of energy depleted vs. energy preserved in each round by the sensor nodes. The energy consumed indicates the energy depleted in data transmitting phase while the energy preserved indicates the energy saved when the nodes were in sleep mode. The study reveals that the a significant amount of energy can be preserved by putting the idle nodes in the sleep mode which results in improved network stability period and decreasing unstable period.

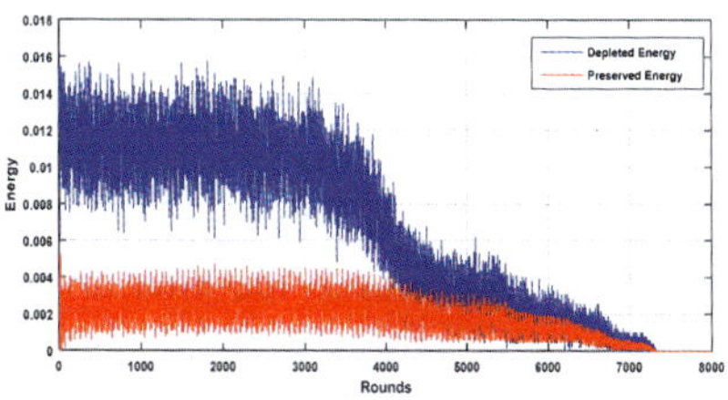

Fig. 1. Depleted energy vs preserved energy.

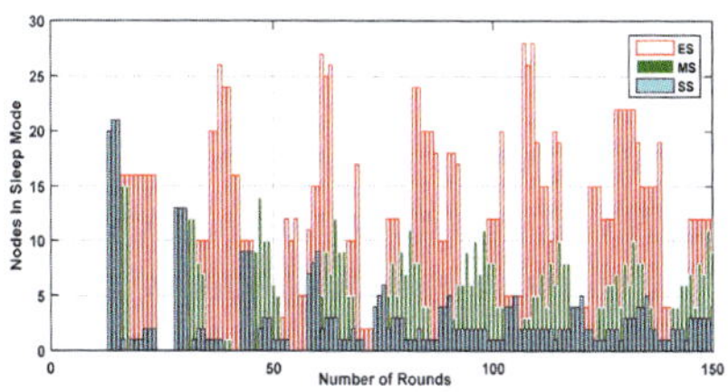

Fig. 2. Multi level sleep scheduling.

The sleep durations of the nodes is calculated according to the predicted inter-arrival time between two packets based on ESA values. Figure 2 represent the proposed method which is based on multi-level sleep scheduling. Due to the enormous data, the number of rounds is set to 150 rounds for the sake of clarity. It can be verified from the figure 2 that the inter arrival rate of the data generated by the nodes is variable and based on this inter arrival rate of data, multi-level sleep scheduling is implemented i.e. Excellent Sleep (ES) refers to the maximum inter arrival rate between two packets, Moderate Sleep (MS) points to the moderate inter arrival rate between the two packets, in which the data generated by the nodes neither high or neither low and Short Sleep (SS) is related to the scenario where the nodes are generating data continuously.

Figure 3 illustrates the total time period of each node spent in sleep mode. The nodes having maximum inter arrival time have minimum energy depletion due to the prolonged sleep and the preserved energy is again consumed in the network operation thus enhancing the overall network lifetime.

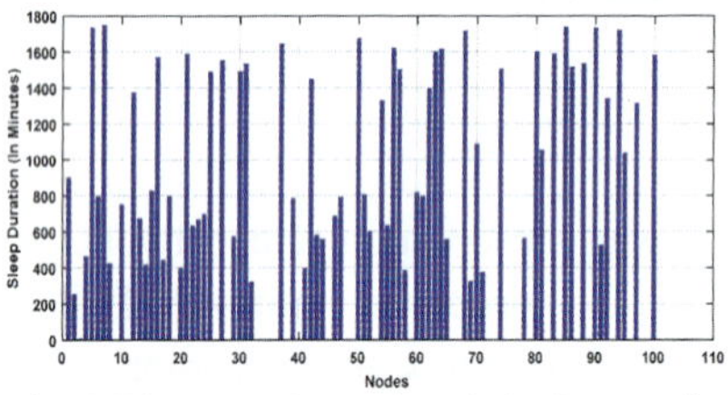

Fig. 3. Time spent by each node in sleep mode.

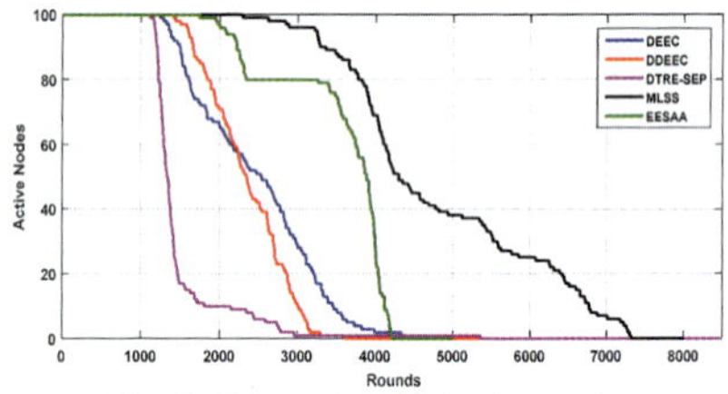

Fig. 4. Comparison of active nodes.

Network stability period is defined as when the first node in the network drains out its energy. It can be conclusively concluded from figure 4 that MLSS takes full advantage of heterogeneity and extends the stability period beyond that achievable and has extend the stable region 106% as compared to DTRE-SEP, 84% as compared DEEC, 62 % DDEEC and 34% to EESAA respectively. It is because that MLSS takes the advantage of the preserved energy from idle mode.

5. Conclusion

In this paper we propose an IAT model for HWSN characterized by ESA. Based on IAT of packets, a multi-level sleep scheduling (MLSS) is proposed in which the each node randomly enters into sleep mode with respect to the inter arrival time between the two packets. The simulation and analysis verified that the network life is enhanced by utilizing the energy which would otherwise have depleted in idle mode and outclass other protocols with the same power in terms of stable and unstable region. The protocol is best suited for delay insensitive environments. In future we intend to extend this technique to delay sensitive environments.

References

1. Shahzad Hassan , Muhammad S.Nisar, and Hongbo Jiang, DTRE-SEP: A Direct Transmission and Residual Energy based Stable Election Protocol for Clustering Techniques in HWSN, International Conference on Communication Software and Networks(ICCSN 2015), (Chengdu, China, 2015)
2. Smaragdakis, I. Matta, A. Bestavros (2004) SEP: A Stable Election Protocol for clustered heterogeneous wireless sensor networks, In: Second International Workshop on Sensor and Actor Network Protocols and Applications (SANPA 2004).
3. Tianle Zhang, Zongwei Luo, Chunlu Wang, Edward C. Wong, A Model for Evaluating Connectivity Availability in Random Sleep Scheduled Delay-Tolerant Wireless Networks, Global Telecommunications Conference (GLOBECOM 2010), (Miami, FL, 2010)
4. W. Ye, J. Heidemann, and D. Estrin, Medium Access Control with Coordinated, Adaptive Sleeping for Wireless Sensor Networks, Networking, Vol. 12 (IEEE/ACM Transactions on 2014), pp 493-506.
5. Rong Zheng, Jennifer C. Hou and Lui Sha, Asynchronous Wakeup For Ad Hoc Networks, in (MobiHoc '03) : Proceedings of the 4th ACM international symposium on Mobile ad hoc networking & computing, (2003).
6. L.M. Feeney and M. Nilsson, Investigating the energy consumption of a wireless network interface in an ad hoc networking environment, Twentieth Annual Joint Conference of the IEEE Computer and Communications Societies. Proc. IEEE, (INFOCOM 2001), (Anchorage, AK, 2001)
7. L. Qing, Q. Zhu, M. Wang, Design of a distributed energy-efficient clustering algorithm for heterogeneous wireless sensor network, ELSEVIER, Computer Communications, (2006).
8. B. Elbhiri, R. Saadane, S. El Fkihi, D. Aboutajdine, Developed Distributed Energy-Efficient Clustering (DDEEC) for heterogeneous wireless sensor networks, I/V Communications and Mobile Network (ISVC 2010), 5th International Symposium on, (Rabat ,2010)
9. T. Shah, N. Javaid, TN. Qureshi, Energy efficient sleep awake aware (EESAA) intelligent sensor network routing protocol, (CoRR 2012), (Islamabad 2012).
10. Bo Huang, Hao, Hui Zhu, Tanabe, Y., Low-Energy Static Clustering Scheme for Wireless Sensor Network, Wireless Communications, Networking and Mobile Computing, 2006. (WiCOM 2006). International Conference on, (Wuhan 2006)

Mobile multi-node recharging in heterogeneous wireless sensor networks

*Shahzad Hassan, Muhammad S. Nisar and Hong-Bo Jiang

School of Electronic Information and Communications,
Huazhong University of Science and Technology,
Wuhan, Hubei, China
**E-mail: shahzad@hust.edu.cn*

The wireless energy transfer technology has evolved as a reliable approach to solve the energy limitation problem for Wireless Sensor Networks. Major work has been carried out to recharge only the static sensor node through wireless charging vehicle. Some of the wireless sensor networks applications require complete mobility and the proposed techniques cannot be applied to mobile sensor networks since the positions of the nodes are changing at regular intervals. This paper extends the prior approach in mobile sensor networks through which the lifetime of the mobile nodes have been significantly enhanced. The charging schedule of the nodes ensure that the nodes never run out energy.

Keywords: Mobile; Multi-Node; Recharging; Heterogeneous; HWSN.

1. Introduction

Energy efficiency is still an open issue and researchers are trying to achieve minimum energy consumption while retaining the anticipated performance. The energy optimization is more complex task in WSNs since it includes not only reducing energy consumption but also extending the network lifetime. Limited battery capacity is major bottle neck in prolonging the network life time. To enhance the network lifetime, abundant research has been published in the recent era [1,2], but regardless of these rigorous efforts, enhancing the network lifetime remains an open issue. Some energy harvesting techniques [3, 4] were proposed but these techniques are strongly dependent on the environment for extracting energy from the environment. The recent development in wireless energy transfer technology (WET) developed by Kurs et al. [5] offers a bright alternative to reenergize these nodes through magnetic resonating coupling. In WSNs the effect of WET is enormous, instead of adopting energy harvesting a node can be efficiently charged through an independent charging source. This

* Corresponding author.

new breakthrough led the approach to a novel standard for WSN; the Wireless Rechargeable Sensor Networks (WRSNs), in which the nodes can either static or mobile.

Motivated by this novel development in WET, specifically we are interested in multi-mobile node recharging in context of heterogeneous clustering [6] through which the network lifespan can be maximized. Heterogeneous wireless sensor networks (HWSN) are composed of two types of nodes, the higher energy nodes, also known as Advance nodes (AD) and less energy nodes also termed as normal nodes (NN). In traditional mobile sensor networks it is very complex to keep the track of network changes triggered by dynamic network topology, continuous node mobility, overhead messages and fixed bandwidth. To inhibit these restrictions, clustering techniques can be used to manage the mobile network into groups of nodes through which the number of control messages could be minimized and mobility can be managed.

2. Proposed Model

2.1. *Energy consumption model*

Due to the introduction of mobility in our protocol, we have modified the energy consumption model used in [6]. In this model, the cost of transmitting and receiving a P- bit packet across a distance can be calculated as

$$E_{Tx}(P, d) = \begin{cases} P.\,E_{Ckt} + P.\,\varepsilon_{fs}\,.\,d^2 \text{ if } d < d_0 \\ P.\,E_{Ckt} + P.\,\varepsilon_{mp}\,.\,d^4 \text{if } d \geq d_0 \end{cases} \tag{1}$$

$$E_{RX=}P.\,E_{Ckt} \tag{2}$$

Where ETX and ERX is energy consumed in transmitting and receiving the packet, ECk is the energy dissipated per bit to run the circuit, εfs is free space and εmp is multipath (εmp) power loss, d is the distance between transmitter and receiver. While d0 is the threshold distance between the nodes and the base station. Equation (1) shows only the energy which is consumed by the nodes in whole communication process, since the nodes are mobile, energy will also be consumed in mobility. The energy consumed by node during mobility is given by

$$E_{ME=}M_E.\,D \tag{3}$$

Where E_{ME} is the energy depleted by a node moving from Point P1 to Point P2, M_E the amount of Energy required to move unit distance and D is the distance travelled by the node. So the total energy (TE) consumed by network will become equal to

$$TE = E_{Tx} + E_{RX} + E_{ME}(\mathrm{D}) \quad (4)$$

2.2. *Cluster head selection/formation*

In cluster head selection process the, node selects itself as a cluster head, if the generated random number is between 0 and 1 and is less than the set threshold value. In our proposed protocol the selection of cluster head is both probability and residual energy based. Initially, the advance nodes have higher probability to become cluster head but after certain rounds the cluster heads are selected on the basis of residual energy. The general threshold for cluster head selection is,

$$T_{SN} \left\{ \left(\frac{P_N}{1-P_N*(r*mod(\frac{1}{P_N}))} \right) * \frac{NE_{RES}}{NNE_{RES}} \qquad if\ S_N \in Y' \right. \quad (5)$$

$$T_{AN} = \left\{ \left(\frac{P_{AS}}{1-P_{AS*}*(r*mod(\frac{1}{P_{AS}}))} \right) * \frac{AE_{RES}}{NNE_{RES}} \qquad if\ S_A \in Y'' \right. \quad (6)$$

Where, r is the round, Y' and Y" are the set of normal and advance nodes that have not become CHs within the last $1/P_N$ and $1/P_A$, rounds of the epoch. On average $n*P_{opt.}$ nodes must become CHs per round per epoch. After the cluster head selection process, the nodes join the nearest cluster completing the cluster formation phase. The energy expended by cluster head and the cluster members is computed by the equations used in [6]. Once the node is selected as cluster head, it will become static in order to ensure connectivity with its members.

2.3. *Energy aware free random way point model*

Generally the random way point (RWP) mobility model is used for mobile networks performance assessment [7]. It is a basic model, which defines the movement pattern of individual nodes by simple terms. In our proposed protocol we have modified the RWP. Since the nodes are free to move in any direction and are energy aware, the nodes will calculate their energy level and the reference energy (RF) required by them to move directly towards the power bank. It decides to take that safe step if and only if the step is taken and the node will still have enough energy to reach back to the power in case it is needed. Movement towards a new point to travel towards to power bank to recharge which is discussed briefly in next section.

2.4. *Basic mechanism*

In our proposed research, the mobile nodes (MN) are energy aware, equipped with the initial energy (IE) and consumes EME (Energy consumed in mobility) per unit step and ETX in communicating with cluster head. Assuming nodes are randomly distributed in an area A, a mobile node MN starts moving in a direction between (0,2π) with a constant velocity V and stop at Point P1 for short interval of time T1. During this interval it computes its residual energy (RE) and its distance (PD) from the nearest power bank (PB) and transmits packet to the cluster head. It then tries to generate a random step, any step it randomly generates it checks whether it is a safe step or not. In case it is a safe step the node will take that step in this round, if the step is not a safe step it will generate another random step that it can possibly take and will repeat the process ten times. If it is unable to find a suitable step in ten tries, it will move towards the nearest PB. A safe step is that step which if taken, the node will still have enough energy to move towards the PB if need arises. At some point the residual energy will become equal to the reference energy, in this case the mobile node will stop packet transmission and will terminate its membership and with the cluster head in order to avoid loss of data. The mobile node will move in a straight line towards power bank to get reenergized. The reference energy is referred to that energy which is utilized by the mobile node to reach the power bank for recharging from its location in the arena.

2.5. *Multi-node recharge cycle*

The recharging cycle (RC) of the mobile node refers to the time period of a node spent while charging. Four power banks with infinite amount of energy are implemented at an equal distance in the given area. The nodes are aware of power bank locations. Once the nodes enter in the charging radius of the power charging, wireless charging takes place. At a given time only five nodes are allowed to recharge at one PB. If a PB has five nodes attached to it, any other node which wants to connect to this particular PB will have to wait for a vacant slot. We have divided the energy consumption of the mobile nodes in five steps i.e. (i) sensing data (ii) communication between member node and the cluster head (iii) mobility (iv) distance traveled by the mobile node to reach power bank.

2.6. *Recharge schedule*

While scheduling we must consider some performance measurement parameters, recharging for the mobile node, the waiting time of the mobile nodes being serviced. The scheduling would be beneficial in that case where the mobile nodes remain active and get recharged before running out of energy. The recharge scheduling cycle can be termed as the interval between the time when the mobile starts charging and leave the power bank. A reasonable recharge scheduling recharges the mobile nodes to enhanced the network lifetime therefore this description characterizes the lasting efficiency of a scheduling well.

2.7. *Packet delivery ratio*

In WSN the packet delivery ratio (PDR) is degraded by packet loss, which is affected by several factors but in case of mobile node, the main reason for packet loss is link failure due to the link failure between node and the base station. In the proposed protocol because of cluster head scheduling the ratio of packet loss is very low. As the PDR is limited by packet loss, this ratio is calculated by the equation used in [12].

$$Packets\ Transmitted = Recivied\ Packets + Packets\ Loss \tag{7}$$

$$PDR = (Total\ Recieved\ Packets/Total\ \ Packets\ Generated) * 100 \tag{8}$$

3. Simulations and Results

To validate our theoretical foundations, we have performed simulations in MATLAB. In simulations, 50 nodes are randomly deployed in a 100*100 area. having both normal and advance nodes. The base station is deployed in the center of the network. Four power banks are deployed at equal distance in the area. The mobile nodes are equipped with extra battery which is used for mobility only.

We have run the simulations for 5000 rounds to check the stability of the network and afterwards for 500 rounds to verify the remaining parameters. Figure 1 shows the number of active nodes in the given rounds. The simulations results verify the benefit of the recharge schedule cycle as after 5000 rounds none of the node has run out of its energy, thus increasing the stability period of the network significantly.

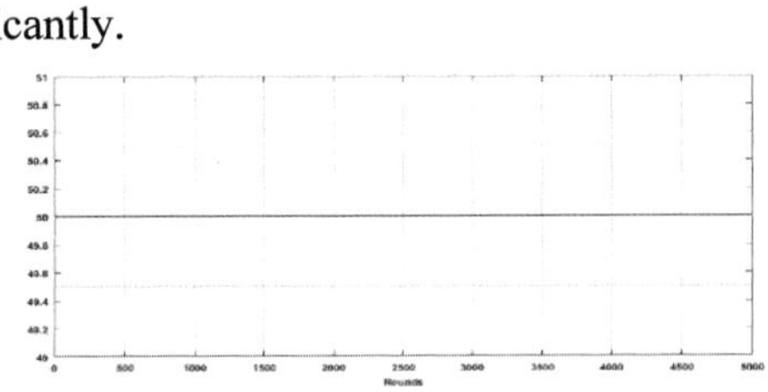

Fig. 1. Active node vs. number rounds.

The PDR in figure 2(a) reveal that, that maximum loss packets in per round is 7%. Due to the cluster head scheduling, the mobile nodes transmit in their assigned slots and cluster head successfully receives the transmitted loss. The packet loss is due to topology changes and since the nodes are mobile. The figures 2(b, c) reveals that the PDR decreases as the number of nodes increases

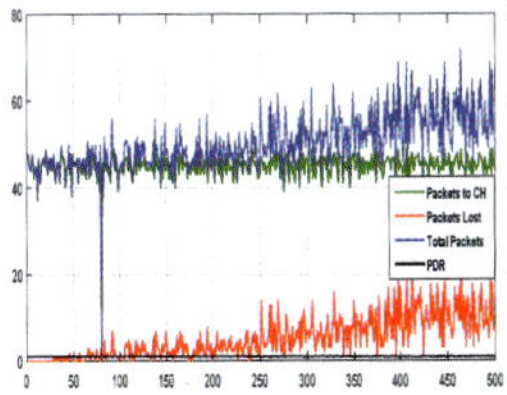

Fig. 2. (a) n=50.

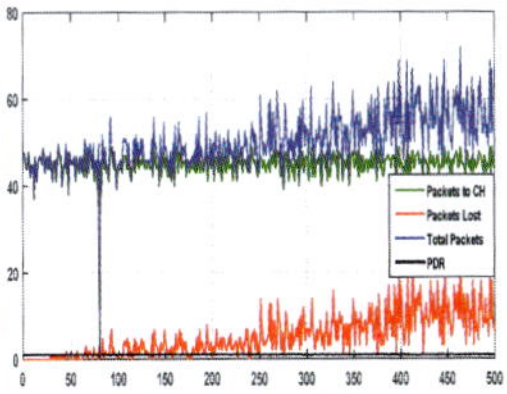

Fig. 2. (b) n=100.

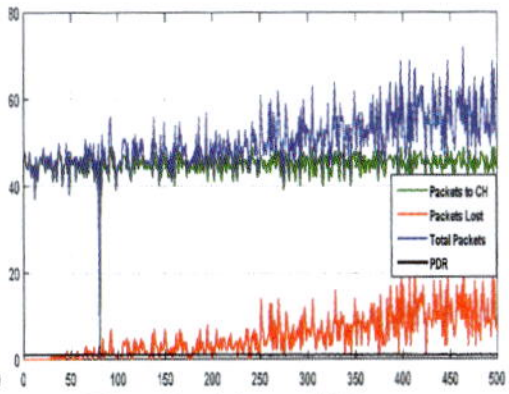

Fig. 2. (c) n=150.

The packets are considered to be lost, when the cluster member moves out of the cluster communication range. Since the node joins the nearest cluster head and may move away in a short while due to which loss of packets occurs. Figure 3 illustrates the connectivity of the cluster nodes with their respective cluster heads. At the start the connectivity was 100% and later it fluctuates according to the situation in each round. The lowest connectivity achieved in our simulations was 75% thus depicting an increased packet loss. However, this loss of packets can be controlled by preventing the nodes to transmit as they move out of range of their cluster head.

Figure 4 represent the energy consumption which is utilized in the transmission of packets from nodes to cluster head and packets forwarded to base station from the cluster head. The energy consumption in the nodes is not uniform because when the nodes becomes cluster head, it depletes energy more rapidly as compared to other nodes and as well as due to the mobility of the nodes. The other reason for this is transmission distance, the energy consumption increased proportionally as the transmission distance between the cluster members and cluster head is increased. Due to the clustering method, the utilization of energy is very low as compared to traditional routing protocols.

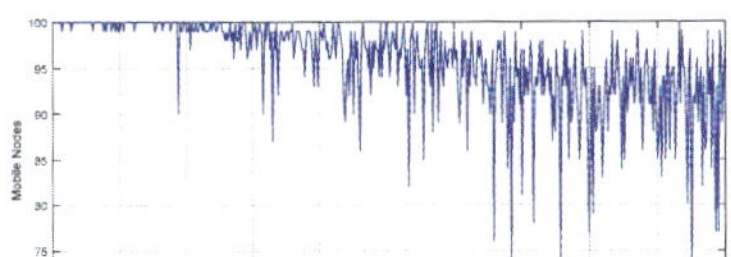

Fig. 3. Connectivity between nodes and cluster head.

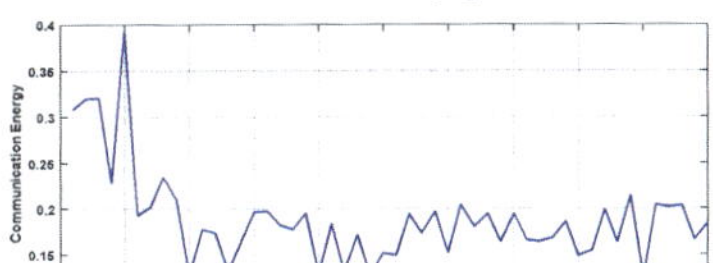

Fig. 4. Energy consumed in communication.

4. Conclusion

In this paper we have proposed a protocol to reenergize multiple mobile nodes simultaneously. Since the nodes are energy aware and when the minimum energy threshold is reached, the nodes travel towards the chargers to reenergize themselves thus eliminating the waiting period for recharging and ensuring that the battery never runs out energy. We have studied a general scenario in which a mobile node enters the recharging radius of the power bank and recharges itself.

References

1. J. Chang and L. Tassiulas, Maximum lifetime routing in wireless sensor networks, vol. 12 ,(IEEE/ACM Trans. Netw.2004), , pp 609–619.
2. Giridhar and P. R. Kumar, Maximizing the functional lifetime of sensor networks, in Proc. (IPSN) Int. Symp. Inf. Process. Sensor Netw., (Los Angeles, CA, 2005)
3. X. Jiang, J. Polastre, and D. Culler, Perpetual environmentally powered sensor networks, in Proc. (IPSN) Int. Symp. Inf. Process. Sensor Netw., (Los Angeles, CA, Apr. 2005)
4. G. Park, T. Rosing, M. D. Todd, C. R. Farrar, and W. Hodgkiss, Energy harvesting for structural health monitoring sensor networks, (J. Infrastruct. Syst. 2008), pp. 64–79.
5. A. Kurs, A. Karalis, R. Moffatt, J. D. Joannopoulos, P. Fisher, and M. Soljacic, Wireless power transfer via strongly coupled magnetic resonances, Vol. 317 (Science 2007), pp 83–86.
6. Shahzad Hassan , Muhammad S.Nisar, and Hongbo Jiang, DTRE-SEP: A Direct Transmission and Residual Energy based Stable Election Protocol for Clustering Techniques in HWSN, International Conference on Communication Software and Networks(ICCSN 2015), (Chengdu, China, 2015)
7. David B. Johnson and David A. Maltz, Dynamic Source Routing in Ad Hoc Wireless Networks, in Mobile Computing, edited by Tomasz Imielinski and Hank Korth, Chapter 5 (Kluwer Academic Publishers, 1996).

Research on an embedded intelligent-unit based on multi-sensor fusion oriented internet

Jian Yang*
School of Electronic & Electrical Engineering, Shandong Jianzhu University, Jinan,250101, China
E-mail:yang8018@sdjzu.edu.cn
www.sdjzu.edu

Combined with the technology of network communication, embedded computer, data processing, device diagnosis and maintenance, Intelligent Maintenance Unit (IMU) is designed for remote monitoring, diagnosing, maintenance and controlling devices where it is embedded into the equipment to be monitored. To enhance the information process and maintenance ability of agile manufacturing systems, we proposed the IMU. The primary architecture and key technique of the Internet-based IMU is based on the research of new local and overseas products. We also constructed remote monitoring and maintenance system with IMU. Key technologies and Internet-based maintenance system integrated with multi–sensors fusion are also analyzed so as to meet the high standards of reliability, networking and intelligence of a monitoring system. Our experimental results have shown that our proposed embedded smart agent provided a sound and reliable e-maintenance mechanism for computer-integrated manufacture (CIM), and sustained the value of products/ equipment during its life-cycle. Finally, the application field and future tendency of IMU-based maintenance technology are presented.

Keywords: Intelligent Maintenance Unit (IMU); Device Diagnosis, E-maintenance; Internet-Based Multi-Sensor Fusion; Embedded Technology

1. Introduction

1.1. *Model design on internet-based embedded smart agent*

Briefly speaking, e-maintenance is a transformation system that enables the manufacturing operations to achieve predictive near-zero-downtime performance as well as to synchronize with the business systems through the use of web-enabled and tether-free (i.e. wireless, web, etc.) information technologies. It integrates information and decisions in data flow (of machine/process level), information level (of factory and supply system level),

* Corresponding author.

and cash flow (of business system level). This transition is dependent upon the advancement of next generation manufacturing practices on "e-factory and e-automation " which is focused on the use of information to make things happen collaboratively on a global basis [1-2].

For the past decade, the impact of web-based technologies has added "velocity' to the design, manufacturing, and aftermarket service of a product. Today's competition in manufacturing industry depends not just on lean manufacturing, but also on the ability to provide customers with total solutions and life-cycle costs for sustainable value. Manufacturers are now under a tremendous pressure to improve their responsiveness and efficiency in terms of product development, operations, and resource utilization with a transparent visibility of production and quality control. Lead times must be cut short to their extreme extent to meet the changing demands of customers in different regions of the world. Products are required to be make-to-order with no or minimum inventory, requiring (a) an efficient information flow between customers, manufacturing and product development (i.e. plant floor, suppliers and designers), (b) a tight control between customers and manufacturing and (c) near-zero downtime of the plant floor assets. The trends in manufacturing and function of predictive intelligence as an enabling tool to meet the needs are clearer with the Internet time coming near [3-6].

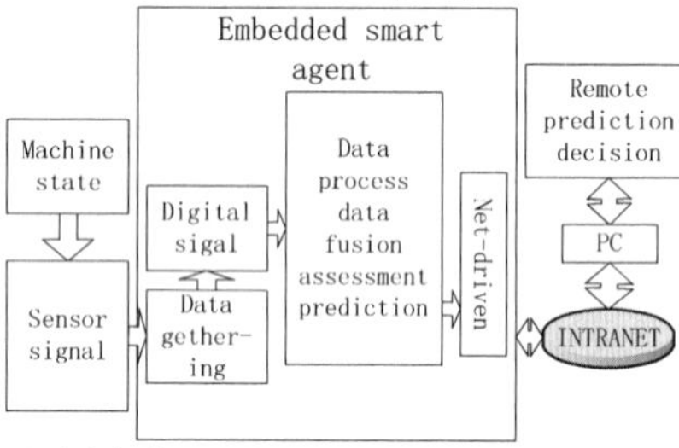

Fig. 1. Model design on internet-based Embedded smart agent.

From analysis above, this paper proposed an internet-based Embedded smart agent model. E-Manufacturing includes the ability to monitor the plant floor assets, predict the variation of product quality and performance loss of any equipment for dynamic rescheduling of production and maintenance operations, and synchronize with related business services to achieve a seamless integration between manufacturing and higher level enterprise systems. Dynamically updated information and knowledge about the capabilities, limits and variation of manufacturing assets for various suppliers guarantee the best decisions for outsourcing at the early stages of design.

Currently, Manufacturing Execution Systems (MES) enables the data flow among design, process, and manufacturing systems. The enterprise resource planning (ERP) systems serves as an engine for driving the operations and the supply chain systems. However, existing structure of the ERP and MES cannot informate (i.e. communicate the information in real-time) the decision across the supply chain systems. The major functions and objectives of e-Manufacturing are: (1) enable an O.H.I.O (only handle information once) environment; (2) predict and optimize total asset utilization in the plant floor; (3) synchronize asset information with supply chain network; and (4) automate business and customer service processes. The proposed e-manufacturing architecture in this position paper addresses the above needs.

To address these needs, an e-Manufacturing system should offer comprehensive solutions by addressing the following requirements: 1) Development of intelligent agents for continuous, real time, remote and distributed monitoring of devices, machinery and systems to predict machine's performance status (health condition) and to enable capabilities of producing quality parts; 2) Development of Internet/Intranet based agent platform that is scalable and re-configurable for data transformation, prognostics, performance optimization and synchronization; Figure 1, illustrates the proposed e-maintenance architecture of Internet/Intranet based agent platform and its elements [7-9].

2. E-Maintenance: Architecture

2.1. *Architecture*

Figure 2, shows an intelligent maintenance system with its key elements. The core-enabling element of an intelligent maintenance system is the smart computational agent that can acquire sensory information and predict the degradation or performance loss (Embedded smart agent), not the traditional diagnostics of failure or faults.

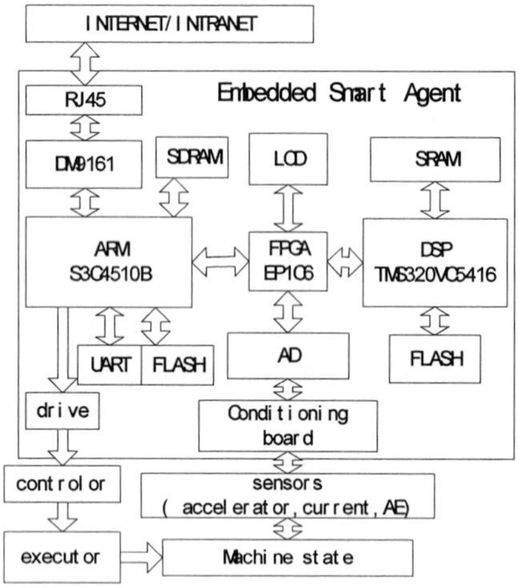

Fig. 2. An internet-based Embedded smart agent.

2.1.1. *Hardware design*

In this paper, we proposed an ARM+DSP master/slave that combine the S3C4510B 16/32-bit RISC micro-controller with the TMS320VC5416 fixed-point, digital signal processor (DSP), in this model, S3C4510B is the cost-effective, high-performance main micro-controller integrated as an embedded smart agent, while TMS320VC5416 fixed-point, digital signal processor (DSP) is the slave processor which deals mainly with the AD converting, data gathering and data processing which include data filtering, features recognizing etc. HPI mode is used to communicating between S3C4510B and TMS320VC5416. An integrated Ethernet controller, the 3C4510B, is designed for use in managed communication hubs and routers. DM9161 is used as net interface adapting circuit which is connected in. Following that circuit designed, is the RJ45 drive circuit, through which the S3C4510B 16/32-bit RISC micro-controller can easily be connected to INTERNET/INTRANET by obeying the TCP/IP communication protocol.

3. Application Case

To meet the demands of high reliability, networking and intelligence of the system, Internet-based maintenance system integrated with multi-sensor fusion of acceleration sensor, current mutual inductance and acoustic emission has also been proposed and implemented here. Signals of acceleration sensor and current mutual inductance are preprocessed to convert into digital signals, then the digital data will be filtered and features—useful information will be extracted from them. Acoustic emission signal is processed by a specialized apparatus called ASCO-P into slowly changed signal named ASL(average signal level) for further proceeding such as feature extraction and state recognition to inspect tool wear, blunt and burn state of the surface of the work-piece, etc. Each input signal is designed in block mode to connect pre-positive circuit in the embedded smart agent that is installed in numerical control machine tool for supervising and diagnosis. Meanwhile different software is tailored to different equipment and different status for supervising and diagnosis.

4. Conclusions

This position paper introduced the e-maintenance architecture of a smart agent-based E-maintenance platform, and outlined its fundamental requirements and elements as well as expected impact to achieve high-velocity and high-impact manufacturing performance. Web-enabled and infotronics technologies play indispensable roles in supporting and enabling the complex practices of design and manufacturing by providing the mechanisms to facilitate and manage the integrated system discipline with the higher system levels such as SCM and ERP. E-maintenance is a major pillar that supports the success of the integration of e-Manufacturing and e-business. The integration among e-Maintenance, e-Manufacturing, and e-Business systems. If implemented properly, manufacturers and users will both benefit from the increased equipment and process reliability with optimal asset performance and seamless integration with suppliers and customers.

In conclusion, an advanced Internet visible and capable embedded smart agent has been implemented in the paper, which has compliant STIM data structure described by IEEE1451.2 standard, while an IP module is presented and adopted herein to instead NCAP to enable the embedded smart agent entering into Internet thus simplifying networking design in a cost-effective way. Adopting embedded Internet technology, the proposed embedded smart agent is feasible for distributed networking application in e-maintenance with advantages of accessibility, integration and expandability, where remote maintenance engineer can transparently monitor and operate the field devices over Internet. The relevant hardware and software are designed and the full working prototype is realized in our work. Experiment has been realized to verity the performances of the proposed embedded smart agent, and the results have been experimentally verified that embedded smart agent is feasible and suitable for distributed networking application under LAN environment. Finally, to meet the demands of high reliability, networking and intelligence of system, Internet-based maintenance system integrated with multi - parameter signals have also been proposed and implemented and results have been experimentally verified that the proposed smart agent-based E-maintenance platform is also feasible and suitable for distributed networking application under LAN environment.

Acknowledgments

This work is supported by Doctor fund of Shandong Jianzhu University(No. XNBS1251) and Key-lab of Intelligent Construction & Building Automation of Shandong Province and partially supported by grant Key-Program of Beijing Natural Science Foundation of Beijing (No. 4031001-1).

References

1. LEE J.Tele-service engineering in manufacturing:challenges and opportunities[J]. International Journal of Machine Tools and Manufacture,1998,38:901-910.
2. SHI J,NI J,LEE J.Research challenges and opportunities in remote diagnostics and system performance assessment[A]. Proceeding of 4th IFAC Workshop on Intelligent Maintenance Systems: IMS'97[C]. Seoul,Korea: IFAC, 1997,21-23.C.Jarlskog, CP Violation (World Scientific, Singapore, 1988).
3. P. Zipkin, 2002, Seminar on the Limits of Mass Customization, Center for Innovative Manufacturing and Operations Management (CIMOM), April 22,2002.
4. Patrick Waurzyniak, 2001, "Moving towards e-factory", SME Manufacturing Magazine, v. 127, n. 5, Nov. 2001, http://www.sme.org/gmn/mag/2001/01nom042/01nom042.pdf,.
5. Patrick Waurzyniak, 2001, "Web tools catch on", SME Manufacturing Magazine, v. 127, n.4, Oct. 2001, (www.sme.org).
6. Rockwell Automation e-Manufacturing Industry Road Map, http://www.rockwellautomation.com L.F. Richardson, Arms and Insecurity (Boxwood, Pittsburg, 1960).
7. M. Koç and J. Lee (June 2002), "e-Manufacturing and e-Maintenance-Applications and Benefits", International Conference on Responsive Manufacturing (ICRM) 2002, June 26-29, 2002, Gaziantep, Turkey.
8. M. Koç and J. Lee (December 2001), "A System Framework for Next-Generation E-Maintenance System", Eco Design 2001, Second

International Symposium on Environmentally Conscious Design and Inverse Manufacturing, December 11-15. 2001, Tokyo Big Sight, Tokyo, Japan

9. Jay Lee, Ahad Ali and M. Koç, (May 2001), "e-Manufacturing- Its Elements and Impact", Proceedings of the Annual Institute of Industrial Engineering(IIE) Conference, Advances in Production Session, May 21-23, 2001, Dallas, Texas, USA

Round-robin differential-phase-shift quantum secret sharing

Zhen-Yuan Sun*, Jian Wang and Chao-Jing Tang
College of Electronic Science and Engineering,
National University of Defense Technology, NUDT
Changsha 410073, Hunan, People's Republic of China
†*E-mail: sunzhenyuan0401@nudt.edu.cn, zhenyuansun@hotmail.com*
http://www.nudt.edu.cn

We put forward a novel quantum secret sharing (QSS) protocol based on round-robin differential-phase-shift scheme (RRDPS), which achieves the distribution of a full secret key to one party and two partial keys to two parties that neither of the two partials alone can decipher the cryptographic message by full key. Making use of the advantages of round-robin differential-phase-shift, the protocol has a high tolerance of bit errors and enjoys a constant amount of privacy amplification and inherent randomness without precise signal disturbance estimation, especially when a finite key effect is considered. In terms of configuration, it is a simple modification, a variable-delay interferometer replacing the one-pulse-delay counterpart, of differential-phase-shift quantum secret sharing(DPS-QSS).

Keywords: Quantum Secret Sharing; Differential Phase Shift; Variable Delay; Randomness .

1. Introduction

Since the first quantum key distribution (QKD) was proposed[1], several significant protocols have also been put forward based on quantum mechanics[2-8]. QKD is considered as the ultimate method to realize private communication securely against eavesdropping and implement future secure communication network[9]. The existing QKD schemes over the past 30 years needed to be set an upper bond for the amount of potential information leakage by monitoring the signal disturbance because of Heisenberg's uncertainty principle[10]. However, a recently proposed RRDPS-QKD can[11-14], surprisingly, bound the amount of leak by its inherent randomness and has a better finite key effect.

Quantum secret sharing (QSS) could be considered as a kind of QKD based on entanglement or polarization[15-20]. A secret message or key is shared by

---

* Corresponding author.

two or more parties, such that it could be reestablished with their collaboration. Ref.21, utilizing weak coherent pulse instead of single photon, presents its merits. Included are simple configuration, fiber transmission and high key creation rate, thus, improving the simplicity and practicality of the scheme.

In this paper, taking advantages of the RRDPS scheme, we make a small change in the Charlie's measurement equipment in the light of configuration of Ref.21. The correctness and inherent randomness of the protocol are also be discussed.

2. Round-Robin Differential-Phase-Shift Quantum Secret Sharing

In this section, a RRDPS-QSS protocol will be described, by which Alice and Bob will obtain partial keys for deciphering while Charlie creates the integrated key for enciphering information. It is achieved by binary phase coding on weak coherent pulse train created by Alice; and a variable-delay interferometer instead of one-pulse delayed Mach-Zehnder interferometer utilized in the DPS-QSS protocol[21], followed by two photon detectors in Charlie's side. Based on this QSS scenario, the schematics of the proposed RRDPS-QSS scheme is displayed in Figure 1, and the specific workflow of the scheme will be explained as follows.

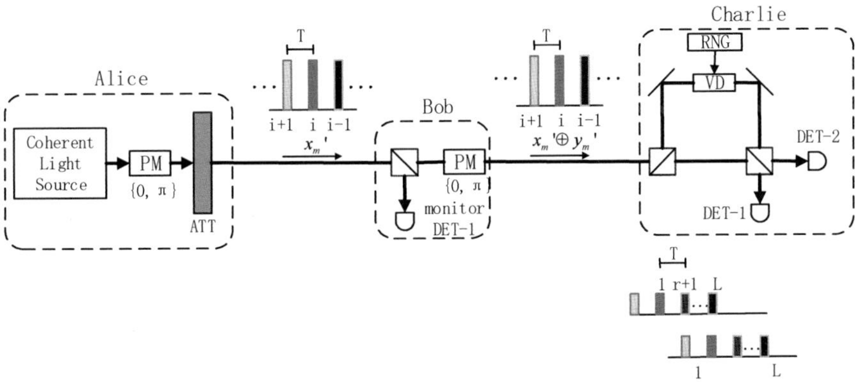

Fig. 1 Configuration of round-robin differential-phase-shift quantum secret sharing. PM phase modulator, ATT attenuator, DET, detector, VD variable delay RNG random number generator.

Quantum Phase:

Step 1. Alice firstly phase-modulates an L-bit weak coherent pulse sequence $s_1 s_2 \cdots s_L$ by randomly choosing phase $\{0, \pi\}$ for each pulse from a laser. The pulse interval ΔT is much shorter than the coherence time. The photon number per pulse is adjusted at less than 1 photon on average, such as 0.1-0.2. Then based on the modulated data, she generates a binary string X while each state is on behalf of a binary variable $x_i, x_i \in \{0,1\}$. To be more specific, if the phase information of the pulse is 0, she sets $x_i = 0$; if the phase information is π, she sets $x_i = 1$.

Step 2. When Bob receives the phase-modulated signal sequence, he splits and detects the received signal by a beam-splitter to monitor the incident count rate with a monitor detector. If the received signal power is too high, he terminates the process; or he modulates the signal again as what Alice has operated by $\{0,\pi\}$ on each pulse and generates a bit string Y , $y_i \in \{0,1\}$ according to $0 : y_i = 0; \pi : y_i = 1$, respectively. Finally, he transmits the L-bit sequence to Charlie.

Step 3. When Charlie receives the phase-modulated signal sequence, initially, he creates a random number $r, r \in \{1, \cdots, L-1\}$ by an independent Random Number Generator (RNG). The he sets the rT as the amount of delay at the Variable Delay (VD), where T denotes the time interval of the incoming pulse.

Step 4. Through the variable delay interferometer, Charlie measures the differential shift between a superposed pair of pulses, which have been split and superposed again and from which a photon is detected. Once detected, he obtains a pair of indices $\{i, j\} \in \{1, \cdots, L\}$ satisfying $j - i = \pm r$ and notes the measurement result of $z'_m = z_i \oplus z_j$ as his sifted key bit, here $\oplus$ represent summation modulo 2. In detail, we assume that he sets $z'_m = 0$ when the detector-1 measures a photon for the differential shift of 0; while $z'_m = 1$ when the detector-2 clicks a photon for the differential shift of π , respectively. Here, noteworthy is that the received signal train pulse rate is smaller than one photon per pulse and the photon is randomly placed, so Charlie will occasionally and randomly detects a photon.

Step 5. One completed, Charlie publicly declares the pair of indices of $\{i, j\}$ though an authenticated public channel. With the position information and their own generated binary string, Alice and Bob records their own sifted key bit. We assume that $x'_m := x_i \oplus x_j$ $\mathrm{y}'_m := y_i \oplus y_j$, respectively corresponding to Charlie's measurement result imposed on the superposed states.

Step 6. In order to get N number sufficient sifted key bits, Alice, Bob and Charlie repeat the above steps more than N times. Then they can conduct error reconciliation and privacy amplification to correct their N-bit sifted key into ideal final key.

Classical Phase:

Step 7. In order to detect the existence of eavesdropper, Charlie, Alice and Bob public randomly some corresponding bits $x_m^{'}$, $y_m^{'}$ and $z_m^{'}$ to test whether the three bits satisfy the condition of $z_m^{'} =: x_m^{'} \oplus y_m^{'}$ and to compute the quantum bit error rate (QBER). If the QBER is in the acceptable range, then continue. Otherwise, terminate the protocol.

Step 8. After casting off the test qubits which have been disclosed before, they execute an improved Cascade error correction as in Ref. [21-23] to correct the errors in the remaining bit sequence. Firstly, they define the number (n times) of iterations to run in accordance with the QBER; divide the sequence into blocks with one bit error on average and same numbers of bits in each block; compute the parity of each block. Secondly, Alice and Bob declare their parities to Charlie, who will check whether the parity of his own block equals to the XOR of other two parties'. If so, turn to next block; otherwise, the block contains odd bits errors. Then they proceed with a binary search to find the error bit. In the next round of iteration, they divide their bit sequence, which have been randomly permuted, into blocks with double sizes.

Step 9. Privacy amplification is then applied to remove the information leakage induced. Then Alice, Bob and Charlie will get final keys X', Y' and Z'.

Step 10. Charlie encrypts his message with his own key; Alice and Bob can decipher the message only when they combine their partial keys.

3. Correctness Analysis

If the RRDPS-QSS is correct, the result key of Charlie is the exclusive-OR (XOR) of Alice's and Bob's partial keys. In the proposed scheme, Alice and Bob both impose their information on the differential phase shift of superposed pair of pulses which are determined randomly by Charlie. When Charlie detects the superposed photon by one detector, his measurement result is decided by the total differential phase shifts. Specifically, if the values of $x_m^{'}$ and $y_m^{'}$ are $\{0,0\}$ or $\{1,1\}$, meaning that the differential phase shifts imposed by Alice and Bob are $\{0,0\}$ or $\{\pi,\pi\}$. So the total phase difference is 0 and we assume DET-1 clicks; if the values of $x_m^{'}$ and $y_m^{'}$ are $\{0,1\}$ or $\{1,0\}$, meaning that the differential phase shifts imposed by Alice and Bob are $\{0,\pi\}$ or $\{\pi,0\}$. So the total phase difference is π and we assume DET-2 clicks. Thus, the measurement result of Charlie will be the XOR of the two parties' bits.

4. Advantages and Security Analysis

The improved RRDPS-QSS could be considered as the combination between the DPS-QSS and the novel RRDPS-QKD. Ref.21 has detailed the merit of security analysis on the betrayal of Alice and Bob and the upper bound of error rate induced by their eavesdropping. This part will focus on the additional advantages and simple external eavesdropping analysis basing on the improvement of the variable delay from the RRDPS-QKD [11], in addition to the ones that have been brought basing on the DPS mechanisms. Included are the security, simplicity and practicality of setup because of weak coherent pulse train replacing the single photon source; fiber transmission due to that information is carried by the differential phase shift other than the phase or polarization which are easily influenced by the environment; high key creation rate by reason of all detected photons contributing to key bits and so on [6-7,12, 21].

4.1. *Privacy amplification*

Privacy amplification is usually used to generate the final secure key bits both in classical and quantum schemes[11-12]. In the proposed scheme, information leakage may be caused by three objects' eavesdropping, Eve, malicious Alice or Bob, the latter two cases of which have been discussed in Ref.21. Here, we focus on the bit errors by Eve who invades in QSS after Bob's operation. The full security proof has derived the asymptotic key length formula in the case of classical light[11], in which the total number v of photons in the L pulses satisfies $\Pr(v > v_{th}) \le e_{src}$; e_{src} is a constant value; v_{th} is an integer.

$$\frac{G}{N} = 1 - h(e_{bit}) - \left[\frac{N_{em}e_{src}}{N} + \left(1 - \frac{N_{em}e_{src}}{N}\right)h\left(\frac{v_{th}}{L-1}\right)\right] \tag{A.1}$$

$$H_{PA} = \frac{N_{em}e_{src}}{N} + \left(1 - \frac{N_{em}e_{src}}{N}\right)h\left(\frac{v_{th}}{L-1}\right) \tag{A.2}$$

H_{PA} contains two parts. The first part can be eliminated by privacy amplification; while the second term applies to $H_{PA} = h(v_{th}/(L-1))$ when $v \le v_{th}$. So H_{PA} depends on the length N of sifted key which has been decided once the scheme finishes without the need of statistical estimation for signal disturbance. It also applies to the case of single photon source that bound the information about sifted key to $h(1/(L-1))$, thus improving the ability of high tolerance of bit errors. At the same time, the characterization of the source and randomness from variable delay become the leading factors in the analysis of finite-key analysis, without the precise statistical estimation of the disturbance. The two advantages really add spices to the existing DPS-QSS scheme.

4.2. *Security stemming from inherent randomness*

Utilizing a weak coherent pulse train in the novel DPS scheme really affects the security of the shared keys [6-8,21], preventing external Eve stealing the full information from the sender. Also, the three parties in DPS-QSS can disclose Alice's or Bob's betrayal by adjusting the sequence of test-bit announcement in the checking stage. Here, we discuss how inherent randomness originated from the simple variable-delay modification will contribute to the security of the RRDPS-QSS scheme.

Let us consider two simple intercept-resend attacks by Eve [11,14]. If Eve uses the same setup with Charlie's, she can master the photon position of incoming train. It is, however, difficult to compel Charlie to declare the same indices $\{i,j\}$ on condition that the amount of delay of Charlie's variable delay is rT. Because the probability of picking up the r by choosing randomly from $L-1$ values is $O(L^{-1})$. If Eve defers his measurement of input signal until he gets the indices $\{i,j\}$ which are obtained from the fake state sent by her, she then can control randomness of variable delay. It is, however, difficult to control the position of the photon in the pulse train because of the uniform distribution over L pulses, thus, making the success probability not overpass $O(L^{-1})$. So, the inherent randomness of the photon position in the source and the amount of delay determines the indices $\{i,j\}$. Information causality also demonstrates the difficulty of guessing the stochastic value that is beyond the ability of Eve [25].

5. Conclusion

In this paper, we put forward a quantum secret sharing protocol based on round-robin differential phase shift scheme, which use variable-delay interferometer instead of one-bit-delay counterpart. Compared with previous one based on differential phase shift, this one restrain the information leakage by differential shift encoding method and random delay. We hope this add to security for QSS scheme in a high bit error rate and short communication time circumstances.

Acknowledgments

This work was sponsored by the National Natural Science Foundation of China (Project No.61101073) and the Graduate Innovation Funds for the National University of Defense Technology.

References

1. Bennett C H. Quantum cryptography: Public key distribution and coin tossing[C]//International Conference on Computer System and Signal Processing, IEEE, 1984. 1984: 175-179.
2. Ekert A K. Quantum cryptography based on Bell's theorem [J]. Physical review letters, 1991, 67(6): 661.
3. Bennett C H, Brassard G, Mermin N D. Quantum cryptography without Bell's theorem[J]. Physical Review Letters, 1992, 68(5): 557.
4. Bennett C H. Quantum cryptography using any two nonorthogonal states[J]. Physical Review Letters, 1992, 68(21): 3121.
5. Huttner B, Imoto N, Gisin N, et al. Quantum cryptography with coherent states[J]. Physical Review A, 1995, 51(3): 1863.
6. Inoue K, Waks E, Yamamoto Y. Differential phase-shift quantum key distribution[C]//Photonics Asia 2002. International Society for Optics and Photonics, 2002: 32-39.
7. Inoue K, Waks E, Yamamoto Y. Differential-phase-shift quantum key distribution using coherent light [J]. Physical Review A, 2003, 68(2): 022317.
8. Waks E, Takesue H, Yamamoto Y. Security of differential-phase-shift quantum key distribution against individual attacks [J]. Physical Review A, 2006, 73(1): 012344.
9. Gisin N, Ribordy G, Tittel W, et al. Quantum cryptography[J]. Reviews of modern physics, 2002, 74(1): 145.
10. Scarani V, Bechmann-Pasquinucci H, Cerf N J, et al. The security of practical quantum key distribution[J]. Reviews of modern physics, 2009, 81(3): 1301.
11. Sasaki T, Yamamoto Y, Koashi M. Practical quantum key distribution protocol without monitoring signal disturbance[J]. 2014.
12. Inoue K. Differential Phase-Shift Quantum Key Distribution Systems [J]. IEEE JOURNAL OF SELECTED TOPICS IN QUANTUM ELECTRONICS, 2015, 21(3): 6600207.
13. Wang S, Yin Z Q, Chen W, et al. Experimental demonstration of a quantum key distribution without signal disturbance monitoring [J]. Nature Photonics, 2015.
14. Koashi M. A protocol of quantum key distribution without relying on information-disturbance trade off[C]//SPIE Optics+ Optoelectronics. International Society for Optics and Photonics, 2015: 950502-950502-7.

15. Hillery M, Bužek V, Berthiaume A. Quantum secret sharing[J]. Physical Review A, 1999, 59(3): 1829.
16. Karlsson A, Koashi M, Imoto N. Quantum entanglement for secret sharing and secret splitting[J]. Physical Review A, 1999, 59(1): 162.
17. Xiao L, Long G L, Deng F G, et al. Efficient multiparty quantum-secret-sharing schemes[J]. Physical Review A, 2004, 69(5): 052307.
18. Zhang Z, Li Y, Man Z. Multiparty quantum secret sharing[J]. Physical Review A, 2005, 71(4): 044301.
19. Schmid C, Trojek P, Bourennane M, et al. Experimental single qubit quantum secret sharing[J]. Physical review letters, 2005, 95(23): 230505.
20. Takesue H, Inoue K. Quantum secret sharing based on modulated high-dimensional time-bin entanglement[J]. Physical Review A, 2006, 74(1): 012315.
21. Inoue K, Ohashi T, Kukita T, et al. Differential-phase-shift quantum secret sharing [J]. Optics express, 2008, 16(20): 15469-15476.
22. Liu X, Zhang B, Wang J, et al. Differential phase shift quantum private comparison [J]. Quantum information processing, 2014, 13(1): 71-84.
23. Brassard G, Salvail L. Secret-key reconciliation by public discussion[C]//advances in Cryptology—EUROCRYPT'93. Springer Berlin Heidelberg, 1993: 410-423.
24. Bennett C H, Brassard G, Crépeau C, et al. Generalized privacy amplification [J]. Information Theory, IEEE Transactions on, 1995, 41(6): 1915-1923.
25. Pawłowski M, Paterek T, Kaszlikowski D, et al. Information causality as a physical principle[J]. Nature, 2009, 461(7267): 1101-1104.

Based on solr distributed classification algorithm of system design for science and technology plan projects

Gui-Lan Cai*
Science and technology innovation monitoring research center,
Guangzhou, 510006, China
E-mail: 61362368@ qq.com

Kai Xu and Ping-Hua Chen.
GDUT, Guangdong university of technology,
Guangzhou, 510006, China
E-mail: 504087493@qq.com

In this paper, we designed a system of scientific and technological project based on classification. Our system uses Check Weight Algorithm to extract keywords and Solr distributed index database of Lucene engine as the core to provide full text indexing and search open source enterprise platforms. This design enables our system to effectively improve the query speed and reduce computational complexity of the traditional query. Based on semantic technology project similarity calculation, it can also hasten the speed of the traditional rechecking system and improve the accuracy of rechecking the effect.

Keywords: Solr; Lucene; Science and Ttechnology Plan; Similarity Calculation; Distributed Systems.

1. Introduction

There are many different kinds of science and technology plan projects and new science and technology plan projects each year, Guangdong province, nearly 20000 items, and continue to increase each year. Data provided by the Guangdong science and technology business management system, for example: eight years from 2007 to 2014, project number is about 100000, which brought challenges to check the heavy work of science and technology plan projects.

The science and technology plan projects in Guangdong province's heavy design to take the whole library check way, namely the new project, necessary and all items in the library than in time, and according to the comparison results determine the situation. The design in the early stages of the project quantity soon, played a good effect, but with the science and technology plan projects of

* Corresponding author.

Guangdong province increased year by year, the design of the limitations of the increasingly exposed. Mainly as follows:

1 Check the project. Guangdong science and technology plan projects check weight system is project scope changes frequently, causing the project often missed, of the need to check the effect of heavy more and not significant, difficult to give managers with scientific, accurate, and the correct reference as soon as possible.2 Part weight content without reference value. Guangdong province science and technology plan projects for rechecking during the course of the full text of the project feasibility study report, feasibility study report in the design of calendar year there are a large number of project background, the unit of basic conditions, personnel information, etc., many projects of similar background, units and personnel information, but the project research contents, key information is completely different. In practice, some repetitive project not project the research content, but the project background, units and personnel information, there is no reference value.3 Check the speed problem. Due to the new project with the library project into alignment, check, check the weight of growing amount of calculation, the system computation is more and more big, the rechecking in recent years, occupy the resources is becoming more and more heavy speed is slower and slower, most of the time can't keep up with the needs of the managers, not only that, due to take up too much system resources, the operation of the system function and other services also caused a certain impact.4 Check owes precision problem. Except for background, units and personnel information mentioned in 2 repeat no reference value, another problem is that owe rechecking precision problem. Mainly displays in the system for some obviously after repeated project than not repeat. Main reason is the content of the same word order have change, or while using different words, but to express the same meaning.

Aiming at these problems, the paper puts forward solutions as follows: first, according to the category of science and technology plan projects are classified, then on the basis of classification for rechecking the project. All relative rechecking during the course of library, this design can greatly increase the rechecking than speed, reduce the amount of calculation. At the same time, the use of advanced distributed technologies, to build a distributed index of the query library, through the solr distributed index of the library management, combined with effective similarity calculation scheme for rechecking.

2. The Flow Chart of the System

System to use lucene for science and technology plan documents and the information index, word segmentation. The index by solr library management, adopts the distributed processing scheme to manage files at the same time. Through machine learning, classifying file, save file after the classification, and then through the similarity algorithm of science and technology plan of similarity calculation. System of the specific process as shown in figure 1:

Below will respectively from classification system, keywords and index build system, similarity of system design scheme of three systems.

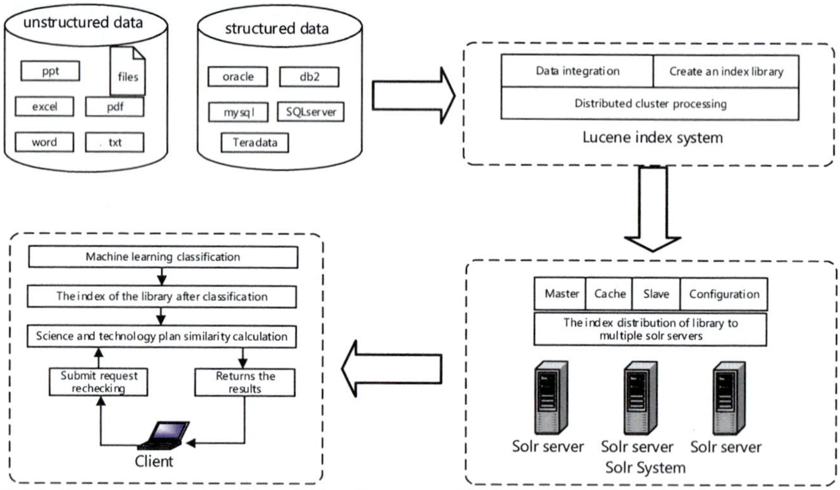

Fig. 1. Flow chart of rechecking during the course of science and technology plan management system in the project.

3. The Classification System Design

Document automatic classification is based on computer category known training samples by machine learning algorithms automatically trained as a classifier, classification of classifier according to stay content automatically determine the category of science and technology.Will be divided into categories of similar content of science and technology to realize automatic integration of similar documents.The way to overcome the manual classification, long cycle, high cost, low efficiency of various defects, effectively reduce the cost, improve the working efficiency.[1]Application of science and technology plan projects generally consists of the following sections: the title, abstract, key words, project introduction, main technical and economic indicators, feasibility study report, etc.Science and technology plan projects is a natural science class.Reference of guangdong province science and technology plan projects commonly used evaluation technology for artificial classification indexing classification number, a total of 35 kinds of technical data, each categories to select 100 data as sample, a total of 3500 data, choose 3000 as the training sample, the rest of the article 500 to test the samples.Using naive bayesian classification method of training data for training.Get classification parameters.Verify the accuracy, if meet the accuracy rates were complete sample training, save training characteristic matrix as a classifier.

Classification indexes as shown in table 1 (excerpts) :

Table 1. Guangdong science and technology plan projects review techniques commonly used classification (excerpts).

The parent encoding	Review technical
110	Computer and software technology
120	Network communication and multimedia technology
130	Electronic devices and components
140	LED technology
150	Advanced manufacturing
160	Satellite and application
170	The new material
180	New energy vehicles

4. Solr Index Management

Solr using Lucene search library as the core, to provide full text indexing and search open source enterprise platform, provide the REST of the HTTP/XML and JSON interface.In the form of Web Ser - vice can be integrated into any programming language is used, with a high degree of scalability, distributed search and index replication.[9]By Posting command is added to Solr document, Solr. The inside of the XML content is:The < field name = "id" > solr < / field >The < field name = "name" > Solr, the Enterprise Search Server < / field >The < fieldname = "manu" > SoftwareFoundation < / field >Solr support rich query syntax, Search field name in the keyword "Search" can use grammar name: Search.Import the data into Solr way also is varied, can use DIH (DataImportHandler) import data from a database, support the CSV, JSON format documents.Binary document such as Word, PDF.Also can call the solrj API to programmatically from define the import.We can also through the browser sends an Http request to operation.Delete the specified data, for example, in the browser to access this address to complete the operation.http://localhost:8080/solr/core1/update?Stream. The body = < delete > the < query > id: solr < / query > < / delete > & commit = trueReal-time text information processing including query, add, update, delete.Query will not modify the index data structure, the other three kinds of operation can create index data changes.Using the Solr shard concept of multiple sets of merging the result set.Solr service is provided by the HTTP service, index combined are also through HTTP requests to merge, this system is improved.Arbitrarily selected grouping a set of concentration, access via TCP/IP protocols, at this time using the Solr provides Embedded Solr Server gets the current set of indexes, and sends a TCP/IP request to the other groups to obtain the index data and indexes of the merge operation, in order to improve the indexing speed.[9].

5. Check Weight Algorithm

5.1. *The keyword extraction*

Application of science and technology plan projects with the characteristics of a lean project text exists some low frequency but important words.In order to reduce the high frequency words inhibitory effect of low-frequency words, USES the nonlinear function of word frequency normalization processing.Chinese text can identify text feature is the general content words, such as nouns, verbs, adjectives, etc.;The interjection, prepositions, conjunctions, such as basic function words in the text has no meaning, it can't be keywords, can bring a lot of to the extraction of keywords.Jia-bin liu et al. Study of a large number of keywords in the database, and after analysis found that the noun or noun phrase accounted for 85% of the total number of words.Importance, therefore, in order to distinguish between different parts of speech of words, the influence of by the part of speech function can give nouns higher ratings, verbs and adjectives.The part of speech weight roughly can be designed according to the following rules.

$$weight = \begin{cases} 1, & \text{noun} \\ 0.6, & \textit{verb or adjective} \\ 0.2, & \text{other} \end{cases} \tag{1}$$

[3]Reference design, word frequency and word length, the part of speech, word position weights were set to 0.3, 0.1, 0.1, 0.5.

5.2. *Science and technology plan similarity calculation*

Title TPSM paragraph statement model. The application of science and technology plan projects title similarity calculation formula is as follows:

$$Sim_t(t1,t2)=\sum_{i=1}^{p} w1_i * w2_{mi} * sim_{imi} + \sum_{j=1}^{q} w2_j * w1_{mj} * sim_{jmj} / 2 \tag{2}$$

The p, q are respectively the application A, B number of keywords in the title, sim_{imi} is the weight of keywords in the application form for A, for A maximum of similarity between keywords.Keyword weight is through the part of speech of words, word frequency and word length, the position of the word, such as statistics, so the semantic similarity calculation give full consideration to the surface layer of the text information and the semantic relations between words.

Application for comprehensive similarity calculation first obtain the degree of similarity between two application project name, the value of two key part of the application project name similarity value between the unknown words from two application project name part of the similarity between the values of and after dealing with the normalized.Formula is as follows: [3]

$$Sim_t(t1,t2)=\frac{sim(tkw1,tkw2)+sim(tuw1,tuw2)}{\sqrt{\sum_{i=1}^{m} w1_i^2} * \sqrt{\sum_{i=1}^{n} w2_i^2}} \tag{3}$$

The m and n, respectively the application of A and B to key words in the main research content and the sum of the number of the unknown words.The same calculation method, can get the degree of similarity between the two application and content similarity.Application for A, B of comprehensive similarity calculation formula is as follows:

$$sim(A,B)=\alpha sim_t(A,B)+\beta sim_c(A,B)+\gamma sim_p(A,B) \tag{4}$$

Among them a, beta, gamma weighting parameters, respectively, and a+ beta + gamma = 1.According to the measured repeatedly experience gives three parameters values are: a = 0.25, beta = 0.25, gamma = 0.5.Comprehensive similarity calculation method fully considers the plan of science and technology application of text semantic information such as words, phrases, surface, which is the title statement model TPSM paragraph.

6. Conclusion

This paper designed on the basis of classification system for science and technology plan projects, effectively improve the query speed, at the same time, combining with lucene engine solr library distributed index, the science and technology plan projects based on semantic similarity computation, effectively solve the slow rechecking during the course of the whole library system, such problems as the effect is not ideal.

Acknowledgments

This paper is supported by National Natural Science Foundation of China(61572144),Guangdong science and technology plan projects (2015 a030401101), ministry of education of Guangdong province production projects (2013 b090500127),project major science and technology project of Guangdong province "integrated management system of science and technology business process reengineering and functional" b080500008 (2012).

References

1. Chiu chun kit, xue-gang hu, a paragraph of word frequency statistics based plagiarism decision algorithm [J]. Computer technology and development, 2009 (4) : 231-233 + 238.
2. The new James ha liu, a Chinese text similarity algorithm based on semantic. Computer and modern [J], 2015 (04) : 6-9.
3. Jian-hai Lin, similarity computing in science and technology project management system of research and application [D], 2014, hangzhou university of electronic science and technology.
4. Luo Hao, science and technology project based on semantic similarity computing research [D], 2013, hangzhou university of electronic science and technology.
5. Hu bin, the research and implementation of science and technology project evaluation experts recommend system [D], 2011, hangzhou university of electronic science and technology.
6. Zhuo qiu, guo-ping tong, danger, a paper similarity rapid detection method based on Spark [J]. Journal of library intelligence, 2015 (11) : 134-142.
7. Warren bishop, hownet, ten thousand party, VIP paper similarity detection system comparative study [J]. Journal of university of book intelligence journal, 2015 (01) : 61-64.

8. Swim static, such as mode of cloud computing systems performance monitoring and evaluation method [J]. Journal of system simulation, 2013 (12) : 2850-2845.
9. Li Shengdong etc, Lucene full-text index efficiency improvement [J]. Journal of shandong university (science edition), 2015 (7) : 79-76.
10. As a tree, LUCENE search algorithm analysis and optimization research [J]. Journal of library journal, 2014 (12) : 17-23.
11. Tang Huajiao, such as distributed parallel based on Lucene index [J]. Computer technology and development, 2011 (02) : pp. 126-123.
12. Yi-fang chang, major national science and technology plan and special supervision and evaluation practice. Science and technology management research, 2015 (8) : 17-20 pages.
13. G.S Alton, aleem walji ong, c. S.Y ang. A vector space model for automatic indexing [J]. Journal of Communications of the ACM, 1975, 18 (11) : 613-620.
14. Wong SKM, Ziarko W, Patrick CN. Wong. The Generalized vector space model ininformation retrieval [C]. ACM SIGIR who lead, 1985, 18 to 25.
15. Sebt Barfrous a. a. A new word sense similarity measure in word - Net [C]. Proceedings of the International, Multi - conference on Computer Science and InformationTechnology, Washinton D C: IEEE Computer Society, 2008:369-373.
16. Monostori K, Zaslavsky A, Schmidt h. the Parallel and distributed document overlap detection on the Web [M]. Applied the Parallel Computing. The New Paradigms for HPC in Industry and Academia. London: Springer - Verlag, London, 2001:206-214.
17. Apache Hadoop. Hadoop [EB/OL]. [2016-01-17]. http://. Hadoop Apache, org.
18. Apache lucene. Lucene [EB/OL]. [2016-01-17]. http://lucene.apache.org.
19. Lin Chieh - Yen, Tsai Cheng Hao, Lee Ching - Pei, et al. The Large - scale logistic regression and linear support vector those using Spark [C]. 2014 IEEE International Conference on Big Data. Washington, DC: the IEEE, 2014:519-528.

An improved approximation formula for the symmetric capacity of 2-dimentional constellations under AWGN channel

Pei Yang*, Li-Qiang Jin and Hong-Wen Yang

Wireless Communication Center, Beijing University of Posts and Telecommunications, Beijing, 100876, P.R. China

*E-mail: *18910257625@163.com*

In this paper, an improved approximation formula is proposed to calculate the symmetric capacity of the 2-dimentional constellations under additive white Gaussian noise (AWGN) channel with the algorithm complexity reduced significantly. The parameters of the formula are evaluated by using the Levenberg-Marquardt (LM) and the Universal Global Optimization (UGO) algorithms. The simulation results show that the analytical solution calculated by the improved formula corresponds with the complex symmetry capacity evaluated by the Gauss-Hermit integration.

Keywords: Symmetric Capacity; Constellation; Channel Capacity.

1. Introduction

The Shannon capacity is defined as the maximum mutual information between the inputs and the outputs of the channel. The maximum for the memoryless additive white Gaussian noise (AWGN) channel is achieved only when the channel inputs are continuous random variables (RV) with Gaussian distribution. Despite the fact that models in [1, 2] are very simple to evaluate, the assumption that the channel inputs are Gaussian distributed is not the case for practical communication systems. For a realistic system, the channel inputs generally take its values from a finite alphabet (constellation) with equal probability, in such cases, the channel capacity refered to as the symmetric capacity. The symmetric capacity is the basic limit of the communication systems and the technology nowadays, such as OFDM and MIMO systems [3]. In order to achieve better channel capacity, many researches works had proposed several methods such as signal constellations in [4], channel polarization in [5] and the joint source-channel coding schemes in [6]. The evaluation of the capacity has drawn much attention in [7], and some tight bounds had been proposed by the

* Corresponding author.

work of [8-10]. The calculation of the symmetric capacity involves an integration that has no closed form solution. In order to calculate the symmetric capacity,the author in [11, 12] proposed a numerical integration by Monte Carlo simulation. An alternative method by applying numerical quadrature formula is presented by [13], where integration in N -dimensions is calculated by Gauss-Hermit Quadrature. A new method by using Groups and Kullback-Leibler distance is given in [14]. Although great works had been done to evaluate the approximate solution of symmetric capacity in [15, 16], but none of them are accurate enough to give precise solution for a real system. The importance of obtaining simple and accurate solution is not only to evaluate the capacity but also to provide solutions to some intractable problems, such as transmit power allocation for fading channels or in a MIMO system.

In this paper, we present a simple and accurate formula for symmetrical capacity problem. The formula is multi-purpose and generally applicable to a wide range of modulations such as QPSK, 8PSK, 16QAM, 64QAM and more accurate over all signal-to-noise-ratio (SNR) of the channel.

The rest of the work is organized as follows. In section 2, the symmetric capacity is demonstrated while in section 3, the general expression is introduced. The numerical results are shown in section 4. While in section 5 the paper is concluded.

2. Symmetric Capacity

Consider a memoryless AWGN channel with input $x \in \Omega$. The output is given by

$$Y = X + Z. \tag{1}$$

where $\Omega = \{a_1, a_2, \cdots, a_M\}$ is the 2-dimensitinal M-ary constellation, Z is the zero mean AWGN. Without loss of generality, we assume that $E\left[|Z|^2\right] = 1$ and $E\left[|X|^2\right] = \gamma$. We also assume that the elements of Ω occurs with equal probability, i.e. $\Pr\{X = a_i\} = 1/M$, $i = 1, 2, \cdots M$. The mutual information between the input and output, which is also referred to as symmetrical capacity, can be expressed as

$$C = H(X) - H(X|Y). \tag{2}$$

where $H(X) = \log M$ is the entropy of X, $H(Y|X)$ is the entropy of X conditioned on Y. Since the conditional distribution of X is given by

$$\Pr\{X=a_i \mid Y=y\}=\frac{\exp\left(\left|y-a_i\right|^2\right)}{\sum_{n=1}^{M}\exp\left(\left|y-a_n\right|^2\right)}. \tag{3}$$

By inserting Eq.(3) into Eq. (2) and perform some mathematical manipulation, we obtain the symmetircal capacity which can be written as

$$C=\log(M)-\frac{1}{M\pi}\sum_{m=1}^{M}\int_{-\infty}^{\infty}e^{-|y|^2}\log\left[\sum_{n=1}^{M}e^{-2\,\mathrm{Re}\left\{y(x_m-x_n)^*\right\}-|x_m-x_n|^2}\right]\mathrm{d}y. \tag{4}$$

It can be seen that Eq.(4) has no closed form expression and can be evaluated only with numerical techniques such as Monte Carlo integration or Gauss-Hermit quadrature [2].

3. Approximation Formula

In this section, we propose an approximation of Eq.(4) as folows:

$$C(\gamma)\approx\log(M)\left(1-\sum_{i=1}^{N}a_i e^{-b_i\gamma}\right). \tag{5}$$

where $N,\{a_i\},\{b_i\}$ are non-negative parameters determined by specific constelation design, and $\{a_i\}$ satisfies

$$\sum_{i=1}^{N}a_i=1. \tag{6}$$

since $C(0)=0$. Note that the approximation formula proposed in [16] is actually a special case of Eq.(5) for $N=1$.

$$C(\gamma)\approx\log(M)\left(1-e^{-b\gamma}\right). \tag{7}$$

The parameters $N,\{a_i\},\{b_i\}$ can be obtained with Levenberg-Marquardt(LM) and Universal Global Optimization(UGO) algorithms. The parameters for typical 2-dimenstional constellations practically in use are shown in Table 1.

Table 1. The parameters for typical 2-dimenstional constellations

Modulations	a	b	N	RMSE
QPSK	1	0.6507	1	3.4287E-3
8PSK	0.5595	0.0801	2	4.3834E-4
	0.4405	0.3776		
16QAM	0.3015	0.7242	2	3.2546E-3
	0.6985	0.1221		
64QAM	0.2496	0.1317	3	1.6163E-3
	0.2439	0.6270		
	0.5065	0.0297		

4. Numerical Results

The curves are shown below where the dotted lines with circles are the real capacity while the solid lines are the approximate capacity evaluated by Eq.(5)

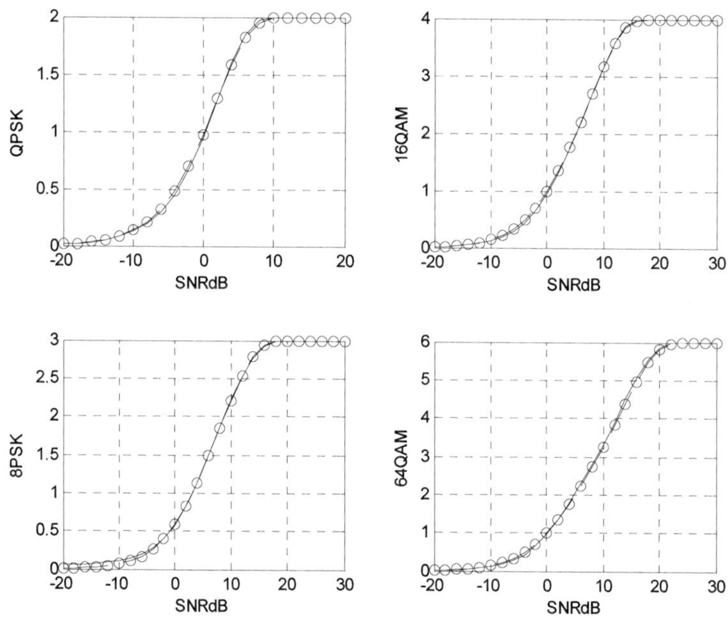

Fig. 1. Exact and approximation capacity for QPSK, 16QAM, 8PSK and 64QAM.

5. Conclusion

In this paper, a simple approximation formula is given to calculate the symmetric capacity under the AWGN channel. The improved formula is multi-purpose and generally applicable to a wide range of modulations such as QPSK, 8PSK, 16QAM and 64QAM. The formula is quite accurate for the calculation of the symmetric capacity under the AWGN channel with the algorithm complexity reduced significantly. Simulation results show that the analytical solution calculated by the improved formula agrees well with the complex symmetry capacity evaluated by the Gauss-Hermit integration over a wide range of SNR with a low RMSE being about 10^{-3}.

References

1. S. L. Fong and V. Y. F. Tan, "Asymptotic expansions for the AWGN channel with feedback under a peak power constraint," *Information Theory (ISIT), 2015 IEEE International Symposium on*, Hong Kong, 2015, pp. 311-315.
2. R. Qian, Y. Qi, P. Tao, W. Wang and J. Yang, "A Simple Approach of Gaussian Approximation for Channel Capacity in Multiple-Antenna Systems," *Vehicular Technology Conference (VTC Fall), 2011 IEEE*, San Francisco, CA, 2011, pp. 1-5.
3. J. Lee and N. Jindal, "Symmetric Capacity of MIMO Downlink Channels," *Information Theory, 2006 IEEE International Symposium on*, Seattle, WA, 2006, pp. 1031-1035.
4. T. Koch, A. Martinez and A. G. I Fàbregas, "The capacity loss of dense constellations," *Information Theory Proceedings (ISIT), 2012 IEEE International Symposium on*, Cambridge, MA, 2012, pp. 135-139.
5. E. Arikan, "Channel Polarization: A Method for Constructing Capacity-Achieving Codes for Symmetric Binary-Input Memoryless Channels," in *IEEE Transactions on Information Theory*, vol. 55, no. 7, pp. 3051-3073, July 2009.B.W. Bestbury, *R*-matrices and the magic square, *J. Phys. A***36**, 1947 (2003).
6. H. Kim, R. Annavajjala, P. C. Cosman and L. B. Milstein, "Source-channel rate optimization for progressive image transmission over block fading relay channels [Transactions Papers]," in *IEEE Transactions on Communications*, vol. 58, no. 6, pp. 1631-1642, June 2010.
7. A. A. Alsugair and R. S. Cheng, "Symmetric capacity and signal design for L-out-of-K symbol-synchronous CDMA Gaussian channels," in *IEEE Transactions on Information Theory*, vol. 41, no. 4, pp. 1072-1082, Jul 1995.
8. E. Baccarelli, "Asymptotically tight bounds on the capacity and outage probability for QAM transmissions over Rayleigh-faded data channels with

CSI," in *IEEE Transactions on Communications*, vol. 47, no. 9, pp. 1273-1277, Sep 1999.

9. E. Baccarelli and A. Fasano, "Some simple bounds on the symmetric capacity and outage probability for QAM wireless channels with Rice and Nakagami fadings," in *IEEE Journal on Selected Areas in Communications*, vol. 18, no. 3, pp. 361-368, March 2000.
10. E. Baccarelli, G. Di Blasio, S. Galli and A. Fasano, "Tight upper and lower bounds on the symmetric capacity and outage probability for bandwidth-efficient QAM transmissions over Rayleigh-faded channels," *Information Theory, 1998. Proceedings. 1998 IEEE International Symposium on*, Cambridge, MA, 1998, pp. 6-.
11. T. F. Wong, "Numerical calculation of symmetric capacity of Rayleigh fading channel with BPSK/QPSK," in *IEEE Communications Letters*, vol. 5, no. 8, pp. 328-330, Aug. 2001.
12. G. Ungerboeck, "Channel coding with multilevel/phase signals," in *IEEE Transactions on Information Theory*, vol. 28, no. 1, pp. 55-67, Jan 1982.
13. P. E. McIllree, "Calculation of channel capacity for M-ary digital modulation signal sets," *Networks, 1993. International Conference on Information Engineering '93. 'Communications and Networks for the Year 2000', Proceedings of IEEE Singapore International Conference on*, 1993, pp. 639-643 vol.2.
14. J. P. Arpasi, "Symmetric channels capacity computation by using groups and the Kullback-Leibler distance," *Computer Science and Electronic Engineering Conference (CEEC), 2014 6th*, Colchester, 2014, pp. 17-22.
15. S. B. Slimane, "Approximation to the symmetric capacity of Rayleigh fading channels with multi-level signals," in *IEEE Communications Letters*, vol. 10, no. 3, pp. 129-131, Mar 2006.
16. W. Li, H. Yang and D. Yang, "Simple Approximation Formula for the Symmetric Capacity," *Wireless and Mobile Communications, 2007. ICWMC '07. Third International Conference on*, Guadeloupe, 2007, pp. 30-30.

Two-hop multi-relay selection for secrecy improvement in broadcast wireless system

L. A. Olawoyin*, Yue Wu and Hong-Wen Yang

Wireless Communication Center, Beijing University of Posts and Telecommunication, Beijing, 100876, Beijing, China.

*E-mail: *lolawoyin@bupt.edu.cn, olawoyin.la@unilorin.edu.ng*

In this paper, we study the opportunistic relay selection for secrecy enhancement in dual-hop network which uses decode-and-forward protocol in the presence of a passive eavesdropper. To deal with this challenge, we consider the selection of relay that has the best instantaneous signal-to-noise ratio to the main receiver as the forwarding node, while another relay with the lowest instantaneous signal-to-noise ratio to the main receiver is considered as the jamming node. The proposed scheme is analyzed in terms of average throughput and secrecy outage probability. Our simulation results show that there is a significant improvement in both secrecy throughput and secrecy outage probability with the proposed scheme.

Keywords: Artificial Noise; Relay Selection; Secrecy Throughput; Secrecy Outage Probability.

1. Introduction

Cooperative communication has been adjudged to be a new paradigm in wireless communication [1]. Due to broadcast nature of wireless systems which make it susceptible to eavesdropping to malicious users, this poses major threat to information security. Traditionally, information secrecy is achieved with the use of cryptography method by exchanging the secret key between the transmitter and legitimate receiver [2]. In recent times, physical layer security (PLS) has become a new research area because it explores the fading characteristic of wireless systems to achieve information security instead of secret key exchange. Wyner in [3] introduced the concept of wiretap channel and this concept was further generalized to many other channels as in [4-7]. One important way to enhance the security is the use of artificial noise (AN) and beamforming [8, 9] where the noise is generated by transmitter which is transmitted in all direction but towards the main receiver channel. In the recent times, opportunistic relay

* Corresponding author.

has been found to be an effective way to increase information secrecy in wireless systems. The authors in [10] proposed amplify-and-forward (AF) cooperative relaying protocol while decode-and forward was discussed in [11] for secrecy enhancement in wireless networks. In this work, we explore the work of [12] where the author suggested the use of suboptimal selection with jamming (SSJ). We propose the selection of antenna with best and worst channel quality to the of main receiver as forwarding and jamming node, respectively. In addition, the power at each relay node is optimize to meet the minimum achieving code rate.

The remaining part of this work is organized as follows. Section 2 presented the system model while the proposed scheme is discussed in section 3. The numerical simulations to justify the work are presented in section 4 while the paper is concluded in section 5.

2. System Model

Consider the system model as shown in Fig. 1. Alice wishes to communicate a secret message to Bob via multiple relays $M_{\text{relay}} = \{1,2,...M\}$. In the presence of passive eavesdropper, Eve. Assuming there is no direct link between Alice to Bob, nor to Eve due to shadow fading effect [13]. The complex channel gain of the link between node $\mathrm{x} \in \{\mathrm{a},\mathrm{b},\mathrm{e}\}$ to the i-th relay node is denoted by $h_{\mathrm{x},i}$, $i = 1,2,...,M$ and the power gain is denoted by $g_{\mathrm{x},i} = |h_{\mathrm{x},i}|^2$. All nodes in Fig. 1 are equipped with single antenna and the channel is reciprocal. Assuming that $g_{\mathrm{x},i}$ for all $i = 1,2,...,M$, $\mathrm{x} \in \{\mathrm{a},\mathrm{b},\mathrm{e}\}$ are independently and identically distributed (i.i.d) and are normalized such that $\mathbb{E}\left[g_{\mathrm{x},i}\right] = 1$.

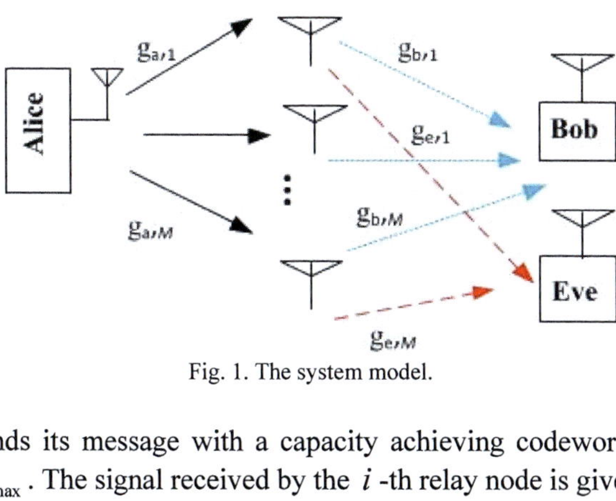

Fig. 1. The system model.

Alice sends its message with a capacity achieving codeword x of rate R with power $P_{\max}$. The signal received by the i-th relay node is given by

$$y_i = \sqrt{P_{\max}} h_{a,i} x + n_i. \tag{1}$$

where n_i is i.i.d. complex Gaussian variables with zero mean and variance σ^2. The i -th relay node can only decode the codeword unless its capacity is greater that the code rate i.e. $\log_2\left(1+\overline{\gamma} g_{a,i}\right) \geq R$, where $\overline{\gamma} = P_{\max}/\sigma^2$. This is also true for Bob and Eve when they are decoding the relayed signals.

3. Proposed Scheme and Performance Evaluation

In order to protect the codeword from being decoded by Eve, the relay node with best channel quality is selected among the relays as the forwarding node, also another relay with worst channel quality is chosen as the jamming node. The index of forwarding and jamming relay nodes are given by

$$f = \arg\max_{i \in \{1,2,\ldots,M\}} \left\{ g_{b,i} \right\}. \tag{2}$$

$$j = \arg\min_{i \in \{1,2,\ldots,M\}} \left\{ g_{b,i} \right\}. \tag{3}$$

The jamming node will broadcast AN w with power $P_{\max}$ while the forwarding node will forward the decoded codeword x with power $P\left(0 \leq P \leq P_{\max}\right)$. The signal received by Bob and Eve is given by

$$y_b = \sqrt{P} h_{b,f} x + \sqrt{P_{\max}} h_{b,j} w + n_b, \tag{4}$$

$$y_e = \sqrt{P} h_{e,f} x + \sqrt{P_{\max}} h_{e,j} w + n_e. \tag{5}$$

where n_b and n_e are respectively, the noise at Bob and Eve which are i.i.d. complex Gaussian variables with zero mean and variance σ^2. The transmit power P is

$$\frac{P}{P_{\max}} = \begin{cases} \dfrac{2^R - 1}{2^{C_b^{\max}} - 1}, & C_b^{\max} \geq R \\ 0, & C_b^{\max} < R \end{cases}. \tag{6}$$

where $C_b^{\max} = \log_2\left(1 + \dfrac{g_{b,f}}{1/\overline{\gamma} + g_{b,j}}\right)$.

The channel capacities of the link between Alice to forwarding node, forwarding node to Bob and forwarding node to Eve can be expressed as

$$C_f = \log_2\left(1 + \overline{\gamma}\, g_{a,f}\right). \tag{7}$$

$$C_{\rm b} = \log_2\left(1 + \frac{P}{P_{\max}} \cdot \frac{g_{{\rm b},f}}{1/\overline{\gamma} + g_{{\rm b},j}}\right) = R. \tag{8}$$

$$C_{\rm e} = \log_2\left(1 + \frac{P}{P_{\max}} \cdot \frac{g_{{\rm e},f}}{1/\overline{\gamma} + g_{{\rm e},j}}\right). \tag{9}$$

Eve will decode the codeword only if $C_e > C_b$ or

$$\frac{g_{{\rm e},f}}{1/\overline{\gamma} + g_{{\rm e},j}} \geq \frac{g_{{\rm b},f}}{1/\overline{\gamma} + g_{{\rm b},j}}. \tag{10}$$

Recall that $g_{{\rm a},i}, g_{{\rm b},i}$ for $i = 1, 2, \ldots, M$ are i.i.d. random variables, and $g_{{\rm b},f} = \max_i \{g_{{\rm b},i}\}$, $g_{{\rm e},f} = \min_i \{g_{{\rm b},i}\}]$, so we have

$$\mathbb{E}[g_{{\rm b},f}] \geq \mathbb{E}[g_{{\rm b},j}] = \mathbb{E}[g_{{\rm e},j}] \geq \mathbb{E}[g_{{\rm e},f}]. \tag{11}$$

In general, the left hand side of Eq. (10) would have less chance because it is larger than the right hand side. Hence the proposed scheme can guarantee a positive secure throughput for arbitrary distribution of $g_{{\rm x},i}, {\rm x} \in \{{\rm b}, {\rm e}\}$, $i = 1, 2, \ldots, M$.

The proposed scheme can work in adaptive rate mode or fixed rate mode. In adaptive mode, the code rate R is adapted to the channel capacity of the two-hop DF relay system between Alice and Bob, i.e.

$$R = \min\{C_f, C_{\rm b}^{\max}\}. \tag{12}$$

The average throughput seen by Bob and Eve can be expressed as

$$\begin{aligned} \overline{R}_{\rm b} &= \mathbb{E}[R], \\ \overline{R}_{\rm e} &= \mathbb{E}[R \cdot I_{C_{\rm e} \geq C_{\rm b}}]. \end{aligned} \tag{13}$$

where $I_{C_{\rm e} \geq C_{\rm b}} \in \{0,1\}$ is an indicator function $I_{C_{\rm e} \geq C_{\rm b}}$ is 1 if $C_e \geq C_b$ and is 0 otherwise. The secrecy throughput is defined as

$$\overline{R}_{\rm s} = \overline{R}_{\rm b} - \overline{R}_{\rm e} = \mathbb{E}\left[R \cdot \left(1 - I_{C_{\rm e} > C_{\rm b}}\right)\right]. \tag{14}$$

In the fixed rate mode, the code rate R is a predefined value independent of the channel condition. Hence outage may occur if the channel capacity is less than R The outage rate seen by Bob and Eve can be expressed as

$$P_{\rm b}^{\rm out} = 1 - \Pr\{C_f \geq R\} \Pr\{C_{\rm b}^{\max} \geq R\}. \tag{15}$$

$$P_e^{out} = 1 - \Pr\{C_f \geq R\}\Pr\{C_e \geq R\}. \quad (16)$$

4. Numerical Simulations

We consider the channel undergo Rayleigh fading where the power gains $g_{x,i}$, $x \in \{b,e\}$, $i = 1,2,\ldots,M$ are i.i.d. random variables with exponential distribution. The numerical results for the throughput defined in Eq. (13) and Eq. (14), are shown in Fig. 2 for $\bar{\gamma} = P_{max} / \sigma^2 = 15$dB. It can be seen that the throughput of Bob channel increases as the number of relay nodes increases while Eve performance decreases monotonically with the number of relay nodes increases. Fig. 3 shows the plot of outage probability defined in Eq. (15) and Eq. (16) as a function of $\bar{\gamma}$ for a given R=1bits/symbol and $M = 4$. We can see that increasing in SNR leads to decrease in the slope of outage curves.

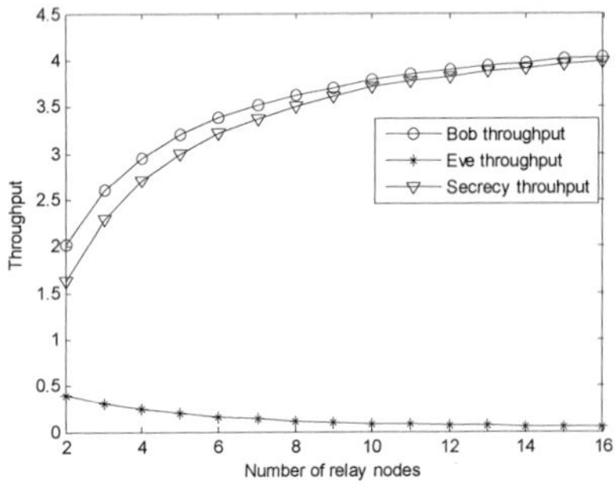

Fig. 2. The throughput.

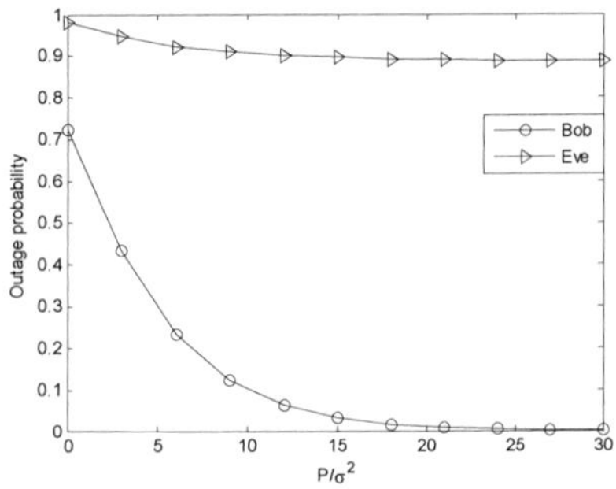

Fig. 3. The outage probability, R=1bit/symbol.

5. Conclusions

In this paper, we propose a scheme which use best relay node to forward user codeword and worst relay as jamming node. The simulation results show that the proposed scheme can improve the secrecy throughput and secrecy outage probability significantly.

References

1. A. Nosratinia, T.E. Hunter, and A. Hedayat, Cooperative communication in wireless networks. *IEEE Commun. Mag.,* vol. 42, 2004. pp. 74-80.
2. B. Mandal, B., S.Chandra, S.Alam and S.S Patra et al. A comparative and analytical study on symmetric key cryptography. *Proc. Int. Conf. in Electronics, Communication and Computational Engineering (ICECCE'14),* (Hosur, India, 2014), pp.131-136.

3. A.D Wyner, The wire-tap channel. *J. Bell System Technical Journal*, vol. 54, 1975, pp. 1355-1387.
4. M.K Hossan, S. Karmakar, and A. Ghosh. On the secrecy capacity of fading Gaussian wiretap channel. *14th IEEE Canadian Workshop in Information Theory (CWIT)*, 2015, pp 36-40.
5. E. Tekin, and A. Yener, The General Gaussian Multiple-Access and Two-Way Wiretap Channels: Achievable Rates and Cooperative Jamming. *IEEE Transactions on Information Theory*, 2008. vol. 54 pp. 2735-2751.
6. T. V. Nguyen, T. Q. S. Quek, Y. H. Kim and H. Shin. Secrecy diversity in MISOME wiretap channels. *IEEE Global Communications Conference (GLOBECOM'12)*, (Anaheim, CA), 2012, pp. 4840 - 4845.
7. G. Zheng, J. Li, K. K. Wong, A. P. Petropulu and B. Ottersten. Using simple relays to improve physical-layer security. *1st IEEE Int. Conf. Communications in China (ICCC)*, Beijing, 2012 pp. 329-333.
8. Goel, S. and R. Negi, Guaranteeing secrecy using artificial noise. *IEEE Transactions on Wireless Communications*, 2008, vol. 7, pp. 2180-2189.
9. N. Romero-Zurita, D. McLernon, and M. Ghogho, Physical layer security by robust masked beamforming and protected zone optimization. IET Communications, 2014. vol. 8, pp. 1248-1257.
10. D.J. Su, S.A. Mousavifar, and C. Leung. Secrecy capacity and wireless energy harvesting in amplify-and-forward relay networks. *IEEE Pacific Rim Conference Communications, Computers and Signal Processing (PACRIM)*, (Victoria, BC), 2015, pp. 258-262.
11. A.I Akin, H. Ilhan, and O. Ozdemir. Relay selection for decode-and-forward cooperative vehicular networks with virtual noise modelling. *13th Int. Conf. in ITS Telecommunications (ITST'13)*, (Tampere, Finland) 2013, pp. 272-277.
12. I. Krikidis, J.S. Thompson, and S. McLaughlin, Relay selection for secure cooperative networks with jamming. *IEEE Transactions Wireless Communications,* 2009. vol. 8, pp. 5003-5011.
13. V. Aggarwal, L. Sankar, A. R. Calderbank and H. V. Poor, Secrecy capacity of a class of orthogonal relay eavesdropper channels, *Information Theory and Applications Workshop, 2009*, San Diego, CA, 2009, pp. 295-300.

Chapter 3

Algorithm and Simulation

Apply classical physical laws to algorithms for analysis of stocks

Xing-Bo Wang[1]
Department of Mechatronics, Foshan University,
Foshan City, PRC,528000
E-mail: wxbmail@msn.com

This article proposes designing algorithms and setting up a mathematical model of stocks' running with classical physical laws. By applying laws of kinetics and dynamics to the motion of stocks, a new model is put forward to predict how a stock is going to rise and fall. Experiments and a large amount of practical facts showed that the model and the algorithms are more accurate and reliable than the present ones, which can be beneficial to investors. This article is valuable for the designing of program software to analyze stocks.

Keywords: Mathematical Model; Algorithm; Stocks; Acceleration.

1. Introduction

Stock buying and selling are important financial phenomena in modern society. Analysis of stock running has early been a special subject in both financial areas and computer technologies. People have to use software to see how stocks run and then decide to sell or buy a stock. To help people to know a stock's information, many theories and approaches have been developed ever since the being of the stock market. For example, Charles Dow developed Dow Theory to identify and signal the change in stock market trends [1], Ralph Nelson Elliott raised wave theory to analyses waves of a stock's run, Nicolas Darvas developed his Box Theory[3], and Chen Hao's Jetton Theory to see a stock's volume and its prices[4].

These theories and strategies have benefit thousands of investors. However they also puzzled thousands of investors because each of the theory deduces a lot of formulas that based on uncertain predictions. Practice of stock market tells investors that, a relatively accurate formula is urgently necessary.

Where does the relatively accurate formula come from? Classic theories cannot be a source because they all rely on statistical approaches and fuzzed

[1] Corresponding author.

judgments. Any engineer knows that a relatively accurate calculation can only come from an analytic approach. Therefore, the needs of a relatively accurate formula for analysis of stocks are a need of an analytic approach to model a stock. Just based on such logic thinking, I have tried an approach to model a stock analytically. This article introduces the outline of my approach.

2. Laws of Stocks' Motion

We set up new models of stock running with classical physical laws. Accordingly, this section first makes a brief overview of some of the classical physical laws including basic laws of motion, Newton's law of motion and the energy laws, and then set up several laws of stock's running.

2.1. *Overview of classical physical laws of motion*

Classical physical laws mainly include the following natural laws.

(1) Basic laws of motion

- Velocity-acceleration relationship: instant velocity v_t of an object that moves by a uniform acceleration a in straight line fits

$$v_t = v_0 + at \tag{1}$$

- Oblique motion: Displacement $d = (d_{hori}, d_{vert})$ of an object that is thrown with a initial velocity v_0 at time t under no resistance except for gravity fits

$$\begin{cases} d_{hori} = v_0^x t \\ d_{vert} = v_0^y t - \dfrac{1}{2} g t^2 \end{cases} \tag{2}$$

Where g is the acceleration of gravity.

A special case of the oblique motion is the vertical projectile motion, in the case of which the formula (2) is rewritten by

$$\begin{cases} d_{hori} = 0 \\ d_{vert} = v_0^y t - \dfrac{1}{2} g t^2 \end{cases} \tag{3}$$

Because the motion's horizontal displacement is zero.

(2) Newton's three laws

- Newton's first law of motion (inertial law of motion): an object either remains at rest or continues to move at a constant velocity, unless acted upon by an external force.
- Newton's second law: The total force F exerted on an object is equal to the mass m of the object multiplied by the acceleration vector a of the object: $F = ma$.
- Newton's third law: when one body exerts a force on a second body, the second body simultaneously exerts a force equal in magnitude and opposite in direction on the first body.

(3) Conservation law of mechanical energy: kinetic energy K of a moving object with m and v being its mass and velocity is $K = \frac{1}{2}mv^2$; potential energy P of the object with mass m and at a height h is $P = mgh$, where g is grail's acceleration. The kinetic energy K can transfer to potential energy P and if there is not any waste of energy the transformation fit $K = P$, namely, $\frac{1}{2}mv^2 = mgh$.

2.2. *Laws of stock running*

We regard a stock's running as a motion of an object. We first give the following definitions.

- Mass of a stock: Floating capital of the stock.
- Action force on a stock: $F = \Delta V$, where ΔV is increment of trading volume V.
- Velocity of a stock's motion: $v = \Delta p$, where Δp is increment of the price in a period of time.
- Acceleration of a stock's motion: $a = \lim_{\Delta t \to 0} \frac{\Delta v}{\Delta t} = \frac{dv}{dt}$, where d is the differential operator. Usually, acceleration of a stock's motion is not a constant one.
- Height of a stock's up or down: magnitude of the stock's waving from a lower price to a higher price, denoted by h or H with h and H being relative and absolute height from a referable price.
- Grail's acceleration: acceleration of composite index, denoted by g, which is a dynamic quantity determined by motion of the grail in a period of time.
- Kinetic and potential energy of a stock: $E = \frac{1}{2}mv^2, P = mgh$

Then we assume that a stock's motion fit the following laws:

(1) Acceleration law:

$$F = ma \tag{4}$$

(2) Velocity-acceleration law:

$$v_t = v_0 + at \tag{5}$$

(3) Inertial law: a stock continues to move at a constant velocity if it is acceleration a vanishes, namely, $a = 0$.

(4) Energy law:

$$\frac{1}{2}mv^2 = mgh \tag{6}$$

In addition, the following theorems are directly deduced from the above laws.

(5) 0-velocity theorem: a stock remains a constant price if and only if it motion velocity is zero.

We call it an initial state when a stock's motion is in zero velocity and zero acceleration.

2.3. *New model of stock's motion*

We use a model of digging ping-pong to describe a stock's motion as follows.

A stock's motion is a process of a walking person's digging a ping-pong ball from a lower altitude to an upper altitude. He first digs and raises the ball to an altitude and then when the ball fall to another altitude he digs a second time; he keeps such digs until the ball is beyond of his reach. Then the ball begins to fall and the guy grabs a step to dig the ball one more time. When the ball falls too lowly, he can no longer dig it; then the ball falls to the ground and declining bounces.

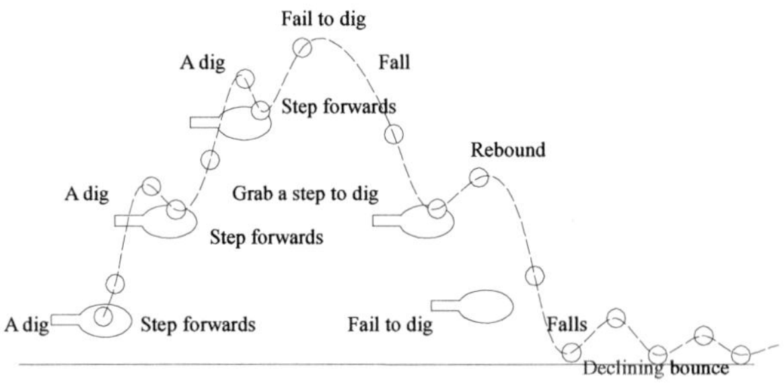

Fig. 1. Model of digging ping-pong.

The model regards each digging action as a vertical projectile motion during which the stock's price keeps rises until its velocity vanishes. When the rising speed turns to be zero, it then begins to fall until another digging catches and punches it upward to a next vertical projectile motion.

By the model, one only needs to know when a digging begins and when the velocity of the gaining stock vanishes if he/she wants to know when a stock begins to up-go and to slip, which is not a difficult computation by definitions of acceleration and velocity and the formulas (4),(5) and (6).

3. Explanation of Stocks' Volume-price Relationship

There is a volume-price octagon to show 8 relationships between volume and price of a stock, as shown in figure 2.

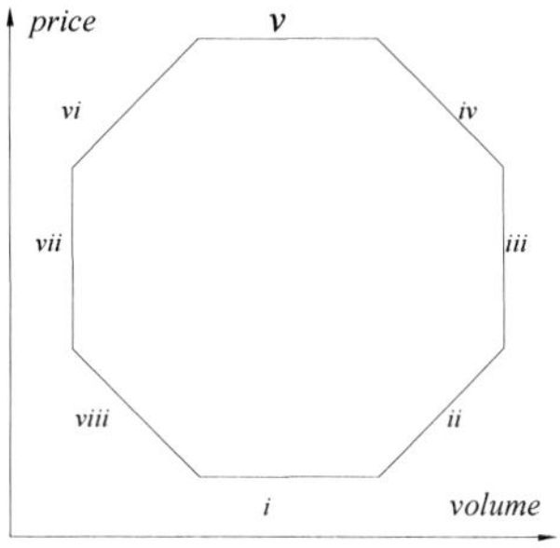

Fig. 2. Volume-price octagon.

In the figure 2, relationship *i* is to express an increment of volume with no increment of price; relationship *ii* is that an increment of volume with an increment of price, relationship *iii* is stable volume with increment of price, relationship *iv* is decrement of volume with increment of price, relationship *v* is decrement of volume with no increment of price, relationship *vi* is decrement of volume with decrement of price, relationship *vii* is stable volume with decrement of price, and relationship *viii* is increment of volume with decrement of price. These volume-price relationships are fundamental for investors to know a stock's running. Now we use our new model to explain these relationships.

(i) The state of the relationship *i* is usually a start period that a stock is gong to move from its initial state that is a 0-velocity state. In this state, the action force F is first not bigger enough to overcome resistance (for example, something like a static friction exerted on a physical object) exerted on it, Hence the stock keeps a 0-velocity state at the beginning. When the action force gradually increases to such a mount that it overcome all the resistance, the stock obtains acceleration and then to run.

(ii) The state of the relationship ii is that a stock begins to run at a speed v that is determined by $v = at$, where a is determined by $F = ma$.

(iii) The state of the relationship iii is to a 0-acceleration motion. Hence the stock keeps a motion of constant velocity by the inertial law.

(iv) The state of the relationship iv is to a motion of negative acceleration. Since the stock has an initial velocity v_0, its instant velocity v_t gradually decreases by $v_t = v_0 - |a|t$ until it turns to 0 and $a = 0$. Since $v_t \geq 0$, the price of the stock still increases.

(v) The state of the relationship v is to a motion of 0-acceleration and 0-velocity. Note that both the velocity v and the acceleration a become zero, the stock run into an initial state. Then it obtains acceleration and velocity with volume's decreasing and into the next stage.

(vi) The state of the relationship vi is to a motion following the state of the relationship v.

(vii) The state of the relationship vii is to a motion like that of the state of the relationship iii except that acceleration of this stage is opposite to the stage iii.

(viii) The state of the relationship $viii$ is to a motion like that of the state of the relationship ii except that velocity and acceleration of this stage are both opposite to the stage ii.

4. Experiment Examples

Based on the previous Medellin approaches, I designed several algorithms to analyze stocks. The most important one is that to a curve of 0-velocity and 0-acceleration that indicates when a stock will fall. The figure 3 and figure 4 are screen captures from software I developed.

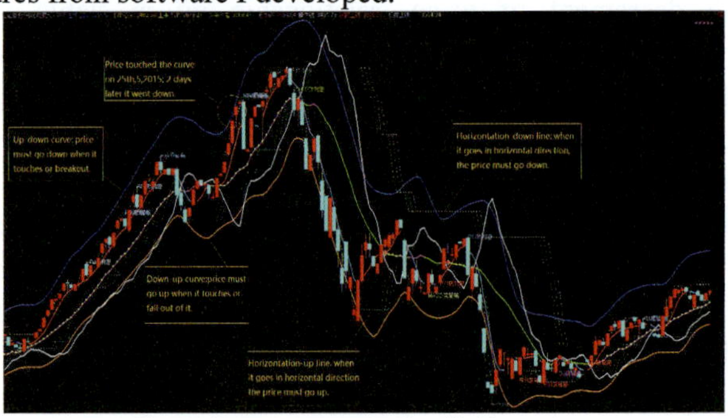

Fig. 3. A screen capture of grail A's motion.

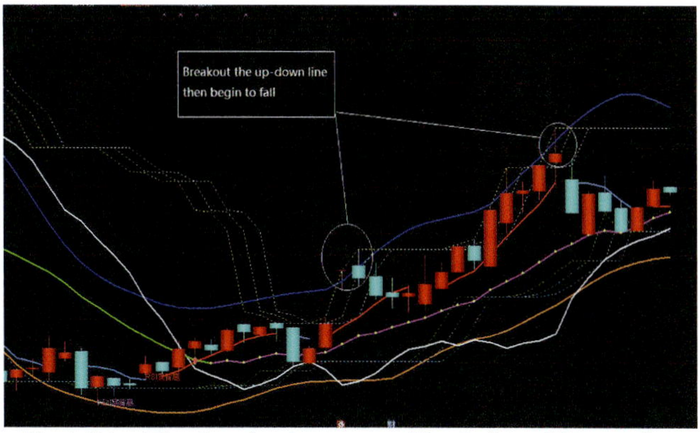

Fig. 4. Price falls after a breakout curve of 0-velocity and 0-acceration.

In the figure 3, the most upper curve is the curve of 0-velocity and 0-acceration. Whenever a stock's price breakouts the curve it must fall in two or three days, as figure 4 shows. For example, on the May 25th of 2015, A-share market A ran out of the curve, I declared at the time that it will fall and three days later, it fell a big fall. The following figure 5 is screen capture of my declaration.

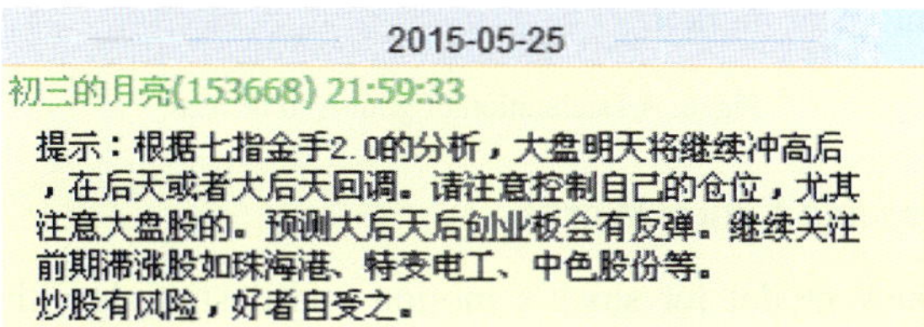

[Translation of the above:]

Professor Wang(153668) 21:59:33

According to the analysis of seven golden fingers 2.0(in Chinese), A-share market A will keep rising tomorrow, but turn to fall the day after tomorrow. So please control your stock position, especially large-cap share. I predict that GEM will rise the day after tomorrow, so please pay close attention to Pre-stagflation stocks such as ZPH, TBEA, NFC and so on.

Fig. 5. Predication of a fall of grail A.

In addition to the curve of 0-velocity and 0-acceration, I also designed many other algorithms to analysis a bear-market or a bull-market. For example, on the Jun 25th of 2015, I warned "A-share market A is turning to be a bear", as shown in figure 5. On the September 25th of 2015, I declared grail A is turning a good market, as seen in http://rc.qzone.qq.com/main?t=0.1261154969688505. Readers can refer my algorithms in my book [5].

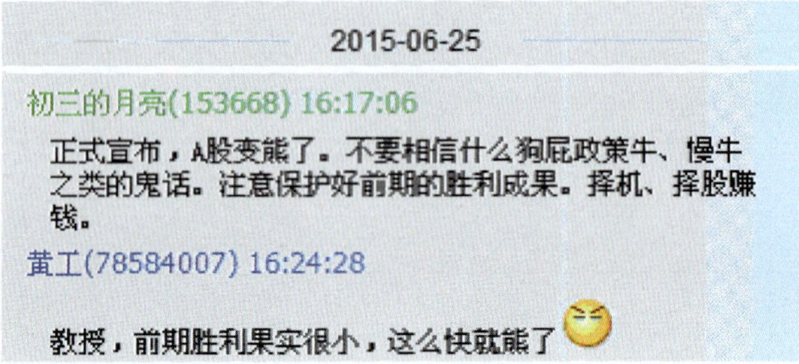

[Translation of the above:]

Professor Wang(153668) 16:17:06

A-share market A is turning to be a bear. Don't believe those nonsense about bull polices and carefully protect your pre-gain.

Engineer Huang(78584007) 16:24:28

I got little achievements in the early stage, but in no reaction to come over, it had turned into bear market.

Fig. 6. A Declaration of grail A to be bear.

5. Conclusions and Future Works

We set up a new model for stock's motion and design algorithms to analyze stock's motion based on classical physical laws. Experiments and practice show that analysis of stock's motion is more accurate and reliable than the classical ones. However in this article, I only present a outline of my work due to limitation of space of the paper. I will introduce more in future occasions.

Acknowledgments

The research work is supported by the National Ministry of Science and Technology under project 2013GA780052, Department of Guangdong Science and Technology under projects 2015A030401105 and 2015A010104011, Foshan Bureau of Science and Technology under projects 2013AG10007, Special Innovative Projects 2014KTSCX156, 2014SFKC30 and 2014QTLXXM 42 from Guangdong Education Department.

References

1. Robert Rhea. The Dow Theory, Snowball publishing, 2013
2. Constance Brown. Mastering Elliott Wave Principle: Elementary Concepts, Wave Patterns, and Practice Exercises,Bloomberg Press,2012
3. Harry Jay.Investment Phrases: Investing Terminology You Need to Know,Createspace,2014
4. Zhang Jian. Analysis and Implementation of Stocks' Distribution Theory. Nankai University, 2005
5. WANG Xingbo. PK A-shares with Seven golden fingers(in Chinese). Guangdong People's press, 2014

An improved recommendation algorithm based on user dynamic interests

Chao Li[1,2], Jun-Min Ye*[1] and Qiang Wang[1], Liu-Yin Huang[1] and Shu Chen[1]

[1]*College of Computer Technology and Science, Central China Normal University,*
[2]*Informatization Office , Central China Normal University*
Wuhan, HuBei, China
*E-mail: * jmye@mail.ccnu.edu.cn*
www.ccnu.edu.cn

Traditional recommendation algorithms not only cannot provide satisfactory real-time recommendation service for Web consumers, but also ignore the shift of user dynamic interests. Thus, this paper introduces the behavior characteristics of user dynamic interests, from which we established the dynamic interest preference extraction model of Web users. The corresponding recommendation algorithm considers the dynamic user interest. Our experimental results show that this method can significantly improve the recommendation efficiency, accuracy and satisfaction of Web consumers.

Keywords: Dynamic Interests; Interest Degree; Interest Degree Model; Recommendation Algorithm.

1. Introduction

The research of recommendation algorithm has been an academic and business focus [1]. The traditional recommendation algorithms based on user interest cannot provide users with satisfying recommendation service for the lack of accuracy and quality[2]. In such background, we designed a recommendation algorithm based on the extraction of user interests by analyzing the behavioral characteristics that can reflect the preference of the customers and extracting those characteristics to build mathematical models. To some extent, the recommendation algorithm in this paper could make up for the accuracy of recommendation when customers' interest shifts dynamically in the traditional

* Corresponding author.

recommendation algorithm. Thus, it can offer satisfactory recommendation service to the users according to their current consumption preferences.

2. Related Researches

Recommendation algorithm is the core of the recommendation system. There have been many algorithms, such as Collaborative Filtering Recommendation algorithm[3], Content-based Recommendation algorithm[4] , Association Rule-based Recommendation algorithm[5] and so on. Considering the scarcity of user-rating data, Deng et al. [6] proposed the Collaborative Filtering Recommendation algorithm. Liang Gong et al. [7] proposed the scene engine. Liang Xiang [8] proposed the recommendation algorithm of Machine Learning and Data mining.

In recent years, recommendation algorithm based on user preference has become a new research hotspot. Yuxiang Li et al. [9] proposed a multi-strategy recommendation algorithm based on users' relationship mining. Guihe He [10] realized the construction of the nearest neighbor set and intelligent recommendation based on users' preference knowledge. Qian Wang, et al. [10] proposed a user preference-oriented collaborative filtering algorithm based on the evaluation of attribute value distribution of user preference. These algorithms enhanced the accuracy and efficiency of recommendation to a certain degree, but ignored the shifting of the users' interests over time, so new projects cannot be recommended in time.

3. An Improved Recommendation Algorithm Based on users' Dynamic Preference

The Mathematical model established on users' dynamic preference is the basis for calculating users' interests in merchandises. So, it is necessary to study users' dynamic interests and related extraction model.

3.1. *Analysis of users' dynamic interest and preference*

Suppose there are four major kinds of action that users may take towards a certain product on a website: Clicking, Storing, Adding to the shopping cart and Buying. According to related surveys, these behavior can reflect users' interest.

•User Preferences are reflected in those four actions towards the product;

•The user preference is dynamic and it is revealed in the users' interest degree. If the users' preference shifts, the degree of the users' interest decreases.

•Users will browse the products that attract them at a higher frequency.

Based on these assumptions, this paper proposes some new perspectives:

•Users are much more prone to the products they have once bought, especially those with high scores of evaluation.

•Users pay a lot more attention to the products that they have stored online than those at which they have just taken a glimpse.

•Users pay much more attention to the products that they have added to their online shopping cart than those which they just have stored online.

•Users' preference towards the products that they have taken actions will continuously decline as time goes by.

3.2. *Model extracting users' dynamic interest and preference*

Define the terms and related criteria:

Definition1. Clicking behavior.

Use $C_{i,j}$ to present the behavior that the user i clicks the product j. Use $C_{i,j,k}$ to present the behavior that it is the kth time that the user i clicks product j. Use t_k to present the time of user i's *k*th click on product j.

Definition2. Purchasing behavior:

Use $B_{i,j}$ to present the behavior that the user i purchases the product j. Use $B_{i,j,k}$ to present the behavior that it is the kth time that the user i purchases product j, And use t_k to record the time that this behavior happens.

Definition3. Storing behavior:

Use $F_{i,j}$ to present the collection behavior of the user i collects product j. Use t_k as the latest moment when the user i collected product j.

Definition4. Adding to Shopping cart behavior:

Use $S_{i,j}$ to present the behavior that the user i adds the product j to his shopping cart. $S_{i,j,k}$ represents that it is the kth time the user i puts the product j into his shopping cart, and use t_k to record the time that this behavior happens.

Definition5. Behavior weight:

W_c is the weight of user's click behavior; W_f is the weight of user's collection behavior; W_b is the weight of user's purchase behavior; W_s is the weight of user's Shopping cart behavior.

Definition6. Time factor:

Time factor T is the time weight which changes as time goes by.

After understanding the dynamic characteristic of users' interests and consumption preferences ,we found user interests follow the following criterion.

•When other factors remain the same, if user i's operations towards product j and product j' are also the same and if the total frequency k that the user has clicked product j is bigger than that (k') of product j', then this user is more interested in product j. Describe this criterion as in Eq(1). The relationship between user interests and storing,buying or putting into the cart is the similar.

$$k > k' \Rightarrow d_{i,j} > d_{i,j'} \tag{1}$$

•When other factors remain the same, if user i's operations towards product j and product j' and their happening time are also the same, and if the total frequency k that the user has clicked product j is bigger than that (k') of product j', then this user is more interested in product j. Describe this criterion as in Eq(2).The relationship between user interests and storing,buying or putting into the cart is the similar.

$$k > k' \Rightarrow d_{i,j} > d_{i,j'} \tag{2}$$

•When other factors remain the same, the weights of a user's operation type differ from each other. The relationship among them is describe as in Eq(3).

$$W_s W_b > W_f > W_c \tag{3}$$

•As the time goes, the time factor T increases with other factors unchanged and time factor changed at different time. Describe this criterion as in Eq(4).

$$t > t' \Rightarrow T > T' \tag{4}$$

To meet the four criteria above, we need define the following parameters:

Define t_0 as the initial time of the given data; define α as the interest shifting factor; defineα_0 as the initial time factor. Describe the criterion as Eq(5).

$$T = \alpha_0 + (t - t_0) * \alpha \tag{5}$$

Define $C_{i,j,k}$ as the behavior that the user i clicks the product j when the user i has clicked the product j k times.

$$C_{i,j,k} = W_c * T_k \tag{6}$$

Define $F_{i,j,l}$ as the behavior that the user i collects the product j at first time.

$$F_{i,j,l} = W_f * T_l \tag{7}$$

Define $B_{i,j,m}$ as the behavior that the user i purchases the product j when the user i has purchased k products.

$$B_{i,j,m} = W_b * T_m \tag{8}$$

Define $S_{i,j,n}$ as the behavior that the user i puts the product j into his shopping cart when the user i has put k products into his or her shopping cart.

$$S_{i,j,n} = W_s * T_n \tag{9}$$

According to formulas and parameters above, we put forward a method to calculate the interest degree of the user i on the product j.

$$d_{i,j} = \sum_{k=1} C_{i,j,k} + \sum_{m=1} B_{i,j,m} + \sum_{l=1} F_{i,j,l} + \sum_{n=1} S_{i,j,n} \tag{10}$$

3.3. *Users' interest degree computing*

Because of extremely sparse data, the neighbor users being found are not accurate and the recommendation accuracy is lower. This paper use Pearson similarity coefficient to evaluate the interest level in the product that users have not browsed before, which increase the number of products which might be mutual interest to users.

Use P_i to represent the product set which user i have never contacted to in the products space set $I_{i,j}$. Use Q_i to represent the set containing the products that have been browsed by the use i, apparently, $P_i = I_{i,j} - Q_i$. The methods used to predict the interest level of user i in the product j are as follows:

a) Search the interest level value of all users who have browsed product m and n in the Products set Q_i, and then use the traditional similarity measurement (Cosine similarity) to calculate the similarity between m and n.

b) Choose top k products of similarity and form a similarity set $M_k = \{Q_1, Q_2, \ldots\ldots, Q_k\}$, where $j \in M_j$, and make $sim(j,Q_1) > sim(j,Q_2), \ldots\ldots > sim(j,Q_k)$

The computational formula this paper uses to predict user i's interest in product j is:

$$P_{i,j} = \frac{\sum k \in M_k \times \mathrm{sim}(j,k) \times D_k}{\sum k \in M_k \times \mathrm{sim}(j,k)} \tag{11}$$

Above, The formula to calculate user i's interest in any product j can be got:

$$D_{i,j} = \begin{cases} d_{i,j} & \text{The user i has browsed the product j} \\ P_{i,j} & \text{The user i hasn't browsed the product j} \end{cases} \tag{12}$$

3.4. ***Creating recommendation set***

In order to establish the model extracting dynamic characteristics of users, a recommendation algorithm based on user's dynamic interest characteristic extraction should be designed. The core idea of this algorithm is to establish users' dynamic interest characteristics extraction model according to consumer's purchase preferences, thus to calculate users' interest score in products. Then using neighbor users' interest score in products to predict target user's interest score in products have not been browsed, so as to gain the interest score in all products. This recommendation algorithm will recommend N products whose interest score is higher than the given threshold to target user.

According the given threshold Dv and computed Di, j, use Di, j which are selected no less than Dv to build recommendation set D. Based on the set D, corresponding recommendation is made.

4. Experiment Simulation

We use experiments to verify the validity of the recommendation algorithm proposed in this paper. And the results have proved that it does have relatively higher accuracy and efficiency even if when users' interest shifts dynamically.

4.1. *Selection of data set and evaluation criteria*

In this paper, we used the Tmall's real behavior data set on the big data research platform of Alibaba-"TianChi" to do the experiment. This data set covered users' behavior logs in the 4 months (from August 15th to April 15th) on Tmall, which includes 884 users, 9531 brands, and 182880 user's behavior records.

We use the optimized data as the training set, and use the Tmall's real behavior data set on "TianChi" in next month as the validation set. Both of the training set and validation set are this experiment's data set, and the evaluating parameters are calculated by both of them. Define the evaluating parameters:

i. Accuracy rate:

$$\Pr ecision = \frac{\sum_{i}^{N} hitBrands_i}{\sum_{i}^{N} pBrands_i} \tag{19}$$

N is the number of predicted users; $pBrands_i$ is the number of predicted brands which user i would buy; $hitBrands_i$ is the intersection of the predicted brands which the user i would buy and the brands user i really bought.

ii. Recall rate:

$$\operatorname{Re} call = \frac{\sum_{i}^{M} hitBrands_i}{\sum_{i}^{M} bBrands_i} \tag{20}$$

M is the number of users who traded successfully; $bBrands_i$ is the number of brands which user i really bought; $hitBrands_i$ is the intersection of the predicted brands which the user i would buy and the brands user i really bought.

iii. F1:

$$F_1 = \frac{2 * P * R}{P + R} \tag{21}$$

4.2. *Experiment results and analysis*

In this paper, we used data sets from section 4.1 and used the improved recommendation algorithm based on users' dynamic interests, designing three groups of tests through adjusting weight parameters.

Test1: Set W_c as 1; Set W_b as 3; Set W_f as 2; Set W_s as 4.

Test2: Set W_c as 1;Set W_b as 3.5;Set W_f as 2.5;Set W_s as 3.5.

Test3: On the basis of test1, W_s increases gradually as time goes.

Table 1. Result of three experiments.

	F1	Accuracy rate	Recall rate
Test1	6.38%	8.76%	5.02%
Test2	6.43%	8.60%	5.13%
Test3	6.39%	8.78%	5.02%

Compared to test1, test2 changed the weight setting. As a result, the accuracy declined a little while the Recall rate and F1 improved. Compared to test1, test3 changed the weight dynamically. As a result, the accuracy and F1improved while the Recall rate remained. The result of F1 in test2 is the highest. This result reflects that the method2's comprehensive effect is the best.

5. Conclusion

Based on the behavioral characteristics of user dynamic interests, this paper considered dynamic user interests and discussed some other characteristics that might affect online users' consumption preferences, such as browsing time and frequency, on the basis of which we established the users' dynamic interest score model. At last, neighbor users' interest score in a certain product which has not been browsed by target user was used to predict his/her interest score and provided users with satisfying recommendation service. The method avoided the low recommendation precision resulted from ignoring user interests and preference change in traditional recommendation algorithm. At the same time, it solved the problems that the nearest neighbor being found may not be accurate enough and the new items cannot be recommended in time.

Acknowledgments

This work is supported by National Key Technology Research and Development Program of the Ministry of Science and Technology of China (Approval number: 2015BAK33B00), the Research Projects of the Social Science and Humanity Fund of the Ministry of Education (Approval number: 15YJA880095) and financially supported by self-determined research funds of CCNU from the colleges' basic research and operation of MOE (Approval number: CCNU15GF003).

References

1. YEONG C,YOON C,SOUNG K. Mining changes in customer buying behaviour for collaborative recommendations [J] . **Expert Systems with Applications**,2004,28(2) : 359-369.

2. HE Gui-he. Research on E-commerce collaborative filtering recommendation algorithm based on user preference mining [J].**Information Science**, 2013, 31(12): 38-42 .

3. PANAGIOTIS S, ALEXANDROS N, APSTOLOSN, et al. Collaborative recommender system: combing effectiveness and efficiency[J].**Expert Systems with Applications**,2007,34(4) : 2995-3013.

4. WENG Sung-shun, LIN Bin-shan, CHEN Wen-tien. Using contextual information and multidimensional approach for recommendation[J]. **Expert Systems with Applications**,2009,36(2) : 1268-1279.

5. LEUNG C W, CHAN S C, CHUNG F, et al. An empirical study of a cross-level association rule mining approach to cold-start recommendations[J]. **Knowledge-based Systems**,2008,21(7) : 515-529.

6. DENG Ai-Lin, ZHU Yang-Yong, SHI Bai-Le. A collaborative filtering recommendation algorithm based on item rating prediction[J].**Journal of Software**, 2003, 14(9) : 1621-1628.

7. GONG Liang, LIU Yi-Jing, GUO Zhi-Jin. Scene engine: The core module of the next generation recommendation system [J].**China Computer Federation**, 2013, 9(3): 14-17.

8. XIANG Liang. The application of machine learning and data mining in personalized recommendation system [J]. **China Computer Federation**, 2013, 9(3) : 8-13.

9. LI Yu-xiang, ZHOU Jie, Xu Bin, LI Bi-Cheng. Multi-Strategy recommendation algorithm based on user relation mining[J].**Journal of Information Engineering University**, 2013, 14(4): 492-498.

10. WANG Qian, YANG Li-yun, YANG De-li. Collaborative filtering algorithm based on rating distribution of attributes faced user preference[J].**Journal of Systems Engineering**, 2010, 25(4): 561-568.

Indoor localization algorithm based on multiple logistic regression analysis and quad-tree with irregular boundary

Tian-Yi Feng[a], Shu-Yi Liu[b], Jing-Kang Yang[*,c], Chen Lu[d], Zhao-Jing Zhang[e] and Dong Liang[f]

Beijing University of Posts and Telecommunications, Beijing, China

E-mail: [a]fengtianyi@bupt.edu.cn, [c]j.yang@se13.qmul.ac.uk, [d]2013213110@bupt.edu.cn, [b]liushuyi@bupt.edu.cn, [e]buptzjj@163.com, [f]liangdong@bupt.edu.cn

When exploiting indoor localization, it is known that there are two most important factors: how accurate and how fast users can obtain the result. Quad-tree algorithm is one of the most creative and novel method that aims to improve the localizing speed and accuracy and is deemed to be more superior to traditional n*n indoor area gridding method. It divides the region into 2*2 grids, and determines one target to be divided into 2*2 grids again and again. With a level to level advancement method, the accuracy will significantly improve. However, the regularly and evenly segmented geometry grids for the Quad-tree algorithm cause particular problems. Experiments show that a boundary error with a low probability of occurring but one that inflicts a lot of damage cannot be ignored. In this paper, we managed to come up with a solution, which is to use the irregular segmenting method based on Multi Logistic Regression to address this error. Also, we further our research to explore how we can determine the threshold of gridding based on the tradeoffs of accuracy and precision, so as to maintain the highest accuracy. Our experimental results have shown that our revised algorithm reveals at least 52% more results that leads to errors within 1 meter, causing the percentages to be up to 80%.

Keywords: Rss Indoor Localization; Quad-Tree Algorithm; Multiple Logistic Regression; Irregular Segmenting; Uneven Segmenting.

1. Introduction

With localization technology widely used in many aspects of military and civil domain, more attention is paid to the mobile localization technology. Nowadays, we gain a lot of benefits from outdoor localization technique of Global Positioning Systems (GPS). Unfortunately, GPS is not available for indoor localization because of the indoor obstacles to satellite signals[1]. However, people spend more time indoors rather than outdoors. And more and more people concentrate on location-based service (LBS)[2], the network operators in the world are ready to apply the potential LBS as the enhanced service which provides user more personal service to satisfy user's need.

---

* Corresponding author.

With this increasing needs of indoor localization, many improved techniques are proposed such as radio frequency identification (RFID)[3], ultra wideband[4], Wi-Fi, Bluetooth[5], and a combination of these[6]. WLAN RSS-based indoor localization method is widely accepted and the traditional WLAN RSS indoor localization method based on the propagation path loss model was proposed to estimate the location. The RSS is affected by the indoor environment factors, such as indoor walls, moving interference which leads to deviating from the path loss model. With overcoming these factors, fingerprint approach arose and have been extensively used in indoor localization, which consists of two phases: offline phase for establishing the fingerprint map, and online phase for estimating the locations. [7]

The above localization methods have the drawbacks of much large memory requirements for the offline fingerprint, especially in the large area. [8] In this paper, we propose a novel Indoor Localization Algorithm Based on Multivariate Logistic Regression Analysis and Quad-Tree with Irregular Boundary. We can change indoor localization into a multi-classification problem.[9] The Quad-Tree can reduce the online complexity while improving the localization accuracy. Combining the Multiple Logistic Regression to produce the irregular boundary can solve the high error rate area's problem by correct the wrong classification of boundary points.

The following are the organization of this paper. Section 1 introduces the traditional indoor localizations and backgrounds. In Section 2, we take an overview of the traditional fingerprinting localization based on RSS and establish the mathematical model of the indoor positioning. In Section 3, we describe the details of the novel proposed indoor localization algorithm and the introduction of multiple logistic regression and decision tree. In Section 4, the simulation results are discussed. Section 5 is the conclusion of this paper.

2. Problem Statement

2.1. *Traditional fingerprinting localization algorithm*

The Traditional Fingerprinting localization based WLAN always divides the area into several uniform squares and uses a terminal to collect received signals strength (RSS) of all access points (APs) they can received in this area. The localization process consists of two-stage phases: the Online Phase and the Offline Phase. The location can be estimated by comparing and matching online information to an offline observation according to max likelihood.

On offline phase, divide the whole area into $I * J$ squares. And the location of reference points(RP) of (i, j)-th can be expressed as:

$$A_{i,j}^{(m)}(\tau) = \gamma_0(\tau = 1,2,\cdots,t,t > 0,m = 1,2,\cdots m) \quad (1)$$

Here, i and j represents the location, γ represents to the RSS in the grid, t represents the total number of sample, m represents the total number of AP.

Therefore, the radio map for the average value of each RP $A_W(\tau)$ can be computed and stored in a database, and it can be denoted as:

$$B_{\mathrm{w*M}} = \begin{pmatrix} \mathrm{A}_1^{(1)} & \cdots & \mathrm{A}_1^{(\mathrm{m})} \\ \vdots & \ddots & \vdots \\ \mathrm{A}_{\mathrm{w}}^{(1)} & \cdots & \mathrm{A}_{\mathrm{w}}^{(\mathrm{m})} \end{pmatrix} \quad (2)$$

And

$$B_W^{(m)} = \frac{1}{q}\sum_{\tau=1}^{q} A_W^{(m)}(\tau) \quad (3)$$

On online phase, by using mobile terminals, present RSS information can be measured in the current location, and it will be recorded as a RSS vector:

$$R_{1*M} = [\gamma^{(1)},\gamma^{(2)},\cdots,\gamma^{(m)}] \quad (4)$$

Compare the online FP γ to the offline FP to estimate the location by using the maximum likelihood algorithm. The estimation of w0 is given as:

$$(i^*,j^*) = \underset{i\in I,j\in J}{argmin}\, Pr\{\gamma|\omega = (i,j)\} \quad (5)$$

2.2. *Quad-Tree Algorithm*

Quad-Tree Algorithm is a modified algorithm of the traditional fingerprinting localization based WLAN.[10] It also consists of two-stage process, but on the online phase, it changes a lot. The offline FP database saves the radio map of 2*2, 4*4, 8*8, ··· $2n*2n$ grids. And we compare the observation vector we collected online by calculate the Euclidean distance. The Quad-Tree Algorithm can decrease the calculated quantity to a great extent and can also reduce localization errors. The steps of Quad-Tree Algorithm are shown as below Figure 2. However, in the Quad-tree algorithm, once the estimation position is divided into the wrong area, it'll not be corrected anymore. It may cause very high error rate on the boundaries and we must adjust the method of dividing the areas to reduce the localization error rate on the boundaries. At the same time, the Quad-Tree cannot find the moment that the splitting can stop appropriately. The unmeaning even incorrect grids may be split. Thus, the Quad-Tree

Algorithm need to be modified. The Algorithm in this paper can provide a novel algorithm based multiple logistic regression with higher reliability and accuracy.

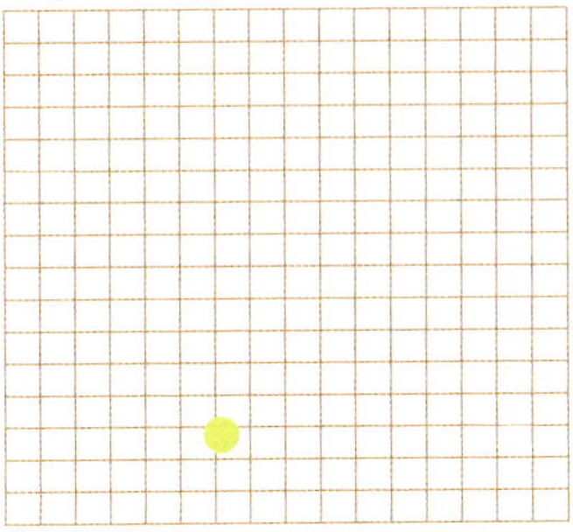

Fig. 1. Uniform grids for localization

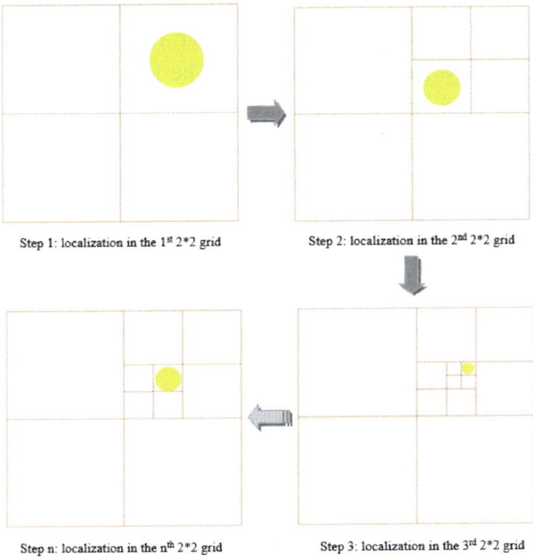

Fig. 2. Specific steps of the Quad-Tree Algorithm.

3. Novel Quad-Tree with Irregular Boundary Based on Multiple Logistic Regression (MLR Quad-Tree)Equations

3.1. *Multiple logistic regression*

Logistic Regression is one of the classical classification methods in statistics. It can be used to solve binary classification and multiple classification.[10]

Firstly, we introduce binomial logistic regression model. Binomial logistic regression model is the conditional probability distribution as following:

$$P(Y=1|x)=\frac{exp(\omega\cdot x+b)}{1+exp(\omega\cdot x+b)} \quad (6)$$

$$P(Y=0|x)=\frac{1}{1+exp(\omega\cdot x+b)} \quad (7)$$

Here, $x \in R^n$ *is* the input, $Y \in \{0,1\}$; is the output, $\omega \in R^n$, $b \in R$ is the parameter, w is the weight vector, b is the polarization. In order to be convenient, we extend the x and w into $w=(w^{(1)},w^{(2)},\cdots,w^{(n)},b)$, $x=(x^{(1)},x^{(2)},\cdots,x^{(n)},1)^T$. Then, the logistic regression model is:

$$P(Y=1|x)=\frac{\exp(\omega\cdot x)}{1+\exp(\omega\cdot x)} \quad (8)$$

$$P(Y=0|x)=\frac{1}{1+\exp(\omega\cdot x)} \quad (9)$$

The value of linear function is closer to positive infinity, the probability value is closer to 1. And the value of linear function is closer to negetive infinity, the probability value is closer to 0. The test described above is binary logistic regression used for two-class classification. We can extend it to multi-nominal logistic regression model used for multiple classification. Assuming the value set of discrete type random variable Y is $\{1,2,\cdots,K\}$, and then, the multiple logistic regression model is:

$$P(Y=k|x)=\frac{\exp(\omega_k\cdot x)}{1+\sum_{K=1}^{K-1}\exp(\omega_k\cdot x)}, k=1,2,\cdots,K-1 \quad (10)$$

$$P(Y=K|x)=\frac{1}{1+\sum_{K=1}^{K-1}\exp(\omega_K\cdot x)} \quad (11)$$

Here, $x \in R^{n+1}, w_k \in R^{n+1}$.

3.2. *MLR Quad-tree system*

The MLR algorithm can solve the multi-classification problems accurately. As a matter of fact, localization is also a classification problem. Localization just classifies the location into $I*J$ classes and estimate the current class as the location. Then MLR algorithm can help us estimate the location as a multi-classification. By processing the RM in SPSS, we found that Multiple Logistic Regression can solve the boundary problems by correcting the wrong classification of boundary points. And then we can reduce the error rate on the boundaries greatly. The Localization process of MLR is also arranged in two-stage phases. In the offline phase, we can use the same RM as the previous part, but we will not classify the region into regular geometrical shape. We should use MLR to do the multi-classification and produce the boundary.

1) Offline phase

Grid the area as 2*2, 4*4, 8*8 •••$2n$ *$2n$ squares and collect RSS $A_{i,j}$. (here n is the last layer)

Using the Multiple Logistic Regression Analysis to process all the data we collect, Re-grid the area and correct the classification especially on the boundaries. (Just like the Fig.4 shows.)

Define the offline radio map(RM) according to the new division.

2) Online phase

a) Collect current RSS matrix $w_{\lambda_k,n} = (w_{\lambda_k,1}, \cdots, w_{\lambda_k,2}, \cdots, w_{\lambda_k,N})$,$\lambda_k$ is the k-th AP set for node *N* obtained by the terminal.

b) Use Quad-Tree algorithm to estimate the location.

i. Start with λ_0 , and the layer k=0.

ii. Read $\Gamma(t_{\lambda_k})$ from RM to determine whether current λ_k can be split. If it is true, skip to step iii; otherwise, skip to step c).

iii. Read fingerprint of λ_k's four sub-nodes from RM and save as $\mathbf{u}_\vartheta$, where ϑ=1, 2, 3, 4.

iv. Calculate the Euclidean distance between $\mathbf{u}_\vartheta$ and $w_{\lambda_k,n}$ as:

$d_\vartheta = \left\|\mathbf{w}^*_{\lambda_k} - \mathbf{u}_\vartheta\right\| = \sqrt{\sum_{j=1}^{N}\left(u_{\vartheta,j} - w_{v_{k,j}}\right)^2}$. Let $\lambda_{k+1} = \text{argmin}\left\|d_\vartheta\right\|$, λ_{k+1} is the current node to be split, and then skip to step ii.

c) Output the localization result: The location estimation is the geometric center of the current cluster q^*, and the estimation uncertainty is defined as half of the longer side of q^*.

3.3. *Uneven segmentation of grids*

In the prior Quad-Tree algorithm, boundary problem is the one matters the most. Which, if happens, will do large harm to the final result and users' experience. However, there exists another issue remained to be discussed, the uneven grids segmentation. Subdividing level represents the number of subdivision of the area. For example, if we divide the localizing room into 2x2 grids, then the level is 1. So what is the highest subdivision level for the localizing area? Is this level a determined number for all the grids in the whole area? It is demonstrated that the curve of precision-to-subdividing level will reach a peak and then starts to decline as the gridding process propelling. For different grid in different sub-dividing level, the peak value is what we pursued to reach the relatively highest localizing subdividing threshold should be raised, which means when we noticed a declination of the result, the subdividing should be ended there, leavings the grid at the lower level.

4. Simulation Results

After putting forward a novel Quad-Tree boundary revising algorithm, simulations need to be done to verifying this model and quantifying the validity and the improvement compared to the prior way in a mathematical way. In the simulated scenario, 14x14 grids and 10 APs were the setting. The simulation is separated into two parts: figure of collection distribution function (CDF) - error curve, both revealing the characteristic of the boundary and the whole area, demonstrating the error value from 0 to 10, and the corresponding percentage of the localization results distributed within. The red line in the figure shows the result of the traditional max likelihood (ML), while the green line represents the situation employing the prior Quad-Tree algorithm. In both the CDF figure, the revise algorithm of ours reveals at least 52% more results has led to errors within 1 meter, and the percentages are up to 80%. Compared to the result of precision improvement in the whole area, it increases more while applying to the boundary.

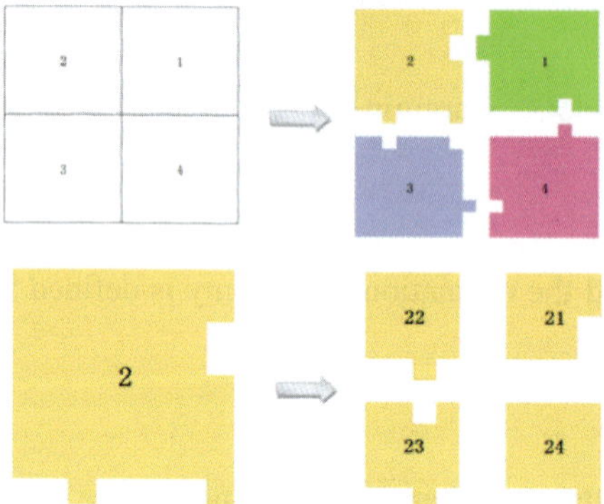

Fig. 3. Grids of the Quad-Tree with Irregular Boundary.

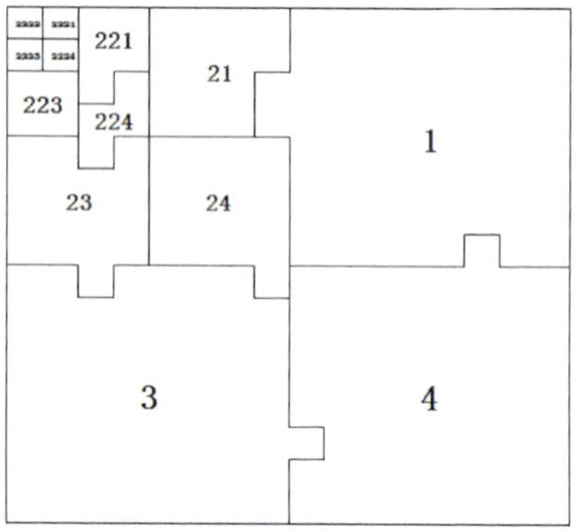

Fig. 4. Quad-tree with Irregular Boundary.

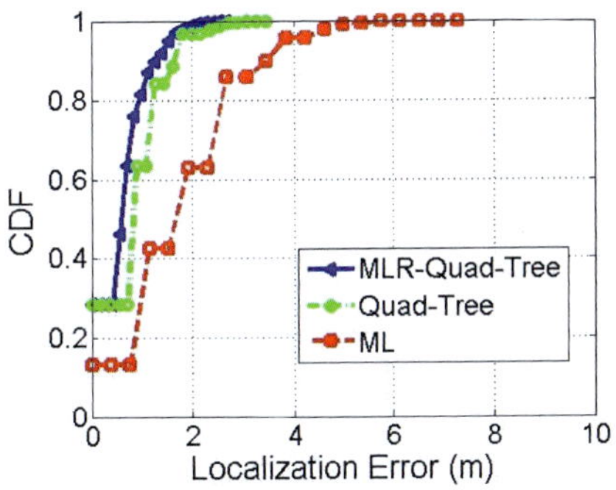

Fig. 5. Simulation results in whole area.

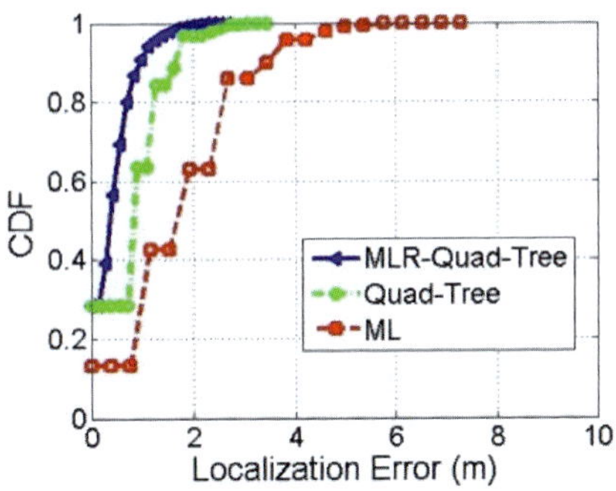

Fig. 6. Simulation results in boundary area.

5. Conclusion

In this paper, we proposed a novel Quad-Tree with Irregular Boundary Based on Multiple Logistic Regression in order to address the boundary problem, which presents as the irregular geometry gridding method. In detail, the grids at the boundary of the prior Quad-Tree achieve the lowest probability of error but the highest damage to the final result. It seldom happens but once can be the worst. When applying irregular geometry gridding, we determined those grid which get error most frequently, and the re-partitioned them into a better group. Instead of revising the error, we think about to adjust to it. When it comes to some sub-level grids that take the half of higher-level one, we leave the higher one undecomposed. These is a way of unevenly segmented grid. The boundary problem of prior Quad-Tree algorithm is preferably resolved by the novel Quad-Tree with Irregular Boundary Based on Multiple Logistic Regression.

Acknowledgments

This work was supported by the Program for Research Innovation Fund for College Student of Beijing University of Posts and Telecommunications.

References

1. Pablo Men´endez, Carmen Campomanes, Krzysztof Trawi´nski, and Jose M. Alonso," topology-based indoor localization by means of wifi fingerprinting with a computational intelligent classifier".
2. M. McGuire, KN Plataniotis, and AN Venetsanopoulos. Data fusion of power and time measurements for mobile terminal location.
3. J. Yin, Q. Yang, and L.M. Ni. Learning adaptive temporal radio maps for signal-strength-based location estimation. IEEE Transactions on Mobile

4. C. Steiner and A. Wittneben, "Efficient training phase for ultra-wideband based location fingerprinting systems".
5. Ling Pei, Ruizhi Chen, Jingbin Liu, Heidi Kuusniemi, Tomi Ten hune n and Yuwei Chen, "Using Inquir y-based Bluetooth RSSI Probability Distributions for I ndoor Positionin g", Journal
6. Ganesh Ananthanarayanan and Ion Stoica. Blue-fi: enhancing wi-fi performance using bluetooth signals. In MobiSys '09, 2009.
7. D. Mohapatra and S. B. Suma , "Survey of Location Based Wireless Services" , IEEE Int. Conf. Personal Wireless Communications.
8. Dong Liang, Zhaojing Zhang and Mugen Peng: Access Point Reselection and Adaptive Cluster Splitting-Based Indoor Localization in Wireless Local Area Networks, Internet of Things Journal, IEEE (Volume:2, Issue: 6)
9. Heng Li, Zhiwen Liu, Xing An and Yonggang Shi: "Multi-classification of cell deformation based on object alignment and run length statistic".
10. Hang li, "Statistical Learning Theory", Tsinghua University Press, Beijing, March, 2012.

Explosion simulation based on the CUDA physical model and particle system

Bai-Yu Zhao* and Ji-Cheng Quan

Aviation University of Air Force,
Changchun, Jilin 130022, China
*Email:*1051464044@qq.com*

The current battlefield situation visualization field of military intelligence is the focus of the study, the visual effects to the commander's decision. According to the physical model of the physical process of battlefield explosion, we proposed to improved simulation model, using Eulerian simulation of instantaneous explosion fireball model, using the explosion of fluid physics volume as a function of spatial location and time, then using the particle system method to simulate the explosion after the spatter model and explosive fluid splash material point in the movement process of various physical quantities with the particle to the variation of the position and time recording, tracking all the fluid particles. The separate explosion simulation process, which can simulate the instantaneous explosion of the overall effect, and could clearly simulate sputtering spatter particles. Experiments show that this method is realistic, can satisfy the real-time simulation of the explosion.

Keywords: Euler; Particle System ; Gpu; Cuda; Real-Time.

1. Introduction

Visual simulation is a comprehensive information technology based on computer graphics, which is widely used in military, railway, aerospace and other fields. The visual effect of the explosion is determined by the physical laws of nature, in the process of explosion will produce flame, smoke, debris and other phenomena. The main method for simulation of the explosion is based on the Navier Stokes equations accurately describe the physical process of explosion, the biggest advantage is that the effect is very realistic, so we propose a segmented simulation method, instantaneous explosion of the overall effect of the grid Euler method to simulate the fire ball, and then Lagrange method based on particle system simulation after the explosion of the spatter. Compared with the original method, this method can simulate the ball of fire and the splash of the particle effectively, which can accord with the needs of most applications.

* Corresponding author.

2. Explosive Effect Modeling Based on Physical Model

2.1. *Navier - Stokes equations*

Explosion is a kind of fluid, the explosive moment due to its strong pressure and high temperature, the gas liquefaction quickly, and to spread around, diffusion process that can be as fluid density has not changed, so you can use the incompressible Navier - Stokes (N – S) equation as the mathematical physics model. Navier - Stokes equations is composed of the following partial differential equation:

$$\frac{\partial u}{\partial t} = -(u \cdot \nabla)u - \frac{1}{\rho}\nabla p + \nu\nabla^2 u + F \tag{1}$$

$$\nabla \cdot u = 0 \tag{2}$$

2.2. *Physical modeling of fire bomb*

Due to explosion leads to a huge ball of fire, the fire ball will spread around slowly, when the fire is out, it would become smoke and splash. Foster and others [1,2] using the equations of incompressible fluid as physical model in 1996, concreting form is as follows:

$$\frac{\partial u}{\partial t} = f - (u \cdot \nabla)u - \frac{1}{\rho}\nabla p + \nu\nabla^2 u \tag{3}$$

The external force term is defined as follows in the model:

$$f = -\beta g_v (T_0 - T_k) \tag{4}$$

β types for thermal expansion coefficient, g_v means gravity effect, To means the temperature value comparative with original point, Tk for the current temperature value. (5) stands for the effect of the force caused by the temperature difference----Thermal buoyancy.

In the process, the temperature is reduced gradually, so the temperature of the transmission formula is:

$$\frac{\partial T}{\partial t} = \lambda\nabla^2 T - (u \cdot \nabla)T \tag{5}$$

λ defines as control coefficient of motion diffusion. Using explicit euler method to solve the N-S equations.

Due to euler method is to calculate quantities of fixed point in the motion of a fluid, such as speed, temperature, density, etc., each fixed point is effected by some external factors such as wind, warm, the fixed point quantities will vary over time when effected. And the energy of the beginning of explosion process gets the largest and highest temperature point, and we can only see the outermost explosion during the explosion effect, so irregular (tensile) grid algorithm is introduced (drawing) to simulate the process of explosion in an instant.

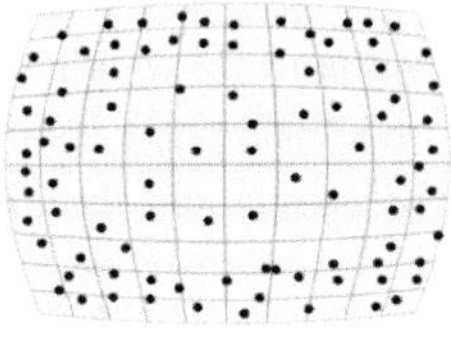

Fig. 1. Irregular grid.

According to above mentioned, it is assumed that the instant of explosion, in which grid middle particle velocity for u, we only need to use iterative method to calculate the particle velocity at every point, in order to avoid numerical dissipation caused by quality changes, using the implicit way to solve the diffusion term in this paper, namely:

$$\frac{u_{i,j,k}^{t+\delta t}-u_{i,j,k}^{t}}{\delta t}=\left(\begin{array}{l}\dfrac{u_{i-1,j,k}^{t+\delta t}-2u_{i,j,k}^{t+\delta t}+u_{i+1,j,k}^{t+\delta t}}{(\delta x)^2}\\+\dfrac{u_{i,j-1,k}^{t+\delta t}-2u_{i,j,k}^{t+\delta t}+u_{i,j+1,k}^{t+\delta t}}{(\delta y)^2}\\+\dfrac{u_{i,j,k-1}^{t+\delta t}-u_{i,j,k}^{t+\delta t}+u_{i,j,k+1}^{t+\delta t}}{(\delta z)^2}\end{array}\right) \tag{6}$$

Here we use Jacobi iteration method for solving, when n times iteration:

$$u_{i,j,k}^{t+\delta t,(n+1)}=\frac{u_{i,j,k}^{t,(n)}-\delta t*v\left(\begin{array}{l}\dfrac{u_{i-1,j,k}^{t+\delta t,(n)}+u_{i+1,j,k}^{t+\delta t,(n)}}{(\delta x)^2}\\+\dfrac{u_{i,j-1,k}^{t+\delta t,(n)}+u_{i,j+1,k}^{t+\delta t,(n)}}{(\delta y)^2}\\+\dfrac{u_{i,j,k-1}^{t+\delta t,(n)}+u_{i,j,k+1}^{t+\delta t,(n)}}{(\delta z)^2}\end{array}\right)}{1+2\dfrac{\delta t*v}{(\delta x)^2}+2\dfrac{\delta t*v}{(\delta y)^2}+2\dfrac{\delta t*v}{(\delta z)^2}} \tag{7}$$

Among them, δ is time step for speed, (u,v,w) is the form of 3d coordinate. Due to the grid is uneven, from what has been discussed above, finally the result is the grid particle velocity field.

2.3. *Explosion spatter model*

2.3.1. *Splash particles motion model*

Splash particles, which caused by missile blast, is subjected to the action of two forces—gravity and wind. Here Lagrange method is used to make fluid as a particle system, through tracking each particle velocity, temperature, location attributes to simulate the movement. Through the thought of literature [4], a wind and a gravity way acted on splash particles is described in this paper.

Particles in x, t, z, three directions add the effect of wind, expressed by the following formula:

$$\begin{cases} u_x = u_{x0} + f_x \\ u_y = u_{y0} + f_y \\ u_z = u_{z0} + f_z \end{cases} \tag{8}$$

Explosive particle motion in the space is following Newton's law, once the particles get initial speed, their spatial location in each period can be represented by the following formula:

$$\begin{cases} x = x_0 + u_x t \\ y = y_0 + u_y t - \dfrac{gt^2}{2} \\ z = z_0 + u_z t \end{cases} \tag{9}$$

Because of reducing calculation complexity, ignoring air buoyancy effect[5].

2.3.2. *The color, transparency, and shape of splash particles*

In the real world, the colors of explosion splash particles are different, so a model, which changes with particle survival time to calculate the color, is introduced. The calculation formula is shown in (10), color is the color of the particle, t shows particle survival time , T said particle life cycle, function is the function to calculate the particle color, the specific calculation is decided by the specific situation.

$$color = function(t, T) \tag{10}$$

In the process of explosion, it can produce a variety of shapes of debris. Particle shape is different with the method of texture mapping. Therefore, a group can be provided with pieces in different shape of texture. When generating debris particles, randomly selecte a texture from a set of texture, then assigned to specific particles, as is shown in type (11), textures is texture arrays, accessing a texture through an index value randomly.

$$\begin{cases} index = random(0, num-1) \\ texture = textures[index] \end{cases} \tag{11}$$

In addition, debris particles in sports also happens spinning. As is shown in type (12), scale shows parameter scale value, two directions and Y scaling respectively, its scope is determined by min and max, angle is the random rotation angle, take z axis as rotation.

$$\begin{cases} scale = size(random(\min, \max), random(\min, \max)) \\ angle = random(0^{\circ}, 360^{\circ}) \end{cases} \tag{12}$$

3. Acceleration simulation based on CUDA

In the simulation of explosive particle phase, major impact of speed is continuously updated of particle attributes. In this paper, CUDA parallel programming is used to accelerate the particles attribute updating; making full use of the GPU parallel computing ability. On the CPU side, through the class instantiation, creating object of splash particle, then the initial properties of particles are calculated by CPU, then passing each particle initial attribute data to GPU. Here we use OpenGL to create pixels in the global memory of GPU buffer object, then CPU makes initialization of all particles attribute data stored in the pixels in the buffer.

In order to update and compute the attribute of each particle parallel, we can map each particle to the each thread in GPU, a thread is responded for a particle attribute update. Using a one-dimensional thread index is accepted. According to the multi-level threading of CUDA programming model, we need a parallelized programs as a grid, then the grid is decomposed into several blocks, each block contains a number of thread. The grid can be divided into one-dimensional, two-dimensional or three-dimensional, block can also be set to a one-dimensional, two-dimensional or three-dimensional.

In the implementation of this article, the block and the kernel function is designed [6] as follows: firstly, set the size of the block and the thread size, secondly, we need design the kernel function, according to the corresponding thread ID thread index, we adopt the one-dimensional block, so the mapping relationship between thread ID and the storage array index can be represented as follows:

$$\text{offset} = threadIdx.x + blockIdx.x \times blockDim.x \tag{13}$$

4. Experimental analysis

Using the method of modeling in this paper, the authors realize the explosion effect, as shown in the figure below Figure.2(A and B).

In order to reduce the workload of computer processing, this paper defines each particle as a spherical using Billoard (billboards) effect Figure.2(C), and each spherical are posted textures that certain Alpha transparency and Figure.2(D and E).

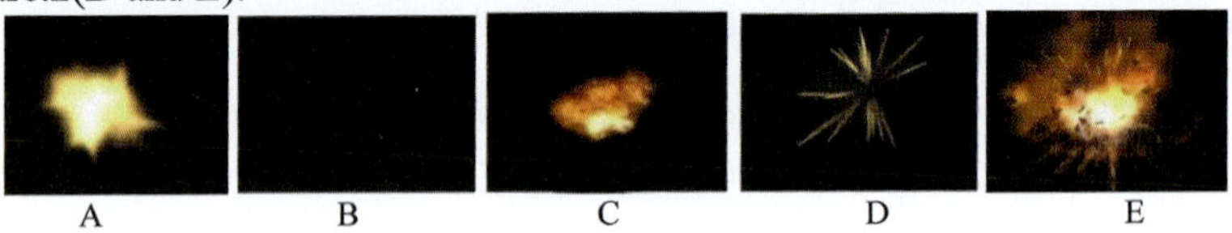

Fig. 2. explosion effect.

5. Conclusions

The model which is based on the physical methods of explosion and splash content based on particle system with the combination of a set of algorithms is introduced in this paper. What's more, use GPU to make acceleration. The algorithm combines particle system and physical model ingeniously, which simulates uncertainty of explosion splash very well. And use CUDA programming, which maintains efficiency in effect but also has greatly improved Next we will research on how to improve efficiency with less particles under the condition without affecting the simulation results. And try to use texture memory to store some of the properties of particles when CUDA programming, in order to accelerate access speed.

References

1. JENSENH, STAM.J, FEDKIM.R: visual simulation of smoke [J]. Computer Graphics:2001. 35,15-22
2. P.CCONSTAINTIN, FOIAS: Navier-Stokes Equations[M]. University of Chicago Press, Chicago, London, 1988
3. JOS STEM. Simulation and control of physical phenomena in computer graphics[A]. Proceedings of 12th Pacific Graphics on IEEE[C] 2004:171-173
4. Yang Pingli, Goofy, Jiang Ling, Huang Shaohua. Proceedings of the conference on the simulation technology on nuclear explosion cloud

dynamic modeling method of [C]//2011. Beijing: "computer simulation" magazine agency, 2011:94-96
5. Zhang Xingkai. Theory and test of blasting dust behavior, post doctoral research. Beijing: University of Science and Technology Beijing.1995.
6. KATRIN FSICHER, GEORG-PETER, OSTERMEYER. Massively Parallel Particle Simulations on Graphics Processing Units with CUDA[J]. Technische University Braunschweig, Institute of Dynamics and Vibrations,2011,pp.873-874

Cognitive student model based on the gray partial correlation

Sheng-Bing Ren*, Di Feng† and Xiao-Nan Chen‡
School of Software, Central South University,
Embedded System and Network Laboratory,
Changsha, Hunan, 410075, *China*
E-mail: **rsb@csu.edu.cn, †feng_di10@163.com, ‡409770352@qq.com*

From the perspective of agile learning, the cognitive student model is an important means to recognize students' knowledge level and make an instructive learning. Due to the existence of several cognitive factors in this model, which can be viewed with strong uncertainty for properties with gray, fuzzy and difficulty quantifying, this paper presents a gray partial correlation method to quantify this model. Firstly, the mutual influence is discussed between any two factors to get the weight vector by calculating the partial correlation coefficient. Then we define the gray classes and white nization weight functions to output gray evaluation weights so as to build comprehensive evaluation model that combines with the weight vector. The model can give a guidance on learning catering exactly to the agile learning.

Keywords: Cognitive Ability; Cognitive Student Model; Agile Learning; Gray Partial Correlation Method.

1. Introduction

Owing to the students' individual differences, different students have different learning status and knowledge level. The definition of agile learning is timely grasping students' learning status, scientifically evaluating knowledge level, then achieving the goal of the tutoring[1]. This new mode is contribute to promoting the notion of learning status, improving learning quality and forming lifelong learning for students[2]. The cognitive student model directly reflects students' knowledge level and provides the basis for further study, thus serving the purpose of agile learning[3].

Cognitive student model is established with the specific technology in the literature[4~6], such as fuzzy analytic hierarchy process and assignment method, but the tutoring cannot be reflected better in above method.

Cognitive factors are affected by each other, different experts have different awareness of these factors. Above all, these factors are gray and fuzzy[7]. In this

---

† Corresponding author.

paper,the grey partial correlation method is gone to build the cognitive student model.

2. Related Theory About Cognitive Student Model

In view of the Benjamin Bloom theory, students' cognitive ability is divided into six factors according to the complexity of intellectual activity: memorization (B_1), comprehension(B_2), application(B_3), analysis(B_4), synthese(B_5) and evaluation(B_6) [8]. Among those factors, memorization and comprehension belong to low ability, while application, analysis, synthese and evaluation belong to advanced ability.

3. Building Cognitive Student Model

3.1. *Application of partial correlation analysis*

In the learning process, students' cognitive ability is not static, any change of certain cognitive factors inevitably affects the other factors. Firstly, the change in lower ability factors is necessary affecting the advanced one, and the change in any advanced cognitive factors is also affecting the other advanced one [9]. The partial correlation analysis is used for discussing the influence or related degree about two factors under conditions of ignoring the influence of other factors. The whole process is completed through the following subsections.

3.1.1. *Determining the expert evaluation matrix*

Limited on individual differences, different experts have different understanding for the importance of cognitive factors. We define scale of marks as the Table 1 shows.

Table 1. Experts scale of marks.

Serial number	Comment remark	Score
1	Extremely important	5
2	Relatively important	4
3	Common important	3
4	Relatively unimportant	2
5	Extremely unimportant	1
6	Between the above two comments	4.5, 3.5, 2.5, 1.5

Supposing scores from k experts about each cognitive factor $B_j(j=1,2,3,\cdots,6)$ is respectively $I_{ij}(i=1,2,3,\cdots,k)$, which can constitute the expert evaluation matrix $I=(I_{ij})_{k\times 6}$.

3.1.2. *Determining simple correlation coefficient*

Simple correlation coefficient is commonly used to inferring whether two factors are relevant to each other. Supposing sample set of score in two cognitive factors is respectively X and Y, $X=\{x_1, x_2, x_3,\cdots, x_k\}$, $Y=\{y_1, y_2, y_3,\cdots, y_k\}$, so the calculation pattern of correlation coefficient is as the formula(1) shows.

$$\rho_{xy} = \frac{\sum_{i=1}^{k} x_i y_i - \frac{1}{k}\sum_{i=1}^{k} x_i \sum_{i=1}^{k} y_i}{\sqrt{\sum_{i=1}^{k} x_i^2 - \frac{1}{k}(\sum_{i=1}^{k} x_i)^2}\sqrt{\sum_{i=1}^{k} y_i^2 - \frac{1}{k}(\sum_{i=1}^{k} y_i)^2}} \tag{1}$$

In the formula(1): ρ_{xy} stands for correlation coefficient about x, y, *and* k stands for number of experts.

3.1.3. *Determining partial correlation coefficient*

Any two cognitive factors should be analyzed with the partial correlation to control the influence of other factors. The general formula of partial correlation coefficient with order n is as the formula(2) shows.

$$\rho_{yx\bullet x_1x_2\cdots x_n} = \frac{\rho_{yx\bullet x_1x_2\cdots x_{n-1}} - \rho_{yx_n\bullet x_1x_2\cdots x_{n-1}}\rho_{xx_n\bullet x_1x_2\cdots x_{n-1}}}{\sqrt{1-\rho^2_{yx_n\bullet x_1x_2\cdots x_{n-1}}}\sqrt{1-\rho^2_{xx_n\bullet x_1x_2\cdots x_{n-1}}}} \tag{2}$$

In formula(2): $\rho_{yx\bullet x_1x_2\cdots x_n}$ stands for partial correlation coefficient about x and y in case of factors $x_1, x_2, x_3,\ldots, x_n$.

Table 2. Interacting of cognitive factors.

Serial number	Factors to be analyzed	Affected by other factors
1	B_1, B_2	nothing
2	B_1, B_3	B_2
3	B_1, B_4	B_2, B_3
4	B_1, B_5	B_2, B_3, B_4
5	B_1, B_6	B_2, B_3, B_4, B_5
6	B_2, B_3	B_1
7	B_2, B_4	B_1, B_3
8	B_2, B_5	B_1, B_3, B_4
9	B_2, B_6	B_1, B_3, B_4, B_5
10	B_3, B_4	B_1, B_2
11	B_3, B_5	B_1, B_2, B_4
12	B_3, B_6	B_1, B_2, B_4, B_5
13	B_4, B_5	B_1, B_2, B_3
14	B_4, B_6	B_1, B_2, B_3, B_5
15	B_5, B_6	B_1, B_2, B_3, B_4

According to the recommendation of related expert, interacting among cognitive factors is as the Table 2 shows.

3.1.4. *Determine weight vector*

Partial correlation coefficient matrix $P=(\sigma_{ij})_{6\times 6}$ is established depending on all the partial correlation coefficient.

In matrix P, $\sigma_{ij}(i,j=1,2,\ldots,6)$ stands for partial correlation coefficient about B_i, B_j.

We can obtain weight value according to matrix P, which is as the formula(3) shows.

$$z_i = \frac{P_i}{\sum_{i=1}^{n} P_i}, \text{ and } P_i = \sum_{j=1}^{n} \sigma_{ij} \tag{3}$$

In formula(3): $z_i(i=1,2,\ldots,n)$ stands for the proportion of factor B_i in the cognitive factors system, namely weight, and $\sum_{i=1}^{n} z_i = 1$. So we can get the weight vector Z which is as the formula(4) shows.

$$Z = (z_1, z_2, z_3, z_4, z_5, z_6) \tag{4}$$

It is necessary to note that the value of the partial correlation coefficient may be negative, which means negative partial correlation, negative numbers are first to take the absolute value before calculating the weight value.

3.2. *Building cognitive student model based on the gray comprehensive evaluation*

Gray comprehensive evaluation is a kind of method based on gray system theory, namely evaluable analysis of incomplete principle combining qualitative and quantitative process. It can model the uncertain system such as evaluation of students' cognitive ability whose factors are gray, fuzzy and hard to quantify, then make a comprehensive evaluation for informational system combining with various gray class degree[10].The steps of gray comprehensive evaluation to cognitive student model is showing below [11].

3.2.1. *Determining the sample evaluation matrix*

Evaluating students' knowledge level cannot do without quantifying the cognitive ability factors. In this paper, we design several quizs, and ensure that each exercise of each quiz tests a cognitive ability, then do statistics for each relative score of ability in each quiz. The formula for relative scores is as the formula(5) shows.

$$f_{ij} = \frac{s_{ij}}{a_{ij}} \times 10 (i = 1,2,3\cdots, m; j = 1,2,3\cdots, 6) \tag{5}$$

In formula(5): m stands for number of quizs, f_{ij} stands for relative score of ability j in quiz i, s_{ij} stands for Actual score of ability j in quiz i, and a_{ij} stands for full score of ability j in quiz i.

The sample evaluation matrix which shows in matrix $F=(f_{ij})_{m\times 6}$ is defined by formula(5).

3.2.2. *Defining gray classes*

In this step, we should define gray classes U, gray class interval V, gray numbersand whitenization weight functions $W(X)$. We define U={bad, medium, fine, excellent}, accordingly V={[2, 4], [4, 6], [6, 8], [8, 10]} in the cognitive student model. According to the method of mixed center-point triangular whitenization weight function, we should extend V around right and left to get the new gray class interval, namely[0, 2] and [10, 12], then we can get the center-points based on this interval,namely the 1, 3, 5, 7, 9 and 11. Then the gray numbers and whitenization weight functions are as shown in formula(6)~(7) by all center-points.

$$W_1(x) = \begin{cases} 0 & x \notin [1,5] \\ 1 & x \in [1,3] \\ (5-x)/2 & x \in [3,5] \end{cases} \tag{6}$$

$$W_2(x) = \begin{cases} 0 & x \notin [3,7] \\ (x-3)/2 & x \in [3,5] \\ (7-x)/2 & x \in [5,7] \end{cases} \tag{7}$$

$$W_3(x) = \begin{cases} 0 & x \notin [5,9] \\ (x-5)/2 & x \in [5,7] \\ (9-x)/2 & x \in [7,9] \end{cases} \tag{8}$$

$$W_4(x) = \begin{cases} 0 & x \notin [7,11] \\ (x-7)/2 & x \in [7,9] \\ 1 & x \in [9,11] \end{cases} \tag{9}$$

3.2.3. *Determing gray evaluation weight matrix*

For cognitive factor B_j, supposing the relative score which tests ability B_j in number m of quizs is respectively f_{1j}, f_{2j}, f_{3j}, $\cdots$, f_{mj}, the whitenization weight about the gray class h(h=1,2,3,4) of factor B_j is respectively $W_h(f_{1j})$, $W_h(f_{2j})$, $W_h(f_{3j})$, $\cdots$, $W_h(f_{mj})$. Supposing M_{jh} is the gray evaluation coefficient about the gray class h of factor B_j, and M_j is the total gray evaluation coefficient, the formula of M_{jh} and M_j are as shown in formula(10)~(11).

$$M_{jh}=\sum_{d=1}^{m}W_h(f_{dj}) \quad (10) \qquad M_j=\sum_{h=1}^{4}M_{jh} \quad (11)$$

The formula of gray evaluation weight r_{jh} is given based on M_{jh} and M_j, which is as shown in formula(12).

$$r_{jh}=\frac{M_{jh}}{M_j} \qquad (12)$$

According to formula(12), we can get the gray evaluation weight matrix $R=(r_{jh})_{6\times 4}$,

3.2.4 *Comprehensive evaluation*

Comprehensive evaluation of cognitive student model is build according to weight vector Z in 3.1.4 and matrix R, which is as shown in vector Q.

$$Q=Z\times R=(q_1,q_2,q_3,q_4) \qquad (13)$$

In formula(13):

Q ---- Comprehensive evaluation vector of cognitive student model.

q_i(i=1,2,3,4) ---- Comprehensive gray evaluation weight about each gray class, and $\sum_{i=1}^{4}q_i=1$.

4. Conclusion

Integrating partial correlation analysis and gray system theory to the cognitive student model is effectively improving the objectivity and reliability of evaluation results, above all, suitting the requirements of agile learning.

Among the student cognitive ability factors, different experts keep a different cognition degree from cognitive factors, and each factor is affected by each other. Partial correlation analysis method is used to calculating the partial correlation coefficient of comparative factors to determine the weight value of cognitive factors in the case of controlling influence from other factors so as to come up with an objective and accurate result.

With the application of gray system theory which determines the gray class and the gray evaluation weight of cognitive factors, comprehensive evaluation model is set up complied with the above weight value. This method could more accurately quantify the cognitive factors.

References

1. P.M. Jansson, Ravi Ramachandran, John Schmalzel, S.A. Mandayam.Creating an Agile ECE Learning Environment Through Engineering Clinics. Vol. 53, Issue. 3(IEEE Transactions on Education, 2010) , pp.455-462.
2. LIU Sheng-tao. On Student Learning Assessment Reform in Universities and Colleges. Vol. 6, Issue. 6(University Education Science, 2013) , pp. 53-56.
3. Karaoli & Dimitriou. Student modeling approaches: A literature review for the last decade. Vol. 40, Issue. 11(Expert Systems with Applications, 2013), pp.4715–4729.
4. WANG Wan-sen,GONG Wen. Research on emotion, cognition and personality student model in E-Learning. Vol. 28, Issue. 11(Application Research of Computers, 2011), pp.4174- 4176 + 4183.
5. HE Guo-qiang, LIU Li-zhen, DU Chao. Design of student model in intelligent learning system. Vol.30, Issue. 10(Computer Engineering and Design, 2009), pp. 2554-2557.
6. YE Xiao-qing. Students Assessment Model based on Fuzzy AHP theory. Vol.28(China Electric Power Education, 2013) , pp.210-212.
7. WANG Lei,HUANG Meng-xing. Research on trust evaluation of supplier based on gray AHP in cloud computing. Vol.30, Issue. 3(Application Research of Computers, 2013) , pp. 742-744+750.
8. Yun-Wu Wu,Kuo-Hua Weng.. Using an analogical thinking model as an instructional tool to improve student cognitive ability in architecture design learning process. Vol.23, Issue. 4(International Journal of Technology and Design Education, 2013) , pp.1071-1035.
9. Stewart, Paulette, Rodney-Wellington, Kerry-Ann. (2014). Moving Beyond Tradition: Technologies Facilitating Students' Cognitive Ability and Modifying Pedagogical Practices of School Librarians in Jamaica and Antigua. International Association of School Librarianship:282-307.
10. Yu Zhang, Jie Ni,Jian Liu, Li-rong Jian. Grey evaluation empirical study based on center-point triangular whitenization weight function of Jiangsu Province industrial technology innovation strategy alliance. Vol.4, Issue. 1(Grey Systems: Theory and Application, 2014), 4(1), pp.124-136.
11. WANG Hua-zhong, QIANG Feng-jiao, HE Bao-cheng. New method of grey assessment based on the improved center-point triangular whitenization weight function. Issue. 8(Statistics & Decision, 2014), pp.69-72.

Application of smart medical based on distributed apriori algorithm

Hai-Feng Lu*
School of Information and Technology, Shanghai DongHua University, 200051,
School of Information and Technology, Shanghai Jianqiao University, 201306,
Shanghai, China
E-mail: 1771097725@qq.com

Fang-Qin Xu
School of Information and Technology, Shanghai Jianqiao University, 201306,
Shanghai, China
E-mail: xufqin2009@126.com

Due to continuous improvement of medical information construction, the scale of medical data is expanding, and its value will continue to increase. However, the traditional storage architecture cannot store such large amounts of medical data. Not only that, the common mining algorithm cannot analyze the value of potential information, largely hindering the development of medical wisdom. According to the concept and characteristics of smart medical, this paper proposes an architecture scheme of a medical cloud platform. Based on MapReduce, our Apriori medical data mining algorithm is conceived by introducing a degree of interest to improve the classic Apriori algorithm and combining with the knowledge of the cloud. The algorithm can ease the development bottleneck encountered by current smart medical. Overall, the proposed technology is more efficient and is able to apply itself more extensively in the field of medical research.

Keywords: Smart Medical; Apriori Algorithm; Map Reduce; Distributed System; Data Mining.

1. Introduction

With the continuous improvement of living standards, people also pay more and more attention to their health. But along with deteriorating of living conditions, the trend of population aging is gradually obvious, these diseases such as diabetes, high blood pressure and heart cerebrovascular have affected many people's daily life. At the same time, the pace of life in modern society becomes faster and faster, and working pressure is growing with each passing day, many young workers are in the sub-health state without self-knowledge, and they are

---

* Corresponding author.

potential patients of these chronic diseases, so modern society has a higher requirement in medical health care. In general, there will be a large amount of data that needs to be preserved in the process of medical service, for example, doctors make diagnoses and give treatment measures, biomedical data of patients as well as their medical history, there are many internal relationships between these data. Therefore, the concept of health service system has appeared in early, and now people just need to combine it with new information technology such as cloud computing, Internet of things and big data, then developing a quicker and more convenient smart medical to reduce pressure on hospital services and provide more comprehensive medical treatment.

Smart medical works for processes of medical research, diagnosis, verification, decision-making by analyzing the data. But the format of medical data is usually a wide variety, which can be text, images, video, voice and other forms. Therefore, the traditional structured database can not satisfy their needs, and new semi-structured and unstructured databases need to get useful information by analyzing a large amount of data. o how to quickly and efficiently excavated important information, and find the potential relationship between these data has become an urgent problem that needs to be solved.

2. Smart Medical

2.1. *The concept of smart medical*

Smart medical is a medical information ecosystem which has high information mobile and sharing, and medical Internet of Things as its core. Through building regional medical information platform of health archives, which has realized the interaction between patients and medical staffs, the interaction between institutions and medical equipment, and gradually formed comprehensive informationization of medical service. Depending on the subject of interactive information, it can be divided into business intelligence hospital services, interactive regional medical services, community / family self-health care services, intelligent remote emergency services and other categories, which are used to satisfy the different users of the service under a variety of environments demand.

According to the definition of smart medical, it usually has the following features:

(1)Interconnectivity

Authorized doctor can view the patient's medical history, medication records, treatment, insurance and other details, and patients can also choose or change doctors and hospitals as needed.

(2)Cooperation

Medical information is no longer in hospitals or areas, and its information warehouse will be changed to be a comprehensive, professional medical sharing platform.

(3)Preventability

Smart medical can quickly and effectively respond to major medical event and lest bungle illness by real-time monitoring the patient's vital signs, effectively analyzing and processing patient's history data.

(4)Universality

Traditional regional medical resources in the health care system are fixed without flow, now smart medical can help achieve the township hospitals and community hospitals seamless connection to the central hospital, so that small hospitals can be obtained expert advice, referral arrangements and online training at major hospitals.

(5)Innovativeness

Doctor make a diagnosis and give a treatment by uploading real-time video, combined with professional recommendations made by other doctors, help to improve the knowledge and process capabilities, to further promote innovation and clinical research.

(6)Reliability

Doctors are able to search, analyze, and cite scientific evidence to support their diagnoses.

2.2. *The composition of smart medical*

Smart medical is an information ecosystem which based on Internet of things, so its system architecture can be evolved from the Internet of Things. Smart medical usually can be divided into intelligent sensing layer, data transport layer, data integration layer, cloud layer, application layer, and intelligent sensing layer corresponds to sensing layer in the Internet of Things, data transport layer and data integration layer correspond to network layer in the Internet of Things, cloud layer and application layer correspond to application layer in the Internet of Things. Its structure is shown in Figure 1.

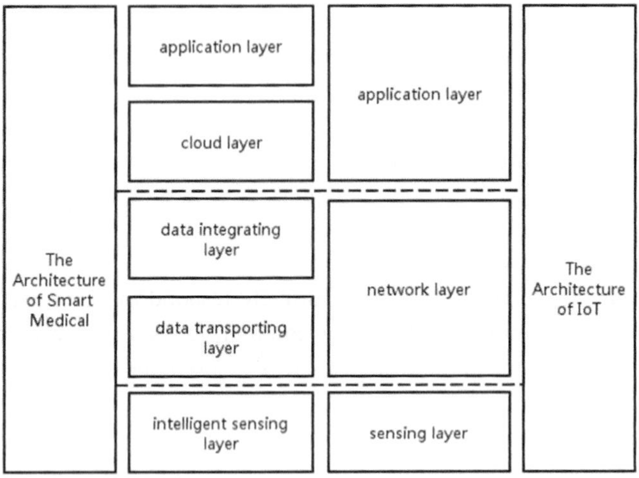

Fig. 1. The corresponding architecture picture between smart medical and IoT.

(1)Intelligent sensing layer

Intelligent sensing layer is mainly used to collect the surrounding environment parameters and transmit data to backbone network by using the short distance transmission technology, so it is mainly composed of sencing devices and network devices. Sensing devices consist of temperature sensors, humidity sensors, heart rate sensors, GPS modules and other common sensors, they collect data and identify objects by joint operation. Network devices build up a network of sense devices, and they will converge monitoring data to the backbone network by using RFID, Bluetooth, infrared and other short-distance transmission technology, in order to archieve transmission and data monitoring of large area. And smart medical is based on this model, through the wireless network will get patient data and sum them to the smart medical data center to establish widely interconnecting between sense body and service body of patient data.

Due to the complex environment in which patients, and data easily causes error and transmission failure by the multifaceted impact, so it should set up an integrated sensor layout. Using wired, wireless, static and dynamic integrated architecture, to archieve key technologies such as automatic access and collaboration, rapid deployment of large-scale sensor, adaptive network, mass data interconnection and sharing, and it is used to satisfy the event of multiple heterogeneous processing and millisecond automatically sense in the health care system, establish a unified standard format of data content, collection methods and convergence. It constructs smart health care system to adapt to different people in different environments.

(2)Data transport layer

Data transport layer is mainly used for data transmission of intelligent sencing layer, which generally acts as a connecting link. It up to ensure the rapid and reliable data transmission to the application layer, down to real-time data for a large number of sensors. Therefore, smart medical data transport layer should be established on the basis of mobile communication network and the Internet, and it connects to mobile communication network and the Internet through a variety of access devices to archieve the function of data real-time transmission. In addition, because the network topology structure that patients use is extremely complex, so data transport layer needs to satisfy isomerization, time-varying and application diversity of nodes to a certain extent, at the same time it should ensure secure access and share of mass data when servicing customers. Therefore, smart medical data transport layer is constructed by data-aware network, transmission network, network application services, on the basis of data-aware security, storage, transport and security access to set up a ubiquitous Internet protocol standard.

(3)Data integration layer

With the explosive growth in recent years of data, which is promoting the development of digital medical information technology and also had some obstacles, such as the fast-growing amount of data, redundant data meaningless and fragmented original relationship data, all of them put forward a severe challenge in the aspect of data storage and use. Therefore, smart medical data integration layer requires to use a more efficient, more accurate and more intelligent way to preprocess massive data, to achieve the objective of the integrated medical sensor real-time observes data. The Data Vitalization is a forward-looking organization and new data processing techniques, through perception, association, existence and other means to achieve self-awareness, self-learning and auto-grow of massive multi-source multi-mode data. So it can solve data restriction brought by the rapid growth to a certain extent through integrating data vitalization into the infrastructure of smart medical, and coupled with depth perception data, large-scale scenario modeling, data efficient sharing, visualization and other key technologies, in order to establish a reliable real-time intelligence medical data processing platform.

(4)Cloud layer

Cloud layer mainly provides a quick safety data processing ability for smart medica, and processing data in the data transport layer into valuable information by using efficient resource management and task scheduling mechanism, and coordinating allocation between various computing resources and smart medical applications to effectively use resources, improving the execution efficiency of applications. Typically cloud computing has features such as reliable in data security, low client demand, easily share data, hardware extension conveviently and so on. So it has obvious advantages in response to a large number of computation and mass storage, so combining it with the smart medical and using high-performance medical network resources to provide necessary services for users in the form of on-demand extension.

The combination of the traditional medical system and cloud computing will greatly reduce the medical cost and improve the utilization of medical resources, at the same time it also will provide the unified management and privacy security of users' personal data, and storings history information and current physical condition of patients in the medical services platform, providing an important reference for doctors to develop diagnostic protocols.

(5)Application layer

Application layer is the depth integration of smart medical and industry expertise, it is usually responsible for problems of coordinating information processing and human-computer interaction, and providing users with information service system that is high availability, high efficiency, high throughput and quick response through integrating and monitoring the effectiveness of information resources in the network. Therefore, the application layer should satisfy several aspects such as scalable architecture, high efficiency of the registration query mechanism, quickly organize information resource allocation mechanism and capability assessment mechanism. In addition, because the application layer achieves real-time interaction and satisfies the needs of different users through the acquisition of the cloud layer analysis and the use of a variety of devices. So this layer can also be divided into two categories in accordance with the form of pattern: One kind is the application layer, which process data to achieve the main function of payment, monitoring, positioning, inventory, forecasting, etc, in order to satisfy needs of different user groups when they are in the respective application scenarios; Another kind is the terminal equipment layer, it mainly provides users with multifunctional interactive interface between various devices to implement the smart medical center thought of people-oriented.

3. Steps and Improvement of Apriori Algorithm

The data mining algorithm of traditional association rule is not very ideal in the process of practical application because smart medical has big data size, high potential value and information complicated structure. With the rapid development of big data and cloud computing, people can develop a solution that is scientific, reasonable and in line with the characteristics of smart medical by improving the traditional association rule algorithms and combining it with the operation of cloud computing model.

3.1. *Steps of apriori algorithm*

Apriori algorithm is put forward by Agrawal and Srikant in 1994, which is typically used for mining association rules of boolean, single-layer and single-demension, using recursive two-stage frequency set thinking and an iterative manner step by step to search frequent itemsets (k + 1) through frequent itemsets k. Because Apriori algorithm find frequent set by scanning the entire transaction set, so the time complexity is very large and it often use two steps of connecting step and pruning step to search for all frequent itemsets.

(1)Connecting step

To find the collection L_k of all the frequent itemsets k, all the frequent itemsets (k-1) is L_{k-1}, it should be firstly used to connect with itself and generate the set of candidate k-itemsets, it as C_k. Assuming that I_m and I_n are members of the L_{k-1}, and regarding $I_i[j]$ as the item j in I_j. Apriori algorithm is used for sorting transaction set D and items of all itemsets in alphabetical order, namely having $I_i[1]<I_i[2]<\ldots<I_i[k-1]$ for item I_i of itemsets (k-1). The L_{k-1} is connected with itself, if satisfied:

$$(I_m[1] = I_n[1])\&\&(I_m[2] = I_n[2])\&\& \ldots \quad \text{(St.1)}$$
$$\&\&(I_m[k-2] = I_n[k-2])\&\&(I_m[k-1] < I_n[k-1])$$

So thinking that I_m and I_n is connected. The result of a connection between I_m and I_n is:

$$\{I_m[1], I_m[2], \ldots, I_m[k-1], I_n[k-1]\} \quad \text{(St.2)}$$

(2)Pruning step

C_k is a superset of L_k, it also means that items of C_k is may not frequent item. By scanning all of the transactions to identify the number of occurrences of each candidate in the items C_k, then determining whether the candidate is the frequent item by comparing with the minimum support threshold min_supp. Here people can use natures of the Apriori to compress C_k and simplify the calculation, that is, all non-empty subsets of any frequent itemset must also be frequent, on the contrary, if nonvoid subset of the candidate is not frequent, then the candidate set is certainly not frequent. These properties may help to remove the candidate set that is not frequent from C_k.

Borrowing connecting step and pruning step, concrete steps of Apriori algorithm are as follows:

(a) Scaning the transaction set D and generating frequent itemsets 1 through comparing with minimum support threshold min_supp;

(b) Generating candidate itemsets (k + 1) by connecting frequent itemsets k with itself;

(c) Calculating the support of each candidate, if the result is or equal greater than minimum support min_supp, then adding frequent itemset (k + 1) and initial state of frequent itemset (k + 1) is empty;

(d) If frequent itemset (k + 1) is the empty set, then frequent itemset k is the maximum frequent itemset, at this time jumping to step (5) to generate association rules; Otherwise, jumping to step (2) to continue the cycle execution;

(e) Generating rules R on the basis of frequent itemsets generated in step (4), namely the association rule X=>Y satisfies support(X=>Y) ≥min_supp and confidence(X=>Y)≥min_conf, the strong association rules that is screened out can actually be applied to various fields.

3.2. *Defects of apriori algorithm*

Apriori algorithm mines useful association rules from mass data by setting the minimum confidence threshold min_conf, however in practice, especially in medical data mining and Apriori algorithm exists disadvantages that are as follows:

(1)The selection of minimum confidence threshold min_conf in the Apriori algorithm is very important, if the choice is not reasonable, it may lead that strong association rules is different from the actual, in this case not only it would not achieve practical significance, but also misleading to some extent, which is particularly important in the medical field, because wrong diagnosis that generated will be very deadly.

(2)Association rules of the Apriori algorithm are mainly determined by setting thresholds of support and confidence, if the support threshold min_supp is set too low, although it can fully mine association rules in the effective, but the overhead of the algorithm is too large, it will increase the time complexity of the algorithm; If the support threshold is too high, it may miss many useful rules and can not guarantee the integrity of the algorithm. The performance of these two cases will reduce Apriori algorithm.

(3)Apriori algorithm traverse the database for many times when it generates candidate frequent itemsets, so when the database is too large, I / O load of the system will increase. As a result, the speed of system processing slow, it greatly increases the time complexity of the algorithm and reduce the efficiency of the algorithm.

3.3. *Improving algorithm by importing the interest degree*

The Apriori algorithm based on support and confidence has many shortcomings, such as a large number of scanning times, the large system load and the generation of wrong association rules. Therefore, this section overcomes the shortcomings of traditional algorithm by introducing interestingness threshold, in order to achieve mining fastly the strong association rules that have high value.

Interest of association rules is mainly used to describe the degree of correlation between item X and item Y in the rule X=>Y. If the interest value in the rule X=>Y is greater, the related degree of X and Y is bigger, and the smaller the degree of correlation. On the contrary, the related degree is smaller. And it mainly plays two roles by introducing a degree of interest: Firstly, it is able to filter the rules with low association rules by using the degree of interest, which further reduces the number of candidate frequent sets, and improves the efficiency of the algorithm; Secondly, it can filter the association rules which have no actual meaning and can easily lead to misleading, so it enhances the practicability of generated association rules[6]. Therefore, this section combined with a variety of interest model and actual needs of medical data, generating a relatively more suitable for medical interest mining model, the formula is as follows:

$$\mathrm{Int}(\mathrm{X} => Y) = \mathrm{Conf}(\mathrm{X} => Y) - \mathrm{Conf}(\overline{\mathrm{X}} => Y) \qquad \text{(Im.1)}$$

$$= \frac{\mathrm{P(XUY)}}{\mathrm{P(X)}} - \frac{\mathrm{P(\overline{X}UY)}}{\mathrm{P(\overline{X})}}$$

$$= \frac{\mathrm{P(XUY)} - \mathrm{P(X)P(Y)}}{\mathrm{P(X)[1 - P(X)]}}$$

Among them, the $\mathrm{Conf}(\mathrm{X} => Y)$ is the confidence of association rules $\mathrm{X} => Y$, which refers to the present probability of consequent Y when antecedent X appears; the $\mathrm{Conf}(\overline{\mathrm{X}} => Y)$ is the confidence of association rules $\overline{\mathrm{X}} => Y$, which refers to the disappeared probability of consequent Y when antecedent X appears. And $\mathrm{P(X)}$, $\mathrm{P(Y)}$and $\mathrm{P(XUY)}$ respectively represent the probability of transaction X, transaction Y and both of X and Y occur in the transaction set D.

The improved interest model not only uses probability to study the correlation of the association rules between the front and the back, but also highlight the correlation is positive correlation or negative correlation in the way of seeking difference. In addition, the interval model evaluates the degree of interest in [-1, +1], when the Int(X=>Y)>0, it refers to the antecedent X shall be positive related to the consequent Y; On the contrary, the negative related. Therefore, interest values can help users understand association between the front and the back, screen out redundant rules and improve the operation efficiency.

4. Distributed Apriori Algorithm

4.1. *Operation mode of MapReduce*

MapReduce is a parallel programming model and the distributed computing framework that processes mass data which packages parallel calculation, local computing, load balance, data distribution and fault tolerance processing, and it provides only a simple and powerful interface, greatly reduces the difficulty of development and deployment. The core idea of MapReduce is "decomposition and merge of tasks", it mainly consists of two processes that are Map and Reduce. Map process refers to all data is divided into M independent data blocks, and then assign to each map a data block. And it is calculated according to a certain calculation process, which will calculate the results in the form of Key-Value and output it, then forming N Key-ValueList according to grouping sort of the Key value. Reduce process refers to the N Key-ValueList as the input, and assigning to the N Reduce functions, then calculating and resulting them. So MapReduce can not be ignored in the analysis and processing of massive data with its two processes.

4.2. *Generation of frequent items*

The generating process of frequent item based on the MapReduce of Apriori algorithm is mainly that the calculation process of Apriori algorithm is distributed to multiple Map to calculate the number of local candidate sets. Finally, global candidate itemsets are generated by using Reduce. Its specific processes are as follows:

(1)Data block

First, the medical data is uploaded to the distributed file system, then the data is divided into N <key, value> data blocks by the working mechanism of MapReduce, and the data blocks are allocated to the M nodes in the cluster.

(2)Map process

Map function processes and scans input data records to generate pair of <key, value>, then taking this to generate frequent 1 set, and then generating a local candidate K itemset C_k for each data block, the generation algorithm is the same as the traditional Apriori algorithm, finally calculating the local support part_supp of local candidate itemsets C_k in the end. It is required to output intermediate key value (I_m, part_supp) in each map program, I_m is the item of local candidate itemset C_k, part_supp is local support of I_m.

(3)Combine function

Combine function is mainly to assist the Reduce function, when the Map function is completed, the local output of the Map function to do a combination and generate local frequent itemsets, then merging the key into r different partition by a hash code, and each partition perform an operation by using the Reduce function, in order to avoid excessive consumption of network resources because the data is too large, resulting in data delay.

(4)Reduce function

Each data node is obtained by using the Reduce function to sum the support degree of the same candidate itemsets in different data blocks, so as to get the global support.

(5)Comparison and fusion

Comparing global support degrees of the local candidate k-itemsets C_k with the minimum support threshold min_supp, in order to get the local frequent k-itemsets L_k. Then fusing local frequent k -itemsets L_k in the r partitions and generating the global frequent k-itemsets SUM_k. Finally, through the continuous iteration to generate the maximum frequent itemset, the algorithm ends.

The flow chart of the frequent items is shown in Figure 2.

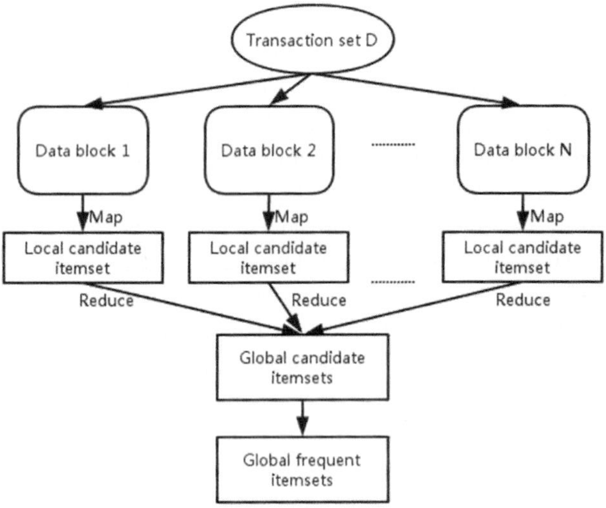

Fig. 2. Flow chart of frequent itemset based on MapReduce.

4.3. *Discovery of association rules*

The steps of using the MapReduce framework to compute the association rules by using frequent item generated are as follows:

(1)The data of each line is divided into a data block and the key-value generated is (line, S_i), the line is the row offset and S_i is the frequent item of this line.

(2)Map function scans (line, S_i) and generate the corresponding association rules (S_i, R_i), and R_i represents the corresponding association rules of S_i.

(3)The output (S_i, R_i) of Map function is to be ruled and store it in HDFS through MapReduce function.

(4)The confidence value and interest value in every association rule are compared with a preset threshold value, and draw the final association rules.

5. Conclusion

This paper improves the shortcomings of traditional Apriori algorithm by introducing the parameter of interest, and gives the design thought of the Apriori algorithm based on distributed combination by combining working mechanism of MapReduce. Compared to the traditional data mining algorithm, the improved algorithm of distributed data mining has the characteristics of reducing the number of original data scanning, multi node parallel computing frequent itemsets and so on, which not only reduces the time consumption, but also saves the computational overhead, and to a certain extent, ruled out the association rules for the diagnosis of misleading, make it more fit the actual situation. Therefore, the distributed data mining technology will be able to play its high efficiency, and has more extensive application in the field of medical information, such as prevention, diagnosis, treatment and analysis.

References

1. Gong T, Huang H, Li P, et al. A Medical Healthcare System for Privacy Protection Based on IoT[C]// Parallel Architectures, Algorithms and Programming (PAAP), 2015 Seventh International Symposium on. IEEE, 2015:217-222.
2. Kldiashvili E. Application of the Cloud Computing for the Effective Implementation of the Medical Information System[J]. International Journal of Natural Computing Research, 2014, 4(3):52-68.
3. Almaolegi M, Arkok B. An Improved Apriori Algorithm for Association Rules[J]. Eprint Arxiv, 2014, 3(1).
4. Hu L H, Ren Y L, He Z L. Improved Apriori algorithm based on classification and database compression[J]. Journal of Chengdu University of Technology, 2015, 42(1):110-114.
5. Qi S, Wong C U I. An Application of Apriori Algorithm Association Rules Mining to Profiling the Heritage Visitors of Macau[M]// Information and Communication Technologies in Tourism 2015. Springer International Publishing, 2015:139-151.
6. Reshamwala A. Improving Efficiency of Apriori Algorithms for Sequential Pattern Mining[J]. Bonfring International Journal of Data Mining, 2014, 4(1):01-06.
7. Zhang J H. Design and implementation of data mining based on distributed computing[J]. Applied Mechanics & Materials, 2014, 644-650:1702-1705.
8. Jula A, Sundararajan E, Othman Z. Cloud computing service composition: A systematic literature review[J]. Expert Systems with Applications, 2014, 41(8):3809-3824.
9. Lin W, Xiao X, Ghinita G. Large-scale frequent subgraph mining in MapReduce[C]// 2014 IEEE 30th International Conference on Data Engineering (ICDE). IEEE Computer Society, 2014:844-855.
10. Grolinger K, Hayes M, Higashino W A, et al. Challenges for MapReduce in Big Data[C]// Services (SERVICES), 2014 IEEE World Congress on. IEEE, 2014:182-189.

An improved genetic algorithm based on k-means clustering for solving traveling salesman problem

Li-Zhuang Tan, Yan-Yan Tan*, Guo-Xiao Yun and Chao Zhang

School of Information Science and Engineering, Shandong Normal University, Jinan, 250014, China

Shandong Provincial Key Laboratory for Novel Distributed Computer Software Technology, Shandong Normal University, Jinan 250014, China

**E-mail: yytan928@163.com*

www.sdnu.edu.cn

Based on k-means clustering, we designed and implemented an improved genetic algorithm(IGA) to find the best solution for the traveling salesman problem. IGA includes three characteristics of classical genetic algorithm(CGA). The first characteristic is the use of k-means clustering method on all cities to reduce the complexity of the problem by dividing the cities into several groups. The second characteristic is the adoption of genetic algorithm in each group for optimizing the sub-path. The third characteristic is the use of evolution operation on the inter-group for integral optimization. We performed many experiments to test the ability of IGA. Our experimental results show that IGA is more effective than CGA, especially for large-scale traveling salesman problems.

Keywords: Traveling Salesman Problem; Genetic Algorithm; K-means; Hybrid Mutation.

1. Introduction

Traveling Salesman Problem(TSP) is a classical combination optimization problem. The TSP was first formulated as a mathematical problem in 1930 and become increasingly popular after 1950[1]. Different types of TSP problems had been studied by the researchers during the recent years[2-7]. TSP has many real world applications in planning scheduling, VLSI routing problems, etc. In TSP, the salesman must be given a closed shortest tour between n cities.

G=(V,E) is a complete undirected graph with a vertex set V={1,2,…,n}, an edge set E. In this paper, we consider the distance from city i to city j is the same as from city j to city i. We define $d(c_i,c_{i+1})$ as the distance traveled between cities. The objective function is as Eq.(1)[3]:

*Corresponding author.

$$\min D = \left(\sum_{k=1}^{n-1} d(c_k, c_{k+1})\right) + d(c_n, c_1) \quad (1)$$

where c_k presents kth and c_1 is first and c_n the last city on the tour. If the coordinates $(c_i x, c_i y)$ of all cities (c_i) that the salesman will visit is known, we can rewrite Eq.(1) as Eq.(2)[4]:

$$\min D = \sum_{i=1}^{n-1} \left(\sqrt{(c_i x\text{-}c_{i+1} x)^2 + (c_i y\text{-}c_{i+1} y)^2} + \sqrt{(c_n x\text{-}c_i x)^2 + (c_n y\text{-}c_i y)^2}\right) \quad (2)$$

The exact algorithms can solve the TSP by a number of steps. But with the scale of TSP becomes large, for example,100 cities with approximately 10155 different solutions. Consequently, many bioinspired algorithms are carried out to obtain accepted solutions for NP-hard problems with short running time, such as ant colony optimization, simulated annealing, genetic algorithm and so on.

Genetic algorithm is an efficacious technique based on natural election for problems with huge search, such as TSP, in which the initial population decides iterations. The aim of GA is to achieve an approximate solution in a large-scale problem thought a couple of steps like selection, crossover, and mutation. Compared with other standard search algorithms, its advantages mainly consist in that the search is conducted using information of a population of tours instead of just one tour[6-8]. Aside from the foregoing content, the GA evaluates the unfit quality of the individual by the numerical value of fitness function, reduces the risk of being immersed in a local optimum when using heuristic algorithms.

K-means algorithm is one of the major algorithms of clustering analysis and a kind of clustering algorithm based on partitioning. The algorithm randomly chooses k points as the initial clustering centers. In order to reduce the complexity of TSP, especially for large-scale problems, we first use the k-means method to cluster cities into k groups. Then, genetic algorithms are carried out both in intra-group and inter-group for optimizing paths. An improved genetic algorithm(IGA) based on k-means clustering is proposed in this paper. In order to test the performance of our proposed IGA, a great quantity of experiments are implemented on IGA, and the experimental results are compared with those obtained by classical genetic algorithm(CGA) both on accuracy and running efficiency.

The remainder of this paper is organized as follows. Section 2 describes the k-means clustering algorithm used in this paper. Section 3 presents the genetic operations adopted in our proposed algorithm. Then in Section 4, experiments, comparing results and discussions are given. Finally, we conclude this paper in Section 5.

2. Clustering Method

The clustering method used in this paper is to convert a large-scale TSP into several small-scale problems by *k*-means. Its main idea is dividing *n* cities into several groups, and using genetic algorithm in each groups to find the shortest sub-traveling route. The use of the clustering is to accelerate the convergence and improve the accuracy of genetic algorithm.

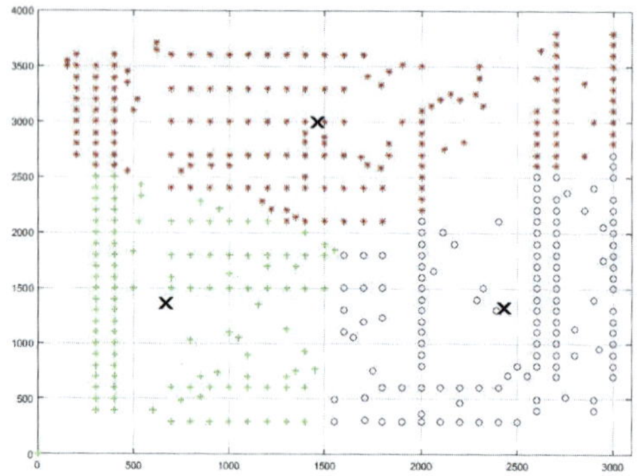

Fig. 1. *k*-means clustering.

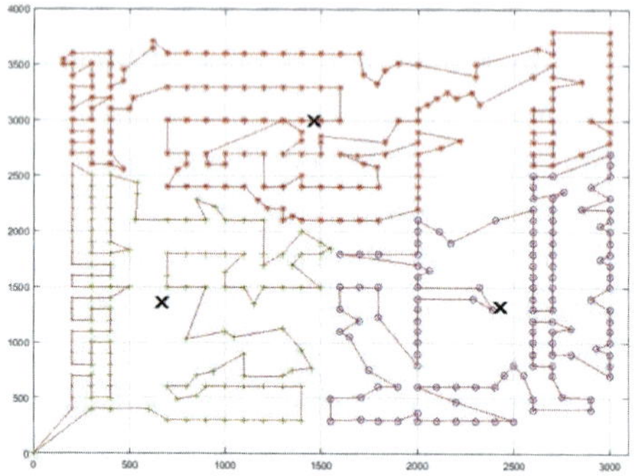

Fig. 2. Example result of clustering formation.

The first step of the clustering algorithm is to form *k* groups by *k*-means clustering with $k = [\sqrt{n} + 0.5]$, each groups is a cluster. We can number the cluster from 1 to *k*.

The second step is to find the shortest loop for the given vertexes, which divided into groups with genetic algorithm. Shown as Figure 2.

The third step is to choose an edge of the shortest loop randomly in each cluster, and deleted the edge and one of the two adjacent vertexes will be the start vertex and other will be the end. Then rewire the start and the end one by one. Shown as Figure 3.

The fourth step is to repeat the third step so as to generate the initial population.

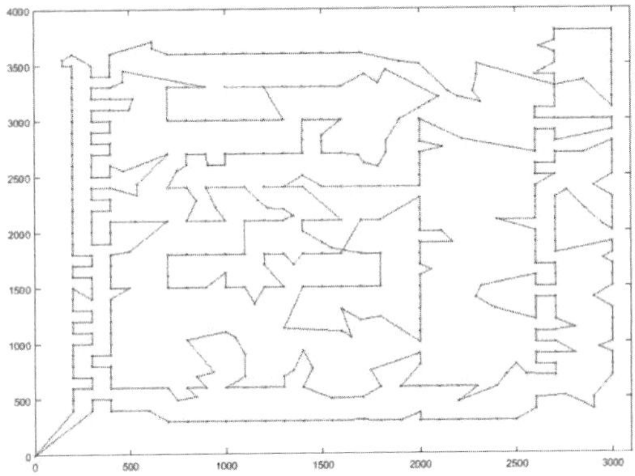

Fig. 3. The initial population.

3. Genetic Evolution

In the lower gradation, the genetic algorithm searches the shortest loop for each cluster. In the higher gradation, we delete the chosen edge from the above loop.

3.1. *Intra-group evolution operation*

Intra-group evolution operation, which applied cluster by cluster, is to find the shortest loop. We use genetic algorithm to optimize the loop during the operation.

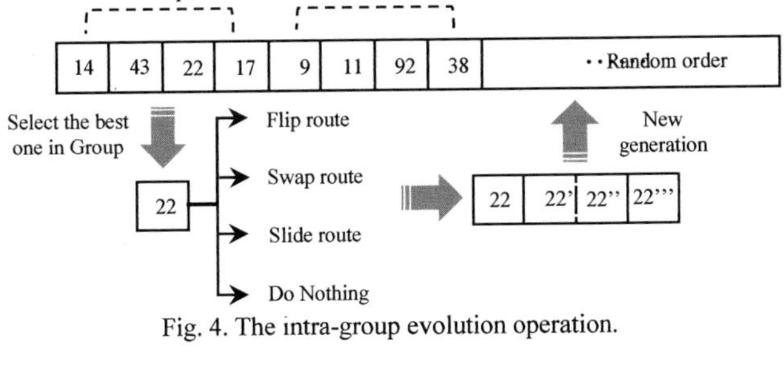

Fig. 4. The intra-group evolution operation.

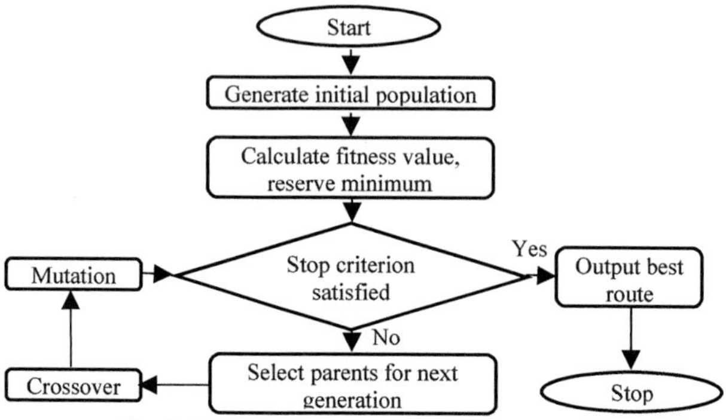

Fig. 5. The overall structure of the proposed IGA.

3.2. *Inter-group evolution operation*

The initial population was formed randomly based on the shortest loop in each cluster. Then it will be optimized by the genetic algorithm.

The population is initialized based on the rewired cluster shortest loop. First, two adjacent vertexes are chosen to be the start and the end vertex. Then a complete chromosome is formed by randomly arranging all the gene section. The next step is choosing the population size and a method for evolution. So far, the generated initial population has been built. Next, we calculate the fitness value of each chromosome and reserve the minimum. Selection is the process of choosing two parents for crossing. Crossover is to explore better solution. It is not performed on every pair of individuals and controlled by probability. Mutation is an operator which is critical to the solution of the genetic algorithm. Finally, the program stops when the criterion are satisfied.

4. Experiments

To investigate the performance of the proposed IGA based on clustering, we test the efficiency of IGA against the classical genetic algorithm (CGA) with MATLAB 2015b. we choose some benchmark test instances from TSPLIB[9]. Two comparing algorithms run 20 times for each test instance.

Table 1. Experimental results obtained by CGA and IGA under the same number of iteration.

TSP question	Population size	Iteration number	CGA		IGA	
			Best	Average	Best	Average
berlin52	50	200	9699.85	10162.13	8345.46	8541.15
	100	200	9427.18	9874.52	8254.63	8401.12
	200	200	8989.61	9474.03	8242.81	8356.11
	500	200	8799.56	9103.03	8163.27	8262.34
	1000	200	8285.21	8749.46	8084.73	8125.18
eil101	100	1000	808.61	837.09	677.02	689.88
	200	1000	767.09	791.43	673.01	684.40
	500	1000	740.58	758.59	668.25	673.64
	1000	1000	726.36	747.84	657.67	670.97
	2000	1000	714.39	722.35	656.62	665.50
bier127	100	1000	159400.49	163414.35	139944.03	145238.93
	200	1000	146106.62	153221.49	131478.22	134265.21
	500	1000	142237.61	148745.09	129671.72	130807.50
	1000	1000	140176.84	142931.92	127900.18	129944.84
	2000	1000	139835.04	142092.88	127052.11	129300.45
kro200A	100	1000	72116.92	74620.79	35230.71	36031.27
	200	1000	67067.89	69609.7	34946.30	35462.12
	500	1000	61672.51	64976.96	33543.25	34443.01
	1000	1000	59956.84	61628.26	32240.51	33296.84
	2000	1000	57244.11	59245.79	31512.77	32809.47

Table 2. Experimental results obtained by CGA and IGA under the same population size.

TSP instances	Population size	Iteration number	CGA		IGA	
			Best	Average	Best	Average
eil101	500	50	1599.41	1610.18	871.03	898.52
	500	100	1150.32	1391.23	728.04	753.72
	500	500	842.28	868.51	680.35	691.64
	500	1000	740.58	758.61	676.61	684.28
	500	2000	679.57	691.21	673.90	679.93
bier127	500	100	237123.17	269132.81	130807.25	137579.35
	500	500	164351.54	172929.27	129555.21	131478.19
	500	1000	142237.61	148745.09	125731.18	129273.77
	500	2000	127663.09	132703.63	123022.83	127551.81
	500	5000	122313.88	130210.18	120077.62	125280.82
kro200A	500	100	180921.11	177608.62	81808.22	89058.73
	500	200	137501.23	139441.09	48858.73	52588.77
	500	500	84559.72	88444.19	38819.12	38230.72
	500	1000	61672.51	64976.96	33909.47	34852.43
	500	2000	47152.68	48342.31	30636.73	31188.28
d657	1000	100	161902.88	176572.27	89979.84	91691.43
	1000	500	159170.02	162237.98	83291.11	86837.84
	1000	1000	138735.24	139983.61	79128.12	82170.33
	1000	2000	123594.08	134340.46	75864.72	76765.91
	1000	5000	109622.85	110753.05	57614.77	58399.32

Table 1 and Table 2 show the experimental results. It can be shown from the two tables that IGA performs much better than CGA. Under the same iteration, IGA can achieve a relatively better solution with a small population size while CGA need a large one. Furthermore, with the same population size, IGA can always solve the TSP effectively, especially on the large-scale test instances, the advantages of IGA are more obvious. As shown in Table 1, take berlin52 for an example, CGA achieves the result 8749.46 with 1000 population size, and IGA just uses population size of 50 to obtain the result of 8541.15. As shown in Table 2, take eil101 for an example, when the iteration number is under 500, IGA obtains a more better results than CGA. So as other test instances.

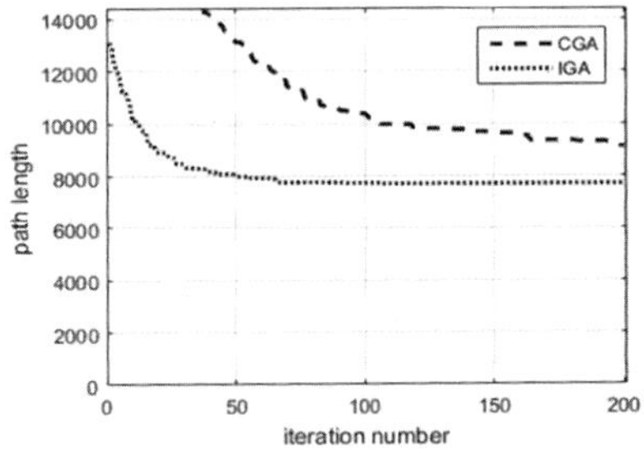

Fig. 6. Convergence curve of berlin52.

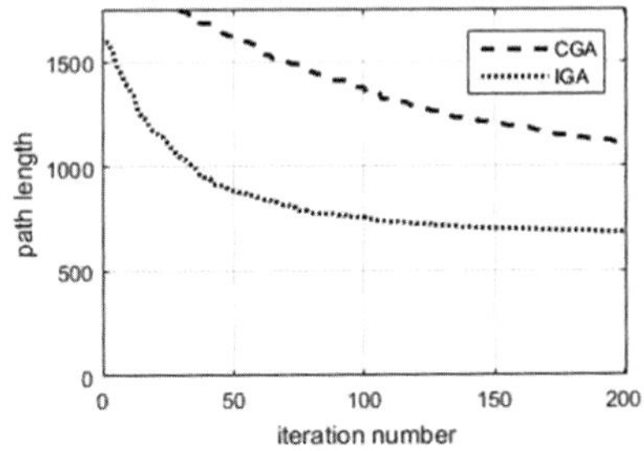

Fig. 7. Convergence curve of eil101.

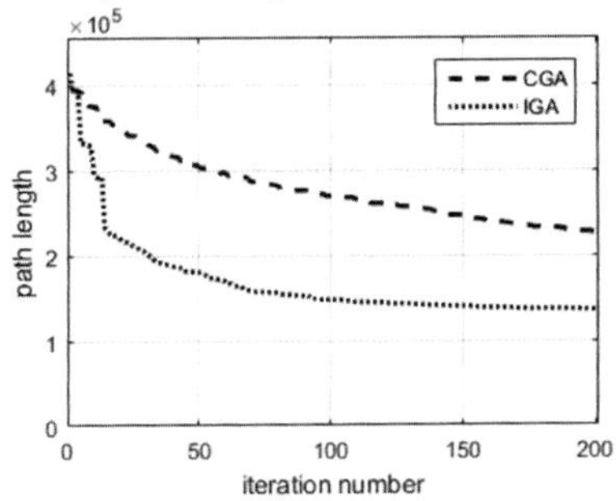

Fig. 8. Convergence curve of bier127.

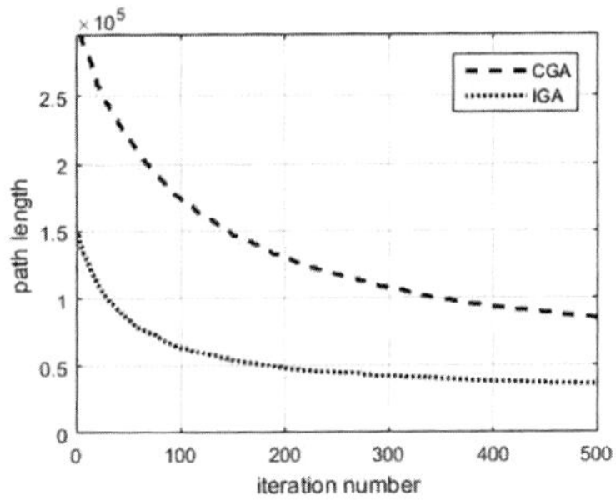

Fig. 9. Convergence curve of kro200A.

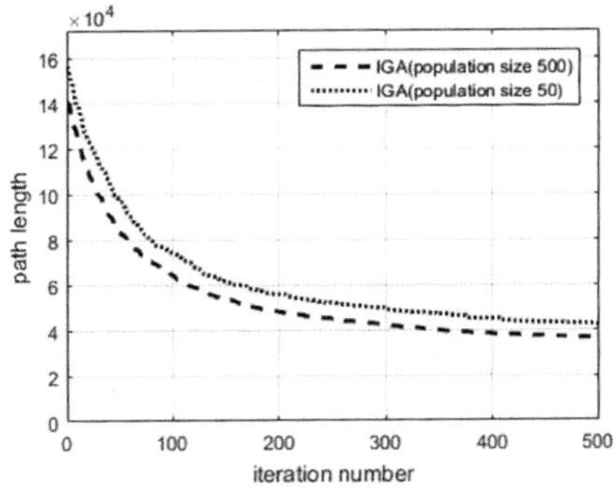

Fig. 10. Comparison between the population size of 50 and 500.

In Figure 6-Figure 9, the convergence curves of the test instances are shown. Obviously, IGA has an excellent convergence and efficiency compared with CGA.

In order to test the parameters sensitivity, we also test the performances of IGA when setting different population sizes. It can be seen from Fig.10 that, the solutions quality and the convergence speeds of 50 and 500 bodies are almost the same. We can get that the proposed IGA is less sensitive to population size. IGA behaves well on solving TSPs, its efficient performance attributes the success to clustering.

5. Conclusion

In this study, we propose a modified genetic algorithm with *k*-means clustering for solving traveling salesman problems. Firstly, k-means clustering method is used on all the cities to reduce the complexity of the problems, through which the cities are divided into several groups. Secondly, genetic algorithm is adopted in each group for optimizing the sub-path. Thirdly, evolution operation is used on the inter-group for integral optimization.

In order to test the performance of the proposed IGA, we run a great quantity of experiments and compare with classical GA (CGA). Experimental results show that, the convergence speed of IGA is faster, and the results obtained by IGA are better than those obtained by CGA. IGA is more effective than CGA, especially for large-scale traveling salesman problems.

Acknowledgments

This work is supported by the Natural Science Foundation of China (No.61401260), Shandong Province Young and Middle-Aged Scientists Research Awards Fund (No.BS2014DX006), and Shandong Normal University Training Programs of Innovation and Entrepreneurship for Undergraduates (No.201501259).

References

1. Samir Maity, Arindam Roy and Manoranjan Maiti, A Modified Genetic Algorithm for solving uncertain Constrained Solid Traveling Salesman Problems, *Computers & Industrial Engineering*, 83(2015).
2. Lin S, Computer solution of the traveling salesman problem. *Bell Syst Technical Report*. (MITRE Corporation,1965).
3. Davendra D. Traveling Salesman Problem, theory and applications.(In Tech, Croatia 2010).
4. Ding-Chao, Cheng-Ye, and HE-Miao, Two-Level Genetic Algorithm for Clustered Traveling Salesman Problem with Application in Large-Scale TSPs. *Tsinghua Science and Technology* 12,4 (2007).

5. Chiranjit Changdar, G.S. Mahapatra, Rajat Kumar Pal, An efficient genetic algorithm for multi-objective solid travelling salesman problem under fuzziness, *Swarm and Evolutionary Computation*, 15 (2014).
6. Yong-Deng, Yang-Liu, and Deyun-Zhou. An Improved Genetic Algorithm with Initial Population Strategy for Symmetric TSP, *Mathematical Problems in Engineering,* 3(2015).
7. Zakir Hussain Ahmed, Improved genetic algorithms for the travelling salesman problem, *Int. J. Process Management and Benchmarking,* 4, 1(2014).
8. TSPLIB,(2015),http://www.iwr.uni-heidelberg.de/groups/comopt/software/ TSPLIB95.

Research on HOG algorithm based on compressed sensing

Xi Chen* and Song Wang†
Computer & Communication Engineering Institute, Changsha University of Science & Technology,
Changsha 410114, China
*E-mail: *chentianjun@163.com, †cswangsong@126.com*

To solve the problem of extraction algorithm of high complexity for large-size images feature, sketch retrieval and compressed sensing theory is fused in this paper. The theory on a small amount of measurements can accurately reconstruct the original signal, thus, HOG algorithm based on compressed sensing is presented. Firstly, the algorithm is needed to preprocess and block in sketch retrieval. Then by means of the compressed sensing principle of block image feature extraction, a small amount of compressed image features are obtained. Finally, we used the weighted distance method to calculate measurement features of similarity, which is then used to implement the image retrieval. Our experimental evaluation which uses two basic benchmarks shows that our proposed algorithm is more accurate in image retrieval than other algorithms.

Keywords: Sketch Retrieval; Matrix Operation; Feature Extraction; Weighted Distance.

1. Introduction

In recent years, with the rapid development of Internet technology and multimedia information technology, multimedia information increased dramatically and the multimedia data is a serious problem in image retrieval. Hand-drawn sketch based on image retrieval [1] (Sketch Based Image Retrieval, SBIR) has become a hot research at home and abroad. Through literature study, found that at present many SBIR methods are computational complexity is high, and feature extraction ,feature matching need long time, feature information is easy to lack of problems, especially for large size image retrieval. Therefore, by researching new image retrieval algorithm to improve existing algorithms for solving the current image retrieval problems, it has a very important significance. The compressed sensing theory shows that ,under certain conditions, by a small number of measurements may accurate signal reconstruction [2][3] the original signal, thus indicating that the measured value of the few can maintain the

† Corresponding author.

structure and sufficient information about the original signal[4][5]. Currently this theory has been investigated in image processing , i.e., digital watermarking, image fusion etc. and achieved initial results. These research ideas can be extended to the field of image retrieval, and compressed sensing theory can be proved theoretically be applied to retrieval.

Based on the above background, the paper studies the basic theory of compressed sensing sketch retrieval technology, and the related theory is applied freehand sketch retrieval process in feature extraction and matching, meanwhile it put forward a new sketch retrieval method. The method of innovation lies in introducing compressed sensing for image feature extraction, using points blocked polynomial deterministic measurement matrix of image is divided into blocks measurement, and solves the problems of traditional image retrieval methods in a certain extent.

2. Compressed Sensing Theory

Compressed sensing, CS, as a new sampling theory, which uses the sparse nature of the signal, and less than beyond the traditional Nyquist frequency signal is sampled to obtain compressed data, then utilize numerical optimization algorithm to reconstruct the original signal.

Suppose that the original signal X : $N \times 1$ in Sparse transform base Ψ : $N \times N$, Conversion coefficients are sparse, so X base Ψ in Conversion coefficients Θ can be expressed:

$$\Theta = \Psi^T X \tag{1}$$

From the above description, Compressed sensing theory is to use the transform matrix and metric matrix to transform the dimensional signal into the observation set, so as to realize the data compression. Compressed sensing theory is to use the transform matrix Ψ and metric matrix Φ to transform the N dimensional signal into the M dimensional observation set Y , so as to realize the data compression. According to the theory of compressive sensing, when the signal is reconstructed, it only needs to be solved:

$$\min \|\Theta\|_0 \; st. \Phi\Theta = \Phi\Psi^T X = Y \tag{2}$$

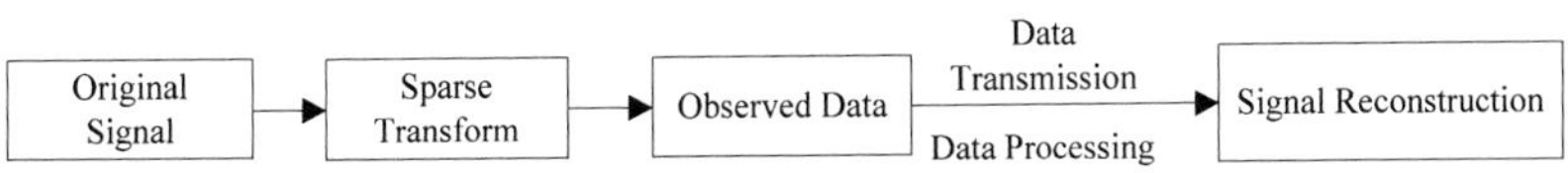

Fig. 1. Framework of CS theory.

3. Sketch retrieval based on Compressed Sensing

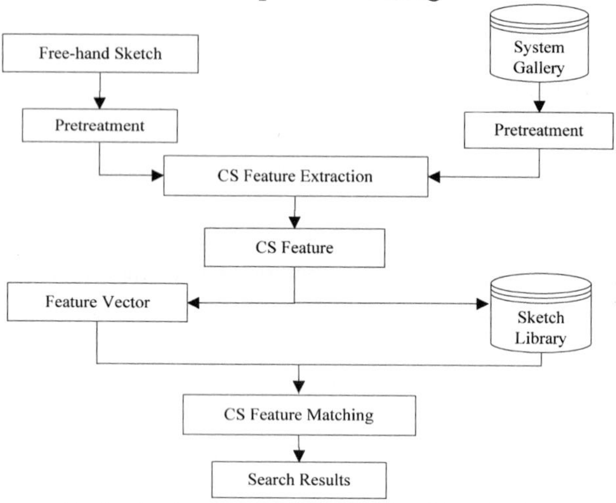

Fig. 2. the Architecture of sketch retrieval based on CS.

Figure 2 is introduced into the theory of compressed sensing in image feature extraction process, which is used to express the cs feature that extracted from the feature vector and sketches feature library, then to match.

3.1. *Sketch retrieval algorithm based on compressed sensing*

algorithm Sketch retrieval algorithm based on Compressed Sensing

Input: 70 Sketch of different complexity;

Step 1 Image library in the image using Canny edge detection method to preprocessing, access to sketch gallery; According to step 2 in 3.2 ,block a image into each sub-block image and calculate the average grey value, and then to count statistical edge information in five kind of edge feature;

Step 2 According to step 4, step 5 in 3.2, extract compression sensing feature of the sketch and image, and get the block measure vector, block difference measurement;

Step 3 According to step 2, step 3 in 3.3, match compression sensing feature of the sketch and image, meanwhile;

Step 4 Finally obtain the top 10 images of similarity, and order the best similarity.

Output: 10 images of the best similarity retrieval;

3.2. *Feature extraction*

Before utilizing compressed sensing extract image feature in SBIR, there is need to preprocess the image, Generally including Cannny edge detection[6]and Berkeley boundary detection[7],In this paper, the algorithm uses Canny edge detection for image preprocessing.

Step 1 The input image (size M× N) is converted to grayscale, Divided into 4×4 to 16 non-overlapping sub-image block, and then divide each sub image block into a number of sub blocks with a certain number of non overlapping blocks.

Step 2 Divide each sub image block into the same 2×2 sub blocks, and digital 0-3 mark 4 subblocks, then Calculate the average value of the gray level in each sub block, and denoted by a_0,a_1,a_2,a_3. Add up edge information in vertical, horizontal, 45 degree angle, 135 degree angle, and non direction. Five kinds of edge features are represented by five kinds of filter coefficients: $f_v(z)$, $f_h(z)$, $f_{45}(z)$, $f_{135}(z)$, $f_{non}(z)$, $z\in[0,3]$,Calculate the corresponding amplitudeits about image block:

$$A=\sum_{z=0}^{3} a_z \times f_j(z),\ j\in\{v,h,45,135,non\} \tag{3}$$

According to statistics, obtaining a five dimensional vector G_j, $j\in\{v,h,45,135,non\}$, the dimension of each image is $5\times 4\times 4$=80.

Step 3 Building block polynimial matrix Φ :

$$\Phi=\begin{bmatrix} \Phi_B & & & \\ & \Phi_B & & \\ & & \ddots & \\ & & & \Phi_B \end{bmatrix} \tag{4}$$

Among Φ_B Orthogonal Gaussian random matrix, the size is 80×256, Φ is matrix($M\times N$) , M=$80\times$ K, N=$256\times$ K.

Step 4 Obtaining the column vector Xj (256× 1) by scanning the j sub block image of an image. Acording to Eq.5,calculating the image perception measurement vector about j image blocks :

$$Y_j=\Phi_B X_j,\ j\in[1,K] \tag{5}$$

Step 5 Where Y_j is a measured value of the j block image, Y_j^* is a retrieved measured value of the j block image, so the Block measurement difference can be written as:

$$U=\left\|Y_j-Y_j^*\right\|_1 \tag{6}$$

3.3. *Feature matching*

In this paper, weighted distance is used to calculate the characteristic matching degree of the image, as shown in Eq.7.The closer the distance between the two images, the greater the degree of similarity ,according to the similarity of the output image from big to small order, we can get the final sketch retrieval results.

Step 1 According to the CS feature extraction step, the feature vector of compressive sensing measurement is extracted.: Y_v, Y_h , Y_{45}, Y_{135}, Y_{non}, Y_v^*, Y_h^*, Y_{45}^*, Y_{135}^*, Y_{non}^*.

Step 2 We can calcualte similarity matching degree between sketch and image database by using the Eq.7:

$$w=w_1\left\|Y_v-Y_v^*\right\|_2+w_1\left\|Y_h-Y_h^*\right\|_2+w_1\left\|Y_{45}-Y_{45}^*\right\|_2+w_1\left\|Y_{135}-Y_{135}^*\right\|_2+w_1\left\|Y_{non}-Y_{non}^*\right\|_2 \quad (7)$$

where w_1, w_2, w_3, w_4, w_5 is weighted coefficient, $\|b\|_2=\sqrt{\sum_{j=1}^{n}\left|b_j\right|^2}$.

Step 3 In addition, given matching threshold $\Omega=0.55$, if $w<\Omega$, so the output result is sorted according to the value from small to large.

4. Experiment

4.1. *Experimental deployment*

In order to verify the validity of the proposed method, we randomly select 70 sketches and the corresponding image as the experimental image database from the literature [8] and [9] standard database used in the standard database. Each sketch corresponds to a class, each class 40, the 70 class contains a total of 2800 related images. The experiments is evaluated from the recall, precision, parallel speedup and efficiency. Precision and recall rate standards for the evaluation standard is commonly used in image retrieval by using Eq.8 and Eq.9:

$$\text{Precision}=\frac{\text{Number of output related image retrieval}}{\text{Total output image to be retrieved}} \quad (8)$$

$$\text{Recall}=\frac{\text{Number of output related image retrieval}}{\text{Total number of related images in image database}} \quad (9)$$

4.2. *Experimental result*

For comparison, here we choose flower and sailing as a sketch to be retrieved, in figure 3. The algorithm and the other three have made a experimental simulation , and each algorithm returns only the top 10 similar images, which is shown in Figure 4, Figure 5.

Fig. 3. The sketch image to be retrieved.

Fig. 4. The retrieval results for different algorithms-flower.

Fig. 5. The retrieval results for different algorithms-sailing.

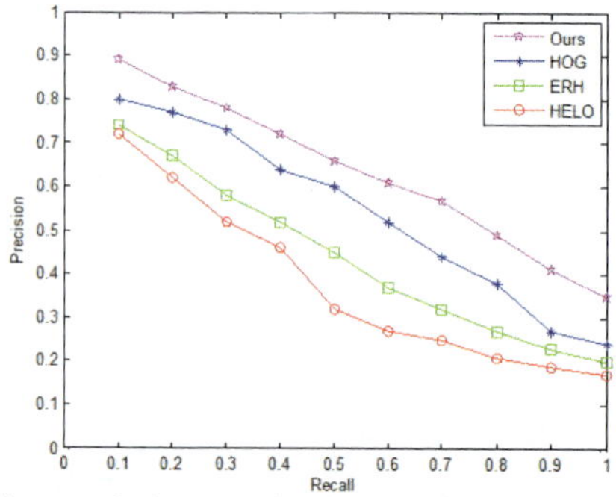

Fig. 6. The comparison results between algorithms using the method of precision-recall.

With the other three algorithms, the proposed algorithm is compared with the precision and recall of the experimental results is shown in Figure 6, shows that ours algorithm of precision and recall rate are higher than those of the other three algorithms.

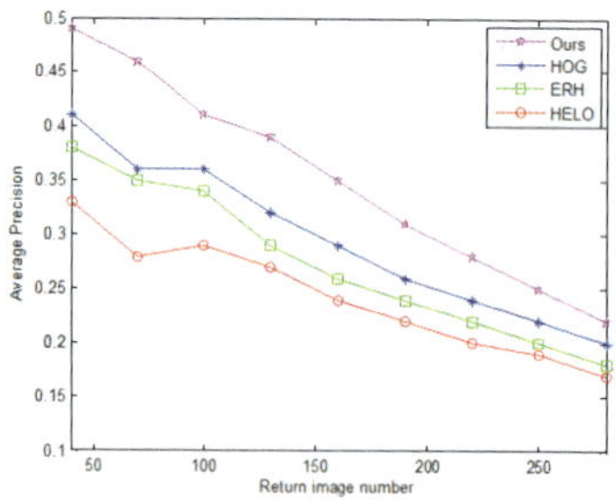

Fig. 7. The comparison of average precision.

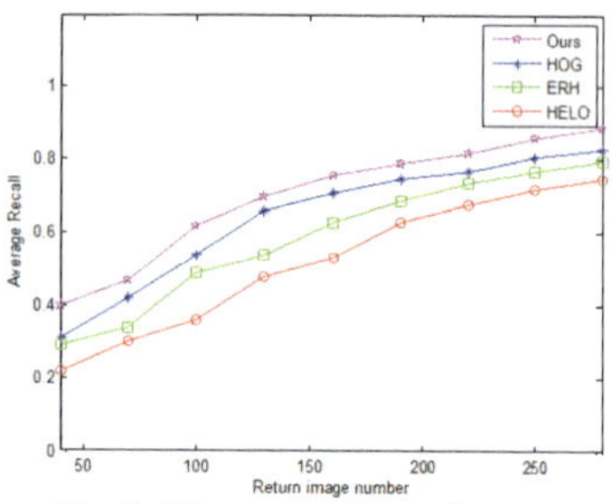

Fig. 8. The comparison of average recall.

From Figure 7,with the increase of returns the number of images, all the average precision of the algorithm are falling, and tend to be the same, but our average precision rate slightly higher than the other algorithms. From Fig.8,with the increase of returns the number of images, all the average precision of the algorithm are raising, but our average recall rate slightly higher than the other algorithms. It is proved that the algorithm is feasible and the image retrieval ability is better.

5. Conclusion

We propose a new SBIR method based on CS, which use compressed sample characteristics and the theory is applied to image feature extraction and sketch. In the process of retrieval of feature extraction and matching application by CS principle, we achieve the purpose of dimension reduction for the feature vector.

Compared with other methods, the proposed method shortens the feature extraction and matching time, furthermore, the algorithm is computationally efficient. Our future work will focus on researching on parallel CS with applications to big data platform.

References

1. Sousa P, Fonseca M J. Sketch-based retrieval of drawings using spatial proximity.[J]. Journal of Visual Languages & Computing, 2010, 21(2):69-80.
2. Key Laboratory of Instrumentation Science amp Dynamic Measurement North University of China, North University of Chi. An Overview of Compressed Sensing Implementation and Application[J]. Journal of Detection & Control, 2014.
3. Ling-Juan Y U, Xie X C. Brief Introduction of Compressed Sensing Theory[J]. Video Engineering, 2008.
4. Zhou Yan, Zeng Fan-zhi, Lu Yan-sheng,et al.. Research on image retrieval method based on compressed sensing[J]. Journal of Zhongshan University: Natural Science Edition, 2014..
5. He Guo-dong, Xie Xiao-juan, Yang Lin-yun, et al. Signal reconstruction based on Compressive Sensing[J]. Radio communication technology, 2014, 40(3): 26-28.
6. Liu G, Fan J. An improved Canny edge detection algorithm[J]. Microcomputer & Its Applications, 2013.
7. Ren X. Boundary Detection[M]// Computer Vision Springer US, 2014:59-64.
8. Eitz M, Hildebrand K, Boubekeur T, et al. Sketch-based image retrieval: Benchmark and bag-of-features descriptors[J]. Visualization and Computer Graphics, IEEE Transactions on, 2011.
9. Hu R, Collomosse J. A performance evaluation of gradient field hog descriptor for sketch based image retrieval[J]. Computer Vision and Image Understanding, 2013, 117(7): 790-806.

Chapter 4

Data Mining and Cloud Computing

Optimizing resource-service sequence based on genetic algorithm for collaborative task systems

Xiu-Yang Lei*, Hai-Bo Li† and Ya-Feng Gao‡
College of Computer Science and Technology, Huaqiao University, Xiamen, Fujian 361021, P.R. China
*E-mail: *leixiuyang@foxmail.com, †lihaibo@hqu.edu.cn*
www.hqu.edu.cn, ‡yafeng.feng@qq.com

In collaborative task systems, resource service sequences can complete more complex tasks and provide more efficient and better services than single resource services. During a business process execution, resource services are used in a certain sequence, which is called Resource-Service Sequence (RSS). A RSS mining and optimization approach based on Genetic Algorithm (RSS-GA) is proposed. In our approach, the degrees of temporal relationship between two resource services are first resolved via historical data. Initial RSSs, which are represented as a temporal constraint matrix, are then obtained. Next, a genetic algorithm is used to optimize the initial RSSs. Lastly, a simulated data set is used to test our algorithm and the experimental results prove that our approach is stable, consistent and effective.

Keywords: Resource-Service Sequence; Genetic Algorithm; Collaborative Task System.

1. Introduction

Collaborative task system (CTS) is used to support effective collaboration across enterprise boundaries. The growing trend of cross-organization collaboration has many causes, such as globalization of markets, the increasing complexity of modern products and services, and the emergence of new information and network technologies. Taking cloud manufacturing platform [1] as a example, it utilizes the web service technology to encapsulate dispersed and diverse manufacturing resources, which are provided by various enterprises and heterogeneous systems, into services. The resource services can be managed and operated in an intelligent and unified way [2].

In the process of cross-organization collaboration, various resource services are used by business processes according to their models. With the business process execution, the temporal relationships between resource services are

† Corresponding author.

formed. Therefore, from a global perspective, mining the temporal relationships can manage and optimize resource services in a sequence way. This is because resource service sequence can provide more efficient and better value-add services [2], and complete more complex tasks in a coarse-grained mode. It is also helpful for enterprises to reduce the selection time of resources, optimize business process models, and improve profits.

If some resource services are used frequently in a certain sequence, which is called the Resource-Service Sequence (RSS). Our problem is to mine and optimize the RSSs from resource services from historical data for CTS. We call this problem RSSMO (Resource-Service Sequences Mining and Optimization).

At present, many researches on relationships of resources for CTS were proposed. A unified multi-granularity resource model was proposed to compose manufacturing resources in a process in [3]. In [4], attributes of QoS were used to compose resources. However these methods were focus on the spatial relationships between resources, but not paid more attention to temporal relationships. In [5], an approach based on dependencies of resources was proposed to cope with the temporal relationships between resources.

Some existing researches have used GA for mining various data sequence patterns, which include mining closed sequential patterns [6], finding sequential web pages [7], and mining negative sequential patterns [8]. These algorithms are efficient to solve specific problems, but not suitable for all problems of sequential patterns mining.

In this paper, we propose a new approach called RSS-GA (Resource-Service Sequences Mining and Optimization based on Genetic Algorithm) to resolve the RSSs for CTS. The RSSMO is an optimizing searching problem. RSSMO can be divided into two stages, mining initial RSSs and resolving the optimization of RSSs. Also, we present a value function to describe the business targets of enterprises, so as to optimize the collaborative capabilities of business processes.

2. Problem Formulation

2.1. *Problem statement*

In CTSs, enterprises collaborate to accomplish complex business processes. A business process is composed of many business activities, which execute in a certain sequence. Then, the resource services which are used by activities can form some sequences as well. A business process can be defined as follows.

Definition 2.1. *Business process BP is a 4-tuples, (id, Activity, $\prec$, R), where id is a unique identifier of a business process, Activity is a set of activities, $\prec$ is a temporal order between activity such that for any activities a_i, $a_j \in$ Activity, $a_i \prec a_j$ indicates that a_i is precursor of a_j, a_j is subsequence of a_i. R is a set of resource services for entire business process.*

With the executions of business process, resource services are used frequently in sequences, RSSs are formed. A RSS can be defined as follows.

Definition 2.2. *A resource-service sequence RSS can be denoted as RSS=$<r_1,r_2,\ldots, r_n>$, where $r_i \in R$, $1<n\leq L$, where L is the maximum length of BP from the start activity to the end activity. For all resource services r_i, $r_j \in$ RSS, if $i<j$, the service time of r_i is earlier than the service time of r_j.*

Our problem is to resolve the set of optimal RSS via mining business historical data using evolutionary algorithm.

2.2. *Strategy of RSS mining and optimization*

The RSS-GA can be divided into two stages: (1) Mining initial RSSs - At this stage, the temporal relationships between any two resource services are resolved by mining historical data, and (2) Resolving the optimization of RSSs – At this stage, we optimize the initial RSSs by genetic algorithm.

For the first stage, a temporal relationship matrix T is built to represent the temporal degrees of any two resource services, denoted as $T(r_i, r_j)$. We adopt an unsupervised statistical clustering method to obtain the temporal degrees, as the values in the matrix, denoted as *freq*($<r_i, r_j>$). Then, we filter the resource services with a threshold min_{of} (minimum occurrence frequency) and min_{vf} (minimum value). The two thresholds are used to keep those temporal relationships which have higher temporal degrees and higher values. Finally, the RSSs are obtained via the matrix.

RSSs should satisfy the requirements of different industrial features [9], therefore, we introduce a function called value function to evaluate RSSs. Value function is defined as follows.

Definition 2.3. *Value function is a function to calculate value of RSS by attributes of resource services, denoted as VF.*

For example, if the manufacturing enterprises are interested in the costs of RSSs, value function can calculate the average unit price of resource services in RSS as the value of RSS.

3. RSS-GA: Genetic Algorithm for RSS

In this section, we describe our genetic algorithm called RSS-GA for resolving the optimization of RSSs for CTS. Firstly, we present our chromosome encoding schema. And then we design the genetic operators. Finally, we give the structure of RSS-GA.

3.1. *Chromosome encoding schema*

In RSSMO, integer encoding scheme is used to represent resource services. A RSS is represented by a chromosome. a resource service in a RSS is mapped into a gene in a chromosome. Given a $RSS=<r_1, r_2, \ldots, r_n>$, it is mapped into a chromosome with n-genes, where $n \leq L$, where L is the maximum length in a business process model. For example, a $RSS=<r_8, r_4, r_2, r_9, r_1>$ is mapped into a 5-gene chromosome <8, 4, 2, 9, 1>.

3.2. *RSS-GA algorithm*

RSS-GA uses genetic operators which includes selection, crossover, and mutation to generate offspring of exiting population. RSS-GA will select individuals regarding to fitness function, which is transformed from value function. The roulette-wheel selection is applied in selection operator.

Then crossover operator takes place. Crossover operator is used to replace some of genes in parent with corresponding genes of other, and then creates two new offspring. In our problem, a single point crossover is applied. Firstly, choosing randomly a crossover position j, and then replacing everything before the point. Crossover can be shown as Figure 1.

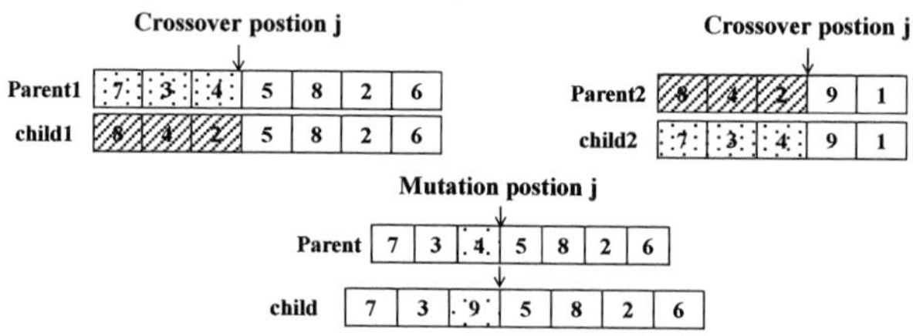

Fig. 1. Crossover and mutation example.

RSS-GA selects individuals regarding to fitness function, which is transformed from value function *VF*. The roulette-wheel selection is applied in selection operator.

The pseudo code of the algorithm is provided in Algorithm 1.

Algorithm 1. RSS-GA

Input: T, the temporal constraint matrix
  VF, value function
  R, RSs set used by business process
  p_c, the crossover probability
  p_m, the mutation probability

Output: $RSSS'=\{RSS_1', RSS_2', \ldots, RSS_n'\}$, optimal $RSSs$

$P \leftarrow$ initialize(R, TCM);
fitness(P, VF); $gen \leftarrow 0$;
while($gen < MAXGEN$) **do**
  $NewP \leftarrow$ select(P); cross($NewP$, p_c); mutate($NewP$, p_m);
  $P \leftarrow NewP$; fitness(P); $++gen$;
end while

4. Performance Evaluation

We evaluate the performance of the proposed approach on a inventory data set, collected from an appliances enterprise. simulated data set in our experiment. The dataset consists of 60,000 items, 30 resource services for 1000 projects, each with 13 attributes, including unit price and service time.

We used the average unit price of resource service to represent the value function of a RSS. Here 7 parameters should be determined: minimum occurrence frequency (min_{of}), minimum value function (min_{vf}), the maximum length of RSS (L), maximum number of generation (MAXGEN), population size ($POPSIZE$), crossover probability (p_c), mutation probability (p_m). In our experiments, we sets min_{of}=0.03, min_{vf}=40, L=10, $MAXGEN$=500, p_c=0.9, p_m=0.08, and $POPSIZE$= [50,100, 150, …, 500]. The performance of RSS-GA is shown in Fig.2, our algorithm has steadiness in mining and optimizing RSSs.

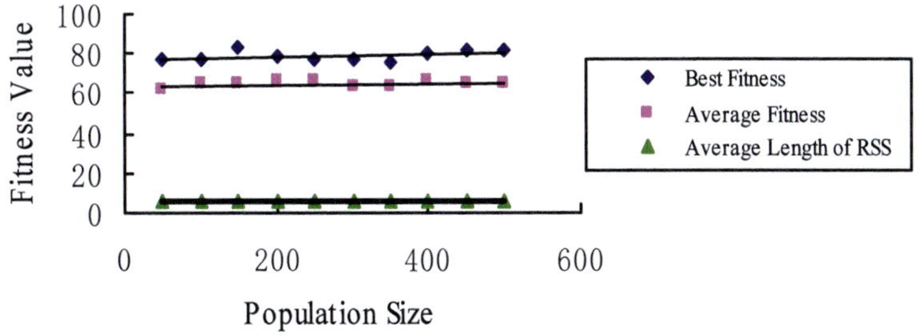

Fig. 2. Experimental results for population size changes from 50 to 500.

5. Conclusions

In this paper, an approach RSS-GA is proposed to mine and optimize resource service sequence for collaborative task systems. RSS-GA can generate various RSSs with different lengths, which for different enterprises business targets. We will focus on more efficient strategies of optimization in future.

Acknowledgments

This work is supported by grant 2015Z125 of the Quanzhou Science and Technology Plan Projects, and grant 1400414003, 1511314028 of the Subsidized Project for Cultivating Postgraduates' Innovative Ability in Scientific Research of Huaqiao University.

References

1. H.B. Li, Approach to multi-granularity resource composition based on workflow in Cloud Manufacturing, *Comput. Integr. Manuf. Syst*, **19**, 210 (2013).
2. G.D. Jules, M. Saadat, and S. Saeidlou, Holonic Ontology and Interaction Protocol for Manufacturing Network Organization, *IEEE Trans. Syst. Man & Cyber. Syst* **45,** 819 (2015) .
3. W. Tan, X. Yao, and X. Liu, A Unified Multi-coarse Manufacturing Resource Model and Its Semantization Based on Extended OWL-S, *Inf. Tech. J* **13**, 1831 (2014).
4. F. Tao, Y. Cheng, L.D. Xu, L. Zhang, and B.H. Li, CCIoT-CMfg: Cloud Computing and Internet of Things-Based Cloud Manufacturing Service System, *IEEE Trans. Ind. Informat* **10**, 1435 (2014).
5. H.B. Li, K. Chen, M.X. Liang, and X.Y. Luo, Composition of Resource-Service Chain for Cloud Manufacturing, *IEEE Trans. Ind. Informat* **PP**,1 (2015).
6. V.P. Raju and G.P.S. Varma, Mining closed sequential patterns using genetic algorithm, in *Proc. IEEE Int. Conference on Advanced Communication Control and Computing Technologies (ICACCCT)*, (Tamilnadu, India, 2014).
7. M. Saravanan and V.L. Jyothi, A novel approach for sequential pattern mining by using genetic algorithm, in *Proc. IEEE Int. Conference on Control, Instrumentation, Communication and Computational Technologies (ICCICCT)*,(Kanyakumari, India, 2014).

8. Z. Zheng, Y. Zhao, Z. Zuo, and L. Cao, An Efficient GA-Based Algorithm for Mining Negative Sequential Patterns, in *Proc. 14th Pacific-Asia Conference on Advances in Knowledge Discovery and Data Mining (PAKDD 2010)*, (Hyderabad, India, 2010).
9. F. Tao, Y. Laili, L. Xu, and L. Zhang, FC-PACO-RM: A Parallel Method for Service Composition Optimal-Selection in Cloud Manufacturing System, *IEEE Trans. Ind. Informat* **9**, 2023 (2013).

GSkiplist: A GPU-method to accelerate key-value database search

Xin Zhao*, Kai Lu, Xiao-Ping Wang and Gen Li
Collage of Computer, National University of Defense Technology,
Science and Technology on Parallel and Distributed Procession Laboratory,
Collaborative Innovation Center of High Performance Computing,
Changsha, 410073, China
**E-mail: zhaoxin10@nudt.edu.cn*
www.nudt.edu.cn

The performance of key-value handling in database is not fast enough in situations with large-scale data sets and queries. Database searching is usually a type of data-thick application, which operates on relatively simple logic when handling requests and can be accelerated on the Graphics Processing Unit (GPU). In this paper, we designed a new data structure named as GSkiplist that can be implemented on GPU effectively. We also described algorithms used in database operations such as inserting, deleting and searching. We chose OpenCL as our programming platform and the experimental results proved that implementing GSkiplist on the GPU can achieve speeds up to 8 times the speed when it is implemented on the CPU.

Keywords: Skiplist; Key-Value Database; OpenCL; GPU.

1. Introduction

Due to the rapid growth of the web and the number of web users, web search engines are faced with enormous performance challenges. Current large-scale search engines are based on data sets of many terabytes, and have to be able to answer tens of thousands of queries per second over tens of billions of pages. To reduce response time, a novel solution is to maximize the quantity of queries at one time taking advantage of GPU's computing feature.

In this paper, we proposed a new approach to make concurrent queries on key-value database using Graphical Processing Units (GPUs). We also redesigned an original structure of skiplist in order to fit better on GPU.

GSkiplist inherited the idea of skiplist[1] invented by William Pugh of the University of Maryland around 1989. It is a viable alternative to balanced trees such as AVL trees or to self-adjusting trees such as splay trees. It uses

* Corresponding author.

probability to ensure balance instead of whirling like in balanced tree, so that it is easier to implement, and with higher efficiency. The time complexity O(logn) is the same, but the constant term of it is relatively smaller. Figure 1 shows a general skiplist data structure.

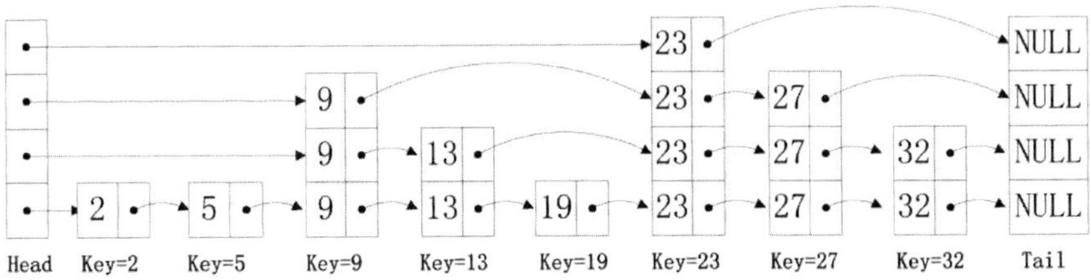

Fig. 1. A traditional skiplist. Each box with an arrow represents a pointer and a row is a linked list giving a sparse subsequence; the numbered boxes at the bottom represent the key for each node. Searching proceeds downwards from the sparsest subsequence at the top until consecutive elements bracketing the search element are found.

Through normalizing the skiplist nodes which is defined as random height in the classical definition, the GPU oriented GSkiplist is more suitable to run on GPUs. Moreover, the retrieve algorithm of GSkiplist map the forward node comparison of GSkiplist node to different GPU threads and handle multiple key-value requests concurrently, therefore the efficiency of GSkiplist is optimized.

2. Approach

In this section, we are going to cover several design key points of GSkiplist including its structure and operations on database.

2.1. *Structure design of GSkiplist*

In traditional definition of skiplist, the height of the node is a random number. So the number of cycle is also random when comparing the backward pointers. That brings two challenges when designing the GPU accelerating arithmetic.

Firstly, the random levels cause branches when GPU executes loop unrolling with multi-threads. Secondly, it is hard to determine a certain step length when searching in a skiplist when the level is random. As GPU is suitable for accelerating geometric calculations such as the rotation and translation of vertices into different coordinate systems, it is necessary to do some change for the existed skiplist data structure.

We come up with a new structure called GSkiplist for the GPU searching in key-value database. The main idea is simple: Fix the height of each node and change the linked list structure to a two-dimensional array, so that it is possible to use fixed number of threads to unrolling the outer for cycle. What's more, the same step length for searching is beneficial for implementation.

Table 1. The basic data structure and simple explanation of GSkiplist.

Name	Type	Explanation
state	bool	Open for activation and close for invalid
key	Key type	Stored a key generated by certain hash function
value	Value type	Related to the database
index	int	An array connected the index with the key
next[index][level]	int	Stored the backward index of the current node
level	int	A tag for different searching threads
randomLevel	int	The height generated when inserting a new node

Apart from the given the definition of node in GSkiplist as Figure 2 above, we still need several global arrays such as findIndex[] and findKey[] to record the connection between index and key, which are both generated or altered during the operations of GSkiplist. The idea of converting the skiplist into a "flat" form can be clearly illustrated by the following figure:

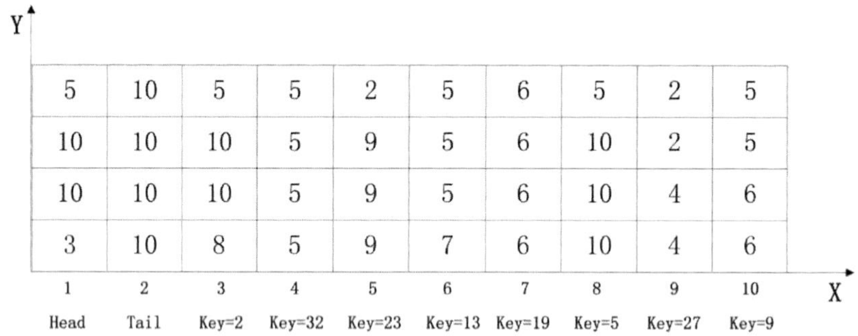

Fig. 2. The schematic diagram of a GSkiplist which is transformed from the skiplist in Fig.1. All nodes are arranged in the form of a two-dimensional array. Each row of Y axis direction represents for a node, so this figure shows a skiplist with 10 nodes for instance. The height of each node is fixed to a certain number. GSkiplist uses the index number of X axis direction to show the next nodes instead of the backward pointers.

For example, in Figure 2, the index number of X axis direction for node which key number equals 23 is 5, its last backward pointer is 2 which means that the last backward pointer of this node is the node with index number 2. We can figure it out in Figure.1 that it is the tail node whose index number is 2.

2.2. *Operations of GSkiplist*

2.2.1. *INSERT operation*

When a new node is coming, it will be added at the end of the two-dimensional array. The index number of it will be N+1 (suppose the current last index is N).

And here is the basic steps of inserting:

Step 1: Initialize the node's information.

Step 2: Sequence the current key array and find the right place for the new node.

Step 3: Change the value of next[][] for the node's previous node according the new node's index and randomlevel.

Step 4: Record the new node's next[][].

2.2.2. *DELETE operation*

When a node is no longer used, the solution is not to delete the node from the index table, but just alter the value of next[][] of its previous node and change the node's state into close.

Step 1: Find the node's index according to global array findIndex[].

Step 2: Change the value of next[][] of its previous node to the current node's next[][].

Step 3: Set the state of the node to "close" and the garbage collection will be implemented afterwards.

This kind of operation is simply handled because we are mainly aiming at improving the searching efficiency and most of these kind of applications concerns little about storage and have little operation such as deleting. In order to make good use of storage, these invalid nodes will be cleared after a certain period considered the balance of time cost and space cost.

2.2.3. *SEARCH operation*

The main idea of searching in GSkiplist is similar to that in skiplist. It begins at the highest level of the header node and follow the references along this level until the value at the node referred to is equal to or larger than X. If not, we switch to the next lower level and continue the search.

Step1: Set a bit array Results[] to tag the searching results.

Step2: N threads executes comparison and stored results into Result[].

Step3: Count the number of 1 in Result[] in order to get the new index and continue searching from the beginning.

The process shows in Figure 3 are taken when one single searching operations begin.

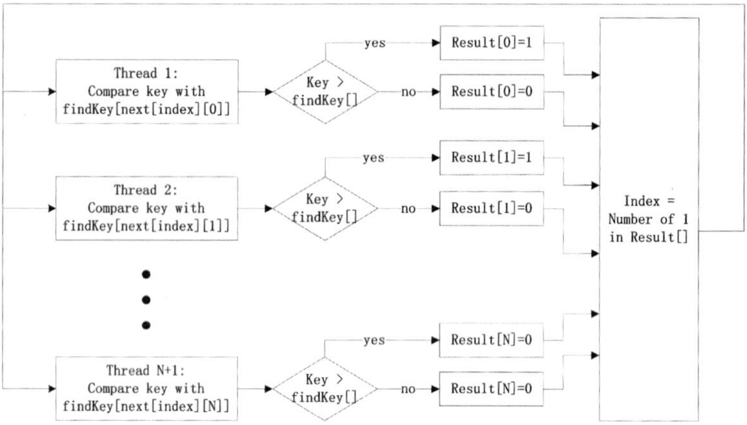

Fig. 3. A general flow chart of one single searching process when using multi-threads in GPU.

The main benefits we can get from the GPU accelerating arithmetic is from the parallel processing ability when large quantities of queries come. When it applies to thousands of queries, the process can be described as follows:

Firstly, use the copy operation in OpenCL to move the generated GSkiplist from main memory to global memory. Then map N × M threads of GPU to the graphic memory. M here represents the number of enquiries being processed at one time, and N represents the number of threads used for one search enquiry. Secondly, using M threads execute M searching enquiries and store the comparison results to a one-dimensional array Result[]. Finally, when all searching process finished, unblock the buffer step and continue next reading value step if Result[] is not null.

It is obvious that we can gain a lot speed up here because we have handle thousands of enquiries at one search time. A detailed experiment results will show the improvement in Section 3.2.

2.3. *Implementation on OpenCL*

We choose OpenCL[2] as the implementation platform because it exploits parallel computation on compute devices by defining the problem into an N-dimensional index space (N = 1,2, or 3).

When a kernel is queued for execution by the host program, and index space is defined. Each independent element of execution in this index space is called a work-item. Each work-item executes the same kernel function but on different data. When a kernel command is placed into the command queue, an index space must be defined to let the device keep track of the total number of work items that require execution.

3. Experimental Results

3.1. *Performance comparison*

The testing platform are set up as follows:

Table 2. A general view of testing platform configuration.

Name	Configuration
GPU	ATI Radeon HD7950, 800MHz, 16 stream processing core
CPU	Intel Core i5-3317U, 1.70GHz

We defined a skiplist sized 1G and also got its corresponding GSkiplist. The data set we used in our experiments is the TPC-H[3] benchmark which is a decision support benchmark suite that provides a set of ad-hoc business queries. The benchmark raised from 10 MB to 100MB, and the results will be showed as follows.

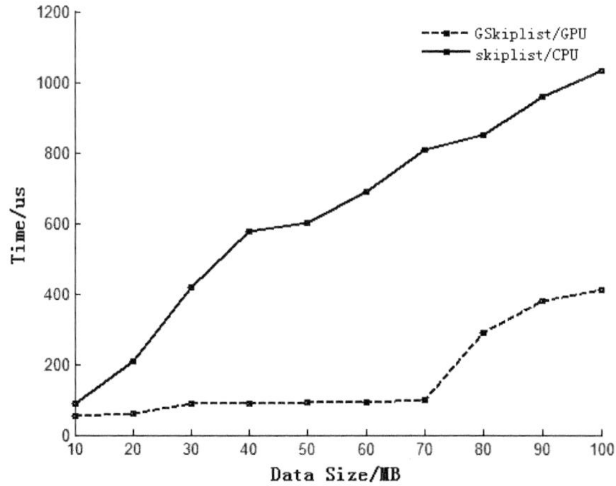

Fig. 4. A general flow chart of one single searching process when using multi-threads in GPU.

Figure 4 shows the search time in GSkiplist in GPU is less than that in skiplist in CPU, and the maximum speedup ratio is 8.

3.2. *Results analysis*

With the increasing quantity in query data, the execution time in CPU is almost linearly growing. The increase of CPU time is due to the multilayer nested loop in searching algorithm.

While there is a different trend in the execution of GPU. At the very beginning, the time was not significantly increased because the GPU has larger memory bandwidth. When data query volume increased to 70MB, the time overhead shows a relatively significant growth due to the limitation of GPU processing core's local cache memory (about 70MB). And cache miss penalty makes a mainly contribution to the time overhead.

4. Related Work

Parallel execution for skiplist: There are two kinds of main solutions of parallel computations for skiplist, one is the tranditional lock-based implementations[4], which is known to offer poor scalability due to high lock contention in the presence of thousands of active threads, the other is a concurrent lock-free implementations done by Fomitchev et al[5].

GPU-related database research: Pioneering early work in GPU database systems was GPUQP, developed by He et al.[6] which has started to migrate to OpenCL. Rauhe et al.[7] designed a multi-level parallelism framework to execute relational queries in GPUs. GPU threads execute the programs adopted from their earlier CPU implementation upon portion of the data. Mostak et al.[8] built a GPU accelerated database system, MapD, specialized for geography map queries.

Parallel Algorithms: There are several parallel algorithms literature such as prefix sums[9], recently studied for GPUs[10,11], and on merging sorted lists in parallel[12].

5. Conclusion

In this paper, we have designed a new data structure GSkiplist which applies skiplist-like searching to GPUs. We focus on applications with large quantity of queries, and can get a performance improvement up to 8X using OpenCL as programming platform. However, our methods are not suitable for databases that need to delete elements frequently. We look forward to further research to optimize the data structure.

Acknowledgments

We thank Hu Hao for his original idea of accelerating key-value database using skiplist in GPU. This work is partially supported by National Science Foundation (NSF) China 61272142, 61402492, 61402486, 61379146, 61272483, by National Hightech R&D Program of China (863 Program) under Grants 2012AA01A301 and 2012AA010901, by program for New Century Excellent Talents in University and by the State Key Laboratory of High-end Server & Storage Technology.

References

1. Pugh W. Skip lists: a probabilistic alternative to balanced trees[J]. Communications of the ACM, 1990, 33(6): 668-676.
2. Khronos Groups. The OpenCL Specification, version 2.0. November 2013.
3. T.P.P Council. TPC Benchmark H (Decision Support) Standard Specification, Revision 2.17.0. http://www.tpc.org/tpch/spec/tpch2.17.0.pdf, 2013.
4. Pugh W. Concurrent maintenance of skip lists[D]. Marylard; University of Maryland, College Park; 1998.
5. Fomitchev M, Ruppert E. Lock-free linked lists and skip lists[D]// *Proceedings of the twenty-third annual ACM symposium on Principles of distributed computing*. Washing ton, DC, ACM, 2004: 50-59.
6. B. He, M. Lu, K. Yang, R. Fang, N. K. Govindaraju, Q. Luo, and P. V. Sander. Relational query coprocessing on graphics processors. *ACM Trans. Database Syst.*, 34(4):21:1-21:39, Dec. 2009.
7. H. Rauhe, J.Dees, K.U. Sattler, and F.Faerber. Multi-level parallel query execution framework for cpu and gpu. *In Advances in Databases and Information Systems*, volume 8133 of Lecture Notes in Computer Science, page 330-343. Springer Berlin Heidelberg, 2013.
8. T. Mostak. An overview of mapd (massively parallel database). 2013.
9. G. Blelloch. Prefix sums and their applications. *In J. H. Reif, editor, Synthesis of Parallel Algorithms*, pages 35-60, 1993.
10. M. Harris. Parallel prefix sum (scan) with CUDA, April 2007. http://developer.download.nvidia.com/compute/cuda/sdk/website/projects/scan/doc/scan.pdf.
11. D. Lichterman. Course project for ECE498, Univ. of Illinois at Urbana-Champaign,http://courses.ece.uiuc.edu/ece498/all/Archive/Spring2007/HallOfFame.html.
12. R.Cole. Parallel merge sort, *SIAM J. on Computing*, 17,1988.

Label-weight adapting (LWA): a method for non-independent multi-label learning

Jing-Kang Yang*, Jiang Jun†, Hao-Ran Lv‡ and Tianyi Wang[d]

International School Queen Mary University of London, London, UK

*E-mail: *j.yang@se13.qmul.ac.uk, †j.jiang@se14.qmul.ac.uk, ‡h.lv@se13.qmul.ac.uk, [d]t.wang@se13.qmul.ac.uk*

Wei-Wei Wang

China Mobile Group GuangXi Co.,Ltd

E-mail: wangweiwei@gx.chinamobile.com

Zhaojing Zhang§ and Liang Dong¶

Key Laboratory of Universal Wireless Communication, Ministry of Education, Beijing University of Posts and Telecommunications, Beijing, 100876, China

E-mail: §buptzzj@163.com, ¶liangdong@bupt.edu.cn

Multi-label learning is one of the most important branches in the data mining field, and many algorithms have been proposed to solve many recent problems such as automatic annotation for multimedia contents, web mining, tag recommendation, etc. However, most of the classic methods have a hypothesis that the labels are independent, which may not hold in every situation, especially in tariff package recommendation where packages are always connected. In this paper, we present a novel algorithm which has a self-adaptive mechanism to implement the correlations between labels. We also analyzed the performance of the framework on tariff package data set. The simulation results show that the novel algorithm achieves better results compared to classical methods in both theory and practice.

Keywords: Label-Weight Adapting; Correlated Label; Multi-Label Learning; Adaptive Learning.

1. Introduction

Multi-label learning, which deals with the task of assigning a new data instance with multiple labels, is one of the important branches in data mining field. It is

---

* Corresponding author.

also a classics problem of machine learning. It has many applications in real world, attract much attention in the recent machine learning literature and help us solve a great number of knotty problems in different fields. In recent years, a lot of people have started to concentrate on studying Multi-label Classification. From 2010 to 2015, there are hundreds of relevant papers published, some use tree-structured and DAGstructured hierarchies to deal with the problem[7], while some other do multi-label classication in parallel tasks[8], with concept drift and class imbalance[9], etc. As the researches move along, more and more researchers pay more attention to use it to solve practical problems.[1] Many algorithms have been proposed to solve a large amount of current problems in practical, for instance, in the field of computer vision, Li used implicit relative labeling-importance information to decide the topic of pictures containing many objects comprehensively. In the field of medicine, human immunodeficiency virus type 1 (HIV-1) inhibitors can be predicted using Multidimensional Bayesian network classifiers[4]. And in addition, Backpropagation for Multi-label Learning is applied to functional genomics and text categorization[6]. Multi-label learning is also widely applied in natural language processing, web mining, rule mining, tag recommendation, etc.

The widely-known binary-relevance method (BR) learns the relevance of each label separately as separate binary problems, however this method has been widely criticized because it has a high possibility to encounter with imbalanced problem[2]. Another typical solution is the one-versus-all method (OVA), a strategy involves training a single classifier per class, with the samples of that class as positive samples and all other samples as negatives. This strategy requires the base classifiers to produce a real-valued confidence score for its decision, rather than just a class label; discrete class labels alone can lead to ambiguities, where multiple classes are predicted for a single sample[1]. However, the OVA has a hypothesis that the labels are independent, which may not hold in every situation. It just solves the problem of classifying instances into one of the more than two classes, but not reveals the correlations between labels when multiple labels are to be predicted for each instance.

In this paper, we present a novel algorithm of Label-Weight Adapting (LWA) which detects the correlations of labels and has a self-adaptive mechanism to eliminate the correlations between labels, consequently ensuring the predicted labels of each instance are independent. To obtain a fixed number of independent labels, we first use a classifier to generate a corresponding value for each label by one-versus-all method, using a chosen algorithm. Secondly, we exploit the correlation parameter of each two labels. By using correlation analysis for the set of labels, we can obtain a covariance matrix representing the correlation of two labels in label space. Additionally, the label with highest adapted score is allocated as the prior choice of the corresponding instance. Till now, initialization is completed. Then the novel LWA algorithm is proposed to decide the target label step by step. In that way, we finally pinpoint a set of target labels. It is noted that with a certain times of iteration, the algorithm will generate a fixed number of optimal labels with independence. Furthermore, we present experiments with dataset of tariff package purchase situation. The results show that in agreement with the real data, the experiment implies that LWA method performs well in prediction of independent labels and it is of significant value.

2. Problem Statement

Although the classic algorithms including One-versus-All and All-Pairs are efficient in solving problems of multi-label classification, the correlation between labels is

Algorithm 1 One-versus-All Method

Input: training set S = $(x_1,y_1),\ldots,(x_m,y_m)$.
algorithm for binary classification A
For each $i \in Y$

$$S_i = (x_1, \delta(y_1 = i)), \cdots, (x_m, \delta(y_m = i))$$

$$h_i = A(S_i)$$

Output: A set of classifiers $H = (h_1, h_2, \ldots, h_m)$.

needed to take into consideration. For instance, when it comes to the matter of music recommendation, we need to think twice of the correlations of music types that suggested to one person because it may not common if people adopt jazz and opera at the same time. From the example we can interfere that the correction rate of prediction may rise if we attempt to increase the correlation among labels towards a single instance. However, sometimes we need to reduce the correlations between labels allocated to one instance for better prediction. For example, when encountering with the tariff package marketing for telecoms enterprise, the packages aims at one customer should not be in strong connection because if A who has accepted a bonus will not take in another similar bonus. Thus, the correlation between labels has both positive and negative influence towards prediction. In the following discussion, we mainly talk about the latter case, where correlation can have a reverse impact on prediction.

There are various approaches to solve multi-label learning problems, and one of the simple and practical methods to tackle multi-label prediction is reduced to binary classification. The two main methods reflecting the thought of binary classification conversion are One-versus-All approach and All-Pair approach. The aim of learning is to find a predictor $h : X \rightarrow Y$, where X is space of features while Y is the space of labels. As for One-versus-All approach, training set is separated into t subsets, with each set S_i = $\{(x_1,\delta_{(y1=i)}),...,(x_m,\delta_{(ym=i)})\}$, where the function $\delta_A = 1$ if the incidence A is true, inversely, $\delta_A = -1$ is held if the incidence A is false. Finally we will obtain a rule set for further implementation, where $H = (h_1,h_2,...,h_i)$. On the other hand, another famous method called All-pairs can also deal with multi-label learning in a simpler way. It will create $\frac{1}{2}n(n-1)$ subset of training set, in the schema of adding $(X_t,1)$ or $(X_t,-1)$ to S_{ij} by matching y_t with y_i and y_j. Consequently, several classifiers will be combined together for prediction on test set. The detailed algorithm with flow chart can be referred in Algorithm 1 and 2.

Algorithm 2 All-Pairs Method

Input: training set S = $(x_1,y_1),...,(x_m,y_m)$.
algorithm for binary classification A
For each $i,j \in Y$, *s.t.* $i < j$ **For** each t =
$1,2,...,m$.

$$if\ y_t = i,\ add\ (x_t, 1)\ to\ S_{ij}.$$
$$if\ y_t = j,\ add\ (x_t, -1)\ to\ S_{ij}.$$

end **For**

$$h_{ij} = A(S_{ij})$$

end **For**
Output: A set of classifiers $H = (h_1,h_2,...,h_m)$.

---

3. Algorithm

In this paper, we aim to propose a novel algorithm for multi-label learning, which highlights the correlations between labels. The algorithm is called "Label Weight Adapting". The input of the algorithm is label correlation and label score derived by classic multi-label classifier. To better illustrate the algorithm, we firstly introduce correlation detection and label scoring method for preparation.

3.1. *Detect the correlations of labels*

To detect the correlation of labels, we define a label space Y spanned by $y_1, y_2, \ldots, y_m$, thus a sample matrix in space Y can be represented by $M = [\alpha_1, \alpha_2, \ldots, \alpha_n]^T = [\beta_1, \beta_2, \ldots, \beta_m]$, where β_j denotes the sample vector of jth label, and α_i denotes the vector containing every label value of i-th instance. We suppose to analyze the correlations between each two labels through evaluating the covariance matrix of M. The covariance of label y_i and y_j is denoted as equation(1). Through the instances we have in the training set, we can use equation(2) to obtain the covariance matrix, further, by simplifying we can easily obtained it by calculating like equation(3). Thus we can obtain the correlation coefficient of $i-th$ label and $j-th$ label easily by referring to either c_{ij} or c_{ji}.

$$c_{ij} = E([y_i - (y_i)][y_j - (y_j)]) \tag{1}$$

$$\begin{aligned} c_{ij} &= \frac{1}{m}\sum_{k=1}^{m}\left[\left(M_{ik} - \frac{1}{m}\sum_{a=1}^{m} M_{ia}\right)\left(M_{jk} - \frac{1}{m}\sum_{b=1}^{m} M_{jb}\right)\right] \\ &= \frac{1}{m}\sum_{k=1}^{m} M_{ik}M_{jk} - \frac{1}{m^2}\sum_{a=1}^{m} M_{ia}\sum_{b=1}^{m} M_{jb} \\ &= \frac{1}{m}a_i^T a_j - \frac{1}{m^2}\sum_{a=1}^{m} M_{ia}\sum_{b=1}^{m} M_{jb} \end{aligned} \tag{2}$$

$$\begin{gathered} \mathrm{C} = \frac{1}{m}\sum_{i=1}^{m}(\beta_i - \beta_0)(\beta_i - \beta_0)^T \\ where\ \beta_0 = (1/m)[\beta_1 + \beta_2 + \cdots + \beta_m] \end{gathered} \tag{3}$$

3.2. *Obtain label scoring vector*

Firstly, we define a classifier based on One-versus-All(OVA) algorithm or Allpairs(AP) algorithm. By operating the classifier on the training data, we obtain a sample matrix M with discrete values. However, to obtain the label scoring vectors, we first need to calculate the probability of every instance of each label. We suppose to choose a few specific algorithms with continuous predicted values, such as Logistic Regression, Support Vector Machine(SVM), Neural Network(NN) and Decision Tree, to predict the probability of instances of labels. In this case, each predicted value will be continuous between 0 and 1.

And we can obtain a probability matrix M with continuous values from matrix M. Secondly, since we realize that different labels may reflect different commercial values in practical work, for example, in package recommendation for Telecoms enterprise, we define a profit coefficient based on profit function model[reference] to reflex values of labels. By calculating the profit of different labels, we obtain a profit vector $p = (\gamma_1, \gamma_2, \ldots, \gamma_m)^T$, where γ_m is the profit value of $m - th$ label[3]. Hence, we can obtain the label scoring matrix by inner product $M = p \cdot M$.

3.3. *Label-weight adapting*

In this section, we propose a novel framework named Label-Weight Adapting (LWA) to obtain a set of $N - 1$ recommended labels by the correlation matrix C of labels and label scoring vector s. It consist of two steps when predict recommended labels: Firstly, initialize $y_l(0)$. Secondly, iterative the following

---

Algorithm 3 Label-Weight Adapting for Single Instance

Input: Correlation matrix $\boldsymbol{c}$ of labels
Label scoring vector $\boldsymbol{s}^{(0)}$
Output: A set of N recommended labels L.
Initialize: L = $\{y_{l(0)}\}$.
Iteration N-1
1: Find correlation coefficient between $y_{l(n-1)}$and other labels, denoted as:

$$\boldsymbol{c}^{(n)} = \left[c_{l(n-1),1}, \ldots, c_{l(n-1),m}\right]$$

2: Calculate the adapted label weight vector $w^{(n)}$, where

$$w_j = f(c_j). \tag{4}$$

3: Calculate the adapted scoring vector $s^{(n)}$

$$s_j^{(n)} = w_j^{(n)} \cdot s_j^{(n-1)} \tag{5}$$

4: Add $y_{l(n)}$ into set L, where

$$l^{(n)} = \arg\max_i \left\{s_j^{(n)}\right\}$$

Algorithm 4 Label-Weight Adapting Algorithm

Input: Correlation matrix $\boldsymbol{C}$ of labels
Label scoring vector $\boldsymbol{S}^{(0)}$
Output: A set of N recommended labels L.
Initialize: L = $\{y_{l(0)}\}$.

Iteration N-1

1: Find correlation coefficient between $y_{l(n-1)}$ and other labels, denoted as:

$$c^{(n)} = \left[c_{l(n-1),1}, \ldots, c_{l(n-1),m}\right]$$

2: Calculate the adapted label weight vector $w^{(n)}$, where

$$w_{ij} = f\left(c_{ij}\right). \tag{6}$$

3: Calculate the adapted scoring vector $s^{(n)}$

$$s_j^{(n)} = w_j^{(n)} \cdot s_j^{(n-1)} \tag{7}$$

4: Add $y_{l(n)}$ into set L, where

$$l^{(n)} = \arg\max_i \left\{s_j^{(n)}\right\}$$

---

steps for N times: find correlation coefficient between $y_{l(n-1)}$ and other labels, and then calculate the adapted label weight vector $w^{(n)}$, where $w_j^{(n)} = f(C_j)$; calculate the adapted scoring matrix $S^{(n)}$, where $S_j^{(n)} = w_i^{(n)} \cdot S_j^{(n)}$. Finally we take $l^{(n)} = \text{argmax}^j \, S_j^{(n)}$ and add $y_{l(n)}$ into set L.

It is necessary to highlight how to obtain the initialized $y_l(0)$. If we implement the formula $l^{(n)} = \text{argmax}_j\{s_j^{(n)}\}$ to derive the target $y_l(0)$, the extreme case may occur that the label with highest score may be mutually exclusive with all the other labels. In the other word, if we choose an initialized label which reversely limits the contribution of other label, obviously, the decision is not optimal. Thus, a developed matrix T is derived in equation (8), which represents the contribution to whole selecting incidence every instance owns. Thus, $l_i^{(n)} = \text{argmax}_j\{t_{ij}\}$.With this step in initialization, the extreme case illustrated above can be avoided.

$$T = S \cdot C \tag{8}$$

4. Analysis of LWA

In the section of Algorithm, we have proposed the novel algorithm 'Label Weight Adapting'. And most importantly, the relation between correlation among labels and adapted weights of labels is presented as a function f, and there are three main chioces for the function f, which is presented in equation(9)(10)(11). It is noted that a in the equation(11) is a positive constant.

$$f(x) = \begin{cases} \dfrac{1}{1-x} & when\ 0 \le x < 1 \\ x+1 & when\ -1 < x < 0 \end{cases}$$

(9)

$$f(x) = log_2 \frac{2}{1-x}$$

(10)

$$f(x) = log_{1+\frac{a}{2}} \frac{a}{1-x} + \left(1 - \frac{a}{2}\right) \quad (11)$$

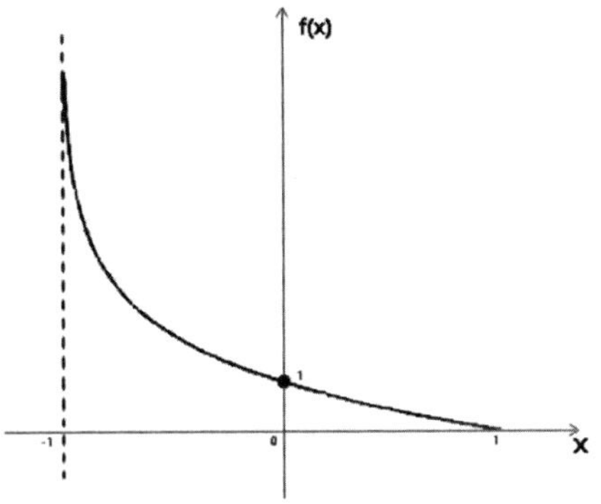

Fig. 1. Plot of functions.

In the following, the theoretical conclusions with respect to LWA algorithm, especially the function of correlation coefficients and adapted weights, are analyzed in a detailed way. Theorem 1 discuss the domain and the shared property those three types of functions have, and in Theorem 2, we analyze the unique property equation(9) has, while in the Theorem 3 we highlight the characteristics of equation(10)(11) and introduce how to implement it in practice. The main propose of this section is to verify the feasibility of the given function *f*.

Theorem 1

The domain of f(x) is $x \in [-1,1]$, since the independent variable x means the correlation coefficients between labels. It is known that when two layers have a correlation coefficient equals 1, the labels are in absolute relations. However, if the coefficient is -1, the labels are of perfect negative correlation. The aim of function f is to calculate the adaptive coefficient upon the label scores of each instance. So, it is acceptable that if one chooses a certain package, another package having closest relationship will not be selected because function overlaps. Further, when $x = \pm 1$, the outcome of $f(x)$ is ∞. Additionally, $f(0) = 1$ means if two labels are independent, one will not affect the other.

$$-\infty < f(x) < \infty$$

$$\lim_{x \to 1} f(x) = -\infty$$

$$\lim_{x \to -1} \mathrm{f(x)} = +\infty$$

Theorem 2

The problem of choosing a suitable function is depended on the property. We provide two usable specific functions. One is a linear function while the other is in logarithmic form. Firstly, we introduce the linear one. The equation(12) presents that if label A and B share same absolute value but inverse sign of correlation with a chosen label C, the corresponding score on A and B is set to adapt in the same rate.

$$f(x) \cdot f(-x) = 1 \tag{12}$$

Theorem 3

As for the logarithmic one, there is a bias on either positive or negative correlation. For example, the corresponding score on A rises more than B if label A and B share same absolute value but inverse sign of correlation with a chosen label C, as the figure(1) shows. In practice, the phenomenon exists when positive and negative correlation need to get different punish factors.

5. Simulation

In this section, we show the experimental results of our proposed framework on the data set of tariff package purchase situation in comparison with the classic multi-label method, Binary Relevance method.

5.1. *Information of data sets*

The data set we used is information of mobile users, who once has subscribed some tariff package. The data set has 0.32 million instances and up to 60 features, including telephone expenses in six months, average call duration, monthly data usage, etc. Most importantly, the data set contains the purchase situation of 20 different tariff packages. Our goal is to verify the advantage of the three-layer framework.

5.2. *Outcome of experiment*

We have compare the outcome of the LWA algorithm with that of Binary Relevance. We used three different data sets which are extracted in different months in order

	June		July		August	
Classic Method (BR)	hit rate	0.6357	hit rate	0.5190	hit rate	0.6142
	cover rate	0.3185	cover rate	0.2613	cover rate	0.3053
LWA	hit rate	0.8111	hit rate	0.7411	hit rate	0.7846
	cover rate	0.6978	cover rate	0.5418	cover rate	0.6058

Fig. 2. The experimental result showing LWA's better performance.

to provide a thorough comparison. The outcome is shown in the Figure 2, which indicates that with the three-layer framework, the accuracy of package recommendation can achieve an improvement. It is necessary to mention that the standard of evaluation uses both average hit rate and average cover rate. Both of them have definition below. For example, for a single instance, it may have 5 correct choices in reality but a learner is set to allocate 3 choice at a time, while the result shows that the learner does 2 correct answers and 1 mistake. Thus, the hit rate is $\frac{2}{3}$ and the covering rate is $\frac{2}{5}$. The term 'average' means those two index is supposed to show the performance of multi-label learner through the outcome.

$$hit\ rate = \frac{number\ of\ correct\ labels}{number\ of\ predicted\ labels}$$

$$cover\ rate = \frac{number\ of\ correct\ labels}{number\ of\ observed\ labels}$$

6. Conclusion

In this paper, we present a novel algorithm called 'Label-Weight Adapting', which has a self-adaptive mechanism to leverage the correlations between labels, ensuring the predicted labels of each instance independent. The experimental results show that LWA method performs well in prediction of independent labels and it is of significant value.

Acknowledgments

This work is supported by Research Innovation Fund for College Students of Beijing University of Posts and Telecommunications (Project No.243).

References

1. M.Zhang and Z.Zhou, *A review on multi-label learning algorithms*, in IEEE Transactions on Knowledge and Data Engineering, VOL.26,NO.8, August 2014.
2. K. Dembczynski, et.al. *On Label Dependence in Multi-label Classfication*,in International Conference on Machine Learning, Haifa, Israel. 2014.
3. Z.Zhang, Regina Wang, *Profit maximization analysis based on data mining and the exponential retention model assumption with respect to customer churn problem*, in ICDM 2015, Altantic city, NJ. 2015.
4. H. Borchani, C. Bielza, C. Toro, and P. Larraaga, *Predicting human immunodeciency virus type 1 inhibitors using multidimensional Bayesian network classiers*, Artif. Intell. Med., vol. 57, no. 3, pp. 219229, Mar. 2013.
5. S. S. Bucak, R. Jin, and A. K. Jain, *Multi-label multiple kernel learning by stochastic approximation: Application to visual object recognition*, in Advances in Neural Information Processing Systems, Eds. Cambridge, MA, USA: MIT Press, 2010, pp. 325333.
6. M.-L. Zhang and Z.-H. Zhou, *Multilabel neural networks with applications to functional genomics and text categorization*, IEEE Trans. Knowl. Data Eng., vol. 18, no. 10, pp. 13381351, Oct. 2006.
7. W. Bi and J. T. Kwok, *Multi-label classication on tree- and DAG-structured hierarchies*, 3rd ed. in Proc. 28th Int. Conf. Mach. Learn., Bellevue, WA, USA, 2011, pp. 1724.
8. E. Loza Menca, *Multilabel classication in parallel tasks*, in Proc. Work. Notes 2nd Int. Workshop Learn. MLD, Haifa, Israel, 2010, pp. 2036.
9. E. Spyromitros-Xious, M. Spiliopoulou, G. Tsoumakas, and I. Vlahavas, *Dealing with concept drift and class imbalance in multi-label stream classication*, 3rd ed. in Proc. 22nd IJCAI, Barcelona, Spain, 2011, pp. 15831588.

Schedule model in a cloud computing based on credit and cost

Yi-Li Zhang*, Jin-Bai Zhang
Nanjing Institute of Industry Technology,
Nanjing, Jiangsu, China
**E-mail: jsnjzhangyl@163.com*

The existing market schedule strategies in a cloud environment mostly seek to maximize income without paying sufficient attention to the credibility of service suppliers. The market schedule model based on credit price and cost computing is proposed for the purpose of income maximization of service suppliers and good credibility among users in combination. The model introduces credit evaluation of service suppliers into the task preemption scheduling mechanism based on cost computing, employing economical concepts including credit evaluation coefficient, credit price, task penalty and income discriminant function. It also uses the discriminant function value of service suppliers' income for decision making in task preemption. The simulation experiment shows that the model achieves income maximization of service suppliers and schedules the tasks with low credibility first.

Keywords: Cloud Computing; Price ; Schedule Model; Credit.

1. Introduction

The market mechanism in the economics can fully activate the individual and realize the reasonable allocation of resources. At present, in the study of cloud computing, Many scholars have introduced the economic model into the cloud computing and establish market Schedule model of cloud computing with the idea of economics. R. Buyya proposed to give a priority to tasks taking less time on resources under the control of user tasks budget and expected completion time. Such a Algorithm fully considered the market benefit issues such as task budget; In the research of market mechanism task scheduling, David E. adopt the concept of opportunity cost in economics, specially emphasizing the influence of opportunity cost on the task scheduling [1].

However, most of these algorithms are in pursuit of income of the service suppliers but rarely consider the credibility of resource suppliers among users.

* Corresponding author.

Considering that the existing market schedule strategy in a cloud environment mainly seeks to maximize the income but without sufficient attention to credibility of service-suppliers, a Schedule Model in a Cloud Computing Based on Credit and Cost (SMCC)is proposed. Economical concepts including credit evaluation coefficient, credit price, task penalty and income discriminant function are introduced to model. The model introduce the credit rating of service party into the user task preemption scheduling mechanism based on cost computing and take the discriminant function value of service suppliers' income as decision-making basis of task preemption, thus realizing the income maximization of service-suppliers and first scheduling its tasks with worse credibility.

2. Relevant Concepts of Credit Price and Task Penalty

In SMCC strategy, the participants in the task scheduling include service-suppliers and service-applicants. Service-suppliers get income by providing users with services but they have to pay users penalty calculated by unit time if exceeding the budget time of users. The users select the service types according to their funding and task features.

Definition 2. 1 The **task penalty** in SMCC model is defined as group of three variables:

$$pena(bugTim, exeTim, \max Tim) \tag{1}$$

The *bugTim* is the estimated time of user's task. The service suppliers need to complete the task before *bugTim* and have to pay the *penalty* beyond the *bugTim*. The *exeTim* is the actual time of executing user's tasks. The *maxTim* is the maximum time that users can stand in order to finish the task. And the *maxTim* normally should not exceed more than one times the *bugTim*, otherwise the contract will be terminated and the *maxPen* has to be paid.

The *penalty* in SMCC model includes *uniPen* in case of $exeTim \geq bugTim$ and *maxPen* in case of $exeTim \geq maxTim$. There are three situations[2]:

When $exeTim \leq bugTim$, namely, complete the task according to the agreed time or ahead of the time, no penalty need to be paid: $pena=0$.

When $bugTim < exeTim < maxTim$, namely, *exeTim* exceeds *bugTim,* penalty need to be paid: $pena=uniPen*(exeTim - bugTim)$.

When $exeTim \geq maxTim$, namely, *exeTim* equals or exceeds *MaxTim*, *maxPen* need to be paid: $pena=maxPen$.

Definition 2.2 The **credit price** is defined as group of three variables:

$$crePri(reso_i, task_i, exeTim), i = 1,...,n \quad (2)$$

Reso is resource marker, task is the marker of user's task, *exeTim* is the time of user's executing tasks. Credit price is user's evaluation of credibility of services by resources-suppliers, it is a variable in dynamic change with the changes of service actions and has memory of previous credibility. The lower the value is, the better the credibility of service-suppliers is among users.[3]

$$crePri = \rho * marPri \quad (3)$$

$$\rho = \eta * \rho_{hist} + \theta * \rho_{now} \quad (4)$$

ρ is the coefficient of credit evaluation, made up of history cumulative value ρ_{hist} and latest evaluated value ρ_{now},$0 \leq \rho \leq 1$;*marPri* is a relative stable value decided by current market price; η and θ are adjustment coefficients, $\eta + \theta = 1$.

According to SMCC model strategy, credit evaluation coefficient of the user task in the first scheduling is: ρ=0.5.

ρ_{now} is evaluation value of user's task after completing the latest scheduling task. The situations are as below:

When $exeTim \leq bugTim$, namely, complete the task according to agreed time or before the agreed time: ρ_{now} =0.

When $bugTim < exeTim < maxTim$, $\rho_{now} = (exeTim - bugTim)/bugTim$.

When $exeTim \geq maxTim$, ρ_{now} =1.

ρ_{hist} is a variable changing with the time. The model adopts time-driven evaluation mechanism. T stands for evaluation period and its value renew once within that period. The formula is as follows[4]:

$$\rho_{hist} = \sum_{i}^{n} \rho_i * \mu^i, i = 1,2,...,n, 0 < \mu < 0.5 \quad (5)$$

The *i* in the formula is the sequence number of credit evaluation after every task is executed, counting from the latest credit evaluation in turns is 0, 1,…, n. The value taken here is: i=1,…,n, i.e., the latest credit evaluation coefficient is not included in ρ_{hist}; μ is the base of history evaluation, $0<\mu<0.5$. Formula(5) shows that the closer to the recent time the coefficient of credit evaluation is, the greater contribution to ρ_{hist} it makes. All user's tasks after service have evaluation value in service-suppliers, which are recorded in credit price list and automatically renewed every period T. Whenever task scheduling, relevant credit evaluation form will be read automatically.

3. Market Schedule Model Based on Credit Price and Cost Computing

The task schedule strategy of SMCC model in a cloud is as follows: if service – supplier receive a new task request while he is executing a user's task, the sunk cost paid by service-suppliers for giving up current task is calculated. At the same time, based on the credit price of current task, the reward for completing new task and the penalty paid by service-suppliers for giving up current task, income value is obtained for accepting new task and giving up current task. By this income value judgment is made about task preempting strategy.

3.1. *Cost computing*

The Sunk Cost is the concept in economics, suggesting cost which cannot be retrieved after being invested. If new task $task_i$ grab current task $task_{i-1}$, the cost which service-supplier shave invested in current task $task_{i-1}$ is *sunCos*.

The foumula of calculating the total $cos_{i,i-1}$ paid for $task_i$ to grab $task_{i-1}$ is :

$$\cos_{i,i-1} = \alpha * sunCos_{i-1} + \beta * pena_{i-1} + \gamma * cre\Pr i_{i-1} \tag{6}$$

The $sunCos_{i-1}$ is the *sunCos* paid in order to give up the current task, *pena* is the compensation paid to users because service-suppliers has not completed user's tasks, *crePrie* is the value of user's credit evaluation of the services provided, i.e., credit price; α,β,γ are adjustment coefficients, $\alpha+\beta+\gamma=1$.[5]

Formula(6) shows user's evaluation of the credibility of service-suppliers. The worse the credibility of service-suppliers $task_{i-1}$, the higher the value of *crePri* and the higher the cost of $task_i$' grabbing $task_{i-1}$, is at this moment, suggesting service-suppliers' intention of preempting to schedule that task so as to diminish the impression of bad credibility among users.

3.2. *Market schedule model based on credit price and cost computing*

Definition 3.1 The **income discriminant function** of SMCC model is defined as a group of three variables[6]:

$$inc(rew, sunCos, pena, cre\Pr i) \tag{7}$$

$$inc_{i,i-1}=rew_i-cos_{i,i-1} \tag{8}$$

The rew_i in the formula(7) is the reward that service-suppliers get in scheduling $task_i$, $cos_{i,i-1}$ is the cost of giving up the current $task_{i-1}$ in order to accept $task_i$. Obviously, the value of income discriminant function is considered the criterion to judge whether SMCC model task preempt or not.

When $inc_{i,i-1} \geq \varepsilon$, current task can be preempted, otherwise preempting fails, ε is the threshold value of current market price.

4. Simulation Experiment and Results Analysis

Simulation experiment: 5 machines, $M=(M_1,M_2,M_3,M_4,M_5)$, each machine provide 20 computing resources, whose computing functions and prices are same; 100 computing resources, $R=(R_1,R_2,\ldots R_i\ldots,R_{100})$ $(i\in 100)$; 100 tasks, $tast=(task_1,task_2,\ldots,task_i, task_n)$ $(i\in 100)$[7].

Parameter description: credit evaluation adjustment coefficient parameter, $\eta=0.4$, $\theta=0.6$; Task preempt cost adjustment coefficient parameter, $\alpha=0.5$, $\beta=0.3$, $\gamma=0.2$. Two groups of experiments are designed in contrast as follows:

The first group of experiments in contrast(called Contrastive Experiment): Select contrastive experiment scheduling model based on cost computing without considering credibility. In the experiment of SMCC strategy(called MSCCC Experiment), select tasks with higher value of credit price$(0.8\leq\rho\leq1)$ in T period i.e., the task of worse service-suppliers' credibility, record these tasks which are successfully scheduled in T period and calculate the percentage of these tasks against the asked tasks during that period; In Contrastive Experiment, based on those tasks of worse service-suppliers' credibility chosen from MSCCC Experiment, record successful scheduling tasks in every T period and calculate the percentage of these tasks against requested tasks during corresponding period in contrastive experiments. The two groups of experiments are shown as Figure 1.

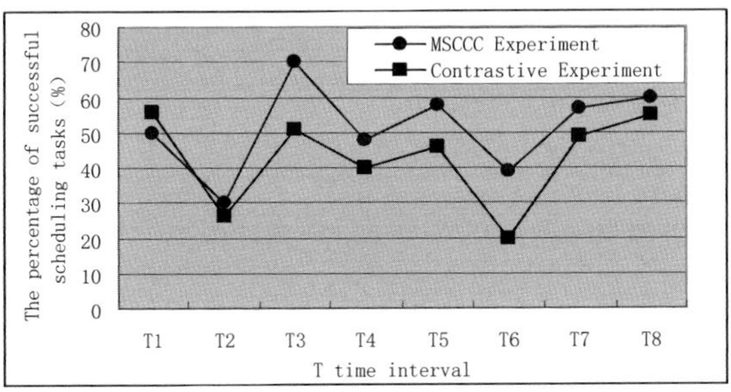

Fig. 1. Simulation experiment of percentages of successful scheduling tasks in different time intervals.

Experiment analysis: As shown in Fig.1, MSCCC Experiment data points are primarily distributed above the experiment data points in contrast and this is because the former adopts a market schedule model strategy based on credit price and cost computing. As is shown in formula(3)(4) and formula(6)(8), the worse the credibility of service-suppliers is, the higher the credit price is, the more its cost of task preempting, so the percentage of the current task continue to be scheduled increases, i.e., service-suppliers preempt to schedule low-trust task so as to diminish the impression of bad credibility among users. However, Contrastive Experiment does not take the credibility of service-suppliers into consideration but pursue maximizing the income of service-suppliers in the process of scheduling, i.e., “income-oriented”, hence tasks of low credibility in MSCCC Experiment do not possess any advantages in the scheduling of the contrastive experiment and their percentage of preempting scheduling tasks in MSCCC Experiment are relatively low.

The second group of contrastive experiments[8]: Select cloud environment market scheduling model experiment which have not adopted credit price and cost computing to prove SMCC strategy so as to maximize the income of service suppliers. Divide task budget and parameter combination of expected completing time into 8 groups with data of combined parameter in every group increasing gradually: G_1(20000,1000), G_2(40000,2000), G_3(60000,3000), G_4(80000,4000), G_5(100000,5000), G_6(120000,6000), G_7(140000,7000), G_8(160000,8000).The relation between the income of service-suppliers, task budget and parameter combination of expected completing time is shown as Figure 2.

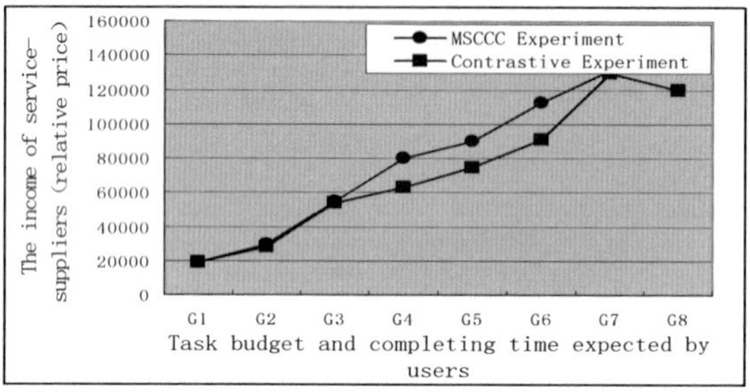

Fig. 2. Simulation experiment of verifying service-providers’ income maximization.

Experiment analysis: As shown in Figure 2, data line of MSCCC Experiment are mainly distributed above that of contrastive experiment and this is because the former adopts a market schedule strategy based on credit price and cost computing. Judging by task income discriminant function and formula(8), as long as income value exceeds any threshold executing task income, preempting current task may occur and the income maximization of service-suppliers scheduling task can be realized. As to contrastive experiment, because the market schedule strategy based on credit price and cost computing has not been adopted in task scheduling, the income of service-suppliers is relatively low.

5. Conclusion

Cloud market schedule strategy based on credit price and cost computing in a cloud environment is an improvement and enrichment to classic market scheduling model proposed by R.Buyya and David E. The model introduce the user's credit evaluation of service-suppliers into the task preemption market scheduling model based on cost computing and take the income discriminant function value of service-suppliers as decision-making basis of task preemption. The market scheduling model first schedules service-suppliers' tasks with worse credibility among users while realizing the income maximization of service-suppliers so as to eradicate their bad impression of "income-oriented".

Acknowledgments

This project is supported by the Major Program Foundation of Nanjing Institute of Industry Technology (Grant No. YK14-04-01)

References

1. Liu Xiaoqian. Research on Data Center Structure and Scheduling Mechanism[D].Hefei: University of Science and Technology of China, 2011.
2. Du Xiongjie, Tang Xuehai. Scheduling mechanism based on service level objective in multi-tenant cluster[J].Journal of Computer Application, 2015 Vol. 35 (4): 944-949.
3. Liu Jin-zhi,Yu Dan,Zhu shuai. Research on new cloud storage service model [J].Application Research of Computers, 2011,28(5):1869-1872.
4. Li Songtaoi. Low-cost cloud storage scheme based on hybrid strategy[J]. Journal of Computer Application , 2014 Vol. 34 (10): 2800-2805.
5. He Huaiwen, Fu Yu.Service layer agreement-aware resource allocation for cloud center profit maximization[J].Journal of Computer Application, 2015 Vol. 35 (6): 1585-1589.
6. Wu Xiu-guo, Wang Min. Minimum-cost Based Data Replication Strategy in Cloud Computing Environment [J]. Computer Science, 2014,41(10):154-159,190.

7. Zhang Chun-yan, Liu Qing-lin.Task allocation based onant colony optimization in cloud computing[J].Journal of Computer Applications, 2012,32(05): 1418-1420.
8. Yang Dan.Cloud Computing Resource Scheduling Based on Improving Firefly Algorithm[J].Computer Engineering ,2015 Vol. 41 (2): 17-20,25.

A method of SQL access logs pretreatment for power business system

Wei-Ei Li, Xiu-Li Huang*, Bo Zhang, Rong Zeng and Yong Li
State Grid Smart Grid Research Institute, China Smart Grid Research Institute, Nanjing, Jiangsu Province 210003, China
**E-mail: huangxiuli@epri.sgcc.com.cn.com*
www.sgcc.com.cn

With the development of smart grid interactive, the database interaction between internal and external information is increasing. In this paper, we proposed a detailed design of a method of SQL access log compression for power business system, so as to respond to the demands of database security audit of the power system. This method can achieve the compression of SQL access log and also has the function of simple classification for SQL logs. The test results show that the method can achieve a large margin compression of SQL access logs and meet the requirements of log preprocessing for the database behavior audit of power business system.

Keywords: Power Business; SQL Access logs Compression; Security Audit; Pretreatment.

1. Introduction

With the application of smart grid information technology and SG-ERP of interactive, power systems business continues to enrich. Interactions with inside and outside increased diversely. For example, power trading system, production system, financial management and control, ERP system, marketing system, there are network data exchange behavior across internal and external borders. To be able to conduct real-time monitoring of various database access information outside the network's users. And auditors are remaindered timely to auditing, tracking and warning various irregularities. Internal and external network security audit database interactions essential boundary [1]. Inside and outside the network boundary database interactions audit includes information collection, pretreatment, data mining and analysis. This is good to promote the smart grid information security.

* Corresponding author.

2. Problems and Research Status

Currently, in the information exchange companies inside and outside between the network boundary, the access behavior is important analysis and auditees. Database holds critical information of the power business systems. For example, user electric energy, marketing business data. So, database are highly attractive for hackers and criminals. It is frequently used to establish the database. Frequent insert, modify, delete and other operations behavior. And there is unrestricted use of the database of time and place. Because the amount of data traffic power system operation log database SQL statement is very large. There is a big difficulty storage analysis. In order to better conduct continuous audit of the database. Electric power business system SQL to access data compression is necessary.

Current research on database audit system has been more widely [2-4]. The related research is less for log compression. In 2010 Hebei Polytechnic University Zhang Chunling, et al designed and implemented WAN heterogeneous database asynchronous data compression based on virtual log synchronization scheme. Compress virtual log by calculating the combined logging. Only retain the log information related to data synchronization. Thereby reduce network traffic and improve the efficiency of data synchronization [5]. In 2010 China Science Institute of Computer Network Information Center gives an efficient compression algorithm of DNS logs. The use of DNS query types of redundancy and DNS query time, IP address and domain name for DNS log compression repeatability [6].

3. SQL Log Compression Method

Database systems operating in the same power business through a web page filling manner. It has to follow a certain pattern. SQL log compression method is through the feature by customizing the user blurring and other alphanumeric input. Create SQL mode database for operational systems. For SQL logs received via mode matching database only record the mode number. Greatly reduce the length of the SQL log record.

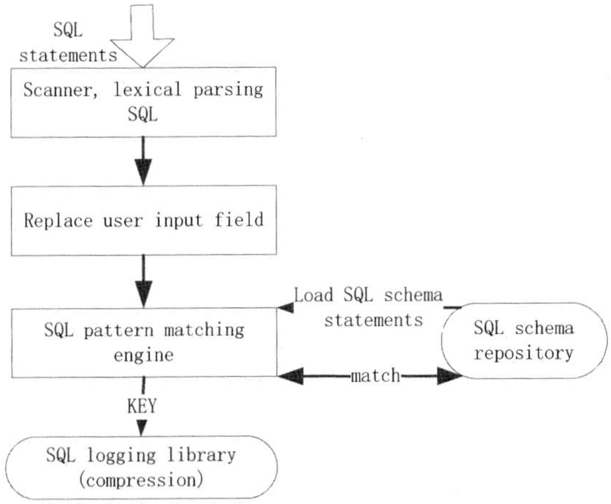

Fig. 1. SQL mode matching method based on log compression.

SQL log compression method is shown in Figure 1. Journal receiver receives the SQL Server log after scanning procedure SQL statement analysis. Form the array of SQL and replace item unrelated. SQL mode matching performed. If the match is successful, returns the number matching mode SQL schema repository and write to SQL logging database. If the SQL mode database fails to match, record mode to the SQL database and write serial number to the SQL database logs.

The database structure is defined as follows:

(1) The application database structure is defined as: {APP_ID: application ID, APP_IP: application IP address}.
(2) SQL mode database structure is defined as: {MODE_ID: mode ID, APP_ID: application ID, SQL: mode statements}.
(3) SQL database log database structure is defined as: {SQL_ID: SQL statement ID, MODE_ID: model ID, TIME: timestamp}.

3.1. *SQL lexical parsing*

SQL lexical parsing refers to SQL statement which is stored in a continuous buffer is broken down to get separate words, and the separate words are organized as a list of word. Detailed parsing process of the SQL is as follows:

(1) We iterate through each character of SQL statements, whitespace as the beginning and end of each word, which is used to parse out each word of SQL statements, and identify the type of each word. At present, in accordance with the provisions, the type of word is divided into as follows: keywords, identifier, number, digital wildcard, string wildcards, arithmetic

operators, logical operators, delimiters, function, manipulate object and other.

(2) The word are stored in the word structure, and multiple words structure form a list. At the same time, the related domain structure(such as type) of the word structure is assigned by the analytical results.

SQL parsing process is shown in Figure 2:

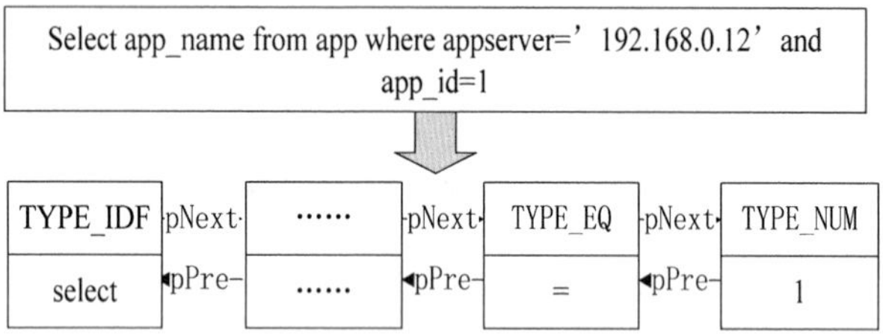

Fig. 2. SQL parsing process.

3.2. *Replace user input*

User input of the SQL statement from application server is constantly changing. It is key point that how to correctly abstract the SQL statement, isolate from the user input and the SQL main statement. A particular wildcard is replaced as part of a user input as far as possible, which is our replacement rule. Then a plurality of similar structures SQL statements can be abstracted as a SQL statement. Example of replacing the user input is as shown in Figure 3.

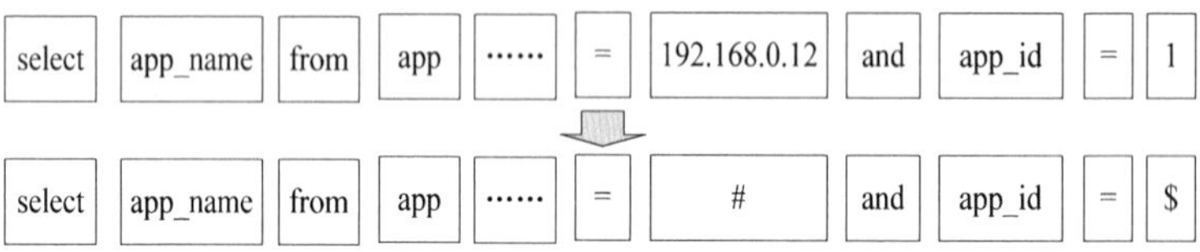

Fig. 3. Example of replacing the user input after SQL parsing.

3.3. *SQL mode matching engine*

SQL mode matching is defined by lexical parsing result to match SQL lexical list pool. Detailed steps are as follows:

(1) SQL lexical list pool is an array of hash, Firstly, the matching process obtain the hash value for lexical lists, and search in the hash array. HASH value calculation method: Hash=Hash<<8+Type (Hash is initialized to 0, Type is the type of word parsed)

(2) Find the corresponding items of an array via HASH value, and then make comparison in the hung lexical list of the array elements. Check for the

exact same lexical list, if any, that the SQL mode already exists, save only the SQL mode number to SQL log table, or that the SQL mode does not exist, after saving mode, and then the mode number is saved to SQL log table in.

SQL mode matching engine detail shown in Figure 4:

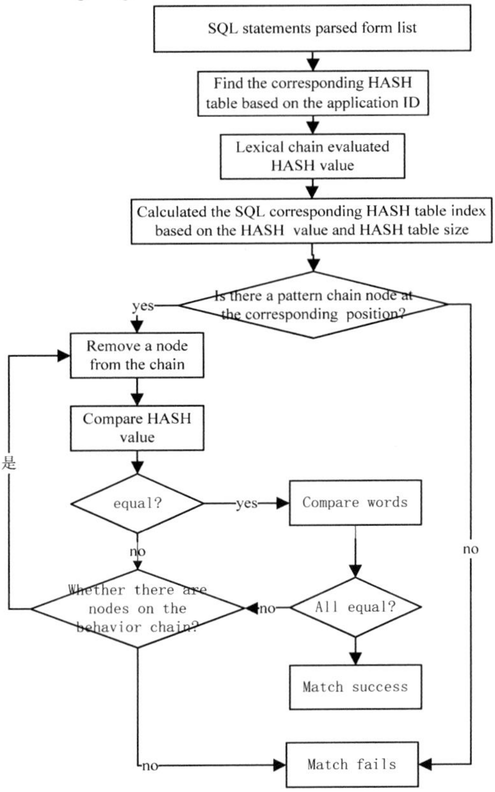

Fig. 4. SQL mode database matching process.

4. Implementation

In the implementation of Power Business Systems SQL access log compression, there are three main modules: initialization module, mode database matching module, the result written modules.

(1) initialization module, it is mainly to initialize the global mutex variable of the system and the lexical scan list corresponding to each business system by reading SQL statement in the mode database.

(2) mode database matching,SQL Logs read by SQL lexical analysis, Users enter replacement and Forming a linked list to be matched with SQL mode in the mode database.
(3) Writing Results, It need to write the log that has already a corresponding SQL mode to the log table and to both write to mode database and also write to the log table.

The system implementation process as shown in Figure 5:

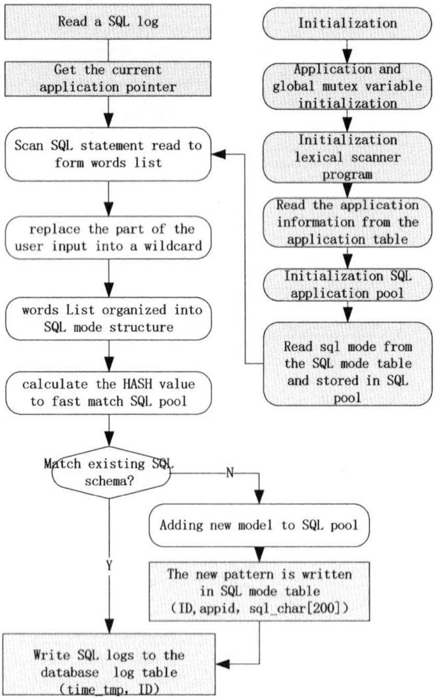

Fig. 5. SQL access log compression method implementation process.

5. Test results

5.1. *Test environment*

Experiment equipment hardware environment: Intel 5500 + ICH10R chipsets, 2 road 4 core cpus. Operating system is Windows 7

Programming languages: C.

Sampling test database SQL statement: real-time data acquisition.

5.2. *Test procedure*

1)Loading SQL log text file.

2)Perform Compression, record the SQL statements number in SQL mode database and logs number in SQL logs database.

5.3. *Test results*

Table 1. Data error comparison between two methods.

SQLLogs	SQL number of mode database	SQL number of log database	Compression ratio	The number of failed
1000	119	1000	11.90%	0
5000	157	5000	3.14%	0
10000	164	10000	1.64%	0
15000	261	15000	1.74%	0
20000	570	20000	2.85%	0
25000	634	25000	2.54%	0
30000	647	30000	2.16%	0
35000	850	35000	2.43%	0
50000	1024	50000	2.04%	0
75000	1103	75000	1.471%	0
100000	1197	100000	1.093%	0

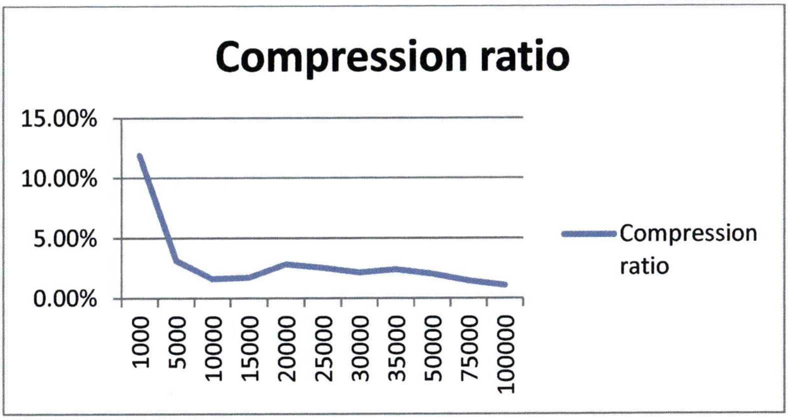

Fig. 6. SQL compression test results.

5.4. *Test result*

In the test result, along with the number of SQL performed by the system increasing, the compression ratio is to 50: 1. During the test, the server CPU, memory and hard drive indicators is detected properly.

6. Conclusion

This paper designs and implements a power business system SQL access log compression method to realize the compression of SQL access log and also have the function of Simple classification for SQL logs. It resolves the problem of SQL statements oversized and hard storage and hard analysis in database audit process, to provide support services for the power system security audit. System test results show that it can achieve SQL access log substantial compression by this method, to the requirements of log preprocessing for the database behavior audit of power business system.

References

1 Yang Bo, *Zhu Weiwei, Du Ninglan, Zhang Xiaodong. Research and Application of Database Auditing Technology for Grid Business Systems* [J]. Electric Power ICT, 2013,11(10):101-103.

2 Helman P, *LiePins G. Statistical foundations of audit trail analysis for the detection of computer misuse*[J]. IEEE Transactions on Software Engineering.1993,19(9):886- 901.

3 Joachim Biskup.*Recent Advances in Intrusion Detection* [M].Heidelberg Springer Berlin,2000:28-48.

4 Wang Yuan Majun. *A Method of Database Secure Audit Based on Intrusion Detection*[J]. Computer Simulation 2007,24(2):33-36.

5 Zhang Chunling, *Lv Zhenyu, Liu Zunfeng. Data Synchronization Solution Based on Virtual Log Compression*[J]. Computer Engineering. 2010, 18(36):67-69.

6 Wang Yanfeng, *Wang Zheng, Yan Baoping. High-efficient DNS Log Compression Algorithm* [J]. Computer Engineering, 2010,15(36):32-35.

Recurrent neural network for forecasting stock market trend

Xin-Kai Xie* and Hong Wang
State Key Laboratory of Intelligent Technology and Systems,
Department of Computer Science and Technology, Tsinghua University,
Beijing, 100084, China
**E-mail: xiexinkai@163.com*

Forecasting of stock market trends has always been an important issue in the field of finance. Neural networks are useful for understanding non-linear relationship between historical data and stock price changes. In this study, recurrent neural network (RNN) is used for time series data. The dataset contains large amount of intraday data from the constituent stock of China Shanghai Shenzhen 300 Index. Using the intraday data, we created and classified different daily features for the RNN model. Instead of using accuracy as a parameter to measure the effectiveness of the model, precision and average profit are used. Our experimental results demonstrate that this RNN model performs well with large amounts of testing data, which can provide useful information for stock market investors.

Keywords: Recurrent Neural Network; Stock Market Forecasting; Time Series Intraday Data.

1. Introduction

Stock market is an unstable and complicate system. High degree of uncertainty, different kinds of noise makes difficult to forecast stock market trend. Therefore, forecasting stock price movement direction is a challenging work.

Recurrent neural network (RNN) can deal with time series features. In RNN, the hidden layer not only get information from input layer, but also get information from last time hidden nodes. It makes the model easier to reveal the information inside the time sequence. Kim S H et al. [1] tried to use RNN to forecast Korea stock index. Saad E W et al. [2] compared RNN with other two neural networks. Güreşen E et al. [3] used genetic algorithm to optimize the parameters of RNN, to forecast Istanbul stock market. Besides optimizing the parameters, Dash and Liu [4] suggested that decreasing the input features' dimension may improve model performance. So using more precise stock data is to better than lots of multi-collinearity features.

---

* Corresponding author.

Although these existing stock market trend prediction systems show effective to forecast on stock market, there are some aspects could be improved. Firstly, some researches have only hundreds of stock market data for training and validation [2][3][5]. NNs usually have great trouble in over-fitting problem, it may occur over-fitting problem when training by only a few data. Secondly, some studies took care about features selection, but these methods select features by subjective factor or fundamental data [6][7][8]. Thirdly, the stock price change during one day may have useful information for forecasting, but most of the studies use only the daily stock data, which cannot reveal intraday information [1][2][3][5][7][8].

This paper focus on the problem above. We use intraday stock data, which have more information than daily stock data. We create a RNN model to forecast stock market trend. The dataset with hundreds of stocks covering more than nine years is used for the model, which has more than 100,000 data for training and testing. It avoid the over-fitting problem.

2. Dataset and RNN Model

2.1. *Dataset*

As the constituent stocks of China Shanghai Shenzhen 300 Index are not always the same 300 stocks, we choose 260 stocks that are always constituent stocks of this index. The dataset covers the period from January 2006 to July 2015. The historical intraday stock data base on minutes, includes trade price, ask price, bid price, and volume.

Using these base intraday data, we create 30 daily features including price change rate, turnover rate, moving average (MA), relative strength index (RSI), industry index change rate over different period of time. The forecast target is stock price change% between ten minutes after opening and ten minutes before the closing.

For each stock, we divided the data into two parts: the first part covers 95% of total data, while the second part is the remain 5% of total data. The first part of data is used for training, i.e., training data. The second part of data is used for validation, i.e., testing data.

In this paper we use features in five days for one RNN model input, so the daily features are merged by every five consecutive trading days. After preparing the stock data, there are 99436 training data and 5215 testing data.

2.2. *RNN model*

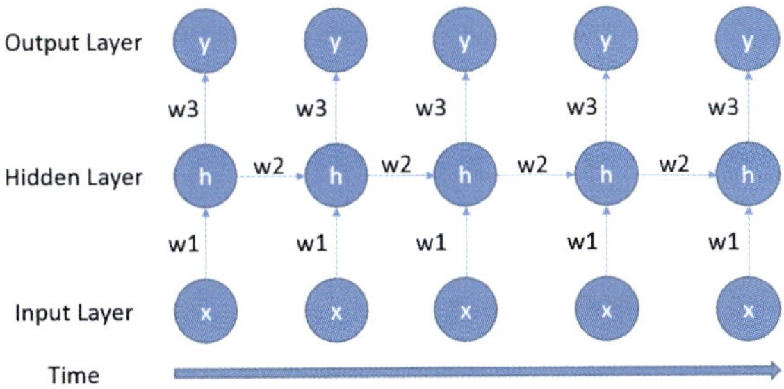

Fig. 1. This is a unfolded structure of RNN.

The RNN structure is shown above. Each circle on the structure refers to one node which contain information in one day. The node of input layer contains 30 daily features. The node of hidden layer contains 100 variables. The node of output layer contains two variables.

Define the ith feature at time t in input node as x_i^t, the ith variable at time t in hidden node as h_i^t, the i-th variable at time t in output node as y_i^t. Define the transition weight matrix from input layer to hidden layer as w1, the transition weight matrix from hidden layer to next time hidden layer as w2, the transition weight matrix from hidden layer to output layer as w3. Define sigmoid function $f(x) = \frac{1}{1+e^{-x}}$ as the node activate function. The forward propagation calculation formula can be written as equations (1) and (2):

$$h_k^t = \sum_{i=1}^{I} w_{1ik} \cdot x_i^t + \sum_{j=1}^{H} w_{2jk} \cdot h_j^{t-1} \tag{1}$$

$$y_k^t = \sum_{i=1}^{H} w_{3ik} \cdot f(h_i^t) \tag{2}$$

The cost function is the mean square of the output layer error:

$$E = \frac{1}{2}\sum_{i=1}^{Y}\left(T_i - f\left(y_i\right)\right)^2 \tag{3}$$

Where T_i is defined by of our forecast target, when training a positive case T_1=1 and T_0=0, while training a negative case T_0=1 and T_1=0. How to classify positive and negative cases will be discussed later. Using gradient descent to train the model [9], the calculation formula of the local gradient δ on the node is written as equation (4) and (5):

$$\delta_{yk}^{t} = \left(T_i \cdot f'(y_k)\right) \tag{4}$$

$$\delta_{hk}^{t} = \sum_{i=1}^{Y} \delta_{yi}^{t} \cdot w_{3ki} + \sum_{j=1}^{H} \delta_{hj}^{t+1} \cdot w_{2kj} \tag{5}$$

So we can adjust the weight matrix using local gradient.

3. Numerical Experiments

3.1. *Data classification*

Data need to be classify into positive cases and negative case for training. A simple idea for classifying the data depends on the sign of the forecast target. When the forecast target is positive, take this data as positive case, otherwise take it as negative case. But one stock may not have any useful information for forecasting when it change only a little bit. Classify these near-zero target into positive cases or negative cases may not help model digging useful information. So here we use two method for classifying the data. Method A, if the absolute value of forecast target is less than a threshold, eliminate this data. Method B, if the forecast target value is greater than a threshold, take it as positive cases, elsewise take it as negative cases.

$$\text{Method A:} \begin{cases} y > \theta & \text{positive cases} \\ y < -\theta & \text{negative cases} \\ -\theta \le y \le \theta & \text{be eliminated} \end{cases} \tag{6}$$

$$\text{Method B:} \begin{cases} y > \theta & \text{positive cases} \\ y \le \theta & \text{negative cases} \end{cases} \tag{7}$$

Where y is the forecast target value and θ (θ>0) is the threshold.

Using Method A, the amount of positive cases are approximately equal to negative cases, but Method B makes the positive cases much fewer than negative cases. So using accuracy to judge the model is meaningless. Instead, we use two technical indicator, precision and average profit to compare models with different parameters. The precision formula is shown below:

$$precision = \frac{true\ positive}{(true\ positive + false\ positive)} \tag{8}$$

Where true positive is that model output is positive and forecast target is greater than zero, while false positive is that model output is positive and forecast target is less than zero.

Besides precision, average profit should be considered. Each time the model predicts positive, trade the stock. It will get money or lose money from stock price change. Despite the transaction cost, the average profit can be calculated. The average profit represent how much an investor can earn by the model. The formula of calculating average profit is:

$$average\ profit = \frac{the\ sum\ of\ profits\ when\ predict\ positive}{the\ number\ of\ predicts\ positive} \tag{9}$$

3.2. *Experimental results and analysis*

There are two experiments by method A with threshold 0.5% and 1%, four experiments by method B with threshold 0.5% to 2.0%. We use an all-buy strategy as baseline. The all-buy strategy, which is a model simply predict all positive. With this baseline, we can get rid of systematic error because the positive cases and negative cases are not exactly 50% to 50%.

Table 3. Experimental results on testing data.

Method	Threshold	Precision	Average Profit
Baseline(all positive)		52.3%	23%%
Method A	0.5%	54.7%	55%%
Method A	1.0%	55.9%	61%%
Method B	0.5%	55.4%	71%%
Method B	1.0%	55.9%	93%%
Method B	1.5%	56.5%	119%%
Method B	2.0%	57.4%	135%%

The testing data are from first half year of 2015. At this period of time, China was in a bull market, so the average profit of baseline is greater than zero. From the table 3, all the models are significantly better than baseline. It proves the effectiveness of the RNN model. Method A is not as good as Method B, a smaller training dataset may be the reason. By Method B, the model is better when the threshold is going up.

4. Conclusions

In this paper, we prepare a large amount of intraday stock data, create features from the data, and training RNN model to forecast stock price change. The RNN model offers the ability to learn non-linear relationships from features and forecast targets. We use precision and average profit to examine whether our model is effective. By experimental results, we find that RNN is effective to forecast stock price change, it is also useful for stock market investors to make an investing decision.

Our study still have some aspects that needs further research. More features can be created from base trade data, it will bring more information. More hidden layers makes the neural network more complicated, which may find more relationships between features and targets.

References

1. Kim S H, Chun S H. Graded forecasting using an array of bipolar predictions: application of probabilistic neural networks to a stock market index[J]. International Journal of Forecasting, 1998, 14(3):323-337.
2. Saad E W, Prokhorov D V, Wunsch D C. Comparative study of stock trend prediction using time delay, recurrent and probabilistic neural networks.[J]. IEEE Transactions on Neural Networks, 1998, 9(6):1456 - 1470.
3. Güreşen E, Kayakutlu G. Forecasting Stock Exchange Movements Using Artificial Neural Network Models and Hybrid Models[M]// Intelligent Information Processing IV. Springer US, 2008:129-137.
4. Dash, M. and Liu, H. 1997, 'Feature selection for classifications', Intelligent Data Analysis: An International Journal, vol. 1, pp. 131-156.
5. Lin Y, Guo H, Hu J. An SVM-based approach for stock market trend prediction[C]// Neural Networks (IJCNN), The 2013 International Joint Conference on. IEEE, 2013:1-7.
6. J. D. Piotroski, "Value investing: The use of historical financial statement information to separate winners from losers," Journal of Accounting Research, vol. 38, pp. 1–41, 2000.
7. Lee W S, Tzeng G H, Guan J L, et al. Combined MCDM techniques for exploring stock selection based on Gordon model[J]. Expert Systems with Applications, 2009, 36(3):6421-6430.
8. Shen K Y, Yan M R, Tzeng G H. An integrated Fuzzy-ANP model for value investing.[C]// Fuzzy Systems and Knowledge Discovery (FSKD), 2010 Seventh International Conference on. IEEE, 2010:982-986.
9. R. Williams and D. Zipser, "Gradient-based learning algorithms for recurrent networks and their computational complexity," Backpropagation: Theory, Architecture, and Applications, M. Chauvin and D. E. Rumelhart, Eds. LEA, 1995, ch. 13.

Text mining of medical insurance policy based on association rules

Li-Li Han
Aerodynamics Research Institute, Aviation Industry Corporation of China, Harbin, 150001, China
E-mail: harbour521@163.com

Since medical insurance is relevant for everyone, it is very important to know how to study the formulation and implement the system. In this paper, I propose a text association analysis method, TACMA, which is based on medical insurance policies by combining the text information of medical insurance policy with the characteristics of semantic relations of the audit knowledge. Compared with the traditional TP-growth algorithm, the TACMA algorithm has improved significantly in mining effectiveness and time efficiency in text knowledge discovery, and can be successfully applied to practical projects.

Keywords: Association Rule; Medical Insurance; TP-Growth Algorithm.

1. Introduction

Text association analysis is one of the applications that association rules [1] take text data as processing targets. The implementation process of which is that extracting a certain pattern from the presentation of text data, and then processing and displaying them to the easily acceptable knowledge for users.

Since the medical insurance is related to everyone's health, how to study the formulation and implementation of the system is very important. The audit text message in the social insurance field refers primarily to the guiding text information which are composed mainly by such Five Insurance[2,3]. This paper provides text association rule classification mining algorithm, which has improved significantly in the mining effectiveness and time efficiency by comparing with the traditional mining algorithm.

2. Basic Definitions

Association rule mining[4] is an important data mining technology in the field of data mining. Concepts, such as frequent item sets and frequent models, mentioned in this technology are also referenced in other data mining technology.

Definition 1. Item: The minimum operational target in the given data set, expressed as i_m.

Definition 2. Item set: The set of one or more items, expressed as $I = \{i_1, i_2, \ldots, i_m\}$.

The association rules mining is one kind of data set in the field of data mining, which is usually stored in the database and manifested in the form of the transaction. At this point, the transaction refers to the form of once operation under a background task, that is, the transaction is item set. The database that storages above transactions is called as transaction database [5].

The association rule generally take the form of $X \Rightarrow Y$. X and Y are items of the transaction database and satisfy the condition of $X \cap Y = \varnothing$. It can be conclude that $X \subseteq I$ and $Y \subseteq I$. Of course, not all formula that is similar to $X \Rightarrow Y$ can become association rules. After discovering the mode of set I, only these formulas that match two parameters named support and confidence can be treated as associated rules.

The support of an item can be defined as the proportion of the appearance of the transaction in the transaction database and extends to the formula $X \Rightarrow Y$. The support in the given data set of which is the proportion of the transaction including transaction X and Y at the same time, shown as in Eq. (1).

$$support(X \Rightarrow Y) = \frac{|\{T \mid X \cup Y \subseteq T, T \in D\}|}{|D|} \tag{1}$$

support $(X \Rightarrow Y)$ is also known as the relative support. In the actual process, it often uses the occurrence number of an item to measure whether it has a lot of data base. At present, the measure standard is called as the support count or absolute support.

Confidence is used to determine reliability of the implication which is composed of items according with the requirements of the support. The measure mode is identifying the probability of the former and the latter items appearing in the existing records at the same time. When the co-occurrence probability is greater, it can be considered that two items under the given data background associate with each other. For the formula $X \Rightarrow Y$, its calculation is shown as in Eq. (2).

$$confidence(X \Rightarrow Y) = \frac{|\{T : X \cup Y \subseteq T, T \in D\}|}{|\{T : X \subseteq T, T \in D\}|} = \frac{support(X \Rightarrow Y)}{support(X)} \tag{2}$$

As mentioned above, whether the formula $X \Rightarrow Y$ can be used as association rules, it need to calculate and judge the values of *support* $(X \Rightarrow Y)$ and *confidence*$(X \Rightarrow Y)$ whether meet predetermined threshold values. Here these two given values are called as the minimum support threshold (*min_sup*) and the minimum confidence threshold (*min_conf*), in which the item matching *min sup* in $X \Rightarrow Y$ is called frequent item sets. But in the actual implementation all frequent item sets matching min_sup conditions are usually obtained first, then the formulas similar to $X \Rightarrow Y$ are formed. It can be discovered by the calculating method of confidence. The value can be easily calculated from the support obtained above. Therefore, the process of finding frequent item sets becomes the key point of association rule mining problems.

3. Algorithm Description

The mining of association rules of medical insurance policy text information based on its classification is taking the association rules mining by the specific business process according above classification. Each section of medical insurance policy is independent. Using the idea of dividing can mine each section of association rules at the same time and the result is expressed as the correspondence between specific business processes and the association rules, meanwhile, each part doesn't need to be aggregated into the overall association rules because the meaning of mining association rules by classification is obtaining more accurate association rules under the specific background.

Though text preprocessing and characterizing, medical insurance policy text information has formed the feature representation and is stored in the database. At the end of the medical insurance policy text information classification, identify each medical insurance policy text information in the database with category is marked according to their business.

On the basis of above database, this paper proposes the text association with classification mining algorithm (Text Association with Classification Mining Algorithm, TACMA). The algorithm is described as follows:

(1) Get all business processes classifications $P=\{P_1,P_2,\ldots,P_n\}$ of the medical insurance policy in the database;

(2) For each classification of medical insurance policy business process P_i, traverse the database respectively, get the characteristic representation of the medical insurance policy text information including record P_i and add contents of the database to the information matrix PM_i corresponding to Pi by taking it as a new line of data;

(3) Get information matrixes $PM=\{PM_1,PM_2, PM_3,\ldots\}$ of all types of business processes of medical insurance policy;

(4) Carry association rules discovery on each information matrix PM_i of category P_i of medical insurance policy business process.

The algorithm first obtains medical insurance policy text information containing each medical insurance policy business process by method *Get_Process_Matrix* and forms the information matrix facilitating to analysis and process.

Then the algorithm carries association rules discovery on the information matrix PM_i of each category P_i of medical insurance policy business process, and obtain the frequent item set from information matrix PM_i. Next, it generates frequent multi-item set and items in which arrange by lexicographic order. Therefore, only two frequent k-1 item sets with different last item need to be paid close attention when generate frequent k item set from k-1 one. At last, we can obtain the association rules from frequent item sets generated above and store rules and their confidence in the association rule set R_i, which is corresponding to the association rules of category P_i mentioned above. Because the confidence of association rule set R_i indicates the meaning of the possibility of appearance of the former and the latter item simultaneously, it is used as a measure of the association strength of association rules.

4. Experimental Design

Aiming at the medical insurance policy project, the host in design environment and effect analysis of the experiment environment of the association algorithm is configured as: operating system Windows 2007, CPU 2.66GHz, memory 4G, database is DB2 V9.1. Compare the mining effect in the characteristic representation of text data of the traditional association rule mining algorithm TP-growth and TACMA algorithm by experiments.

Experiments adopt the relevant literatures in the field of basic social medical insurance as the experimental data, mainly including " New rural medical scheme" and "Urban residents' basic medical insurance". After a preliminary text processing, these materials are divided into 102 smaller documents and adopt as the source of audit information in experiments.

5. Experimental Results and Analysis

Through distinguishing classifications to mine association rules, TACMA algorithm obtains 439 association rules after the summary. TP-growth algorithm obtains 221 association rules. By analyzing the contents, it can be conclude that the number of association rules obtained from TACMA algorithm is more than TP-growth algorithm on each category, which explains that the mining effect is better by adopting classification association rule, as shown in Figure 1.

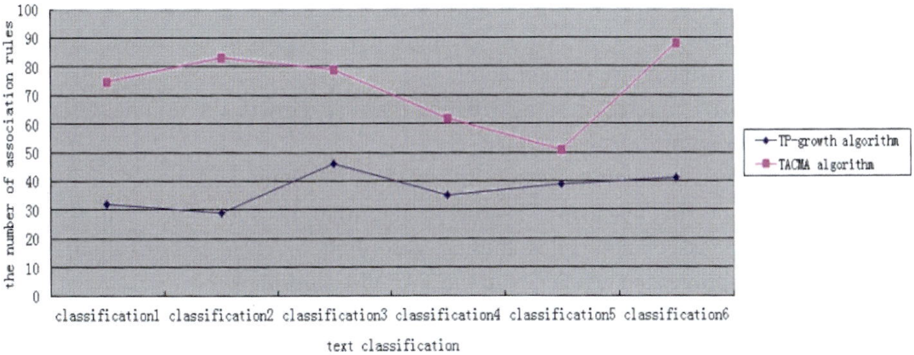

Fig. 1. The comparison of association rules mining.

There are some differences on the confidence between these two methods, which mainly reflect on that the confidence of association rules obtained from TACMA algorithm is higher than that of TP-growth, as shown in Figure 2.

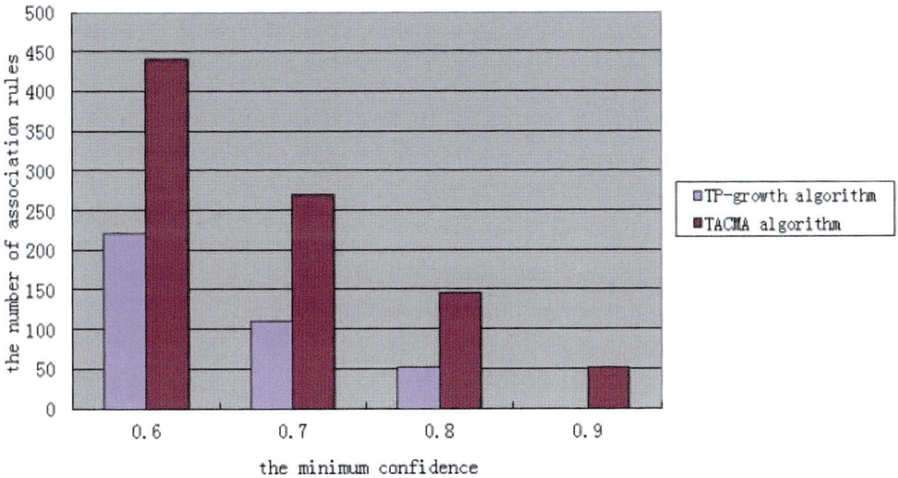

Fig. 2. The comparison of different confidence.

There is no association rule with the minimum confidence 0.9 in the algorithm TP-growth, on the other hand, three are 63 association rules in TACMA algorithm. At the same time, the number of association rules that meet other minimum confidence in TP-growth algorithm is also less than the corresponding TACMA algorithm. It is obvious that association rules obtained by TACMA algorithm show stronger correlation along with the meaning of the specific knowledge type, while the algorithm TP-growth has weak correlation expressed in the entire field of audit.

TACMA algorithm is also better than TP-growth on mining efficiency, shown in Figure 3.

In the same amount of audit information, the mining efficiency of TACMA algorithm is higher than TP-growth algorithm because TACMA algorithm carries on association rules mining according to each classification, which greatly improves the efficiency of the TACMA algorithm.

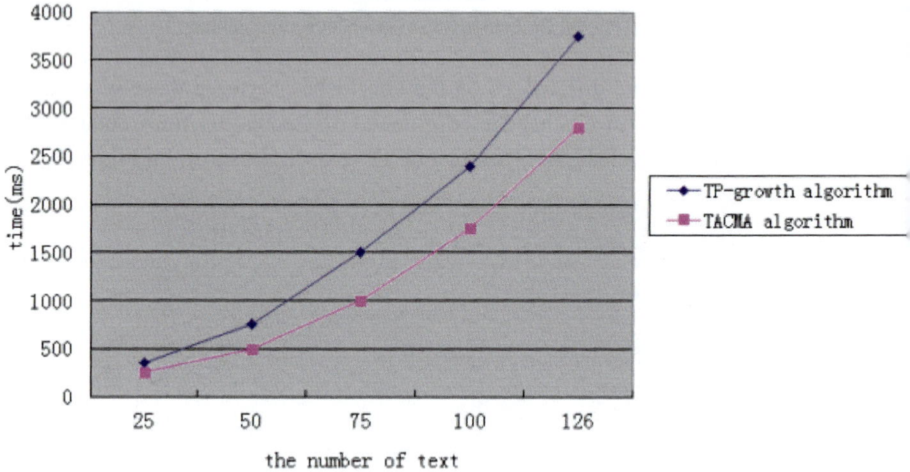

Fig. 3. The comparison of mining time.

6. Conclusion

This paper mainly provides TACMA text association rules classification mining algorithm aiming at the medical insurance policy text, and. Through large amount of experiments, the paper compares the effect of the traditional association rules mining algorithm TP-growth and TACMA algorithm on characteristic representation of text and proves that the TACMA algorithm has obvious advantages on the effectiveness and time efficiency.

Acknowledgments

This work is supported by the Fundamental Research Funds for the Central Universities under Grant No.100603.

References

1. J.M. Korn, F.G. Kuruvilla, S.A. McCarroll, et al. Integrated genotype calling and association analysis of SNPs, common copy number polymorphisms and rare CNVs. Nature genetics, 2008, 40(10): pp. 1253-1260.
2. Liu Quan-Ying. Social Medical Insurance and Commercial Medical Insurance. Financial Theory and Practice, 2004, (04). pp. 93-99.
3. Liu Y. Reforming China's urban health insurance system. Health Policy, 2002, 60(2): pp. 133-150.
4. R. Agrawal, T. Imieliński, A. Swami. Mining association rules between sets of items in large databases//ACM SIGMOD Record. ACM, 1993, 22(2): pp. 207-216.
5. T. Uno , T. Asai, Y. Uchida, et al. An efficient algorithm for enumerating closed patterns in transaction databases//Discovery Science. Springer Berlin Heidelberg, 2004: pp. 16-31.

The research and application of data visualization based on D3 in the field of healthcare

Xin-Liang Xu* and Lian-Jun Chen

Donghua University, Shanghai Jian Qiao University, Shanghai, China

**E-mail: 807998774@qq.com*

www. http://www.dhu.edu.cn/, www. http://www.gench.edu.cn/

With the development of computer and database technology, people have a multitude of ways to access information. However, the rapid increase of big data information has created new problems for people. Due to the vast amounts of monitoring data and ineffective extraction of intrinsic data, more and more information are ignored. There is demand for user-oriented growth in data analysis, which is a technology that presents data in a more graphical way for easier visualization. This data visualization technology will impact users greatly and aid them in making decisions in the future. By combining the D3 visualization library with practical applications in the field of health care, data visualization can be used to effectively explore health care research.

Keywords: Data Visualization; D3 ; Healthcare.

1. Introduction

Since the 21st century, big data is the hottest topic talking about by the people, how to get the most effective data and information to help their decision-making has become a new challenge. Compared with the traditional mainstream relational database, big data belonging to non-relational databases are multi-state, large high-dimensional data, dynamic and heterogeneous. These huge amounts of data are diverse information assets, its need for new processing mode to have a stronger decision-making power, insight found force and process optimization capabilities of massive and high growth rates [1]. At present, China's traditional medical system is facing serious challenges in the era of big data, how fast the acquisition and pre-processing of data, then through the appropriate management and analysis, dig out useful information, which will eventually contain text, images, audio, video, and sensor detection system data.

---

* Corresponding author.

Showing diversified relations of data efficiently and quickly, human decision-making to facilitate data visualization has drawn increasing attention.

2. Data Visualization

2.1. *Data view of data vitalization*

Data visualization is a wonderful combination of art and technology, large data sets and graphically represent by using data analysis and development tool found unknown information processing [2]. Here are talking about the type of data, such as a typical numerical data, as well as text-based data, and there are some atypical data, including symbols, images, audio, video, etc., we pass the original data by using visual elements to express that modes which can express the attributes associated with the data itself Properly. Data visualization is the so-called data in large databases or data warehouse data visualization, lift people out from the relational data tables or scattered data, discarding the middle of the process of observation and analysis, let people see the association of data in a more intuitive way.

The basic idea of data visualization is setting each data unit as a primitive element, in visual rendering process, using computer graphics and image processing technology, the style of high and low, left and right, position, size, color, etc., the arithmetic operation of sorting, classification perspective, etc., converting the data into a acceptable form to the people, so amount of data images represent in front of people. Furthermore, indicate the value of each attribute data with the form of multi-dimensional, it's convenient to help people observed data from different angles, allowing users to have a deeper realization and analysis, thereby providing decision-making role.

2.2. *Related technologies*

2.2.1. *SVG*

SVG (Scalable Vector Graphics), also known as Scalable Vector Graphics, is an XML-based standard, high-quality, flexible, easy and open vector graphics description language [3]. SVG graphics has dynamic and interactivity, you can use SVG rendering scalable, multi-format, high-quality vector graphics.

In addition to the characteristics of SVG vector image is created based on XML, it also has characteristics supporting for vector filter. Follow the text format, SVG images and animations are also text, its images can be generated dynamically by programming language, such as JavaScript, java and other programming language, this feature plays an important role on the database tab.

DOM (Document Object Model, Document Object Model) is fully supported, allowing dynamic configuration scripts or programs stored files, style, content. Because SVG is a text format, so you can operate cross-platform and get out of the flat limitation, avoiding data diversity and other issues when transferring relevant external data.

Because SVG has its unique advantages, it is suitable for network development and compatible with other industry standards, has attractive prospects in the application of wireless devices and image search. you can implement Web graphics publishing, generating graphics dynamically which provide the foundation for data tables and image maps.

2.2.2. *D3*

D3 refers to a data-driven documents (Data-Driven Documents), is essentially a JavaScript library, strict compliance with Web standards, to operate through the data file, compatible with major browsers, can provide a strong visual component, allowing users to manipulate DOM by data-driven. Its advantages are as follows:

① data can be bound together with the DOM; ② data conversion and mapping are independent of each other; ③ concise code; ④ large number of layout; ⑤ based on SVG, zoom with no loss of accuracy.

D3 process does not involve accessing, analyzing, filtering, and mining to a range of data, more scope is in Represent, Refine, Interact [4]. Implementation Mechanism is the combination of sucked data in the database with HTML, SVG, CSS, let the initial data in three-dimensional space form a compact data system, which reflects the relationship between the data and ensure visualization of abstract data. In the visualization process sources of data is from the author, in particular visualization process, D3 uses a chain syntax to achieve a good extraction and application of linear data in the database.

For example:D3.select ("body") append ("p") text ("hello world"); // add text to the P element.

D3 supports a variety of data formats, such as JSON, XML, CSV, HTML, etc.. data and documents set as element nodes, which should link to relational group. Using the powerful built-in layout force, the full name of Force-Directed Graph (force Roadmap), to make visualization process treatment after loading the data. The elements used in the process are divided into two categories, one is a visual chart elements, including the X-axis and Y-axis and a line graph of the number of shape; the other is the background train system design parameters, including the size of the area graphic canvas, visualization activity time, table and other parameters to ensure the authenticity and accuracy of visual graphic information.

D3 is not a frame of processing data, it mainly analysis data from the mass of two-dimensional data table, summarizes the relationship between the data network, connects all the data in the form of three-dimensional information and presents graphics to the user, thus ensuring data relationships fully reflects [5].

3. The Analysis and Application of Data Visualization in Healthcare

In recent years, Malaysia's Muhammad, who study EHR visualization system, from the first perspective of the doctor's ,combine graphical, visualize the abstract and efficient electronic medical record data to people, and obtain critical information, thereby strengthening the management of patients. Using technology of data mining visualization, the American company Vysis and Neuro Medical also make this study, through its research and development of new drugs, analysis on the protein extraction and studies on the drug side effects, side effects of drugs was found lastly.

For the elderly health situation in china, someone has been proposed a three-tier wireless medical sensor network. However, with technological development, four sensor networks will have a good application scenarios, its structure is shown below in Figure 1:

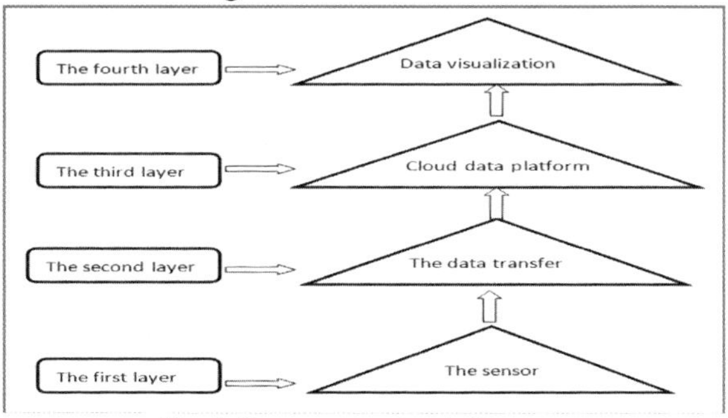

Fig. 1. Four sensor network diagram.

Here the definition of four sensor network, the first layer is made of some medical sensors, mainly include wearable devices; the second layer provides reliable data transmission, universal application of high-speed networks, data transmission node via 4G, Wi-Fi, Bluetooth technology and other means to send data to the data center; the third layer is through cloud data centers of telemedicine platform to analysis; The fourth layer presents the results of analysis data through visualization technology, and make predictions and solutions feedback to the patient.

3.1. *Health monitoring*

The most common test items ECG, oxygen saturation, body temperature, heart rate, blood pressure, how to capture data and real-time data sent to the monitoring center have become the key element. The equipment used most in current market is a smart bracelet, it can record the user's daily survival exercise, sleep and diet, real-time data. The latest bracelets have been added monitoring heart rate, blood oxygen, temperature, blood pressure. Unexpected situations are a major health hazard of life, monitoring movement or activity is essential for patients, after auto-sensing motion sensor analysis, demand will be sent to the monitoring center. There are also a series of doses of drugs, dietary laws and other monitoring, patient will be protected in time, first layer plays the major role in four sensor network.

3.2. *Transmission of health information*

The rapid rise of wireless mobile communication technology, its maneuverability favored four-sensor networks. 4G network of satellite communication technology and the rapid development of wireless networks are free to switch between transmission of health information which was able to transited in time, we believe that in the near future, Bluetooth, 4G, Wi-Fi as well as fiber optic communications can have tighter integration degree.

3.3. *Deployment of wisdom medical center cloud platform*

The central cloud platform of medical wisdom system in accordance with four-layer network builds the architecture, namely: layer of information acquisition, layer of storage and transport, layer of information analysis, layer of presentation of data and provides data for visual feedback of the decision-making level, the system structure Figure 2 is as follows:

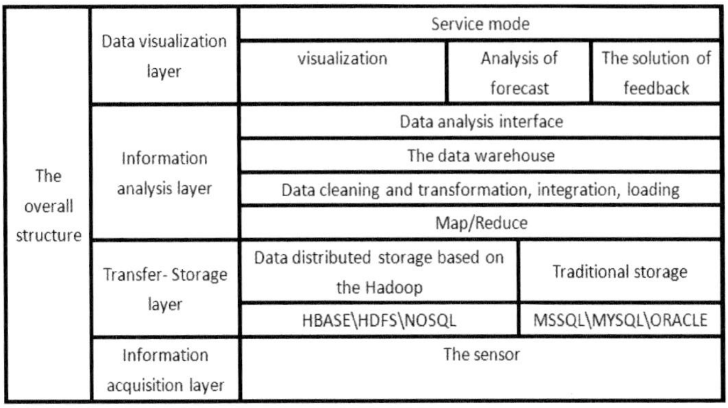

Fig. 2. System block diagram.

By the existing health information system, we can not service all business sectors, also can not meet the data analysis of the cloud, thus establishing a comprehensive information collection layer to obtain medical information. Because acquisition layer is used in the form of a sensor, so mobile wireless technology ensure the platform to collect information in time. Since the advent of the era of big data, consider the fact that relational database and non-relational databases coexist, the high consistency of structured data stored in traditional databases, multiple types of unstructured, low consistency-based data stored in big data distributed systems based Hadoop[6].

3.4. *Visualization of health data*

Results of information analysis layer processing, through data analysis interface, using JSON (JavaScript Object Notation) lightweight data interchange format, reading from the database dynamically. JSON is a subset of JavaScript, a fully language-independent text format, but its habits is similar to C family, precisely this feature, making it easy for machines to parse and generate [7]. The structure of data After anglicizing by large data platform generally is {key: value, key: value, ...}, key is the object's properties, value is the corresponding attribute value, type the attribute value can be numbers, strings, arrays, objects. Combing objects and arrays two structures can form a complex data structure. Processing in the data visualization can be divided into the following steps:

(1)Reading JSON data dynamically

The basic format of JSON is: { name:[{"111","aaa" },{"222","bbb"}, {"333","ccc"}] }

However, in a dynamic interactive referencing of JSON data, you can customize the data formats, such as:

[num:[{“number”}],healthy:[{time:{“yy-mm-dd”}, rate:{“nn”}, tem:{“tt”}, bp: [{“pa” }], bo: {“bl” } }]];

Defined data formats, “num” refers to the patient's medical card number, “healthy” refers to condition of health, time refers to the system time, rate refers to heart rate, tem refers to temperature, “bp” refers to the blood pressure value, “bo” refers to oxygen values in blood. Load the exchanged data from the file healthy.json, using d3.json requesting file healthy.json, the visual syntax is:

d3.json(“healthy.json”,function(error, root)){};

Content of healthy.json file will be saved in the variable root, after the console output can be seen as an array where each item holds key information.

(2)Before drawing

Select the appropriate graphics to effectively present data information, such as four-sensor networks, to show personal physical condition of the patient, we will choose a bar chart, stack charts, line charts, pie charts from various different angles, intuitive rendering patient monitoring health status, and for overall health, we can visualize the data from the map, pie two ways.

(3)Setting

Selecting the appropriate graphic representation of data, conversing the initial data. Setting appropriate visitor is base of data converting, we can create a graphic layout when visitor finished . For example here we must first draw bar, the statement of initial data setting visitor shown as this:

Var stack=d3.layout.stack().values(function(d)){ return d.healthy ;}.x(function(d){ return d.time;}).y(function(d){ return d.rate;}).z(function(d) { return d.tem;}).m(function(d){ return d.bp;}).n(function(d){ return d. bo;});

(4)After setting

Add the <svg> </ svg>, <g> </ g> and other elements to complete the corresponding layout. We set time as the abscissa, and the ordinate as the corresponding range, after loading the data corresponding to displayed on the layout, such as blood pressure, heart rate, oxygen equivalent. When the mouse moves to the corresponding position, the corresponding range can be displayed directly to the user in the form of text.

(5)Data presenting

In addition to the stack effect represented by a rectangle, we can also create a regional builder, to dynamically generate stacked charts, we can see the trend of the monitoring data, it’s easier for people to make intuitive decisions. Line charts here also has a good application, applying D3 features and marking some abnormal data by the special color, abnormal physical condition can be feedback to patient in a certain time, such as heart rate over time is higher than normal or lower than normal.

Also for the condition of overall health, we chose the map layout, a unique advantage of map is that we can see the distribution of data for certain infectious diseases, through data analysis layer in four sensor networks, health data will be distributed on the map using luminous figure expression, certain health monitoring data by color and brightness of the light-filled visualize whose elements complex operations than the general graph, first define the map projection and geographical path builder, request a map file "china.json", then import the data file, you can see the trend of certain out-breaking disease and infectious diseases on the map, and finally help people make early prevention and response measures.

4. Summary

On the basis of D3,this article make a explosion to the wisdom of health care in view of the data visualization. More specific form of monitoring over health data can be unfolded to ensure that the medical center cloud platform fully grasp the physical condition of the patient, through the data map presented, prevention of infectious diseases and responding measures play a decision-making role in the practical application. Thus, the visual analysis has distinct significance [8]. D3 visualization library as a data display technology, has good prospects in market demand ,and has gotten a great deal of attention, I believe both applications in all aspects of medical wisdom, or Internet traffic data visualization and other social groups in society, D3 also has its own unique charm, and better serve humanity.

References

1. China Big Data. The six secret of using big data well[EB/OL]. http://www.thebigdata.cn/YeJieDongTai/29051.html, 2016-01-29.
2. Kan Liu, Xiao Zheng Zhou, Dong Yu Zhou. Research and development of data visualization [J]. Computer Engineering,2002- 8,28(8): 1-2.
3. Yan Jie Feng. Expression of multi-scale spatial data based on SVG [A].2006.
4. Duboeuf. Vitamin D3 and Calcium to Prevent Hip Fractures in Elderly Women — NEJM. New England Journal of Medicine(J), 2015.
5. C Mao, M Wadleigh, GM Jenkins, YA Hannun, LM Obeid. Interactive and Multivariate Choropleth Maps with D3 [J]. Cartographic Perspectives 2015, 272(78):57-76.

6. O Aisling, D Jurate, RD Sleator. 'Big data', Hadoop and cloud computing in genomics[J]. Journal of Biomedical Informatics, 2013, 46(5):774-781.
7. Fei Yan,Hao Li. Improve transmission of AJAX JSON data. China High Technology Enterprises,2008.
8. Julie Steele, Noah lliinsky. Beautiful Visualization[M]. BeiJing: Mechanical industry press, 2011- 6:7-37

Chapter 5

Computer System and Application

Efficient concurrent virtual machine scheduling for Xen hypervisors

Po-Jen Chuang* and Chih-Yuan Chou

Department of Electrical and Computer Engineering, Tamkang University, Tamsui, New Taipei City, Taiwan 25137, R. O. C.

**E-mail: pjchuang@ee.tku.edu.tw*

This paper presents a new CPU scheduler to facilitate concurrent applications in Xen. The new scheduler uses the completely fair scheduler to reduce the gap between concurrent and non-concurrent virtual machines (VMs), and to balance physical CPU time distribution and resource utilization. It uses the red-black tree to shorten the virtual CPU (VCPU) lookup time, achieve efficient task scheduling and to reduce the runtime. To assist with concurrent VCPU synchronization and scheduling, we set a concurrent waiting queue which can handily pick up marked VCPUs for execution. Extended simulation runs are conducted to evaluate the performance of our proposed scheduler and the simulation results showed that our scheduler outperforms existing schedulers in reducing the runtime for both concurrent and non-concurrent applications.

Keywords: Xen Hypervisors; Concurrent Virtual Machines; Scheduling; Performance Evaluation.

1. Introduction

In virtualization practices [1], we establish fully functional virtual machines (VMs) to execute operating systems and applications like a real computer, and therefore need to allocate available physical resources – efficiently – to form the desired VMs. This paper presents a new CPU scheduler over Xen [2] to secure better hardware resource allocation and task scheduling, and shorten the CPU runtime especially in handling concurrent applications. The ultimate goal is to generate less overall runtime and better overall performance than existing schedulers. To achieve the goal, our new scheduler uses the Completely Fair Scheduler (CFS) [3] to reduce the gap between concurrent and non-concurrent VMs. The CFS mechanism can improve the concurrent application problem because it can balance the CPU time distribution and upgrade efficiency. We meanwhile adopt the Red-Black (R-B) Tree structure [4] to replace the run

* Corresponding author.

queue structure of previous schedulers and, by doing so, successfully shorten the virtual CPU (VCPU) lookup time to reach more efficient task scheduling and save overall runtime. To assist concurrent VCPU synchronization and scheduling, we use a simple approach to mark VCPUs and pick up marked VCPUs for handy execution by a proposed concurrent waiting queue. Extended simulation runs are conducted to evaluate the performance – in terms of runtime – of our new scheduler and the existing Credit [5] and Turbo [6] schedulers. The results show that, in contrast to previous schedulers, our new scheduler yields notably reduced overall runtime and enhanced efficiency.

2. The Proposed New Scheduler

As mentioned, our scheduler brings in the CFS mechanism, which is designed to provide fair CPU scheduling, to reduce the gap between concurrent VMs and non-concurrent VMs. We employ it to solve the run queue problem and reduce the runtime. To obtain the virtual runtime (vruntime, an important concept in CFS) of VCPUs, we increase or decrease – based on their weights – a fixed part of the runtime. Vruntime can lead to fair scheduling. It reflects the overall CPU time a current task has consumed and will be a proper measure to help decide if to remove the current task from execution in order to start a new task. We can attain the vruntime of a VCPU by the following formula:

$$\text{vruntime(delta\_exec\_weighted)} = \text{deltaTime} * \frac{\text{Default\_Weight}}{\text{Weight}}$$

In this formula, deltaTime indicates the time between two consecutive vruntime computations. Default_Weight is a predetermined fixed value. As we can see, when the weight of a VCPU grows, the ratio (Default_Weight/Weight) drops and so does the vruntime. When the weight of a VCPU declines, both the ratio and vruntime rise. We then input the obtained vruntime into the R-B Tree to work out task scheduling. All VCPUs that attain vruntime values will be scheduled as follows: The one with the smallest value will get to the leftmost side of the tree and be picked up for the next execution. After having the vruntime of all pending VCPUs, we then sort out their execution priority by performing the R-B Tree algorithm. The R-B Tree [4] is a self-balancing binary search tree, able to guarantee searching in O(log n) time – n is the total number of elements in the tree. It conducts the insertion operation according to the "key" values (i.e., the vruntime values in our scheduler) and moves the smallest key value (the VCPU with the shortest vruntime in our scheduler) to the leftmost side of the tree. By such a practice, it helps us attain the sequence of task scheduling efficiently. That is, it helps us decide the scheduling priority of pending VCPUs – based on their vruntime values – and attain the execution sequence. We can always pick up the VCPU at the leftmost side of the tree, which has the shortest vruntime, for next execution.

As both Credit [5] and Turbo [6] schedulers set concurrent VCPU scheduling by the concurrent task run queue which requires considerable VCPU lookup time, we propose to use a different concurrent waiting queue to obtain the concurrent run queue in a more efficient way. We first use a flag to mark each concurrent VCPU awaiting execution, and before inserting VCPUs into the R-B Tree to work out the running schedule, we will sort out all VCPUs by the marked flags and put them into two groups: unmarked general VCPUs or marked concurrent VCPUs (CON-VCPUs). After the sorting, we insert the general VCPUs into the R-B Tree and put the CON-VCPUs into the concurrent waiting queue (con-wait-queue). We then involve a set of triple parameter indexes (PCPU, VCPU and Domain) to denote those CON-VCPUs waiting to enter the concurrent waiting queue. Of the three indexes, PCPU will lead a CON-VCPU to the concurrent waiting queue of a proper PCPU, and Domain will take it to the proper position in the waiting queue. We then build a one-dimensional array of global variables con_bit [Domain ID] to monitor all CON-VCPUs and decide their bit positions according to index VCPU. Thus, selecting CON-VCPUs from our con-wait-queue becomes an efficient O(1) operation.

Table 1. Scheduler comparisons: key features.

Scheduler	Credit [5]	Turbo [6]	Ours
PCPU's queue operation	1.Linear queue – **Run queue** 2.Time complexity **O(n)**	1.Linear queue – **Run queue** 2.Time complexity **O(n)**	**1.R-B Tree + Vruntime** 2.Time complexity **O(log n)**
Solving concurrent applications	NO	1.Tag priority of Turbo 2.Trace Turbo tags - **O(n)**	**1.con-wait-queue** 2.Time complexity **O(1)**

For clear reference, we list the key features of the three target CPU schedulers, Credit, Turbo and Ours, in Table 1. Recall that our new scheduler makes use of the vruntime and R-B Tree to establish task scheduling. Compared to the original linear queue operation in the other two schedulers which requires O(n) execution time, our scheduling operation takes only O(log n) – a notably reduced amount. In solving the concurrent application problem, the Turbo scheduler uses turbo flags to mark CON-VCPUs and looks up all marked

CON-VCPUs to pick them up for execution. To complete the concurrent execution, our scheduler puts each waiting CON-VCPU in the concurrent waiting queue and starts their execution only when all of them are ready in the queue. That is, we take only O(1) to get all CON-VCPUs in the concurrent waiting queue to be executed by the PCPU. Different from the Turbo and our schedulers, the Credit scheduler has no mechanism to deal with this concurrent application issue.

3. Experimental Evaluation

Experimental evaluation is conducted to evaluate the performance of the Credit scheduler [5], the Turbo scheduler [6] and our new scheduler (Ours). In the simulation, we implement NAS Parallel Benchmark [7] NPB-LU.4.A (briefed as LU) on one VM and gradually increase the working VM from 1 to 8 VMs. Each VM has 4 VCPUs and performs a NPB program. When the number of VMs increases to 8, there will be a total of 32 VCPUs in the system – waiting to get the resources of 4 PCPUs and the 8 VMs will execute a total of 8 NPB programs. We then carry out the experiment to check the performance of the three schedulers and collect the results of runtime in Figure 1 for comparison.

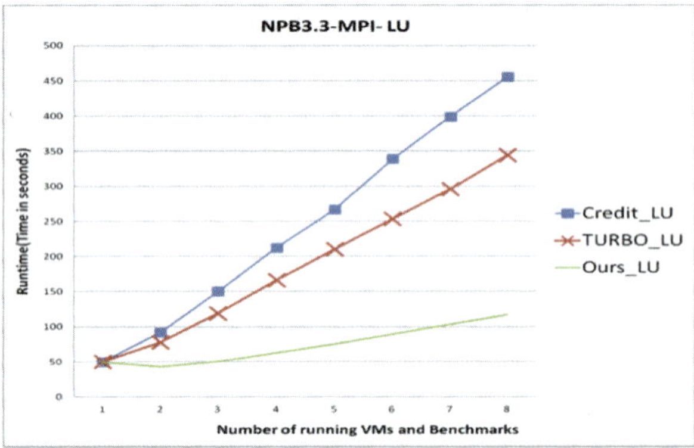

Fig. 1. Performance comparisons under Benchmark LU.

Note that when VMs are put into work under Benchmark LU, they will execute concurrent applications which require involved CPUs to communicate with each other in order to get the needed synchronization information and based on the information to undertake the operation. Concurrent applications like this may constantly consume a significant amount of waiting time as CPUs must spend time waiting for the synchronized act. From the results, we can see the Turbo scheduler needs less runtime than the Credit scheduler because it uses the priority tag to mark VCPUs awaiting concurrent execution and sort them out for the next execution, thus reducing time consumption due to synchronization as with the Credit scheduler. Our new scheduler takes the shortest overall runtime. That is, we take the least time to finish task execution and result in the best efficiency. We outperform the other two schedulers in runtime mainly because our efficient task scheduling can turn over more balanced CPU time distribution, less VCPU lookup time and better synchronization on concurrent VCPUs.

4. Conclusion

Our new CPU scheduler employs the CFS mechanism and the concept of virtual runtime to produce more efficient task scheduling, fairer PCPU time distribution and higher efficiency than existing schedulers. It adopts the R-B Tree, instead of the original run queue structure, to shorten the VCPU lookup time, and builds a simple approach to mark and install VCPUs into an efficient concurrent waiting queue to reach better synchronization on concurrent VCPUs. Simulation results exhibit that, in contrast to previous schedulers, the new scheduler can constantly reduce the runtime for both concurrent and non-concurrent applications.

References

1. Virtualization, http://en.wikipedia.org/wiki/Virtualization.
2. D. Chisnall, The Definitive Guide to the Xen Hypervisor, Prentice Hall, Nov. 2007.
3. I. Molnar, Scheduler core and Completely Fair Scheduler [CFS], http://lwn.net/ Articles/230501/.
4. L. Turbak, Red-Black Trees, http://cs.wellesley.edu/~cs231/spring01/red-black.pdf.
5. Credit Scheduler, http://wiki.xensource.com/wiki/Credit_Scheduler.
6. C. Y. Tseng, S. C. Lin, L. C. Chen, and H. K. Chung, The performance improvement and evaluation of an enhanced CPU scheduler in virtualized environment," in *Proc. 2011 International Conference on e-Commerce, e-Administration, e-Society, e-Education, and e-Technology* (*e-CASE & e-Tech 2011*), Jan. 2011, p.3107-3123.
7. NAS Parallel Benchmarks, http://www.nas.nasa.gov/publications/npb.html.

Rotating laser precision machining computer control system

Tao Wang[1,3,*], Long-Fei Ma[1, 3], Bo Zhang[1, 2], Sha Han[1, 3],Ya-Peng Hu[1, 3], Xin-Chao Zhao[1, 3] and Zhuo Wang[1, 2]

[1]*College of mechanical engineering, Hebei University of Technology, Tianjin 300130, China*

[2]*College of mechanical and electrical engineering, Harbin Engineering University, Harbin, Heilongjiang,150001, China*

[3]*Wuxi Jin Tianyang laser Electronics Co. Ltd., Jiangsu, Wuxi 214192, China*

[*]*E-mail: wtao_1@163.com*

This paper studies the laser rotating precision fuzzy PID computer control system, using laser technology combined with computer intelligent control technology, the control technology of stepper motor control method based on fuzzy PID control algorithm of fuzzy computer, PID computer, PID computer fuzzy controller is designed for stepper motor, stepper motor built computer control system also, the stepper motor control system was analyzed by MATLAB computer simulation in MATLAB/Simulink environment to build the stepper motor fuzzy PID computer control model, the implementation of the MATLAB computer simulation was established based on the laser rotary precision machining machine computer intelligent control system, with the elimination of the stepping motor start and stop. The swing, running well, precision rotating laser processing.

Keywords: Computer Intelligence; Laser Processing; Fuzzy Control; PID Computer Control; Rotary Process.

1. Introduction

The fuzzy PID computer control system of laser rotating precision machining is a science and technology which combine the laser processing with the computer intelligent control, and belongs to the advanced manufacturing technology of the computer intelligence[1].

In rotating laser precision processing system, stepper motor to start and stop, with a swing, influence the quality of products, this paper investigates laser rotary precision processing computer fuzzy control system, by means of laser technology, proposed by the fuzzy PID algorithm of computer control of stepper motor control method to eliminate the step motor at start and stop with the swing, improve the processing precision of laser rotation and establish a rotating laser precision machining machine computer intelligent control system, laser rotating intelligent precision machining.

2. Fuzzy Self-Tuning PID Computer Control

Based on computer PID control principle, the fuzzy control principle and fuzzy computer PID control principle, by the fuzzy PID control computer analysis and Study on the theory, the modern theory of fuzzy control, PID control computer control theory optimization with rotating laser precision processing step for machine computer control, stepper motor design fuzzy PID computer control rules, the computer PID control and fuzzy control combined, integrated both advantages can get higher control effect of the system, construction of step motor fuzzy PID controller computer[2].

Using fuzzy self-tuning PID computer control, the structure diagram is shown in Figure 1.

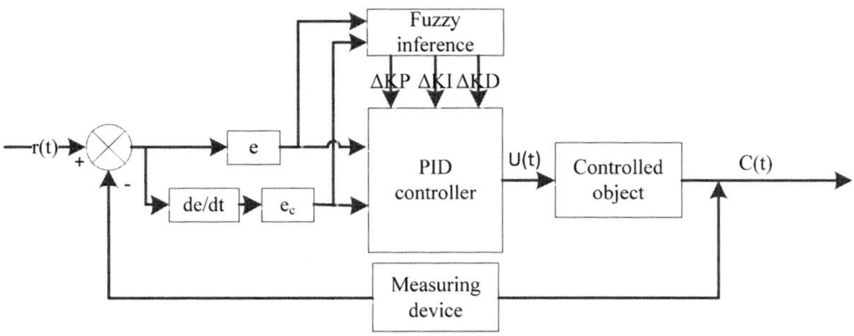

Fig. 1. Fuzzy PID computer controller structure diagram.

The design steps of the closed loop fuzzy PID computer controller of the stepper motor:Stepper motor close loop fuzzy PID controller design steps: first, the establishment of a fuzzy inference system and the system input position deviation e and deviation change ec, three output respectively ΔK_P, ΔK_I, ΔK_D.

Establishing fuzzy control rules, according to the influence of PID computer control parameters on the system output, different stages require different control parameters. The basic principles of e and ec, K_P, K_I, K_D and are the following[3][4]:

(1)When the deviation e is large, increasing K_P to speed up the system response and reduce the steady state error and improve control precision. At the same time take smaller K_I and K_D.

(2)when the deviation e and ec medium larger, for the overshoot and ensure the response speed, K_I appropriate value, K_P, K_D take a smaller value.

(3)when the deviation is small, should be appropriately increased K_P and K_I values in order to make the system has good steady-state performance, and take appropriate K_D to avoid output in response to a predetermined value in the vicinity of the oscillation and improve the anti-interference ability. In particular to ec is small, K_D should take a larger value, when ec is large, K_D smaller values.

Respectively parameters ΔK_P, $\Delta K_{I,}$ ΔK_D fuzzy control rules, as Table 1~3 shows.

Table 1. Rules of fuzzy control of ΔK_P.

e \ ec	NL	NM	NS	ZO	PS	PM	PL
NL	PL	PL	PM	PM	PS	ZO	NS
NM	PL	PL	PM	PS	PS	ZO	NS
NS	PM	PM	PM	PS	ZO	NS	NS
ZO	PM	PM	PS	ZO	NS	NM	NM
PS	PS	PS	ZO	NS	NS	NM	NM
PM	PS	ZO	NS	NM	NM	NM	NL
PL	ZO	ZO	NM	NM	NM	NL	NL

Table 2. Rules of fuzzy control of ΔKI.

e \ ec	NL	NM	NS	ZO	PS	PM
NL	NL	NL	NM	NM	NS	ZO
NM	NL	NL	NM	NS	NS	ZO
NS	NL	NM	NS	NS	ZO	PS
ZO	NM	NM	NS	ZO	PS	PM
PS	NM	NS	ZO	PS	PS	NM
PM	ZO	ZO	PS	PS	PM	PL
PL	ZO	ZO	PS	PM	PM	PL

Table 3. Rules of fuzzy control of ΔK_D.

e \ ec	NL	NM	NS	ZO	PS	PM	PL
NL	PS	NS	NL	NL	NL	NM	PS
NM	PS	NS	NL	NM	NM	NS	ZO
NS	ZO	NS	NM	NM	NS	NS	ZO
ZO	ZO	NS	NS	NS	NS	NS	ZO
PS	ZO	ZO	ZO	ZO	ZO	ZO	ZO
PM	PL	NS	PS	PS	PS	PS	PL
PL	PL	PM	PM	PM	PS	PS	PL

Figure 2 is using Mamdani inference results, when e=1.14, ec=0.646 respectively corresponding to step into motor, lose step several steps, namely given value difference is a misfit formed step different diversification values. It corresponds to the parameter of PID valueΔK_P=-0.542. ΔK_I=0.118 ΔK_D=0.036.

The rule view of the output surface as showed in Figure3, you can observe the input e and ec on the output of the impact.

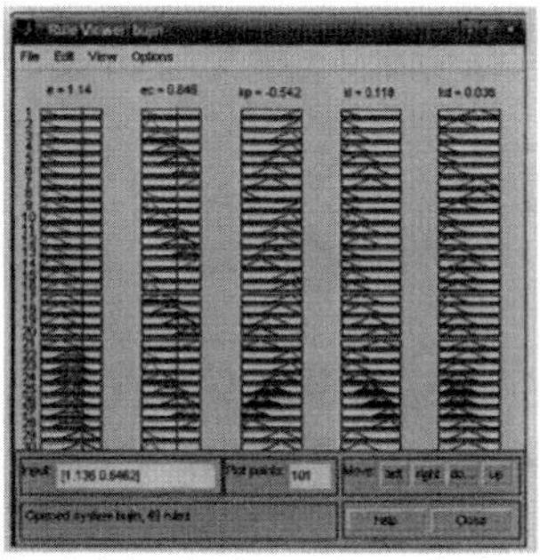

Fig. 2. Mamdani uses the method to control the inference process.

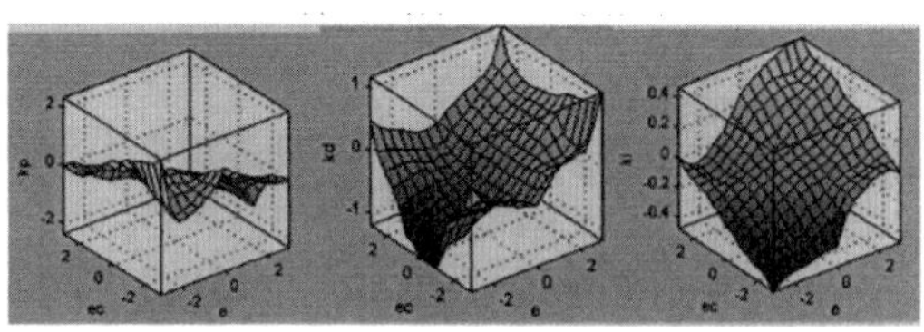

ΔK_P rule ΔK_I rule ΔK_D rule

Fig. 3. Output surface rule viewers.

Mainly of stepper motor control system, the control algorithm is discussed. Through the basic principle of PID computer control, fuzzy control basic principle and basic principle of the fuzzy PID computer control, fuzzy PID computer control algorithm for stepper motor control is applied to the system, and design the step motor based on Fuzzy PID computer controller.

3. MATLAB Computer Simulation of Control System

In order to test the control effect of the algorithm, the control system is simulated by MATLAB software, and the control strategy is analyzed by simulation, and then the relevant parameters are adjusted to achieve the best control effect.

3.1. *A mathematical model of the stepping motor is established*

In regardless of the stator pole and the end of the magnetic flux leakage, regardless of the permanent magnetic circuit of magnetic flux leakage, neglects the eddy current and magnetic flux leakage effect, ignoring harmonic component of stator coil inductance, ignoring the influence of saturation, the two-phase hybrid step into the motor voltage balance equation [9].

$$\mu_a = Ri_a + L\frac{di_a}{dt} - k_m\omega\sin(N_r\theta)$$
$$\mu_b = Ri_b + L\frac{di_b}{dt} - k_m\omega\sin(N_r\theta) \qquad (1)$$

Formula, μ_a , μ_b and i_a , i_b,, respectively, A, B two phase voltage or current; R for the winding resistance; L for the winding inductance; for the counter electromotive force coefficient; for the motor speed; the number of rotor teeth.

The torque equation of the motor is:

$$T_e = J\frac{d\omega}{dt} + B\omega + T_L = -K_m i_a \cos(N_r\theta) \qquad (2)$$

In the formula, Te is the electromagnetic torque; J is the moment of inertia; B is the viscous friction coefficient; TL is the load torque.

By the formula 1 and 2, the mathematical model of two-phase hybrid stepping motor is obtained:

$$\begin{cases} \frac{d\theta}{dt} = \omega \\ \frac{d\omega}{dt} = [-K_m i_a \sin(N_r\theta) + K_m i_b \cos(N_r\theta) - B\omega]/J \\ \frac{di_a}{dt} = [\mu_a - Ri_a + K_m\omega\sin(N_r\theta)]/L \\ \frac{di_b}{dt} = [\mu_b - Ri_b + K_m\omega\cos(N_r\theta)]/L \end{cases} \qquad (3)$$

3.2. *Establishment of a computer simulation model of a fuzzy self-tuning PID system*

In under the environment of MATLAB / Simulink graphical modeling, the use of system simulation basic modules to build the fuzzy PID parameters are adjusted on line by fuzzy self-tuning PID system computer simulation model, and built a simulation model of the conventional PID computer controller, can control results in the two kinds of control methods in the same operating environment, can be better compared to two control algorithm of control effect.

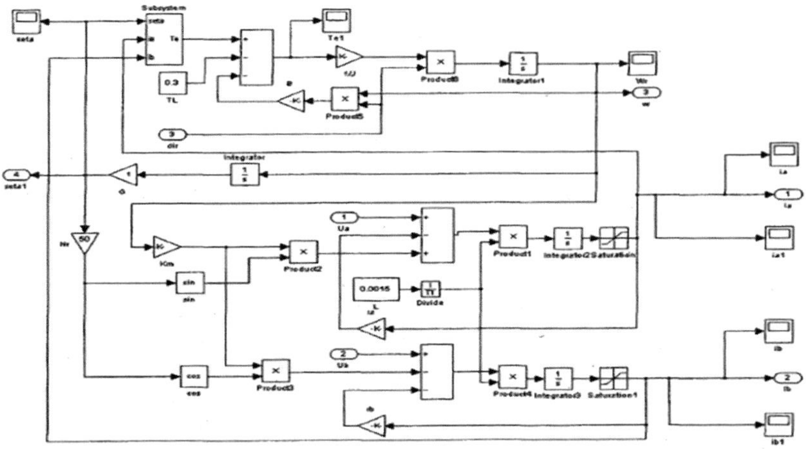

Fig. 4. Simulation model of stepping motor in Simulink.

3.3. *Computer programming*

The normalized parameters are converted to the actual PID parameters and the deviation of the PID e algorithm programming.

```
Function [out]=Fuzzy PID (u1,u2,u3,u4)
Persistent error_1 error_2
Ts=0.001                    %Sampling time
Kph=u2;                     %Setting parameters KP, KI, KD acquisition
Ki=u3;
Kid=u4;
Error=u1;                   %Parameter deviation e acquisition
Error= (error-error_2) /ts;           %Discrete PID from calculation
Error_1=error_1+error*ts;
Out=amp*error +cd* error+ki*error_1;
Error_2=error;
```

3.4. *The computer control systems of the stepping motor are built-in*

Construction of the stepping motor of the computer control system. Under the environment of MATLAB / Simulink to build step into motor fuzzy PID control computer model. Finally, through the analysis of the simulation results can be obtained by a stepping motor fuzzy PID computer control, can make improvement step into motor lose step lost step, oscillation, and can be greatly improve the system response speed, quick start and stop quickly, improve the laser marking machine with rotary processing efficiency is more stable in the process.

4. Computer Simulation and Experimental Results Analysis

Stepper motor PID computer control position response curve, as shown in Figure 5(a), stepper motor fuzzy PID computer control position response curve shown in Figure 5(b). In experimental results show, dated on behalf of the actual output of the simulation curve, data2 standard curve representing the desired output. Contrast Figure 5(a) and (b) can draw computer PID fuzzy control is better than PID computer control has the following several aspects: 1. System response time is to speed up, 2.The system overshoot is significantly reduced. The oscillation amplitude is smaller and in reaching the equilibrium position, fuzzy computer PID control curve and the desired output standard curves coincide, 3.The time to reach the equilibrium position of the system is significantly reduced.

In Figure 5(a) system, the time is longer when the system is close to the equilibrium position, and the system in Figure 5(b) is obviously shorter than that of the system which is close to the equilibrium position and tends to be stable.

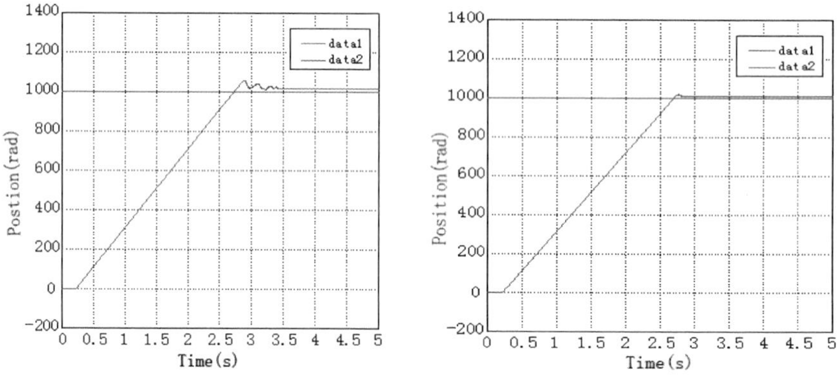

(a)Response curve of PID computer control position (b)Response curve of PID computer control position

Fig. 5. Contrast diagrams of stepper motor position output.

Figure 6 is two control system output speed comparison chart, from two the figure can be seen, in the initial stage, PID computer control starting speed is about 100rad/min, less than the start speed fuzzy PID control computer is about 150 rad/min. in the search process, when close to the target position, the motor speed decreased rapidly, fuzzy PID computer under the control of the stepping motor speed can be more stable close to 0, while the conventional PID computer control, a larger oscillation in this period, and the stable time is relatively long. When the deviation is larger, the highest speed fuzzy PID computer control system to limit, the position deviation decreases rapidly. The position deviation is small, the speed decay rapidly, to eliminate the overshoot.

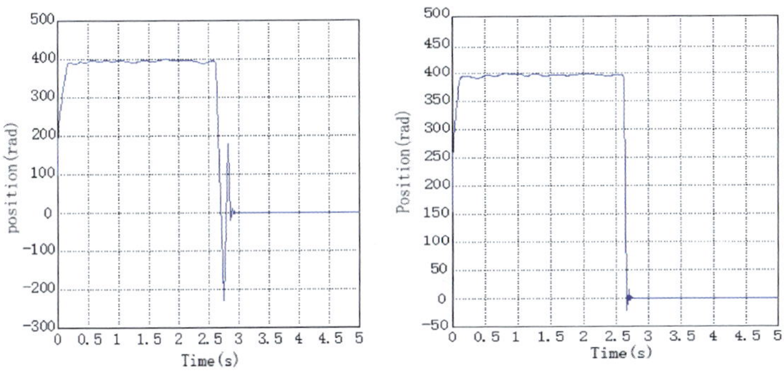

(a) Output speed of PID computer control system (b)Fuzzy PID computer control system output speed

Fig. 6. Stepper motor output speed comparison.

The laser marking machine actuator - stepper motor for improved closed-loop control, and intelligent control, the response speed of laser marking machine in rotating precision machining process faster, reduce the system overshoot response[10]. The stepper motor control of the fuzzy PID controller design of the computer, start fast, stop quickly, and effectively reduce the oscillation phenomenon, when the motor out of step when the fuzzy PID controller can make the computer system quickly restored to normal working condition, and on-line correction of stepper motor because of losing step, error caused by oscillation, eliminated with the stepper motor to start with stop the swing, the laser marking machine is more stable in the rotation process, higher precision, and improve work efficiency[11].

5. Conclusion

This paper studies the laser rotary precision machining of the fuzzy control system of PID computer, using laser technology combined with computer intelligent control technology, the control technology of stepper motor control method based on fuzzy PID control algorithm of fuzzy computer, PID computer, PID computer fuzzy controller is designed for stepper motor, constructed stepper motor control computer analysis of MATLAB system, computer simulation of stepper motor control system was carried out, built in the MATLAB/Simulink environment of stepping motor fuzzy PID computer control model, the implementation of the analysis of the MATLAB computer simulation was established based on the laser rotary precision machining machine computer intelligent control system, with the elimination of the stepping motor start and stop. Swing, realized the high precision laser machining, good operation.

References

1. Lutz Cleemann.Eureka:Eurolaser.Proc.SPIE[C],1020,1988.
2. MEI Sui-sheng.Application of laser processing in electronic industry[J].Laser and infrared,1999,24(1):5-8.
3. XU Qing-ren.Development and application of laser technology in foreign countries[J].International Aviation,1998(03).
4. LIU Jian-dong.Robustness analysis and design of digital PID control system[J].Journal of North China Electric Power University,2003,30(6):31-33.
5. Escuer A, Jarabo S.New theoretical model based on the application of the energy conservation principle for erbium-doped silica fiber lasers[J]. Journal of Modern Optics, 2005, 52(5): 655-657.
6. Cheok A D.Use of Fuzzy Logic for Modeling ,Estimation,and Prediction in Switched Reluctance Moter Drives[J].IEEE Trans on Industrial Electronics.1999,46(6):1207-1224.
7. YE Run-yu.Design and application of an adaptive fuzzy PID controller[J].Journal of Fujian University of Technology,2006,4(6):818-821
8. LIU Sheng,XU Xin-xi,LIU Zhi-guo,et al.Application of fuzzy PID control in pressure control of BSL-3 Laboratory[J].Automation and Instrumentation,2006,(4):47—50.
9. XIAO Yun-mao.Research on the control technology of stepping motor based on fuzzy PID [D]. Hangzhou:Master Thesis of Zhejiang University of Technology ,2008.
10. WANG Xi-jun,JIANG Jun, SUN Fu-dong et al. The application of MAX7219 in the display driver chip[J].Automation Technology and Application，2009,28(10):121-123.
11. Miller.T.J.E .Motor design optimization using speed CAD software[C].IEE Colloquium(Digest),1999,2(14):5-9.

The design of microcirculatory hemodynamic detection system based on Android

W. Jiang
School of Physics and Electronic Information Engineering, Qinghai University for Nationalities, BayiZhong Road 3
Xining, Qinghai, 810007, China
E-mail: my_jiangwei007@163.com

L. Yang*, J. Su, Y. Li and X X. Yang
College of Life Science and Bioengineering, Beijing University of Technology, Pingleyuan, 100, Chaoyang, District
Beijing, 100124, China
**E-mail: yanglin@bjut.edu.cn*

Cardiovascular disease has become one of the most serious diseases which not only endangers human health, but also causes death. The microcirculatory hemodynamic parameters contain abundant cardiovascular information. Through hemodynamic detection, cardiovascular health status can be understood. Results: The data measured by the fingertip pulse wave detection was transmitted to the mobile phone via Bluetooth. The data was obtained at the phone of the Android platform and finally calculated to the relevant microcirculatory hemodynamic parameters. Using this program, users can easily learn about their physical condition through their mobile phones. This equipment allows users to pay attention to their own health at home and improve their health consciousness. Not only that, it also makes the allocation of medical resources more efficient.

Keywords: Android; Microcirculation; Hemodynamic.

1. Introduction

In today's society, medical instruments are moving towards the trend of high integration. Wearable medical devices have become a major development direction. [1-2]Therefore, the real-time monitoring of physiological parameters and information analysis are particularly important.[3-6] With the development of science and technology, medical instruments from the hospital and laboratory gradually go into the home, in order to achieve people understanding their

*Corresponding author.

physiological state. [7-10] In this context, medical devices based on the mobile terminal come into being.[11-12]

Hemodynamic monitoring is based on the laws of physics, combined with the concept of physiology and pathology, to make a quantitative, dynamic and continuous measurement and analysis of the regularity of blood movement in the circulatory system. By measuring the pulse wave, the parameters of blood flow dynamics were calculated together with other physiological parameters. [13-15]

This research will combine hemodynamics monitor and wearable devices. Pulse wave is measured through reflective type sensor. The signal is transmitted to the phone through Bluetooth module after the signal processing circuit. Data is computed to microcirculatory hemodynamic parameters.

2. Method

The analog circuit was used to achieve the voltage regulator, PPG signal acquisition, amplification and other work. Digital circuit is mainly MCU and its peripheral circuit. The microcontroller drives the PPG sensor to work by controlling the electric level. The signal is collected by the PPG sensor and sent to the pretreatment module for processing. The processed signal was A/D converted through the microcontroller and transmitted through Bluetooth. The whole system is powered by 5V and 3.3V. The system block diagram is shown in Figure 1.

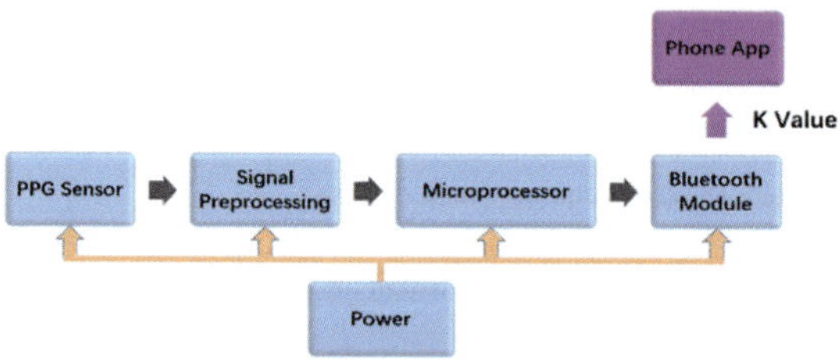

Fig. 1. The system block diagram.

2.1. *Analog circuit*

The PPG detecting circuit was composed of a light emitting diode and a light sensitive transistor. The light emitting diode is used to send out the light of a certain frequency into the finger, which is absorbed and reflected by the blood and tissue of the finger. The residual light signal is received by the photosensitive transistor and converted to electrical signal.

The pulse wave signal has the characteristics of low frequency, low amplitude, high noise and strong randomness, so the amplifier AD623 is used to form a differential amplifier, which is about 100 times. The amplified signal were sampled by ADC which is in the microcontroller and powered by 3.3V. The power supply voltage of AD623 is chosen as 3.3V, and the power supply mode of single power supply is adopted. The signal amplification circuit is shown in Figure 2.

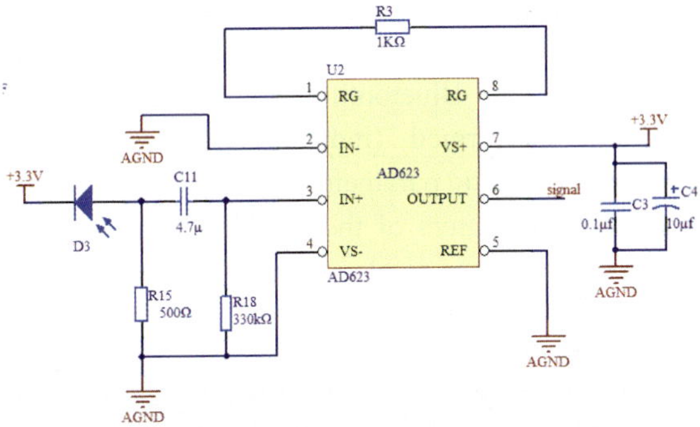

Fig. 2. The signal amplification circuit.

The circuit is powered by USB interface. Due to design, most of the circuit module uses 3.3V power supply, a 5 ~ 3.3V DC to DC conversion was needed. For the consideration of ripple rejection, ASM1117-3.3 was selected as the conversion chip. The typical value of the ripple rejection ratio of ASM1117-3.3 can reach 75, which is much higher than that of the same type of chip.

2.2. *Digital circuit*

Based on the system operating speed, ADC resolution, power consumption, STM32F103RBT6 was used as a microcontroller. PA0 was used as the A/D converter I/O port, and coupled with the AD623 signal output directly through the resistance. PC0 port was reserved for the manual sentence I/O port and connected through a switch circuit. Single chip produced interrupt through the PC0 port state changes, and simply processed the pulse waveform after A/D converting. PA7 port was used to generate the PWM waveform to drive the infrared light emitting diode. PA8 and PA9 port were reserved for the serial port to send data. NRST was the reset port, which can reset the microcontroller.

2.3. *Serial communication (Bluetooth)*

HC-05 master slave Bluetooth module was used, which all functions were controlled by the AT instruction set. Power on the KEY will be set to VCC, then the module can enter the AT command state. After that, the parameters were adjusted, returned OK as an answer if success, connected the circuit to send signals to the phone application.

2.4. *Software analysis based on Android*

Basic information sheet was used to input user's basic physiological parameters: height, weight, systolic blood pressure (SBP), diastolic blood pressure (DBP), etc. Click NEXT to enter the Bluetooth sheet. In the Bluetooth sheet device connectivity test was performed (indicates success). Click NEXT into Calculation sheet. Hemodynamic parameters were calculated using K values and the basic physiological parameters of the user input in the background. Click NEXT to enter the Result sheet to display the results.

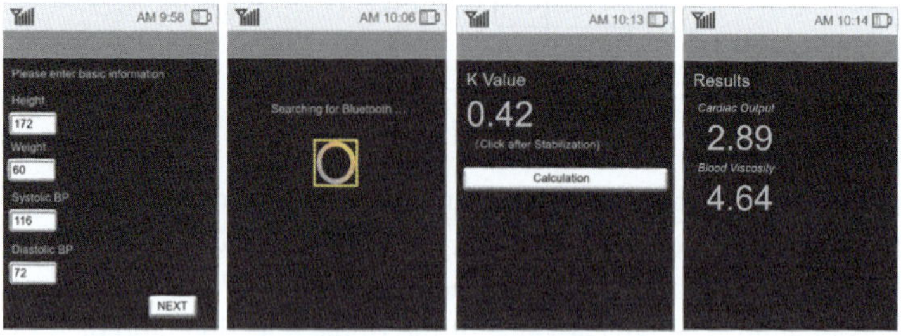

a b c d

Fig. 3. Interfaces (a. Basic information sheet. b. Bluetooth sheet. c. Calculation sheet. d. Result sheet.)

3. Results

In this paper, a blood flow dynamic detection based on Android was designed. The hardware transmitted the data from the sensor to the mobile phone via Bluetooth. The application is used to calculate the physiological parameters of the blood flow dynamics. The following points were mainly completed:

- The physiological pulse wave signal was extracted by using the PPG sensor.
- HC-05 Bluetooth module was used to send the processed signal to the mobile phone.
- The software was programmed based on the Android platform. In the mobile phone, the signal was processed and calculated. Finally, the

microcirculation hemodynamic parameters: cardiac output, blood viscosity were calculated.

4. Conclusion

Through mobile phone users can easily learn about their physical condition. Due to slow disease characterization was first reflected in the hemodynamic information. This equipment can let users at home can always pay attention to their own health, to improve the health consciousness. At the same time, it also makes the allocation of medical resources more efficient.

At the same time the device function is not perfect. There are many problems. The following sections can be forecasted:

- Due to the small size of the device circuit, it can be embedded in other family monitoring equipment, for the prevention and treatment of chronic diseases.
- The circuit embedded into the daily necessities, which can play the role of monitoring and warning in daily life.

Acknowledgments

This work was supported by Project supported by Beijing Municipal Commission of Education Foundation [PXM2016_014204_500079], Project of Excellent talent of Organization Department of Beijing Municipal Party Committee [2014000020124G044] and Project of Excellent talent "RiXin" of Beijing University of Technology.

References

1. C. Zhou, J. Feng, J. Hu and X. Ye, Study of artifact-resistive technology based on a novel dual photoplethysmography method for wearable pulse rate monitors. *J Med Syst.* **40**, 56 (2016).
2. A. Alzahrani, S. Hu and V. Azorin-Peris, A comparative study of physiological monitoring with a wearable opto-electronic patch sensor (oeps) for motion reduction. *Biosensors (Basel).* **5**, 288 (2015).
3. R.H. Thiele, K. Bartels and T.J. Gan, Cardiac output monitoring: a contemporary assessment and review. *Crit Care Med.* **43**, 177 (2015).
4. Z. Zhang, Photoplethysmography-based heart rate monitoring in physical activities via joint sparse spectrum reconstruction. *IEEE Trans Biomed Eng.* **62**, 1902 (2015).

5. E.M. Suess and M.R. Pinsky, Hemodynamic monitoring for the evaluation and treatment of shock: what is the current state of the art? *Semin Respir Crit Care Med.* **36**, 890 (2015).
6. K. Matsumura, P. Rolfe, J. Lee and T. Yamakoshi, Iphone 4s photoplethysmography: which light color yields the most accurate heart rate and normalized pulse volume using the iphysiometer application in the presence of motion artifact? *PLoS One.* **9**, e91205 (2014).
7. W. Wang, S. Stuijk and G. de Haan, Exploiting spatial redundancy of image sensor for motion robust rPPG. *IEEE Trans Biomed Eng.* **62**, 415 (2015).
8. W. Wang, S. Stuijk and G. de Haan, A novel algorithm for remote photoplethysmography: spatial subspace rotation. *IEEE Trans Biomed Eng.* **2015**, (2015).
9. T. Zunyi, M. Sekine, T. Tamura, M. Yoshida and C. Wenxi, A chair for cuffless real-time estimation of systolic blood pressure based on pulse transit time. *Conf Proc IEEE Eng Med Biol Soc.***2015**, 5118 (2015).
10. K.Y. Chin and R B. Panerai, A new noninvasive device for continuous arterial blood pressure monitoring in the superficial temporal artery. *Physiol Meas.* **34**, 407 (2013).
11. F. Guede-Fernandez, V. Ferrer-Mileo, J. Ramos-Castro, M. Fernandez-Chimeno and M.A. Garcia-Gonzalez, Real time heart rate variability assessment from android smartphone camera photoplethysmography: postural and device influences. *Conf Proc IEEE Eng Med Biol Soc.* **2015**, 7332 (2015).
12. X.Y. Zhang, L. Yang, J. Wang, Q. Zhao and A.K. Qiao, The design of wound area measurement software based on android operating system. *Intelligent Control and Automation, (WCICA'2014)*, (Shenyang, China, 2014)
13. A.R. Wang, L. Yang, C.Y. Liu, J.X. Cui, Y. Li, X.X. Yang, S. Zhang and D.C. Zheng, Athletic differences in the characteristics of the photoplethysmographic pulse shape: effect of maximal oxygen uptake and maximal muscular voluntary contraction. *Biomed Res Int.* **2015**, 752570 (2015).
14. T.T. Yan, L. Yang, S. Zhang, Y.M. Yang and X.W. Li, Detecting differences in volume pulse wave parameters among fingers and toes in four different postures. *Biomed Mater Eng.* **26(s1)**, 389 (2015).
15. K. Li, S. Zhang, L. Yang, Z.C. Luo and G.X. Gu, The differences in waveform between photoplethysmography pulse wave and radial pulse wave in movement station. *Biomed Mater Eng.* **24**, 2657 (2014).

The computer control system design and implementation of food production line

Cheng-Jun An[*]

Jinan Vocational College,
Jinan Shandong China,
E-mail: 2002acj@163.com

Due to slow adjustment process of production parameters of food production line control system, production efficiency is slowed down, impacting food production schedules. Thus, in order to improve the control efficiency of food production line and to ensure rapid development of the food industry, this paper focuses on the food production line of computer control system. The production line of the thousand layer cake is analyzed in this paper, where the hardware and software aspects of food production line computer control system are delved into for detailed design and analysis. Research results show that improvements in design of the food production line computer control system can effectively improve the efficiency of food production and management.

Keywords: Food Production Line; The Computer Control System; Industrial Ethernet.

1. Introduction

On food production line control system is analyzed, the most is one of the traditional manual or semi-automatic Way to give priority to, its production efficiency is low. In the production process parameters adjustment process often takes a long time, thus had a serious impact on food production schedule. Therefore, to strengthen the research and application of computer control system for food production line, to ensure that the production line to replace the product, the operator simply operation can quickly adjust the related parameters of the whole production line, in order to satisfy the demands of food production, has become the focus of the work are needed in the field of food.

2. The Food Production Process

For example Thousand layer cake production, , the process is as follows: in the feeding mouth into the already stirring dough, by folding rolling after range

* Corresponding author.

crisp cake machine for the first operation, and the noodle press and fold repeatedly pressing and fold the crisp cake machine, and then under the function of shaping cutter, cascading down good chunk face were cut step by step, after the transition machine processing, by the separator has been well overlying cutting face isometric partitioning, sugar in spray machine after spraying salt or sugar to taste, production line in molding process to complete. The overlying crust at a certain distance, puts its 50 m long thousand layer cake baking furnace for expanding hair, styling, dehydration, mature, coloring, such as operation, and the four period of baking temperature zone (up and down two floors, a total of eight zones), to control adjustment with the face in the temperature range of the moisture evaporation, finally, baked product thousand layer cake is made, and after finishing, packaging net machine into the link [1].

Example: Semi-automatic small food production line production equipment 300-500 kg/h

Process: raw material preparation to mix, transportation and conditioning—— extrusion—— drying——filter—— oil —— the flavor—— cooling —— (packaging)

Device configuration: mix machine, feeding machine——conditioning—— extrusion machine to send wind machine, seven layers of the oven to hoist, injection machine to high and low——vibrating screen——cooling drum machine——packaging machine

3. Food Production Line Computer Control System Hardware Design

3.1. *Based on industrial Ethernet network design*

Put two industrial control computer at the scene of the thousand layer cake production workshop, is used as the operator station and engineer station respectively, and thus help site operation personnel to realize real-time monitoring of workshop equipment and production lines, and to adjust the parameters of the production line. At the same time, put the PC in production management staff in the office, make the management personnel in a timely manner the production condition of field production line and food storage, query and supervise lay a foundation for its video marketing. Three states computer by Siemens industrial Ethernet network, so as to exchange data, considering the production site and the management office of distance is far, to the scene of the two machines (operator station and engineer station) in the form of optical fiber is connected with the production management computer, and with the aid of the TCP/IP Ethernet communication mode to realize data exchange [2]·

3.2. *Based on PROFIBUS DP network middle network design*

PROFIBUS DP network communication wiring is simple, fast transfer rate and communication distance, etc, in the field condition for food production line shall be fully considered, on the basis of the introduction of 2 sets of S7-300 PLC, and 315-2 DP as the system CPU, each PLC are equipped with other related module, and one of them in PLC in the front of the thousand layer cake production technology and the oven link (PLC1), another PLC placed in the bottom section of the video inbound links (PLC2) separation and packing. Establish PROFIBUS DP network system: set CPU5611 communication card operator station at the scene, the S7-300 CPU in a definable communications defined as PROFIBUS mouth, and take the bus connector to connect PC and PLC is shielded twisted-pair cable, among them, the PLC1 is responsible for the production line after the separation section of food and packaging inventory control, while PLC2 is responsible for the detection and control of baking furnace, including eight temperature zone temperature detection and the closed loop control of electric heating system. At the same time, the S7-200 data from the station and the upper machine, rolling machine, fold cake machine process parameters) exchange, so as to help operators and managers determine the stacked layers of the face skin fold cake machine and the running speed of the machine itself[3].

3.3. *The underlying network design based on RS485*

Based on RS485 network on the bottom of the intelligent instrument, each from the stand of S7-200 PLC will with EM277 PROFIBUS communication module as the basis, thereby ditto the host S7-300 PLC to control the machine data exchange, and the related control command and control parameters to be received, and the working state of the field devices and the change of actual parameters shall be returned to [4]. In addition, through PLC S7-300 CPU two communication mouth with TD200 human-computer interaction with the the frequency converter, in order to display and modify parameters to fold cake machine speed control purposes.

4. The Food Production Line Computer Control System Software Design

4.1. *Upper unit configuration software*

Selection of Siemens WINCC configuration software as the system development platform, using its powerful database function and function of graphical monitoring picture design, computer control system to carry on the design of food production line. First of all, to set up of the process parameters involved in the production process. Thousand layer cake production line, the design of the need for its versatility and the production efficiency to be fully considered, based on computer control automatic production line, for all process parameters are set up and adjust automatically, and therefore need to use the computer to the new process parameters sent to the bottom of the programmable logic controller, which is finished by PLC at the bottom of the parameter adjustment for all kinds of equipment, while reducing artificial labor intensity, improve the efficiency of all kinds of parameter configuration. Secondly, in terms of production line monitoring picture, because the PC monitor screen in the simulation of actual working condition of food production line, therefore, with the aid of WINCC industrial automatic drawing function of the library, based on the production process, will be the actual working condition of the thousand layer cake production line, according to the simulated on computer can realize the actual production parameters and the monitoring and management of the relevant information such as [5]. Finally, PC by providing temperature stick figure and temperature rise curve, to help managers to the appropriate length within 8 temperature in the oven temperature control conditions for query, supervision and control, and on this basis, to track the quality of food, at the same time, improve the production process, to improve the quality of food.

4.2. *PLC software programming*

In this paper, the design of food production line PLC part of computer control system is made up of PROFIBUS and RS - 485 bus system constitute, in PROFIBUS bus system, its program mainly includes two S7-315 DP master station programming, respectively responsible for package sorting and the oven temperature detector and control, and also responsible for PC and from the stand of S7-200 PLC data exchange. From the stand of S7-200 PLC via EM277 PROFIBUS bus, host S7-300 PLC by V register as a data exchange area to implement the data communication, and through EM277 with PLC data exchange from the station. On RS485 bus system program, through the WINCC special Variable-frequency Drive control instruction package, which will be laminated machine and oven and so on the control parameter of a given frequency to frequency converter, at the same time, the current state of the Variable-frequency Drive parameter feedback from the station to S7-200 PLC, and then through the grating ruler to test the analog displacement, the incoming from the stand of PLC, after comparing with the given value, to control the step motor, the adjustment of the rolling thickness of thousand layer cake.

5. Conclusion

Based on the process analysis of thousand layer cake production line, and respectively on the production line at the top of the computer control system of the network, middle, and the underlying network industrial Ethernet and PROFIBUS DP network and RS - 485 network hardware design part, and the computer control system of production line of upper unit configuration software and PLC programming software part has made the detailed analysis. Research results show that the designed by food production line of computer control system can effectively improve the efficiency of food production and management, plays an important role for realizing the fine management of food production and meaning.

References

1. Li Benhong, Chen Xiaojun. Food packaging automatic production line based on PLC and touch screen control system design [J]. Mechanical and electrical engineering, 2014,01 (12) : 9-13.
2. Li Guoping, Chang Sheng' ai, Wang Na, etc. Sawing and milling machining production line of computer network control system design [J]. Journal of shan dong building materials, 2011 01 (05) : 38-40.
3. Jin Fenglian. Food packaging production line control system based on CAN bus design [J]. Journal of packaging engineering, 2012,01 (23) : 98-100.
4. Qiang Minghui ,Zhang Jing' e,Zhang Wei. Copper cathode automatic packaging production line control system based on PLC [J]. The design and implementation of electrical automation, 2011,01 (11) : 69-71.
5. Yang Mingjuan. Condoms dipping production line computer control system design [J]. Chinese business (second half), 2012,01 (5) : 328.F.Benhamou and A.Colmerauer (eds.), *Constraint Logic Programming,Selected Research* (MIT Press, 1993).

Extended operation and permission model based on RBAC

Zhi-Hong Liang, Wen-Hong Li, Xing-Ping Sun* and Qing-Yi Chen
School of Software, Yunnan University,
Kunming, Yunnan, 650091, China
** E-mail: sunxp@ynu.edu.cn*

Jun-Hua Hu and Shu-Qiong Sun
Key Laboratory for Software Engineering of Yunnan Province Republic of China,
Kunming, Yunnan, 650091, China

In this paper, an extended permission is introduced. In the permission model, the permission is split into the permission component by analyzing the relationship between the permission and the operation. Quantified permission component is spilt from permission, and the permission component is calculated. The size of the permissions of the two roles is then compared. The size of permissions is compared before authorization, and the authorization operation is performed only if the granted permission is greater than the current permission. This reduces the workload of both the authorized operation and the administrator.

Keywords: RBAC; Accessing Control; Resource Model.

1. Introduction

With the rapid development of Internet technology and network, information security is becoming more and more important in the government and enterprise. It is the issue that current industry and academia must to study and solve how to ensure that the government and the enterprise information security. The current information security technology mainly includes encryption technology, identity authentication, access control, intrusion detection, risk analysis and assessment, and so on. With the system scale and complexity increasing, any single information security means has its limitations. A variety of information security technology is used to ensure the security of information.

From the 90s of the last century, RBAC (role based access control) began to receive welcome, and has been developed as the main popular model[1] of access control in the current stage. The central notion of RBAC is that

* Corresponding author.

permissions are associated with roles, and users are assigned to appropriate roles. This greatly simplifies management of permissions. Each role has their own permissions to access the certain resources, when grant permissions to the role, the size of permission of the role is unknown. Sometimes the permissions assigned to the role are the same or smaller than the current permissions of the role. So, this will result in repeated permissions granted to this role. At the time of authorization, first of all, the current permissions of the role should be compared with the permissions which will be granted to this role. only when the permissions which will be granted to this role are greater than the current permissions of this role , it is necessary to grant the permissions to the role.

2. Operational Model

2.1. *Definition of operation model*

Definition 1 Operation: The behavior to access the protected resources in the system, denoted as Operation[2].

Definition 2 Operation Attribute: Operation Attribute is used to describe the features of Operation, denoted as Operation Attribute. For example, the name of the operation, the relative higher operation and what kind of specific operation, are used to describe operating characteristics, these are all attributes of the operation.

Definition 3 Abstract Operation: The abstract operation does not operate on the resources that are protected by the system. Abstract operation is simply an abstract description of the specific operation and the description of the common attributes of the specific operation.

Definition 4 Operation Bundle: A set of specific operations, denoted as Operation Bundle. The number of specific operations in Operation Bundle is usually determined according to the specific circumstances. Some Operation Bundle have only one specific operation, Some Operation Bundle has a lot of specific operations. These specific operations are put together is in order to correspond to resources of the same type. Operation Bundle which is correspond to the users of different level is different. the number of specific operations which is contained in Operation Bundle of high level roles is larger than the number of specific operations contained in Operation Bundle of low level roles.

2.2. *Representation of operation*

(1)Operation

Operation={operation1, operation2, operation3, …, operationi},

operationi represents specific operation. All the specific operations have the operation name and operation description. For each operation can be represented as operationi={ operationAttribute1, operationAttribute2, operationAttribute3, ⋯, operationAttributei }, For each of the operationAttributei represents an attribute of the operation.

(2)Operation Bundle

OperationBundle={operationBundle1, operationBundle2, operationBundle3, operationBundlei}, Each Operation Bundle is made of a finite number of specific operations. For each Operation Bundle operationBundlei={ operationi }{ operationi+1 }{ operationi+2 }{ operationi+k }, I which is in OperationBundlei is a specific number, the size of k is finite, the value of k is according to the specific operations which is needed by Operation Bundle .

2.3. *Relationship between operations*

(1) Relationship between operations

The operation is based on the specific resource type. If the type of resource is different, the operation on the resource is also different. For example, the operations on the file resource include read, write, edit, delete, and so on. There is a relationship between these operations. For example, the editing operation of the file covers the reading operation of the file. This is because that if contents of the file is not be read ,the contents of the file can not be edited. From the above statement, the following conclusion can be got.

Conclusion 1: the inclusion relationship exists between operations, the operation of the high level include low-level operation implicitly.

The level of operation is relative. For example, the level of editing file is higher than the level of reading file. However, the level of deleting file is higher than the level of editing file. This is because that the operation of editing file covers the operation of reading file and the operation of deleting file covers the operation of editing file implicitly. From the above statement, the following conclusion can be got.

Conclusion 2: The inclusion relationship between operations is transitive.

Symbol is defined to represent a relatively high level of operation symbol. operation1operation2 represent that the level of operation2 is higher than the level of operation3.

If operation1operation2 and operation2operation3, according to the transfer, operation1operation3 can be got. This shows that the level of operation1 is higher than the level of operation3 relatively.

A method to determine the level of operation is proposed. A tree structure for storing operation level is set up. When store the tree structure of operation, both of the current operation and operations which are higher than the current operation should be stored. If there is a high level of operation, the current high level operation ID should be stored. If there is no high level of operation, the location of the storage of high level operations will be stored in -1. IsHaveHighOperation (String operationId) is a function used to determine whether the current operation has high level operation . OperationId is the ID of the current operation. Firstly, the function IsHaveHighOperation(String operationId) query the value of the operation table and judge whether the value is -1. If the value of the query is -1, the current operation does not have relatively high level operations. If the value of the query isn't -1, the current operation has relatively high level operations. The function IsHaveHighOperation (String operationId) return all ID of high level operations. According to the value of the return of the ID to check the operation table, the corresponding operation name can be got. The operation name which got from the operation table should be compared with the operation name which are in the permission component. whether the operation queried from the operating table is in the right component can be judged by comparing the operation name. If the operation is in the permission component, the current operation can be judged as a low level operation and delete the current specific operation. If the current operation is not in the permissions table, the current operation does not have advanced operations, and then to determine whether the next specific operation is a low level of operation, until the list of specific operations that correspond to the rights component is traversed.

(2) The relationship between Operation Bundle

According to the definition of the Operation Bundle, the Operation Bundle is a collection of the specific operation on the same type resource. the number of specific operation in the operation bundle which is corresponding to the different level user is different. With the increase of the permission level of the user role, the number of specific operations which are corresponding to the operation bundle of this type resource will be changed. There is a certain relationship between these Operation Bundles, with the improvement of the level of permission ,the number of the specific operation in the corresponding Operation Bundle will increase . In order to reduce the operation of adding specific operation to the Operation Bundle, we can use the inheritance relationship here. an Operation Bundle can be created when the user's role has the lowest permission. With the level of the character is raising, a new Operation Bundle is created, and the new Operation Beam is created by inherit the original operation. Some of the new specific operations may be added to the new Operation Bundle. Thus, the specific operations of the new operation bundle are composed of operations which are inherited from the original operation bundle and some new operations. As shown in the following The relationship between Operation Bundle figure.

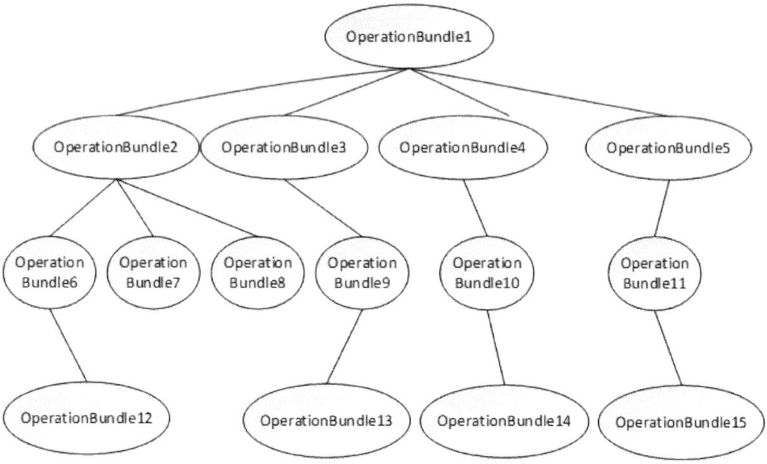

Fig. 1. relations between OperationBundle.

Figure 1 shows the relationship between the Operation Bundle. There is a role1 who is corresponding to the operationBundle1. The operations which is corresponding to the resource1 is contained in the operatioinBundle1. There are other roles such as role2, role3, role4, and role5,the permission level which is corresponding to these roles is higher than the level corresponding to role1. The operationBundle2 which is corresponding to the role2 can be made of the operations inherited from operationBundle1 and some new operation. the way how to form operationBundle2 is same with the way how to form operationBundle3, operationBundle4 and operationBundle5. Similarly, operationBundle3, operationBundle4 and operationBundle5 correspond to role3, role4, and role5, respectively. likely, operationBundle6, operationBundle7, and operationBundel8 are formed by adding some specific operations on the basis of the operationBundle2. operationBundle12 is formed by adding some specific operations on the basis of the operationBundle6.

At the same time , the conclusion got from the relation between the operation bundles is that the number of specific operation in the lower level operation bundle is smaller and the number of specific operation in the higher level operation bundle is larger. That also to say that the number of specific operations in the operation bundle which is the root node is the least and the number of specific operations in the operation bundle which is in leaf node is the largest. So the permission of the role who is corresponding to the root node is highest and the permission of the role who is corresponding to the leaf node is lowest.

3. Permission Scheme

In the previous chapter section, we introduced the operation . Access is for one or more protected resources operation license in the system[3-5].In this chapter mainly focuses on permissions, quantified permissions, computing privileges and permissions. At the time of calculating permissions, computing access rules are put forward. When comparing permissions, more access rules are put forward to determine the rights of the two characters in size.

3.1. *Authority*

In RBAC model, introduces the role between users and permissions. First, the user get the role and the permissions granted to the role .Indirectly through this way the permissions granted to the user. From here we see that role is the link between permissions and users.

3.1.1. *The description of the authority*

In the system, we assign permissions to roles, to give the user a role, the user can obtain certain operations through role permissions. Only the user get the indirect authority, the user can operate the resources in the system. The user's operation range is depending on the user's authority. In general, Permissions is an abstract concept, Authority is some specific operation of protected resources. From what has been discussed above, Permissions can be represented as binary group < operation, resources>in the model, the resource represents the protected object in the system, and the operation is a kind of specific access method [2].

3.1.2. *Representation of authority*

In the system, all authorities sets constitute the whole authority model. In the Permission Scheme, the authority can be expressed as a set.

Permission={permission1, permission2, permission3, ···, permissioni}, For any permissions can represent the operation of the resource. We can use the binary group to represent it, permissioni=<operationi, resourcei>, Resourcei represents the resource to be operated on, and the operationi represents a collection of specific operations. Operationi represents a collection of specific operation, in the other word, it have one or more specific operations, To analyze the relationships between the authority and operating, we split any one of the rights permissioni into smaller permissions and make the authority component and the specific operation of the corresponding. for any permission,

Permission={permission1, permission2, permission3, ..., permissioni}, PI represents the specific authority component, specific authority and the corresponding specific operation is corresponding.

3.2. *Quantifying Permission*

Quantifying permission is mainly to access component quantitative for a special value. These specific authorities and the corresponding specific operations are corresponding. When defining the system, We give each specific operations different values, which are set by the system designers According to the particular case, to represent the level of specific operation . And Then assigned the values to the permission components directly, as permission components value.

3.2.1. *Quantifying Permission*

For any permission, permissioni={p1, p2, p3, ..., pi}, PI represents the specific authority component. The corresponding operation set of the permissioni is Operationi={operation1, operation2, operation3, ..., operationi}, and the operationi represents the Specific operation. and these specific operations are corresponding with the permission components. When defining the specific operations, we assigned a numerical value for each specific operation respectively, the specific value, which we called The value of the operation, is represented by operationiweight which meaning the specific operation value of the ith. If the specific operation of the permission component P1 is corresponding with the operation1, then We assign the specific operation -- operation1 to the permission component p1.in the other word, |p1|=operationiweight. We can transform the value of the specific operation into permission component values in the way we said above.

4 Summary

In the paper, it introduces the permission model, the description and representation of the permission, how to quantify the permissions and relations between rights and specific operations. Permission to split into the permission component, the permission component is corresponding to the specific operation, the value of the specific operation is granted to the permission component. The focus of this paper is the calculation and comparison of permissions. The method of calculating the permission is proposed in the calculation of permission and the method of comparing the permission is proposed in the comparison. By comparing the size of the permissions, to determine the permissions of the relationship. Only when the current permissions of the role is less than the permissions which will be granted to the role, the permission which will be granted to the role can be granted to the role, so that it reduces the number of authorization for the role.

Acknowledgment

This work is funded by Yunnan Provincial Department of Education Fund for Scientific Research of major special projects under Grant No. ZD2014001, and by the Open Foundation of Key Laboratory of Software Engineering of Yunnan Province under Grant No. 2015SE102, 2015SE204.

References

1. Jiang Zenghu,Chert Maohua .Access ControI ModeI Based on Repeated-Role and PRI Task[J], Network Security Technology & Application 2007
2. BAO lian-chengl ,ZHAO jing-bo1,2 A survery on access control teachnology[J], Electrical Drive Automation, 2006,28(4):1-5
3. ZHANG Rui-qingu1,2,SHU Jianl1,CAI Kel1,LIU Lin—lanl1 Improved RBAC model based on organization structure[J], Computer Engineering and Design 2009,30(23)
4. S.Sandhu,E.Coyne,H.Feunstein,Role-Based Access Conrol Model[J] IEEE Computer,1996,29(2):38-47
5. Ravi Sandhu.Rational for the RBAC96 family of access models.Procecdings of the 1th ACM Workshop on Role-based Access Control ,ACM1997

Performance analysis and the staffing optimization for a multi-skill call center in M-design based on queueing model method

Chun-Yan Li*

School of Economics and Management, Yanshan University, Qinhuangdao 066004, Zhijiang College of Zhejiang University of Technology, Hangzhou, 310024, China

E-mail: llccyy1980@126.com

This paper focuses on a queuing model of a multi-skill call center in M-design. In this model, there are two types of customers and three groups of servers who have different skills. We obtained the state-transition rates by using results from an M/M/c/c loss queuing system. We then established equations for the steady-state probabilities of the system. Following that, we obtained the computational formula for the service level and established the staffing calculation model for the optimal number of agents in each group. Subsequently, we adopted the implicit enumeration method to find the solution, and through the numerical examples, we investigated the effects of model's parameters on the results.

Keywords: Multi-Skill Call Center; Queuing Model; Steady-State Probabilities; Service Level; Optimization.

1. Introduction

A call center is defined as a service system in which agents or servers serve callers or customers over the telephone, fax, email, etc. This comprehensive information service system relies on computer telephone integration technology. Examples of call centers include emergency centers, information centers, telephone banking, telephone shopping, and ticket reservation centers. The growth of the call center industry has been observed world-wide, providing significant employment in many countries. Call centers are becoming increasingly important in the global business environment. Correspondingly, as the importance and complexity of modern call centers grow, there is a proliferation of literature relating to them, typically focusing on queuing models. In a queuing model of a call center, the call agents and calls correspond to servers and customers, respectively. For important related surveys, we refer to

* Corresponding author.

Koole and Mandelbaum [1] and Gans et al. [2]. An introduction to staffing problems with relevant bibliographic references can be found in Aksin et al. [3].

Most of the early literature on queuing models of call centers assumes that call centers consist of a single type of incoming call and a single pool of homogeneous agents. Such call centers are referred to as single-skill call centers. However, as telecommunications and information technology have advanced over the years, call centers often handle several types of calls distinguished by the required skills for delivering service. Accordingly, multi-skill call centers have emerged and have recently been studied in the literature. A multi-skill call center handles several types of calls, and each agent has a selected number of skills. The agents are distinguished by the set of call types they can handle. A typical example is an international call center where incoming calls are in different languages, see Gan et al. [2].

Perry and Nilsson [4] considered a multi-skill call center with two classes of calls that are served by a single pool of agents. They determined the required number of agents and an assignment policy to satisfy a target for the expected waiting times of callers. Such multi-skill call centers are referred to as V-models or V-designs ac-cording to the canonical designs presented in Garnett and Mandelbaum [5]. Bhulai and Koole [6] proposed scheduling policies and showed that the policy is optimal for equal service time distributions. Gans and Zhou [7] also studied the same V-design model using a linear programming approach. They obtained results for the case of unequal service rates. For other related work, we refer to [8], [9], [10], [11] and the references therein.

Örmeci [12] studied a dynamic admission control for a multi-skill call center in M-design where there are two classes of calls and three stations: one dedicated to each class, and one shared station. He showed that serving a call in its assigned station, whenever possible, is optimal. In this paper, we study an M-design model for a multi-skill call center by using a queuing model. We focus on the performance analysis and optimization for this M-design model of a multi-skill call center. We develop a queuing model and use results of an M/M/c/c queuing system to calculate the transition rates of the system. Then, we establish equations for the steady-state probabilities of the model. Moreover, we consider the calculation of the service level and an optimization of a staffing problem.

The rest of the paper is organized as follows. In Section 2, we describe the M-design model for a multi-skill call center. In Section 3, we obtain the state-transition rates by using results of an M/M/c/c queuing system. Then, we establish equations for the steady-state probabilities of the system. In Section 4, we obtain the computational formula for the service level and present a staffing problem. Section 5 concludes the paper.

2. System Model

In this paper, we study an M-design model for a multi-skill call center where there are two types of calls and three groups of servers. A diagram of the model is shown in Figure. 1.

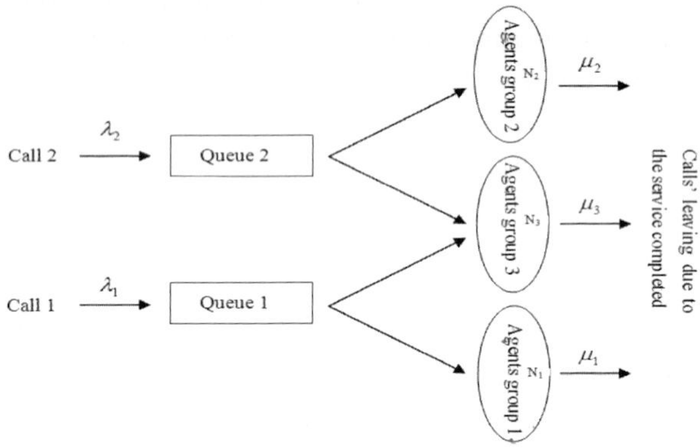

Fig. 1. Diagram of a queuing model of the multi-skill call center in M-design.

- Arrival process: There are two types of calls (or customers). The calls of Type 1 and Type 2 arrive according to a Poisson process with rates λ_1 and λ_2, respectively. Arriving calls are lined in two queues. Queue 1 and Queue 2 consist of calls of Type 1 and Type 2, respectively. There are infinite waiting spaces for both queues.
- Service process: There are three groups of servers (agents). Group 1, Group 2 and Group 3 consist of N_1 servers, N_2 servers and N_3 servers, respectively. Group 1 and Group 2 are specialized servers who can only serve customers of Type 1 and Type 2 calls, respectively. Group 3 is made up of flexible servers who can serve customers of both Type 1 and Type 2. The service times of servers in Group 1, Group 2 and Group 3 are all exponentially distributed with parameters μ_1, μ_2 and μ_3, respectively.
- Routing policy: A arriving customer of Type 1 (or Type 2) have priority to be served by a server in Group 1 (or Group 2) if there are free servers in Group 1 (or Group 2) and free servers in Group 3. If all servers in Group 1 (Group 2) are busy, the customer will be serviced by a free server in Group 3. If all servers are busy in Group 3, the customer must wait in Queue 1 or Queue 2.

- Queuing discipline: A free server in Group 1 (Group 2) serve the waiting customers in Queue 1 (Queue 2) according to a First-come First-served (FCFS) discipline, and a free server in Group 3 serve the waiting customers in Queue 1 or Queue 2 according to FCFS discipline. If all servers in Group 1 and Group 2 are busy and there are waiting customers both in Queue 1 and Queue 2, a free server in Group 3 will select randomly (with equal probability) a customer of Type 1 and Type 2 for service.

3. The Steady-State Probability

In this section, we firstly define the states of the model. Then, we derive the state-transition rates by using results of an M/M/c/c queuing system. Finally, we obtain equations for the steady-state probabilities of the system. It is assumed that the system is stationary.

3.1. *The state space*

The M-design model has three skill groups. Each group has three states: an idle state (denoted by 1), that being the case where at least one agent is idle; a busy state (denoted by 2), the case where all agents in the group are busy and there no calls waiting for service rendered by this group; and an overload state (denoted by 3), the case where all agents in the group are busy and there is at least one call waiting for service by this group. Theoretically, the system has 27 states, and all states are listed as follows:

D={(111), **(112)**, **(113)**, (121), (122), **(123)**, **(131)**, (132), **(133)**,
(211), (212), **(213)**, (221), (222), **(223)**, **(231)**, (232), **(233)**,
(311), (312), **(313)**, **(321)**, (322), **(323)**, **(331)**, (332), **(333)**}.

However, due to the routing policy assumed in Section 2, the 15 states above marked in boldface do not exist. Therefore, the state space actually consists of 12 states, which are given by

$$E = \{(111),(121),(122),(132),(211),(212),(222),(221),(232),(312),(322),(332)\}$$

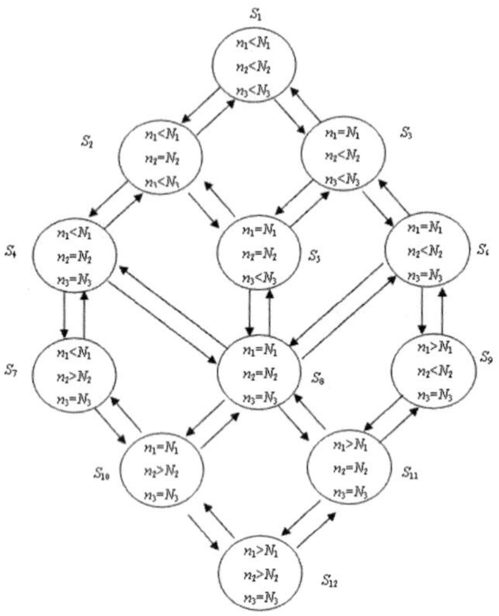

Fig. 2. Diagram of the state-transition.

Let $n_1(n_2)$ be the number of customers waiting for service including those being serviced by servers of Group 1 (2), and $n_3$be the number of customers being serviced by servers of Group 3. The i th state in E is denoted by S_i, i=1, 2, …, 12. The diagram of the state-transition is shown in Figure. 2.

3.2. *The calculation of the state-transition rates*

The transition of states occurs due to either the arrival of a call or the completion of a service.

(i) The state-transition due to the arrival of calls.

Consider the state S_1, the trigger for the transfer from state S_1 to state S_3 is a call of Type 1. If a call of Type 1 arrives at the epoch when n_1Error! Reference source not found. N_1-2, then the state S_1 will not be changed. If a call of Type 1 arrives at the epoch when $n_1=N_1$-1, then the transition will occur from S_1 to S_3. Thus, we can obtain the transition rate $q_{1\text{-}3}$ from state S_1 to state S_3 as follows:

$$q_{1-3} = \lim_{\Delta t \to 0} \frac{P_{S_1,S_3}(\Delta t)}{\Delta t} = \lambda_1 P(n_1 = N_1 - 1) \tag{1}$$

where $P_{S_1,S_3}(\Delta t)$ is the probability that the state process is in state S_3 at time $t+\Delta t$ provided that the process is in state S_1 at time t, and $P(n_1 = N_1 - 1)$ is the probability that there are N_1-1 calls of Type 1 needing to be serviced by the agents in Group 1.

Note that if the process is in the state S_1 then the number of calls of either Type 1 or Type 2 is less than the number of the agents either in Group 1 or Group 2, i.e., $n_1<N_1$ and $n_2<N_2$. In this case, each queue behaves like an M/M/c/c loss queuing system. Thus, using the results of the M/M/c/c loss queuing system, we have

$$P(n_1 = N_1 - 1) = \frac{1}{(N_1 - 1)!}(\frac{\lambda_1}{\mu_1})^{N_1-1} \Bigg/ \sum_{j=0}^{N_1} \frac{(\frac{\lambda_1}{\mu_1})^j}{j!}. \tag{2}$$

Similar analysis gives the other transition rates q_{i-j} caused by the arrival of calls which are given as follows:

$$q_{1-2} = q_{3-5} = q_{6-8} = q_{9-11} = \lambda_2 P(n_2 = N_2 - 1),$$
$$q_{1-3} = q_{2-5} = q_{4-8} = q_{7-10} = \lambda_1 P(n_1 = N_1 - 1),$$
$$q_{2-4} = \lambda_2 P^1(n_3 = N_3 - 1),$$
$$q_{3-6} = \lambda_1 P^2(n_3 = N_3 - 1),$$
$$q_{5-8} = (\lambda_1 + \lambda_2) P^3(n_3 = N_3 - 1),$$
$$q_{4-7} = q_{8-10} = q_{11-12} = \lambda_2,$$
$$q_{6-9} = q_{8-11} = q_{10-12} = \lambda_1,$$

where

$$P(n_2 = N_2 - 1) = \frac{1}{(N_2 - 1)!}(\frac{\lambda_2}{\mu_2})^{N_2-1} \Bigg/ \sum_{j=0}^{N_2} \frac{(\frac{\lambda_2}{\mu_2})^j}{j!}. \tag{3}$$

$$P^1(n_3 = N_3 - 1) = \frac{1}{(N_3 - 1)!}(\frac{\lambda_2}{\mu_3})^{N_3-1} \Bigg/ \sum_{j=0}^{N_3} \frac{(\frac{\lambda_2}{\mu_3})^j}{j!}. \tag{4}$$

$$P^2(n_3 = N_3 - 1) = \frac{1}{(N_3 - 1)!}(\frac{\lambda_1}{\mu_3})^{N_3-1} \Bigg/ \sum_{j=0}^{N_3} \frac{(\frac{\lambda_1}{\mu_3})^j}{j!}. \tag{5}$$

$$P^3(n_3 = N_3 - 1) = \frac{1}{(N_3 - 1)!}(\frac{\lambda_1 + \lambda_2}{\mu_3})^{N_3-1} \Bigg/ \sum_{j=0}^{N_3} \frac{(\frac{\lambda_1 + \lambda_2}{\mu_3})^j}{j!}. \tag{6}$$

(ii) The state-transition due to the completion of a service of a call.
Consider the state S_2, the trigger of the transfer from state S_2 to state S_1 is due to a service completion of a call of Type 2. If the state process is in state S_2, then all N_2 agents are busy. Thus, the transition rate from state S_2 to S_1 is $q_{2-1}=N_2\mu_2$.

Similar analysis gives the other transition rates $q_{i\text{-}j}$ caused by a service completion which are given as follows:

$q_{2-1} = q_{5-3} = q_{8-6} = q_{11-9} = N_2\mu_2$,

$q_{3-1} = q_{5-2} = q_{8-4} = q_{10-7} = N_1\mu_1$,

$q_{4-2} = q_{6-3} = q_{8-5} = N_3\mu_3$,

$q_{7-4} = q_{10-8} = (N_2\mu_2 + N_3\mu_3)P(n_2 = N_2 + 1)$,

$q_{9-6} = q_{11-8} = (N_1\mu_1 + N_3\mu_3)P(n_1 = N_1 + 1)$,

$q_{12-11} = (N_2\mu_2 + 1/2N_3\mu_3)P(n_2 = N_2 + 1)$,

$q_{12-10} = (N_1\mu_1 + 1/2N_3\mu_3)P(n_1 = N_1 + 1)$,

where the probabilities of $P(n_1 = N_1 + 1)$ and $P(n_2 = N_2 + 1)$ can be obtained by using the results of the M/M/c queuing system which are given as follows:

$$P(n_1 = N_1 + 1) = \frac{1}{N_1 \cdot N_1!}\left(\frac{\lambda_1}{\mu_1}\right)^{N_1+1} P_0^1, \tag{7}$$

$$P_0^1 = \left[\sum_{j=0}^{N_1-1}\frac{\left(\frac{\lambda_1}{\mu_1}\right)^j}{j!} + \frac{N_1\left(\frac{\lambda_1}{\mu_1}\right)^{N_1}}{N_1!\left(N_1 - \frac{\lambda_1}{\mu_1}\right)}\right]^{-1}, \tag{8}$$

The probability $P(n_2 = N_2 + 1)$ is similar to $P(n_1 = N_1 + 1)$, it is the probability that there are N_2+1 customers in the M/M/N_2 queuing system with arrival rate λ_2 and service rate μ_2.6

3.3. *The equations for the steady-state probabilities*

Let $P_i, i = 1,2,...,12$ be the steady-state probabilities of the state process. Then, we can obtain the equations for the steady-state probabilities of the system as follows:

$P_1(q_{1-2} + q_{1-3}) = P_2q_{2-1} + P_3q_{3-1}$,

$P_2(q_{2-1} + q_{2-4} + q_{2-5}) = P_1q_{1-2} + P_4q_{4-2} + P_5q_{5-2}$,

$P_3(q_{3-1} + q_{3-5} + q_{3-6}) = P_1q_{1-3} + P_5q_{5-3} + P_6q_{6-3}$,

$P_4(q_{4-2} + q_{4-7} + q_{4-8}) = P_2q_{2-4} + P_7q_{7-4} + P_8q_{8-4}$,

$P_5(q_{5-2} + q_{5-3} + q_{5-8}) = P_2q_{2-5} + P_3q_{3-5} + P_8q_{8-5}$,

$P_6(q_{6-3} + q_{6-8} + q_{6-9}) = P_3q_{3-6} + P_8q_{8-6} + P_9q_{9-6}$,

$P_7(q_{7-4} + q_{7-10}) = P_4q_{4-7} + P_{10}q_{10-7}$,

$$P_8(q_{8-4}+q_{8-5}+q_{8-6}+q_{8-10}+q_{8-11})=P_4q_{4-8}+P_5q_{5-8}+P_6q_{6-8}+P_{10}q_{10-8}+P_{11}q_{11-8}$$
$$P_9(q_{9-6}+q_{9-11})=P_6q_{6-9}+P_{11}q_{11-9},$$
$$P_{10}(q_{10-7}+q_{10-8}+q_{10-12})=P_7q_{7-10}+P_8q_{8-10}+P_{12}q_{12-10},$$
$$P_{11}(q_{11-8}+q_{11-9}+q_{11-12})=P_8q_{8-11}+P_9q_{9-11}+P_{12}q_{12-11},$$
$$P_{12}(q_{12-10}+q_{12-11})=P_{10}q_{10-12}+P_{11}q_{11-12},$$
$$\sum_{i=1}^{12}P_i=1.$$

All the steady-state probabilities can be obtained by solving these equations. However, the calculation of these probabilities are very cumbersome. In next section, these probabilities are calculated numerically by using Matlab software.

4. Optimization Problem

In this section, we first consider the calculation of the service level. Then, we consider the staffing problem.

4.1. *The calculation of the service level*

The service level is defined as the percentage of the serviced calls in a given fixed waiting time, denoted by P/T. Actually, the 80/20 principle is a general rule, that is to say at least 80 percent of the calls should be serviced within a 20 second waiting time. We can derive the service level using the steady-state probabilities.

Let P_{sl}^1 and P_{sl}^2 be the probabilities that the call of Type 1 and Type 2 is serviced in a fixed time T_1 and T_2, respectively.

Consider a call of Type 1. Calls of Type 1 form a queue when the process is in states S_9, S_{11}, S_{12}. It can be seen that the service rate for calls of Type 1 in each state of S_9 and S_{11} is $N_1\mu_1+N_3\mu_3$, and the service rate in state S_{12} is $N_1\mu_1+N_3\mu_3/2$, Thus, we get the probability P_{ns}^1 that a call of Type 1 cannot be serviced in time T_1, as follows:

$$P_{ns}^1=P_9\sum_{i=k_1}^{\infty}P(n_1=i)+P_{11}\sum_{i=k_1}^{\infty}P(n_1=i)+P_{12}\sum_{i=k_2}^{\infty}P(n_1=i) \tag{9}$$

where

$$k_1=N_1+N_3+[T_1(N_1\mu_1+N_3\mu_3)] \tag{10}$$

$$k_2=N_1+1/2N_3+[T_1(N_1\mu_1+1/2N_3\mu_3)] \tag{11}$$

Similarly, we get the probability P_{ns}^2 that a call of Type 2 cannot be served in time T_2 as follows:

$$P_{ns}^2 = P_7 \sum_{i=k_3}^{\infty} P(n_2 = i) + P_{10} \sum_{i=k_3}^{\infty} P(n_2 = i) + P_{12} \sum_{i=k_4}^{\infty} P(n_2 = i) \tag{12}$$

where

$$k_3 = N_2 + N_3 + [T_2(N_2\mu_2 + N_3\mu_3)] \tag{13}$$

$$k_4 = N_2 + 1/2N_3 + [T_2(N_2\mu_2 + 1/2N_3\mu_3)] \tag{14}$$

Then we can get $P_{sl}^1 = 1 - P_{ns}^1$ and $P_{sl}^2 = 1 - P_{ns}^2$.

The probability $P(n_1=i)$ $[P(n_2=i)]$ in Eq. (9) [Eq. (12)] is the probability that there are i customers in the M/M/N_1 [N_2] queuing system with arrival rate λ_1 [λ_2] and service rate μ_2 [μ_1] which is given in [13].

In the actual call center, generally classified according to the scale of the agents number, different countries have different standard. In China, The size of the call center can be divided into small (below 50 agents), medium (51-200 agents), large (more than 201agents). Our calculation formula of the service level is suitable for various scale of call center. Because this paper adopted the division of state set method, there are 12 state set, and 12 steady-state probabilities, the solution of the steady-state probability has nothing to do with the agents number N, So theoretically, we can solve any scale call center. There are some numerical results of the service levels P_{sl}^1 and P_{sl}^2 for different scale call centers in table 1. Because this model is multi-skill call center, the scale of the call center is the sum of the total number of three agent groups. Let N denote the total number of the agents.

Table 1. The numerical results of the service levels P_{sl}^1 and P_{sl}^2.

N	N_1	N_2	N_3	P_{sl}^1	P_{sl}^2
36	11	16	9	0.8075	0.9551
36	11	14	11	0.8028	0.6677
36	11	15	10	0.8200	0.8983
150	65	65	20	0.9961	1.0000
150	64	56	30	0.9876	0.7170
150	62	57	31	0.9175	0.9373
215	91	101	23	0.5967	0.6374
215	93	102	20	0.9099	0.8671
215	100	101	14	0.9984	0.7615

From the table we can observe that, the service level calculation formula can calculate any scale of call center, and at the same time, for multi-skill call center, the total number of agents is the same, but every agent group number is different, the service level of call center also has a great influence. So we have to find the optimal numbers of every agent group, to achieve the service level, and at the same time the cost of the system is minimal.

4.2. *Optimization of a staffing problem*

Let C_1, C_2 and C_3be the costs for each server in Group 1, Group 2 and Group 3, respectively. In order to minimize the cost, we try to find the optimal number of servers N_1, N_2 and N_3 subject to the constraint conditions. The optimization of a staffing problem can be expressed as follows:

$$\begin{aligned} \min \quad & C_1N_1 + C_2N_2 + C_3N_3 \\ s.t. \quad & P_{sl}^1 \geq \alpha_1, \\ & P_{sl}^2 \geq \alpha_2, \\ & a_i \leq N_i \leq b_i, \\ & a_i, b_i, N_i \in Z^+, i = 1,2,3. \end{aligned} \tag{15}$$

where α_1 and α_2 are the given service rates of the call of Type 1 and the call of Type 2, Z^+ denote a set of positive integers.

This is a nonlinear integer programming problem, the objective function is a linear function, there are three variables N_1, N_2 and N_3need to be sure. Constraints are inbound calls to satisfy a certain service level, that is to say, the service level of call 1 is greater than or equal to α_1, the service level of call 2 is greater than or equal to α_2. The calculate formula of service level see above (11) and (14), from the formula we can see, the constraint is highly nonlinear, Since there are only three variables, we can solve this problem very well by the traditional method. We use the implicit enumeration method, by MATLAB software, to find the optimal numbers of every agent group.

The implicit enumeration method was proposed by Balas E in 1965, it is compared with the exhaustive method (complete enumeration method). Its basic characteristic is only need to investigate part of the variable combinations of the problem. In the process of solving, according to the existing information automatically abandon many variable combinations that may not be the optimal solution, thus greatly reduce the workload. So this method is also known as part of the enumeration method. The implicit enumeration method only need to compare a small part of the variable combinations of the objective function, then can obtain the optimal solution and optimal value.

The key of using the implicit enumeration method to solve the staffing problem is to reduce the number of enumeration. According to this model, in order to reduce the number of enumeration, first we use the constraint conditions that the service level of call 1 and call 2 are greater than a certain value to eliminate some variable combinations. When computing, the variable combinations test in turn with these constraints. If constraints did not pass, get rid of this set of variables. Then from the satisfy combinations to find out the combination of the minimum cost, then the combination is the optimal number of agents we want to find.

The specific algorithm is described below:

(1) Initialize the parameters,

(2) Give all variable combinations of N_1, N_2 and N_3,

(3) Use constraints to get rid of the combinations that does not satisfy the constraints,

(4) Then from the satisfy combinations to find out the combination of the minimum cost.

In order to verify the above algorithm's effectiveness for solve the staff problem, There are some numerical results in the table 2, the table below shows the optimal number of agents in the cases of different agent group costs. The parameter settings are as follows: λ_1=5, μ_1=0.5, λ_2=4, μ_2=0.3, μ_3=0.2, T_1=20, T_2=30, $\alpha_1=\alpha_2$=0.8.

We can see from the table 2, the costs of the agent group are different, at the same time, meet the requirements of the service level, the number of agents of each group are also different.

There are some data of a certain logistics company call center in the table 3. In the logistics company, the agent group 1 serves the new customers, λ_1 and μ_1 representative the arrival rate and the service rate of the new customers respectively. The agent group 2 serves the registered customers, λ_2 and μ_2 representative the arrival rate and the service rate of registered customers

Table 2. The numerical results of the optimal N_1, N_2 and N_3.

C_1	C_2	C_3	N_1	N_2	N_3	cost
10	10	10	11	15	6	320
10	60	10	11	15	6	1070
10	100	10	13	14	10	1630
20	100	10	12	14	12	1760
60	100	10	11	14	15	2210
100	100	10	11	14	15	2650
10	10	60	11	15	6	620
10	10	100	11	15	6	860

Table 3. The numerical results of the optimal N_1, N_2 and N_3.

λ_1	μ_1	λ_2	μ_2	μ_3	N_1	N_2	N_3	cost
26	0.8	48	2.5	1.3	35	20	5	1500
22	0.7	67	2.2	1	34	32	5	1840
41	0.8	87	2.8	1.3	54	32	5	2240
58	0.6	83	2.3	1.3	100	37	5	3310
38	0.5	75	2.3	1.1	79	34	5	2800
45	0.5	85	2.6	1.3	94	34	5	3100
39	0.6	76	2.4	1	68	33	5	2550
44	0.7	64	1.6	1	66	41	5	2750
24	0.5	52	2.5	0.8	51	22	5	1880
23	0.5	52	2.2	1	49	25	5	1930
28	0.5	41	2.4	0.8	59	18	5	1920
20	0.7	84	1.2	0.8	31	72	5	2980

respectively, the agent group 3 can serve both new customers and registered customers, the service rate is μ_3. We selected the data of 12 time intervals to calculate the optimal agent numbers. Here we assume the cost of the three agent groups are C_1=20, C_2=30 and C_3=40 respectively, and T_1=20, T_2=30, that is call 1 should be serviced within a 20 second waiting time, call 2 should be serviced within a 30 second waiting time.

The numerical results is obtained in the cases that the costs of the three agent groups are not much difference, we can see that, in this case, The logistics call center demand for the third agent group is not very great.

5. Conclusions

This paper has studied a queuing model of the M-design multi-skill call center. We have obtained the transition rates of the states and then established equations for the steady-state probabilities of the system. We have derived the computational formula for the service level and presented an optimization model of the staffing problem to calculate the optimal number of the agents in each groups. In this work, we studied an exponential model in a multi-skill call center. A further extension for future research would be to study non-exponential models or models with impatient customers.

Acknowledgments

This work is supported by the National Natural Science Foundation of China (No. 71071133).

References

1. Koole G., Mandelbaum A. Queueing models of call centers: an introduction. *Annals of Operations Research*, 113: 41-59 (2002).
2. Gans, N., Koole, G., Mandelbaum, A. Telephone call centers: tutorial, review, and research prospects. *Manufacturing & Service Operation Management*, 5(2): 79-141(2003).
3. Aksin, Z., Armony, M., Mehrotra, V. The modern call-center: a multi-disciplinary perspective on operations management research. *Production and Operations Management*, 16: 655-688(2007).
4. Perry, M., Nilsson A. Performance modeling of automatic call distributors: assignable grade of service staffing. Proceedings of the 14th International Switching Symposium, 294-298(1992).
5. Garnett, O., Mandelbaum, A. An introduction to skills-based routing and its operational complexities. Teaching Notes, Technion. (2001).
6. Bhulai, S., Koole G. A queueing model for call blending in call centers. *IEEE Transactions on Automatic Control*, 48: 1434-1438(2003).
7. Gans N., Zhou Y. A call-routing problem with service-level constraints. *Operations Research*, 51 (2): 255-271(2003).
8. Gurvich I, Armony M., Mandelbaum A. Service-level differentiation in call centers with fully flexible servers. *Management Science*, 54 (2): 279-294(2008).
9. Armony, M, Maglaras C. On customer contact centers with a call-back option: customer decisions, routing rules, and system design. *Operations Research*, 52 (2): 271-292(2004).
10. Armony, M., Maglaras C. Contact centers with a call-back option and real-time delay information. *Operations Research*, 52 (4): 527-545(2004).
11. Jouini, O., Dallery, Y., Aksin, Z. Queueing models for full-flexible multi-class call centers with real-time anticipated delays. *Production Economics*, 120: 389-399(2009).
12. Örmeci, L. E. Dynamic admission control in a call center with one shared and two dedicated service facilities. *IEEE Transactions on Automatic Control*, 49 (7): 1157-1161(2004).
13. Gross, D., Harris, C. M. Fundamentals of Queueing Theory, 2nd edition. Wiley, New York(1985).

A resource management system design for malware behavior detection

Jian-Guo Jiang, Zhi-Song Liu*, Min Yu and Chao Liu
Institute of Information Engineering, Chinese Academy of Sciences,
Key Laboratory of Network Assessment Technology, Chinese Academy of Sciences,
Beijing, 100000, China
**E-mail: liuzhisong@iie.ac.cn*
www.iie.ac.cn

With the development of the dynamic analysis algorithm and behavior detection method, it has become increasingly important to design an effective system to manage and control behavior detecting resources. This paper introduces a design and architecture of a resource management system to manage the virtual environment for malware behavior detection, including its resource management algorithm, network design, system architecture, et al. The design introduced here focuses on the high standards placed on performance, reliability and scalability of malware behavior detection system. From the results obtained from both experimental simulations and real world implementation, this system has proved to be competent and fulfil those requirements.

Keywords: Malware Analysis; Virtualization; Resource Management System.

1. Introduction

The continued rapid growth of malware has attracted the attention of information security experts, as well as public. The number of new malware discovered in 2015 is beyond 140 million,[1] which is even greater than the number of all malware found before 2013. Meanwhile, Advanced Persistent Threat becomes a reality from rumor[11] and the number of them is increasing rapidly. These overwhelming, highly targeted, heavy obfuscated malwares make a new challenge to malware analysis system.

[2] and [3] carried out a novel work towards the automatic malware analysis by its behavior. These systems detect behavior of the executable samples, and try to determine whether it is a malware by their behavior characteristic. Newer research like [10], introduced more practical method to automatically detect behavior in virtualization environments. However, none of these researches give

* Corresponding author.

special attention to the management of the virtualization resources used for behavior detection in large scale system, which is very important when we need to handle huge number of samples.

This paper introduces a virtualization environment management system optimized for malware behavior detecting. In compare to general purpose virtualization management system, the system used for malware behavior detecting has higher requirement in stability. With carefully designed management algorithm and system architecture, as well as virtual environment and virtual network, we try to implement a system to manage large number and multiple type of virtualization based malware behavior detection units with high scalability, performance and robustness.

In the first part of this paper, The resource organization method and sample process procedure are introduced. Next, we will look into the core of the system, a data structure to organize the resources and an algorithm to allocate them to processing task. We then introduce the system implementation and network design to achieve high scalability and robustness. Finally, several tests and their results of the system are provided.

2. System Overview

2.1. *Organization of resource*

We utilize the existing virtualization based behavior detection technology and program as the behavior detecting unit (BDU). A BDU is a virtual machine with a program running inside responsible for detecting behaviors out of a single sample. BDU is the building block that the system need to manage.

APT malware with targeted attack characteristic could pretend to be a normal code in normal environment, and only perform malicious behavior in specific type of environment. The management system should be able to simulate various types of environment, like those with different OS version and different application installed. To simulate multiple type of virtual environment with different characteristic, we use a two-layer structure to organize BDUs: BDUs form the first layer of the structure, providing various OS versions; snapshots in the virtual machine associated with a BDU form the second layer of the structure, providing different user configuration simulation.

2.2. *Process procedure*

Figure 1 illustrates the logic design of the whole system. The process procedure is as follows:

(1) Log collector fetches a sample from external system and determines which type of BDU should this sample run in.
(2) Log collector sends a request to task assign unit for a certain type of BDU.
(3) Task assign unit looks for an idle BDU with the core management algorithm, reverts it to the requested environment snapshot, and returns it to log collector. If there is no idle BDU, the log collector will hold until task assign unit find a new or released idle BDU.
(4) Log collector sends the sample to BDU, BDU executes the sample for a specified time and return the behavior log to log collector.
(5) Log collector returns the behavior log to external system for further process, and reminds the task assign unit to release the BDU.

Some malware may attack and then sabotage the virtual environment it running in. This event is commonly seen in those malware that do buffer overflow to system service. This system sets a system service in each BDU to send heartbeat packet to task assign unit in a specific period. When the BDU is sabotaged, heartbeat service will be halted, and task assign unit then could detect the abnormal in this BDU. Task assign unit would report this event to external system, as well as revert the BDU to a clean snapshot, or just shut it down.

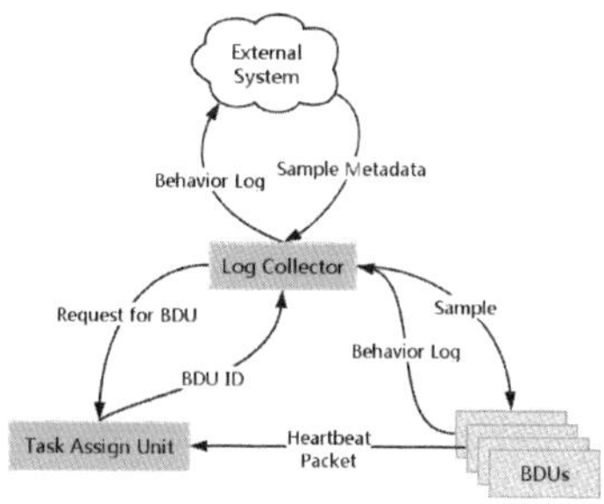

Fig. 1. System architecture.

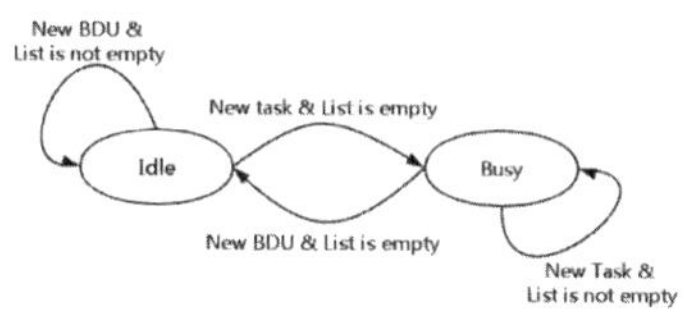

Fig. 2. The state transition of one allocation linked list.

3. Algorithm for Resource Management

3.1. *Resource management algorithm*

Resource management algorithm is the core part in the task assign unit. It is responsible for allocating resource ((BDUs) or manage the pending request ((tasks) from log collector.

The system uses several special designed double status allocation linked list to organize and manage tasks and resources. Every allocation linked list stand for a specific type of BDUs or the pending tasks waiting for this type of BDU. Double status allocation linked list could be in one of the two following status any point in time:

- Idle. In this status the linked list contains only idle BDUs
- Busy. In this status, the linked list contains only pending tasks waiting for BDU
- We define the empty allocation linked list as idle, in such time, all the BDUs are allocated to tasks, and there are no pending tasks. The state transition diagram is shown in Figure 2. Whenever log collector request or release a certain type of BDU, task assign unit will check an internal hash table to look for the corresponding allocation linked list.
- When the log collector requested for a BDU, task assign unit find out the corresponding allocation linked list:
- If the allocation linked list is in busy state, task assign unit will append this task to the end of the linked list. This case is illustrated in the upper diagram in Figure 3.
- If the linked list is in idle state and it is empty, task assign unit will set the state into busy, and add the task to the linked list. If the linked list is not empty, task assign unit will remove the last BDU in the allocation linked list, allocate the BDU to this task and send it back to log collector. This case is illustrated in the lower diagram in Figure 3.

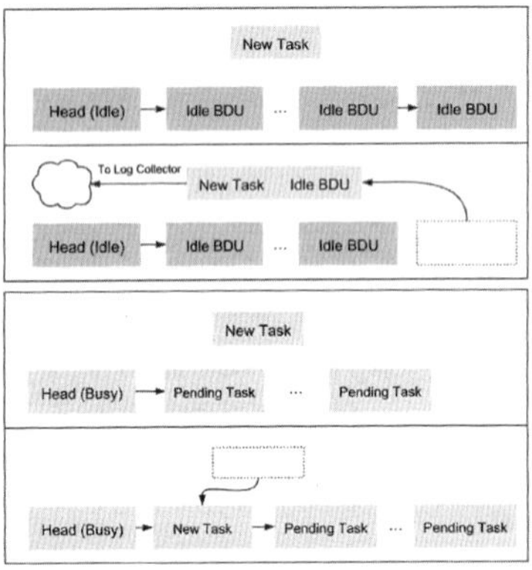

Fig. 3. The modification of allocation linked list when new task arrive.

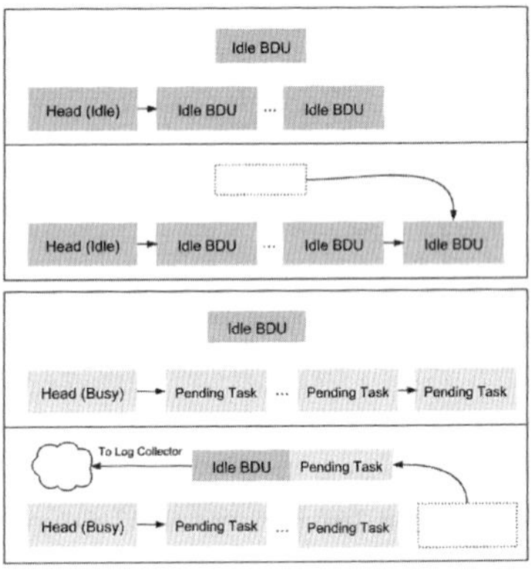

Fig. 4. The modification of allocation linked list when idle BDU is found.

BDU become available to more task when one of the following event occurs:

(1) Task assign unit enumerates a new BDU or
(2) A log collector releases a BDU after a completed task or

(3) A BDU continuously fails to transmit several heartbeat packets.

In these cases, the task assign unit will check the corresponding allocation linked list:

- If the state of the allocation linked list is busy, task assign unit will pop a pending task from the head of the linked list, allocate the BDU to this task and send it back to log collector. This case is illustrated in the lower diagram in Figure 4. If this operation causes the linked list become empty, set it into idle state;
- If the allocation linked list is in idle state, task assign unit will append this BDU to the end of the linked list as illustrated in the upper diagram in Figure 4. If the length of this linked list is long enough, task assign unit could shut down this BDU to reduce system cost.

The algorithm has the following characteristic:

(1) Cooperate with heartbeat monitoring thread, this algorithm could report and revert the damaged BDUs.

(2) For a particular type of BDUs, the task distribution policy is First-come, first-served (FCFS), ensuring the fairness of scheduling.

(3) For a particular BDU, the allocation policy is Last-come, first-allocate, improving the locality of reference and thus improve the throughput of the whole system, especially when the system is running under low load condition.

(4) Algorithm would automatically ensure that when system is running under low workload condition, most of the idle BDU would be shut down. When system is running under high workload condition, the algorithm would not need any additional system cost to switch from low workload condition.

4. Network Design and Implementation Details

4.1. *Reliable network design*

Malware may attack the nearby network by ARP or DDoS attack, which will cause a halt in nearby network transmission. In this system, we place all the BDUs to a separate VLAN, so that the management system would not be interrupted even if network attack is undergoing in one of the BDUs. On the other hand, some malware, especially the backdoor or Trojan program, will only have malicious behavior if there is an Internet connection. To improve the ability of analyzing this type of malware, this system setups a pfsense software router to ensure that every BDU have a real Internet connection. Network design is illustrated in Figure 5.

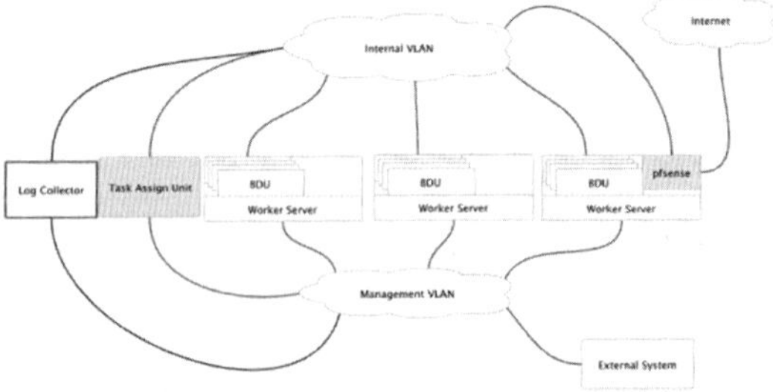

Fig. 5. Network design.

4.2. *Implementation details*

This system is implemented based on Xen Server 6.2, with virtual environments of Windows XP SP3, Windows 7 64/32bit operating system. Most of the system components are scalable across multiple servers, including log collector, Xen host server and BDUs on them.

In this design, task assign unit is an obvious single point of failure, making its performance and robustness critical to the entire system. Task assign unit is implemented by C++, with an RPC service implemented with libevent. Heartbeat monitoring thread uses atomic operator to maintain the heartbeat timeout of every BDUs. This design compromises some performance for a more elegant design of the entire algorithm. It is proven by experiments that the performance cost of this trade-off is limited to 5%.

5. Test Results

We did several tests to this system design to evaluate the overall throughput on different system workload condition. The hardware used as virtualization platform is a Dell 1U server, with 16 core Intel Xen E5620, 8GB system memory, running XenServer 6.2. The task assign unit is running on Mac mini 2014 Late with Intel Core i5, 8GB memory, running Mac OS X 10.11.2

The test method is:

(1) Reboot the XenServer host
(2) Use different number of dummy log collectors to require BDU from task assign unit, and release it back to task assign unit after a specific mount of time.
(3) Randomly stop some heartbeat generator service running in BDU virtual machines, simulating the event of virtual machine failure.

Figure 6 shows the test results, x-axis represents the load factor (number of log collector divide number of cores on server), y-axis is the average time cost of a single request, including time cost of reverting snapshots and data structure manipulation. To illustrate the performance improvement, we implement another similar system as a reference, with a round-robin scheduling algorithm.

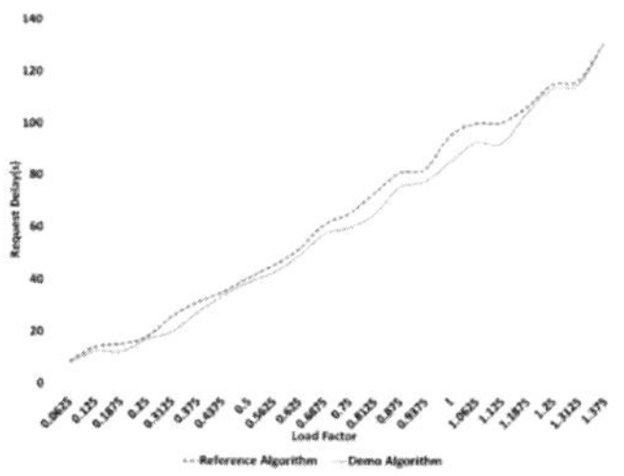

Fig. 6. Comparison of two allocating method.

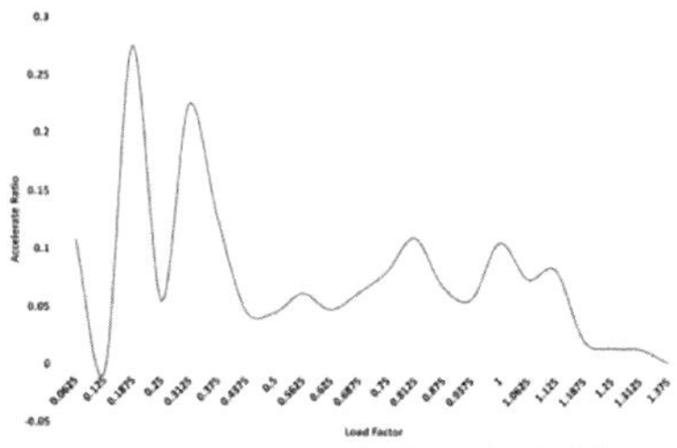

Fig. 7. Accelerate ratio in different load factor.

For a better view of the advantage of this algorithm, Figure 7 illustrates the accelerate ratio by the ratio of the time difference between reference implementation and demo implementation of our algorithm, to the time cost of the reference implementation. It is obviously that in low load factor, this algorithm has a accelerate ratio around 10% to reference implementation. And in all conditions, this algorithm, if is not better, has the comparable performance to reference implementation.

On other tests, if the transmission of heartbeat packet is stopped by manually terminating the heartbeat generating service, this system could report this event to log collector and revert the BDU as promised, proving that this design could handle the damage of BDU properly.

On a larger scale test, we build a behavior detecting system with 524 BDUs distributed in 22serversr. In this system we achieve a throughput of 6,000 samples per hour on average, 11,000 samples on peak. This system has run continuously for 54 days without any breakdown, despite the highly dangerous samples running in the system. We also carry out a benchmark on task assign unit. The result shows that the task assign unit can handle more than 5,000 samples per second.

6. Conclusion

Based on the review of the system design and the analysis of the test result, we believe that this system design is highly scalable, robust and fit for managing large number of malware behavior detection resources. Eliminating the single point of failure, increasing the stability of the task assign unit would be the primary work of this project in the future.

Acknowledgments

This work is supported by Strategic Pilot Technology Chinese Academy of Sciences (No. XDA06010703), Young Scholar Foundation of Institute (No. 1104005704), and Key Lab of Information Network Security, Ministry of Public Security (No. Y6Q0011104).

References

1. AV-TEST. *Malware Statistics & Trends Report* (AV-TEST, 2016),`https://www.av-test.org/en/statistics/malware/`
2. Bayer U, Kruegel C, Kirda E. *TTAnalyze: A tool for analyzing malware*(na, 2006).
3. Dinaburg, A., Royal, P., Sharif, M., & Lee, W. (2008, October). *Ether: malware analysis via hardware virtualization extensions*. In Proceedings of the 15th ACM conference on Computer and communications security (pp. 51-62). ACM.
4. Gionta, J., Azab, A. M., Enck, W., Ning, P., & Zhang, X. (2014, August). *DACSA: A Decoupled Architecture for Cloud Security Analysis.* In CSET.
5. Eliseev, E. Y., & Yablokov, V. V. (2015). U.S. Patent No. 8,978,141. Washington, DC: U.S. Patent and Trademark Office.
6. Shi, H., Alwabel, A., & Mirkovic, J. (2014). *Cardinal pill testing of system virtual machines.* In 23rd USENIX Security Symposium (USENIX Security 14) (pp. 271-285).
7. Dong, Y., Zhang, X., Dai, J., & Guan, H. (2014). *HYVI: A hybrid virtualization solution balancing performance and manageability*. Parallel and Distributed Systems, IEEE Transactions on, 25(9), 2332-2341.
8. Gandotra, E., Bansal, D., & Sofat, S. (2014). *Malware analysis and classification: A survey*. Journal of Information Security, 2014.
9. Branco, R. R., Barbosa, G. N., & Neto, P. D. (2012). *Scientific but not academical overview of malware anti-debugging, anti-disassembly and anti-vm technologies.* Black Hat.
10. Kruegel, C. (2014, August). *Full System Emulation: Achieving Successful Automated Dynamic Analysis of Evasive Malware*. In Proc. BlackHat USA Security Conference.
11. Sophos . *Security Threat Report* (Sophos, 2014). https://www.sophos.com/en-us/medialibrary/pdfs/other/sophos-security-threat-report-2014.pdf.
12. Ali M Q, Ashfaq A B, Al-Shaer E, et al. Towards a science of anomaly detection system evasion[C]. Communications and Network Security (CNS), 2015 IEEE Conference on, 2015: 460-468.

Combining static and dynamic analysis for the detection of malicious JavaScript-bearing PDF documents

Min Li, Ying Zhou*, Min Yu and Chao Liu
Institute of Information Engineering, Chinese Academy of Sciences,
Key Laboratory of Network Assessment Technology, Chinese Academy of Science,
Beijing, 100000, China
**E-mail: zhouying@iie.ac.cn*
www.iie.ac.cn

With the increase in popularity of Portable Document Format (PDF) documents and increasing vulnerability of PDF users, effective detection of malicious PDF documents becomes a more and more significant issue. In this paper, we proposed a method for the detection of JavaScript-bearing malicious documents and established the prototype detection system. We de-obfuscate the JavaScript code extracted from PDF documents through static analysis, and emulate the code execution during dynamic analysis. The combination of static and dynamic analysis in our approach makes the detection immune to obfuscation. Our experimental evaluation shows that our method can detect a broad range of malicious PDF documents and markedly enhance the detection accuracy with an acceptable overhead.

Keywords: Malware Detection; Malicious JavaScript; PDF Documents.

1. Introduction

Accordingly, the Microsoft Security Intelligence Report[1] shows that the prevalence of exploits target document readers and editors rose sharply from 4th quarter last year. As the report shown, the PDF documents are becoming one of the most successful attacking vectors gradually because of their ubiquitous-ness and rich functionalities.

In the following we briefly introduce the most relevant ones about our work. In 2011, P. Laskov and N. Šrndić present PJScan[2], which is based on lexical analysis of JavaScript code and use machine learning to automatically construct models from available data for subsequent classification of unknown data. PJScan has made a great progress, but it still exists some shortcomings: first, the PDF object extraction may be incomplete only with the PDF parser Poppler[3]; second, the extracted JavaScript code may be obfuscated by the

* Corresponding author.

attackers so that extract static characteristics directly from them is not representative. CWSandbox[4] is the most prominent tool in dynamic analysis. It launches the Adobe Reader to load the suspect PDF document literally in an emulated runtime environment and detect malicious behavior by monitoring system calls and modifications, however, the problem with CWSandbox and other dynamic tools in general is that they all have an overhead time expense. As fully static method and fully dynamic method both have advantages and disadvantages, combining the advantages of them is gradually becoming the mainstream method for malicious PDF analysis.

In this paper, we provides the following contributions: First, design the de-obfuscate JavaScript code embedded in PDF in static detection stage and rise the static detection accuracy; Second, design a multi-level detection system with combining static and dynamic detection methods which is easy for deploying on personal pc; Last, combining dynamic emulated detection with static detection with previous JavaScript de-obfuscation in an effort to balance the detection effectiveness and performance overhead. Our aim is to develop efficient detection system suitable for deployment on end-user system.

The remainder of the paper is outlined as follows. The architecture of our detection system and the methodology used in its specific components is presented in section 2. In section 3, we present data corpuses and do experimental evaluation with them. Conclusions and future work are discussed in section 4.

2. PDF Essentials and JavaScript

Before presenting the technical details of our methods, we briefly summarize the main features of the PDF standard and present the techniques used for the embedding of JavaScript. According to PDF references[5], the PDF document's body consists as a hierarchy of objects linked together to describe pages, multimedia, annotations, etc. The logical structure of the PDF document is a tree-like structure and the catalog dictionary is the root of the structure.

Before execution, the embedded JavaScript code has to be included in the PDF document's action dictionary. Such dictionaries have keyword "/S" and may have values "/JavaScript" and "/Rendition", the former must and the latter may have the keyword "/JS" where are important location for the search for JavaScript code.

The "/JavaScript" and "/Rendition" keywords may be found at the following locations:

- The Catalog dictionary's "/AA" entry may define an additional action specified by a JavaScript action dictionary,

- The Catalog dictionary's "/OpenAction" entry may define an action which is to be run after the document is opened.
- The document's name tree may contain entry "/JavaScript" that maps name strings to document-level JavaScript action dictionaries for the execution after opening of a document.
- The document's Outline hierarchy may contain references to JavaScript action dictionaries.
- Pages, file attachments and forms may contain references to JavaScript action dictionaries.

3. System Design and Implementation

In this section, we introduce the design and implementation of our system in detail. The overall architecture of our prototype system in shown in Figure 1.

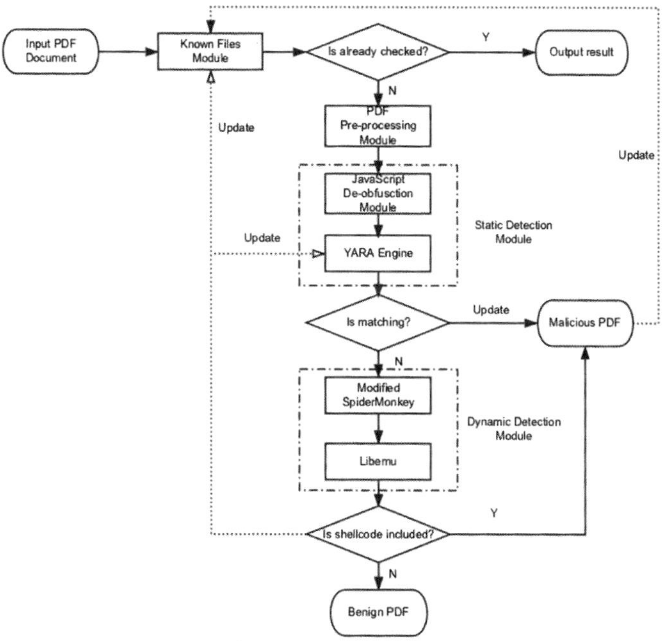

Fig. 1. System architecture.

On the whole, our prototype system consists of four main components, including the known files module, PDF pre-processing module, static detection module and dynamic detection module. For an input PDF file, the known files module will first determine whether the file is detected previously. Detection will be ended if the input file recorded in this set, otherwise the PDF preprocessing module will retrieve all the JavaScript code in the document and begin static detection. In this stage, the prototype system first de-obfuscate the JavaScript code and extract the signature used to compare with the YARA[6] signature set to decide the document is malicious or benign. If the known files module and the static detection module can't give a confidence result, the emulation-based dynamic detection module will start. This module has a modified JavaScript interpreter to run the extracted code in an emulated environment and record all the strings generated in procedure. Then these strings will be detected whether to include ShellCode.

The detailed description of the design and implementation of each module is as follows.

3.1. *Known files module*

This module contains a set which record the files' MD5 value and a corresponding calculation part. In order to increase the detection efficiency, files detected previously will be recorded as MD5 value, so that this module can detect a file benign or malicious quickly if its MD5 value has been recorded in the set.

3.2. *File pre-processing module*

The pre-processing of document includes file parsing and JavaScript extraction. Reliable extraction of JavaScript code from PDF document is itself a major challenge. In our survey, attackers may obstruct the document parsing and JavaScript extracting process with the following ways:

- Deliberately omit some objects in the cross-reference table.
- Replace some letters with hexadecimal digits in some keywords, e.g. Change "Catalog" to "C#61t#41log".
- Use some keywords which don't look right, but are actually legal, e.g. Replace "endobj" with "objend".

Based on the facts above, none of the previous tools for static analysis of PDF documents, e.g. PDFID7, PDF Dissector8, were able to provide full coverage of the possible location of JavaScript code in PDF documents. In this module, we have developed a parsing module with the origin JSUNPACK9 and two improvements as follows:

- In addition to using the cross-reference table, we also use the relationship between the object references chain, so as to parse all the PDF document objects.
- Normalize all the object name in the form of a string to solve the problem of confusion of the object names.

3.3. *De-obfuscation and static detection*

This module is static detection module with the components of de-obfuscation module and YARA engine. We did research on the commonly used obfuscation techniques and present the corresponding de-obfuscation schemes.

First of all, in view of the confusion method by adding unnecessary comments in the JavaScript code, we use the following method to delete them: traverse all the strings in the JavaScript, if any string includes chars "/*", we delete the substring of which begins with "/*" and ends with "*/". Second, for the technique of string partition, we use the following method to restore: traverse all the strings in the JavaScript, if any string include char "+", the char before which is '"', and the char after which is '"', we delete the three chars to merge them.

After de-obfuscation of JavaScript, we can start static detection with higher accuracy. In this module, we use YARA to build the signature set of malicious code. There is a real rule example in Figure 2.

```
Rule ExampleRule
{
    strings:
        $my_text_string = "text here"
        $my_hex_string = " {E2 34 A1 C8 23 FB}
    condition:
        $my_text_string or $my_hex_string
}
```

Fig. 2. Rule example in YARA engine.

In this example, "ExampleRule" is the keyword of the rule, identification of a signature string is included in "strings" and the usage condition of the rule is included in "condition". Due to YARA is a common used signature set engine for malicious signature recognition, in order to adapt to the malicious PDF detection, we have made the following improvements: Use the direct functions about the malicious PDF document vulnerability described in the National Vulnerability Database (NVD) as the identification string to establish the signature set. And for the contents of the "condition", we also use the description about the vulnerabilities or other related variants in NVD.

3.4. *Emulation-based dynamic detection*

This module is the last detection component of our system. In this module, firstly, we run the extracted JavaScript codes with a modified interpreter. In our system, we choose to use a modified SpiderMonkey[10]. SpiderMonkey is a JavaScript engine which is an open source project maintained by the Mozilla Foundation and it currently provides JavaScript support for Mozilla Firefox. Although the original version of SpiderMonkey can execute the code directly, it cannot retrieve the ShellCode. Because the real ShellCode (in the form of op-code) will be saved in a continuous memory area as a string variable during the execution of the JavaScript, we modified the engine to monitor the memory area of all the string variables and record them.

The second stage in this module is detecting the op-code formed ShellCode. We utilize Libemu[11] to achieve this goal. Libemu is a ShellCode detection and analysis library written in C language. It detects shellcode by searching for GetPC code, which is used in a lot of ShellCode to find the actual value of the program counter[12]. We use the function "emu_shellcode_test()" in the library to detect whether ShellCode is included in the op-code. If ShellCode is included in the embedded JavaScript, undoubtedly the PDF document detected is a malicious document.

4. Evaluation

In this section, we present the experimental evaluation result of our prototype system. We implement the prototype system on a virtual machine with Intel Core i7-4510U CPU @ 2.00GHz, 4 GB Memory and Ubuntu 12.04 OS. We collect 43 PDF files from the Internet, including 31 malicious PDF documents and 12 benign PDF documents. We will evaluate the system from two aspects, effectiveness and performance.

4.1. *Effectiveness*

Firstly, we use these samples to test the detection system's effectiveness, the results are shown in Table 1. As Shown in Table 1, the false-positive is zero. That's partially because benign documents rarely contain JavaScript code or the included JavaScript code are rarely involved with sensitive operations. Among the 31 malicious samples, however, only 29 (93.5%) samples are identified as malicious. After analyzing the remaining 2 samples artificially, we found one of them uses too complex obfuscation techniques beyond our de-obfuscation scheme, which indicates we need add more robust varieties of de-obfuscation schemes to cover more kinds of obfuscated PDF documents, the other one contains JavaScript and ShellCode, but our prototype system didn't detect them correctly.

Table 1. Effective test results.

	Malicious Samples	Benign Samples	Sum
Detected Malicious	29	0	29
Detected Benign	2	12	14
Sum	31	12	43

4.2. *Performance*

To evaluate the overhead time of our detection system, we have tested the time of detecting the samples. The results are shown in Table 2. As shown in Table 2, we finished detecting all the 43 samples in 1.5s if we only open static detection module and the average processing time for each sample is 0.035s. Then we open both the static detection module and dynamic detection module, the detection process is finished in 275s and the average processing time for each sample is 6.4s. As we had expected the emulated dynamic detection incurred significant overhead, however, this overhead is comparable to other works that use pure static detection, our prototype strikes a balance between effectiveness and performance. And our system can choose both static detection module and dynamic detection module or only choose static detection module when deploying on the end-user's system, so we can choose higher accuracy or lower overhead according the user's demand.

Table 2. Time of detecting files in different modules.

Situation	Overall Time(s)	Average Time(s)
Only Static Detection	1.5	0.035
Both Static and Dynamic Detection	275	6.4

5. Conclusion and Future Work

In this paper, we present a new detection system with static and dynamic detection methods targeting to JavaScript-based malicious PDF documents. With multilevel detections, our work provides an efficient system with high accuracy at an acceptable overhead. It is also suitable and convenient for deploying on end-user system. For future work, we plan to further develop the file pre-processing module in order to adapt more obfuscation techniques and collect more malicious samples' signatures so as to enrich our static detection signature sets.

Acknowledgment

This work is supported by Strategic Pilot Technology Chinese Academy of Sciences (No. XDA06010703), Young Scholar Foundation of Institute (No. 1104005704), and Key Lab of Information Network Security, Ministry of Public Security (No. Y6Q0011104).

References

1. Microsoft. The Rise in the Exploitation of Old PDF vulnerabilities. https://blogs.technet.microsoft.com/mmpc/2013/04/29/the-rise-in-the-exploitation-of-old-pdf-vulnerabilities/
2. Laskov P, Šrndić N. Static detection of malicious JavaScript-bearing PDF documents[C]//Proceedings of the 27th Annual Computer Security Applications Conference. ACM, 2011: 373-382.
3. Poppler is a PDF rendering library based on the xpdf-3.0 code base https://poppler.freedesktop.org/
4. Willems C, Holz T, Freiling F. CWSandbox: Towards automated dynamic binary analysis[J]. IEEE Security and Privacy, 2007, 5(2): 32-39.
5. PDF Reference. http://www.adobe.com/devnet/pdf/pdf_reference.html, 2016.
6. YARA: The pattern matching swiss knife for malware researchers https://plusvic.github.io/yara
7. M. Barreno, B. Nelson, A. Joseph, and J. Tygar. The security of machine learning. Machine Learning, 81(2):121–148, 2010.
8. C.-C. Chang and C.-J. Lin. LIBSVM: a library for support vector machines, 2001. Software available at http://www.csie.ntu.edu.tw/~cjlin/libsvm.
9. D. Canali, M. Cova, G. Vigna, and C. Kruegel. Prophiler: a fast filter for the large-scale detection of malicious web pages. In International Conference on World Wide Web (WWW), pages 197–206, 2011.
10. MOZILLA DEVELOPER NETWORK. https://developer.mozilla.org/en-US/docs/SpiderMonkey/1.8.8#Platform_support
11. x86 ShellCode Emulation, https://www.honeynet.org/project/libemu, 2016
12. Lu X, Zhuge J, Wang R, et al. De-obfuscation and Detection of Malicious PDF Files with High Accuracy[C]//System Sciences (HICSS), 2013 46th Hawaii International Conference on. IEEE, 2013: 4890-4899.

The design of intelligent remote meter reading system based on mechanical water meter

Lei Jing[1,*], Lai-Lei Jiao[2,†] and Zeng-Kang Wang[1]

[1]*Wenjing College of Yantai University, China*

[2]*Powerise Wireless Power Technology Co. Ltd. China*

E-mail: *jinglei1980@yeah.net, †jiaolailei@tom.com*

Focusing on the defects of the traditional artificial meter reading system, the paper designed a set of intelligent remote meter reading system. The system has core components such as collector and concentrator which can cut the cost of meter copying and enable water management to be concentrated, reliable and convenient by using correct networking solution and communication modes. With mobile apps that connect to public network, users can also view their water consumption and cost data. This paper will introduce the hardware architecture and key modules of the intelligent remote meter reading system. The measurements made by the proposed system are not only accurate, but installation and maintenance are also relatively easy, which meet the demands of system design.

Keywords: Remote Water Meter Reading System; Concentrator; GPRS Communication.

1. Introduction

With the rapid development of the city's modernization process, all kinds of defects of manual meter reading are exposed day by day. For example, the meter reading efficiency is low, the cost is high, the measurement is affected by human factors and cannot be read in a unified time point nor be monitored by Water supply enterprise, be managed reasonably etc. In order to solve the above problem, we designed a set of intelligent water meter reading system which can not only read meter data in real time, but also monitor the whole water use system remotely, save the cost greatly and solve the data lag program.

*Corresponding author.

2. Structure of System

2.1. *Structure*

The mechanical water meter is used widely due to its accurate and reliable measurement and the low price. While the current electronic water meter is relatively expensive, it will take more money and time if you want to carry out the transformation of the old district. In addition, the mechanical water meter has longer life, and can be used for at least six years, that means there is no need to invest to replace all of them. So the design of intelligent water meter consists mechanical part and electronic part.

The system mainly includes mechanical water meter, water quantity collector, collector / concentrator unit, master station software etc.. As shown in figure 1.

The remote water meter reading system is designed with four or five layers, which contain a base mechanical water meter at the bottom. The second layer is a water quantity collector which can record the flow data measured by the double control switch sensor connected on the base meter and upload it to collector unit through the M-BUS. Considering the advantages of convenient installation, the first and the two layer should be able to form an independent collector, so the design of collector should be considered to reduce the power consumption, and can use battery power. The third layer is the unit collector, which is between the collector and the network and can be used to collect the data of the water quantity in the local area, and to detect and report the state of the water meter. Of course, the collector unit can be equipped with GPRS communication module and connected to the public network. If so, each unit can receive sets of data from about 64 house. If there is a one thousand-family's district with five residential units which have six floors and every floor lived two families, it would use seventeen GPRS upload modules to cover all the seventeen residential buildings. Because the GPRS data transmission costs associated with the data flow, the subsequent network solution cost is comparatively high, and the pressure of the network server will be relatively large. To solve above problems, we can add a concentrator as the fourth layer, and the quantity of concentrator depends on the dimensions of the district .It just have little difference from unit collector, and can also connect with unit collector through the small area wireless and upload the data through GPRS, usually a one thousand-family's district only need one. The fifth layer is the master station system. If the site is not big, the unit collector layer can be saved, and use the concentrator acquisition content data from water quantity collector through M-BUS bus directly. In this paper, the network solution is designed for just this kind of situation.

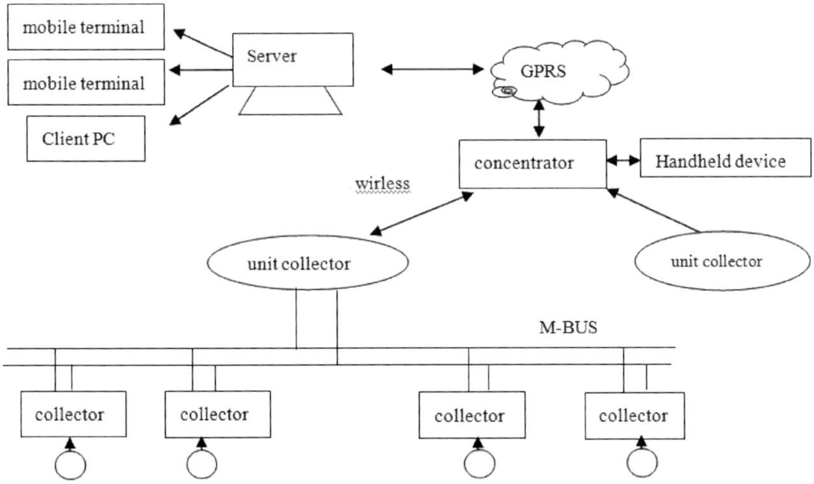

Fig. 1. Structure of remote meter reading system.

2.2. *Choice of communication mode*

The unit collector and water quantity collector are connected via M-BUS, M-BUS is a kind of specified bus standard designed for the data translation of domestic instrument in Europe which was used widely in residential intelligent meter reading due to its two wire structure independence of polarity, topological structure, advantages of convenient installation and low cost. The biggest advantage of M-BUS is that the power was offered by unit collector or concentrator in the process of communication ,that's why the water collector only powered by lithium battery, which ensure that the independent work can continue even the upper power is lost and life data is not influenced by data communication.

Data transmission between the unit collector and concentrator can use technique like 485 bus, power line carrier or small area wireless (ZigBee or 433MHz) etc. In order to guarantee the quality of communication, the design of dual communication interface is adopted. Wired part use the 485-BUS, 485-BUS can effectively prevent the existence of common mode interference due to its differential transmission. At the same time, the 433MHz wireless module is used as a redundant channel. And in the design of added frequency modulation measures to ensure the quality of communication in case the automatic frequency modulation is failed.

2.3. *Design of water quantity collector*

The design of MSP430F4152 MCU with low power consumption is adopted in the water quantity collector. There are five main parts, as shown in Figure 2. Each collector has a unique address number, which can be configured, read and stored. A collector can do the job of data acquisition and statistics, and can store the data acquired from mechanical water meter, and transfer these data through M-BUS bus to the unit collector and concentrator. If a fault occurs, the failure time can be recorded. In addition, it also has battery power alarm function.

The MCU will periodically gather water flow data, and be waked up when the concentrator summon data, but most of the time it's in a low power mode.

The use of precise on-off switch sensor which is key component determines the accuracy of the meter reading data. Installation of the sensors need to first install needle in the mechanical meter, the position of the needle should be on one dial of the meter pointer, and keep about 0.5mm-1mm distance from the glass surface of water meter without touch of mechanism.

The principle of the sensor is shown in Figure 3. A and B for two magnet, when the water meter pointer shaft drives the needle to rotate, magnet A's magnetic induction makes the switch be in a normally open state. In this moment the state of switch will never be changed even magnet A swing arbitrarily at the critical point, unless magnet B rotating in place than the switch turn in the state of off and produced a signal pulse.

The on-off switch sensor has a state self holding function, and high sensitivity, strong stability, long transmission distance. It's not easy to damage, when the mechanical water meter is too old to be used, the sensor can also secondary use, saving the cost of the system.

The software is divided into three modules, namely, water quantity detection module, communication module and data storage module. The procedures for the design of the communication module follow the M-BUS remote meter communication protocol which is general consented in the industry.

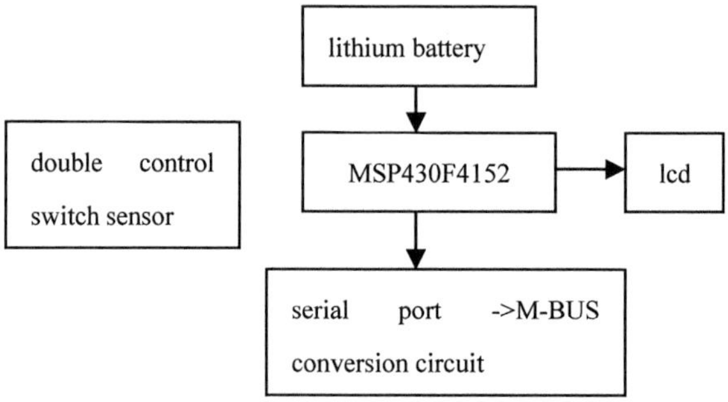

Fig. 2. Structure of water quantity collector.

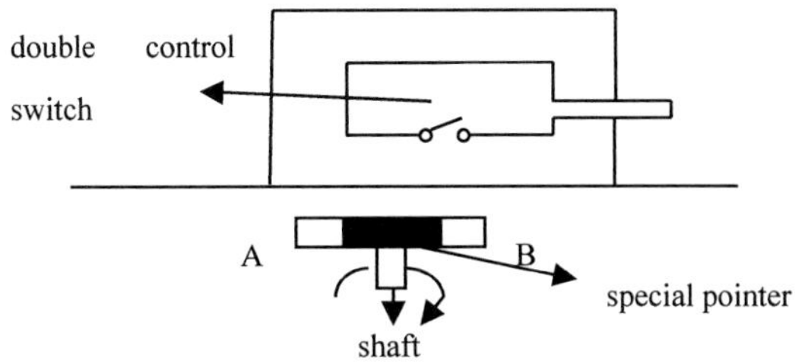

Fig. 3. schematic diagram of double control switch sensor.

2.4. *Design of concentrator*

The concentrator has M-BUS bus interface, 485 interface and GPRS communication function. It can automatically receive data from water quantity collector with M-BUS communication function, and also can expansion multiple M-BUS interface according to the site neatly. At the same time, the data will be uploaded through GPRS function to the monitoring center station after being packaged ,query for the staff. It also support for data storage, active report, fault alarm and other functions. Using 220V power supply.

The concentrator uses STM32F101 main control chip, C/OS-II operating system to complete the design. The hardware architecture is shown in figure 4.

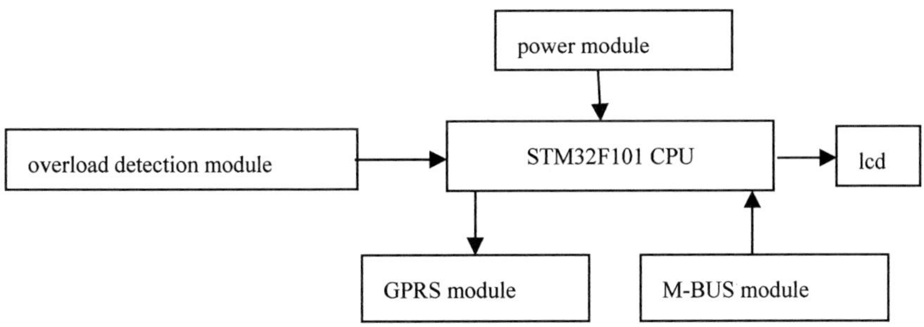

Fig. 4. Structure of concentrator.

The GPRS module selects the SIM900A which is specifically design for China and India market of dual band GSM / GPRS module, works at band EGSM 900MHz and DCS 1800MHz can automatically search the two bands, and can also set the band through AT command .

3. Summary

From now on, there are two districts are installed and using the intelligent remote water meter system designed by the scheme of this paper. And both of the two system's measurement is accurate, installation and maintenance are also relatively easy, basically meet the demand of system design.

References

1. Z Zhang, G Chen, J Li, The research on digit recognition algorithm for automatic meter reading system, World Congress on Intelligent Control & Automation, 2010
2. K Li, J Liu, C Yue, Remote power management and meter-reading system based on ARM microprocessor, Conference on Precision Electromagnetic Measurements Digest, 2008

The theory research on portable monitoring system model of the environment of banana growth

Ling-Ling Wang[1, 2, a], Yu-Ping Li[1, 2, b *], Hong-Xia Luo[1, 2, c] and Ji-Hua Fang [1, 2, d]
[1]*Institute of Information and Technology, Chinese Academy of Tropical Agricultural Sciences, Danzhou, China,*
[2]*Key Laboratory of practical research on tropical crops information technology in Hainan, Danzhou, China.*
E-mail: [a]zishi-010@163.com, [b]xxsrdny@163.com, [c]120081008@163.com, [d]fangdil@163.com*
http://info.catas.cn/

Banana, as an important-part of the fruit project in China, is a pillar industry of tropical agriculture, which is also a major source of farmers' income. The change in the field environment has a great effect on the production of bananas. The key environment factors and characteristics which affect banana high efficient cultivation, are yet to be resolved. We realized that some environmental changes may affect banana yield, especially natural disasters such as typhoon or drought. A kind of mathematical model which can reflect the relationship between banana cultivation and field environment should be built. Based on the problem, we will systemically research the monitoring model between banana growth and field environment and focus on the key factors. We will analysis monitoring model characteristics in order to reveal the change law of the environment parameters of banana growth. The basic data and framework can be provided for banana expert system. Results in this thesis will set up the foundation for serving relevant government offices, farmers and consumers.

Keywords: Field Environment; Data Acquisition; Mathematical Model; Banana.

1. Introduction

Banana cultivation in China is mainly distributed in tropical and sub- tropical regions of South China, Hainan province is one of the major banana production area. The climate is complex and varied with a longer summer. The change of the production condition and environment has a great effect on banana's production and price. Major climatic disasters in Hainan, including drought, rainstorm, flooding, low temperature, and typhoon, affect banana production

* Corresponding author.

greatly. Typhoon Nesat, Haiyan and Rammasun all affect the major banana production area seriously in the previous five years.

In this study, environment factors and characteristics will be analyzed, under the influence of typhoons and droughts, the relational model between banana growing conditions and environment factors will be explored. The work of this subject will provide a solid foundation for environmental control in banana production. It is especially important that it will benefit for the natural calamity reduction.

2. Current Status and Study Trends

2.1. *Information collection*

The research around information collection on agriculture weather and soil had been so much. The research of Information collection of field environment at home and abroad mainly is in two areas: one is the agriculture weather factors research, including[1-3] canopy temperature, air temperature, humidity, atmosphere pressure, light intensity, wind speed, direction detection and carbon dioxide concentration. The other is the soil factors research, including[4] soil moisture, moisture rate, nutrient, PH and electrical conductivity.

2.2. *Data acquisition equipment*

In recent years, the researches of data acquisition equipment are concentrated on sensors and smart mobile terminal, including[4] data collector with the wireless data transmission in Japan, data collector using the wireless broadcasting station in American, data collector using the remote wireless transmission technology in the Netherlands. China and other nations also have the similar data-collecting device[5-6], such as data collection and transmission of wild land, mobile intelligent surveillance, soil wireless strain sensor[7-11].

2.3. *The research on environment information of banana growth*

The studies are focused on technology of drip irrigation of integral control of water and fertilization, and banana growth period of nutritional requirements. Banana variety, plant density, soil property, weather condition and water and fertilizer supplying can affect the growth, development yield and quality of banana tree seriously. Scholars have focused on the nutritional properties, ways of the fertigation, the dynamic process of nutrient accumulation, realizing the coordinated and all-round development of the banana crop, soil, fertilizer and water.

3. Key Technique and Method

3.1. *Key agricultural environment factor determining*

The paper adopts one-factor experiment to confirm key factors based on the data collected from different environments of banana growth and development. This paper generalizes a method for class discrimination with the purpose of formulating dynamic random models starting from sample information. Expression method is shown in Eq. (1).

$$y_{ij} = \mu + \tau_i + \varepsilon_{ij} (i = 1,2,\ldots,a;\ j = 1,2,\ldots,n) \tag{1}$$

μ is public parameter; τi is treatment effect. εij is random error.

$$\varepsilon_{ij} \sim N(0,\sigma^2)$$

SST is the sum of the squares of total deviation as shown in Eq. (2).

$$SS_T = \sum_{i=1}^{a}\sum_{j=1}^{n}(y_{ij} - \tilde{y}_{..})^2 = n\sum_{i=1}^{a}(\tilde{y}_{i.} - \tilde{y}_{..})^2 + + \sum_{i=1}^{a}\sum_{j=1}^{n}(y_{ij} - \tilde{y}_{.})^2 \tag{2}$$

SSt is the sum of the squares after processing as shown in Eq. (3). SSe is experiment error as shown in Eq. (4)

$$SS_t = n\sum_{i=1}^{a}(\tilde{y}_{i.} - \tilde{y}_{..})^2 = \sum_{i=1}^{a}\frac{y_{i.}^2}{n} - \frac{y_{..}^2}{N} \tag{3}$$

$$SS_e = \sum_{i=1}^{a}\sum_{j=1}^{n}(y_{ij} - \overline{y}_{i.})^2 \tag{4}$$

The methods for calculating mean square and error will used. Key factors important factor can be confirmed applying F Inspection law.

3.2. *The mathematical model research on banana effected by field environment*

The relationship between banana developing and field environment is determined through one-dimensional linear regression as shown in Eq. (5).

$$y_i = a + bx_i + \varepsilon_i (i = 1,2,3,\ldots,n) \tag{5}$$

ε_{ij} is random error.

$$\varepsilon_{ij} \sim N(0,\sigma^2)$$

The optimal model coefficients of a and b is determined according to the principle of minimum value of theoretical value and observation. Expression method is shown in Eq. (6).

$$\sum_{i=1}^{n}(y_i - \hat{y}_i) = \sum_{i=1}^{n}[y_i - (a + bx_i)] \tag{6}$$

The functions of two variables of a and b is is shown in Eq. (7).

$$Q(a,b) = \sum_{i=1}^{n}[y_i - (a + bx_i)]^2 \tag{7}$$

The limit value of a and b will determined using the least square estimation method as shown in Eq. (8).

$$\begin{cases} \dfrac{\partial Q}{\partial a} = -2\sum_{i=1}^{n}\left[y_i - (a + bx_i)\right] = 0 \\ \dfrac{\partial Q}{\partial b} = -2\sum_{i=1}^{n}\left[y_i - (a + bx_i)\right]x_i = 0 \end{cases} \tag{8}$$

The methods for calculating mean square and error will used. The multiple linear regressive model will be established applying significance test. Based on banana growth analysis theory, banana growth characteristics under different field environment can be analyzed.

4. The Design of Portable Monitoring System

Portable monitoring system can be designed based on the model above. The hardware of the system includes TINY210 development board and sensors. The information acquisition system uses JAVA. The data includes wind velocity and direction, light intensity, air humidity, air temperatures and soil moisture. The functions of the system include data collection, field production and management. The general concept and scheme of designing is as shown in Figure 1.

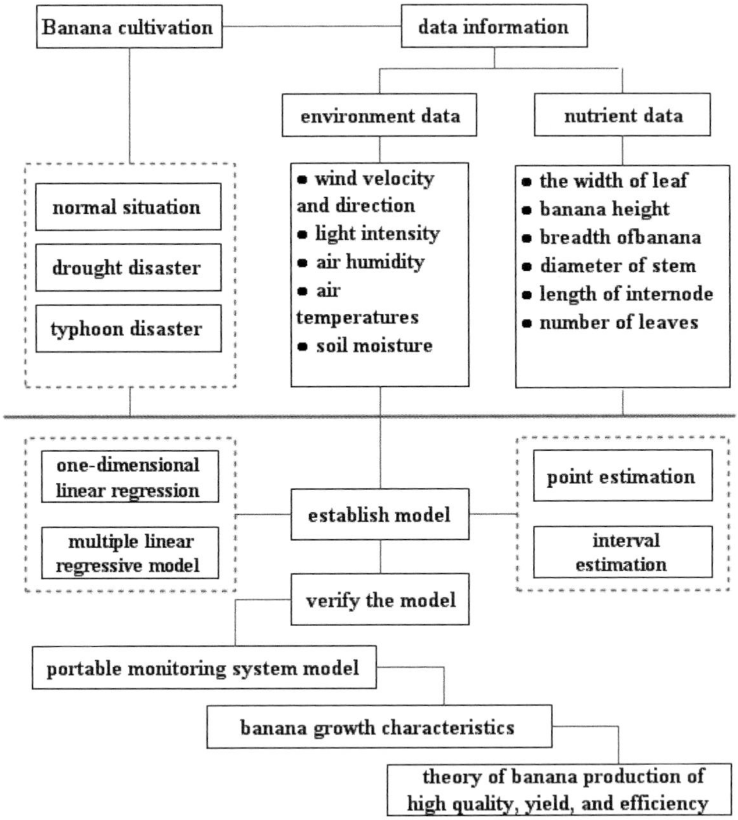

Fig. 1. The general concept and scheme of designing.

5. Summary

Agricultural application of information and computer technology is becoming a focus of agricultural research. The paper presents the solution of the portable monitoring system on banana growth. These findings may widen agricultural thinking in environmental disaster and can be used as the technical pattern for avoiding typhoon and drought disaster in banana producing areas.

Acknowledgments

This work is sponsored by the Project of Technology Extension and Training of Tropical Crops of Ministry of Agriculture in 2016.

References

1. Lingling Wang, Kun Lu, Hongxia Luo, Jihua Fang. The Key Technology Research on Information Acquisition about Tropical Crop Production Environment. Advanced Materials Research Vols. 989-994 (2014) pp 3452-3455.
2. Lingling Wang, Hongxia Luo, Jianhua Cao, et al.The design of Intelligent Monitoring System on Tropical Greenhouse Crop Production Environment. Advanced Materials Research Vols. 1073-1076 (2015) pp 530-534.
3. Jianhua Cao, Lingling Wang, Hongxia Luo, et al. A Research on Information Management System About the Field Environment of Tropical Crop, Advanced Materials Research Vols. 1073-1076 (2015) pp 3013-3016.
4. Wang Bingli, Zhang Ping, Xiong Gang. Farmland Meteorological Information Monitoring System Based on GPR S. Journal of Agricultural Mechanization Research, 2015(1):105-108.
5. Peets S, Mouazen A M, Blackburn K, Kuang B, Wiebensohn J. Methods and procedures for automatic collection and management of data acquired from on-the-go sensors with application to on-the-go soil sensors. Computers and Electronics in Agriculture, 2012, 81:104-112.
6. Nikkila R, Wiebensohn J, Nash E, Seilonen I, Koskinen K. A service infrastructure for the representation, discovery, distribution and evaluation of agricultural production standards for automated compliance control. Computers and Electronics in Agriculture, 2012, 80:80-88.
7. Sorensen C G, Pesonen L, Bochtis D D, Vougioukas S G, Suomi P. Functional requirements for a future farm management information system. Computers and Electronics in Agriculture, 2011, 76(2): 266-276.
8. Sorensen C G, Pesonen L, Fountas S, Suomi P, Bochtis D, Bildsoe P, Pedersen S M.A user-centric approach for information modelling in arable farming. Computers and Electronics in Agriculture, 2010, 73(1): 44-55.
9. Sorensen C G, Fountas S, Nash E, Pesonen L, Bochtis D, Pedersen S M, Basso B, Blackmore S B. Conceptual model of a future farm management information system. Computers and Electronics in Agriculture, 2010, 72(1): 37-47.
10. Nikkila R, Seilonen I, Koskinen K. Software architecture for farm management information systems in precision agriculture. Computers and Electronics in Agriculture, 2010, 70(2): 328-336.
11. Chen Haitao. The development of image acquisition and processing system based on 3G for irrigation area. Northwest A&F University. 2014.

Study on the operation quality assessment of P2P platform based on factor analysis

Lu-Ming Yang*

Electronic Commerce Department, Yunnan University,
Kunming, Yunnan, China
E-mail: 392469456@qq.com
www.ynu.edu

Ya-Na Wu and Liang Shan†

Kunming, Yunnan, China
E-mail: †249247589@qq.com

As the medium for direct interaction with various participants, internet-based Peer-To-Peer (P2P) lending platforms play a vital role. To assess their operational quality in a relatively objective way, we established a theoretical basis and came up with an indicator system to evaluate the operational quality of P2P platforms. This was done by consulting local and foreign documents and by collaborating with researchers on current developments. We also analyzed real-time data of 75 internet-based P2P lending platforms via factor analysis and drew the conclusion that the indicator of website history and scale is the biggest contributor. Thus, we proposed relevant suggestions to regulatory institutions, internet-base lending platforms and relevant investors.

Keywords: Empirical Research; P2P Platform; The Evaluation Index.

1. Introduction

Internet finance emerges with the continuous developing of the Internet technology in our country, and P2P Internet lending platform is one important pattern. The research and discussion about P2P platform play an important role on the developing of Internet finance in our country, which are the hot spot on Internet finance.

In this background, for researching deeply P2P platform both in theory and in practice, this article, first build theory basis and attain the evaluation index system of P2P platform by consulting Sino-foreign literatures; then process the

---

* Corresponding author.

date via factor analysis based on real-time date from 75 P2P platforms; lastly to table a proposal according to gotten conclusion.

This article which is based on organizing gotten literature aims at building evaluation index system focusing on P2P Internet lending platform operation quality, at the same time using the real-time date of P2P Internet lending platform in our country to starting empirical analysis, and getting the webs' ranking, finally offering a thought for evaluation of P2P Internet lending platform's operation quality.

2. Literature Review

In recent years, P2P platform has caught relevant scholars' eyes. After referring literatures, obviously finding that relevant scholars' research mainly focus on following several aspects.

2.1. *The research about lending platform' risk control*

Xiaoguang Wu and Yi Cao (2011) evaluates UK P2P lending platform Zopa and USA Prosper's operation performance, and evaluates its risk factors which are possibly caused in the process of operation[5]. Xiaoguang Wu (2012) via relevant research points out that P2P risk control measures in our country easily lead to borrowers' false information transaction and put forward risk control measures to protect clients' rights and interests[8]. from the perspective of borrowers , Jiaxian Qiu, Zhangxi Lin and Binjie Luo (2012) collect the dates of PaiPaiDai, start empirical analysis for the reasons of affecting successful lend and draw relevant conclusions[3].

2.2. *The research about lending platform's operation and transaction process*

Yixian Mo (2011) [6], Zhongjin guo and Haixia Lin (2013) [9] use P2P Internet lending platform which is famous in our country as an example and describe the characteristics and processes of our country'P2P internet lending platform. S. Berger and F. Gleisner (2009) discuss the intermediary role of USA P2P internet lending in the electronic marketplace and analysis succinctly that investors invest by purchasing the bill issued by Prosper platform[1].Alexander Bachmann and Alexander Becker (2011) review systematically the literatures about P2P's research, especially P2P Internet lending platform's process and the factors that influence financing[2].

2.3. *The research about P2P platform's evaluation*

Haifeng Guo, Xiao Chen (2015) use 50 Internet lending platforms in our country as an example, evaluate P2P Internet platform's comprehensive competitiveness [11]. Xu Ri, Jingjing Feng (2014) make up an lending platform credit rating evaluation model based on Fisher discriminance and verify by the empirical analysis that Fisher discriminance's misjudgment rate is just a twelfth in the Internet lending platform credit rating prediction[10].Yuhang Li (2015) uses principal component method to analyze 9 index factors that influence P2P platform'ability and make a relevant conclusion after clustering the analysis result[13].Jin Nie (2015) discuss and analyze synthetically 27 P2P platforms in our country using Grayscale Relation Method and based on it he put forward some suggestions about P2P web' construction[12].

To sum up, we can find that relevant scholars' relevant researches about P2P platform mainly focus on platform' operation, transaction process and risk control and so on. Although recent two years some scholars start researching P2P platform evaluation model's build, as a whole relevant empirical research is not enough and not deep. So this article based on 75 P2P platforms' real-time date, starting from 8 evaluation indexes, makes up an evaluation system of P2P platform operation quality and dig it deeply using factor analysis for the purpose of making a certain degree of supplement for the existing research and promote certainly P2P Internet lending platform's develop and perfection.

3. Research Object Selecting and Evaluation Index System Building

3.1. *Research object selecting*

This article's research object selecting is based on "2014 Internet lending platform developing index ranking report" issued by 2014 China internet lending industry annual report. This article selects 100 web platforms mentioned in the report as the report object, removing the platforms whose date is missing, finally left 75 research objects.

3.2. *Evaluation index system selecting*

Knowing practically P2P platform web's characteristic, web developing current situation and P2P platform's transaction characteristic, and making use of the existing literature's evaluation index, this article select 8 evaluation indexes in all: the amount of users per million ever looking through this web, average daily IP amount, reverse link number, search engine record amount, accumulative unpaid money, investor number, loan amount, established time. Every index's source is as shown in the table1.

Table 1. Resource of index's acquisition.

Evaluation index	Resource
The amount of users per million ever looking through this web	Junjun Li[4], Zhiping Hou [7]
Average daily IP amount	Zhiping Hou [7]
Reverse link number	Zhiping Hou [7]
Search engine record amount	Author
Accumulative unpaid money	Yuhang Li[13]
Investor numbe	Haifeng Guo[11]
Loan amount	Haifeng Guo[11]
Established time	Yuhang Li[13], Haifeng Guo[11]

4. P2P Platform Evaluation's Empirical Analysis

4.1. *Date resource and acquisition*

According to the differences between every evaluation index, this article's date mainly comes from Alexa, ChinaZ and Online Lending House, the date of date collection is November 29th,2015.

The relevant information of this article's evaluation index is as shown in the Table 2.

Table 2. Relevant information of evaluation index.

Evaluation index	Index meaning	Date source	Measure method
The amount of users per million ever visit this web	The amount of users per million who install Alexa toolbar to visit certain web.	Alexa	Average amount in recent 3 years
Average daily IP amount	The meaning is that in recent 3 years the amount of IP which average daily visits certain web.	Alexa	Average amount in recent 3 years
Reverse link number	The meaning is that a instruction of new document, this instruction can be used to point out a new relevant document and fill the old one instead od updating every old document .	Alexa	The whole amoun so far
Search engine record amount	Baidu record amount,reflect how P2P platform's content is plentiful.	ChinaZ	The whole amoun so far
Accumulative unpaid money	The meaning is that the whole money that platform not return to investor.	Online Lending House	Date in recent 30 days
Investor numbe	The meaning is the number of investors invested in this platform.	Online Lending House	Date in recent 30 days
Loan amount	The meaning is how many times the platform borrow money.	Online Lending House	Date in recent 30 days
Established time	The days from the date Internet lending platform established to November 2015.	Online Lending House	Date using month as Alternate units

4.2. *Factor analysis*

4.2.1. *KMO and bartlett check*

First, use KMO and Bartlett check for date. KMO 's statistic is 0.655, it is greater than 0.6,Bartlett spherical check's chi-square value is 392.586, significance level is 0.000,it is less than 0.001,so this article's date is suitable for factor analysis.

4.2.2. *Total variance explained*

Knowing from the score that the former three factors' Eigenvalue are all greater than 1,cumulative contribution of variance accounted to 80.656%, they can greatly reflect every index's condition , so this article selects the former three common factors instead of quondam index.

4.2.3. *Factor rotating*

For making factor have naming interpretation, this article uses varimax to rotate it, rotation converge after five times.

As is shown in table 3 ,the first common factor F1 synthesize searcher engine record amount, loan amount, established date, it reflects the web's history and scale, the second factor reflects synthetically average daily IP amount, the amount of users per million ever visit this web, reverse line number, it reflects the web's popularity index, the third factor reflects synthetically accumulative unpaid money, investor number, it reflects web debts repayment and investment condition.

Table 3. Rotated component matrix.

	Component		
	1	2	3
Search engine record amount	.942	.084	-.027
Loan amount	.828	.129	.439
Established date	.656	.364	.045
Average daily IP amount	.151	.922	.156
the amount of users per million ever visit this web	.176	.893	.230
Reverse line number	.523	.607	.274
accumulative unpaid money	.085	.103	.929
Investor number	.135	.369	.752

4.2.4. *Factor analysis result*

Three common factors were extracted by factor analysis of the communist ,and according to its contribution rate, can be calculated the weight of their representtatives. W1=50.840/80.648=0.6304, W2=16.791/80.648=0.2082, W3=13.018/80.648=0.1614, comprehensive score:Sn=0.6304×F1+0.2082×F2+0.1614×F3. Through the above calculation, can calculate a P2P platform score of each factor and final standings, as is shown in table 4.

Table 4. P2P platform web factor analysis and ranking (partly).

Platform name	F_1	F_2	F_3	Sn	Comprehensive ranking
Paipai loan	4875536.19	-1396515.16	-1245749.03	2503175.18	1
Niwo loan	509070.63	-386236.42	646649.57	344400.23	2
Hongling venture capital	404233.70	-478721.29	1028509.18	325883.22	3
Easy loan web	561486.87	-185899.63	-74914.73	294794.26	4
Benefit web	-479579.96	-1504453.75	5111231.21	266556.56	5
…	…	…	…	…	
Chujinsuo	-470.13	-2234.11	8227.18	654.89	73
Wan brothers	-717.93	-2416.70	8272.13	471.18	74
Xiaofu finance	-23.65	-717.19	3095.81	367.05	75

4.2.5. *Empirical result analysis*

As is known in the table 4, Paipai loan has the highest comprehensive score and F1 numerical value .Next is Easy loan net, Niwo loan, it indicates that this three webs 's established histories are quite long and their scales are quite great. From F2, Xiaofu finance's score is highest, indicating its popularity is quite high, it may be concerned with that So far it has established for nearly 18 months, it has made loans over 140 million and never be overdue and no bad debt. From F3's score and ranking, Benefit web, Lujinsuo, Yilong loan rank top three, it indicates that this three platforms have an advantage on debts repaying and investors.

After analysis, we can see,F1,F3 and comprehensive ranking mainly show a positive correlation trend, indicating that to a certain extent, the web's history is longer, the scale is greater, the repaying ability is better, the comprehensive ranking is higher. But we can find that Paipai loan's ranking is still first in F1 when it ranks last in F3, it may be concerned with Paipai loan's location purely as an intermediary agent. It didn't offer any guarantee and payment on account for all loans so that invest risk will higher for investors and investors' number will be partly influenced, leading to the low trading volume.

Comparing comprehensive ranking with F2, it is easy to find that F2 and comprehensive ranking mainly show a negative correlation. But it is not enough to illustrate the platform's average daily IP number, the amount of users per million ever visit this web, reverse line number are lower, the comprehensive ranking is better. Because its history , scale and other indexes is lower than other platforms whose ranking is not great, it usually uses high interest rate and other ways to attract investors to invest at the beginning of the establishment, so that it can attract people's eyes and gain more popularity.

4.2.6. *Relevant suggestion*

As an important role of Internet finance, P2P Internet lending platform attract tremendous attention in recent years. From the current developing situation of our country's P2P Internet lending platform ,it still exists many problems. Our country's credit investigation system is not complete, in this context, every Internet lending platforms should enhance personal platform construction and improve operation quality. Basing on it, we put forward the following suggestions.

As an important role of Internet finance, P2P Internet lending platform attract tremendous attention in recent years. From the current developing situation of our country's P2P Internet lending platform ,it still exists many problems. Our country's credit investigation system is not complete, in this context, every Internet lending platforms should enhance personal platform construction and improve operation quality. Basing on it, we put forward the following suggestions.

5. Conclusion

This article, based on Sino-foreign relevant literature, via selecting 8 indexes concerning with P2P Internet lending platform, using factor analysis to evaluate 75 P2P Internet lending platforms operation quality, finally extract three principle component factors. This article 's relevant conclusion can be used to refer for investors, at the same time , relevant P2P Internet lending platform can find personal advantages and disadvantages through horizontal comparison so as to develop intensively.

References

1. S. Berger, F. Gleisner.Emergence of financial intermediaries in elect-ronic markets: The case of online P2P lending[J].BuR Business Resear-ch Journal, 2009,5(2):39-65.
2. Alexander Bachmann,Alexander Becker,Daniel Buerckner,et al.Online Peer-to-Peer Lending:A Literature Review[J].Journal of Internet Banking and Commerce,2011,16(2):1-18.
3. Jiaxian Qiu, Zhangxi Lin, Binjie Luo.Effects of Borrower-Defined Conditions in the Online Peer-to-Peer Lending Market[J].Web-Enabled Convergence of Commerce, Work, and Social Life Lecture Notes in Business Information Processing Volume 108, 2012, pp 167-179.

4. Junjun Li, Haimin Chen.E-Commence Web Site Evaluation Based on Factor Analysis and Correspondence Analysis.[J].information science,2008,26(8):1252-1256.
5. Xiaoguang Wu, Yi Cao. The Regulation Strengthening of the P2P online Lending Platform. [J].southern finance,2011,(4):32-35.
6. Yixian Mo. Internet lending Sino-foreign theory and practice research literature summarize. [J].finance theory and practice,2011,(12):101 104.
7. Zhiping Hou. Analysis on Application and Research of Factor and Clustering in E-commerce Website Evaluation. [J].science and technology manage research, 2011,(18):144-147.
8. Xiaoguang Wu. The clients' protection of the rights and interests in P2P Internet lending platform. [J].finance theory and practice,2012,(2):54-57.
9. Zhongjin Guo, Haixia Lin. Internet lending credit mechanism research-make an example of Paipai loan. [J].modern manage science,2013(5):90-92.
10. Xuri Feng, Jingjing Zhang.P2P Internet lending platform credit ranking evaluation model based on Fisher discriminance. [J].finance theory and practice, 2014,(11):51-56.
11. Haifeng Guo, Xiao Chen.A Study of the Comprehensive Competitiveness Evaluation of P2P Network Lending Platform.[J].finance forum,2015(2):12-23.
12. Jin Nie, Linshuang Hu.P2P Internet lending platform evaluation discusses based on line analysis. [J].library science research,2015(13):37-45.
13. Yuhang Li, Shaomo Xia. The Principal Component Clustering M ethod for Evaluation of P2P Platform. [J].value project,2015,(13):180-182.

An efficient tables memory access optimized algorithms for CAVLC decoding

Xiao-Bo Lin and Hui-Lin Wang

Department of Computer Science and Engineering, HanShan Normal University, Chaozhou ,Guangdong, 521041,China,
E-mail:18927131317@qq.com

In order to solve the problems of high table memory access, an efficient table memory access optimized algorithms are presented for CAVLC decoding in H.264/AVC in this paper. The contribution of this paper rests that we use hash table query technology in table look-up to lower high table memory access by reducing high table memory access in H.264/AVC. Specially, after finding the internal relationships between code length and numbers of 0 in code prefix, we use hash table query technology in table look up to reduce lots of table memory access. The simulation results show that our proposed schemes based on hash query can lower about 25% table memory access for CAVLC decoding compared with TLSS method.

Keywords: CAVLC Decoding; Table Look-Up; Hash Query; Memory Access.

1. Introduction

H.264/AVC is the latest international video coding standard, which has been developed by ITU-T and ISO/IEC [1]. It has been widely adopted in many related video communication aspects and has greatly improved the compression ratio and video quality. Context-based adaptive variable length coding (CAVLC) as the entropy coding tool of the H.264/AVC baseline profile, has improved the video compression ratio, at the same time has also increased the power consumption and hardware cost of the decoder. Thus how to develop a CAVLC decoding method with a low-power design that also meets the demands of real-time CAVLC decoding has become our major issue at present.

Table look up is a very power-consuming operation, which consumes most power of CAVLC decoder [2]. It could seriously affect the efficiency of CAVLC decoding. Looking-up variable length tables have become a very important problem for CAVLC decoding in H.264/AVC baseline profile. During the process of CAVLC decoding, CAVLC decoding needs to decode five syntax elements. Three in five syntax elements, Coeff_token, Run_before and Total_zeros, need be decoded by looking up the variable length tables, while the rest of them, Level and sign of TrailingOnes(T1s), are decoded by the regular arithmetic operations without using the looking-up of variable length tables. So high table memory access will be required to find desired codeword from variable length tables and great mount of storage space will be required to store these variable length tables. It is well known that table memory access is an important performance and power bottleneck in embedded systems especially for small-size multimedia applications [3]. As the conventional method of Table Look-up by Sequential Search (TLSS) needs a great lot of table memory access to spend on every decoded codeword due to complete table look-up for the desired codeword, it cannot meet real-time decoding application requirement. In order to lower the table memory access, many software optimized decoding methods have been developed for the purpose.

For example, Moon et al. [4] proposed a new table look-up method based on integer arithmetic operations to reduce some table memory access. Lu et al. [2] proposed an entropy decoding algorithm to decrease table memory access than the original algorithm in the H.264 reference software; Lee et al. [5] developed new codeword structures, look-up tables and searching methods to improve table look up operation for the CAVLC syntax elements. Kim et al. [6] proposes some other integer arithmetic operations to reduce table memory access. Heng et al. [7] merged all codeword tables into one table and organized the table into sub-tables to drop table memory access. Wang et al. [8] presents a novel low-cost high-performance CAVLC decoder for H.264/AVC to save the CAVLC decoding memory access. Uchihara et al. [9] proposed a new efficient software CAVLC decoder architecture based on level length extraction to reduce decoding memory access, and so on.

In this paper, we propose a table look-up algorithm based on abased on hash query for CAVLC decoding. The achievement of the algorithm rests that it uses hash table query to solve the problem of high table memory access for CAVLC decoding. The simulation results show that our proposed scheme can lower about 25% table memory access for CAVLC decoding without degrading video quality.

The rest of this paper is organized as follows. In Section 2, the principle of CAVLC decoding is introduced. The proposed decoding method is presented in Section 3. The simulation result of proposed scheme compared with TLSS methods is presented in Section 4. In Section 5, we give some conclusions.

2. The Principle of CAVLC Decoding

In baseline profile of H.264/AVC, the CAVLC and Exp-Golomb codes are used as two entropy decoding method. Exp-Golomb codes are used to decode indication information and other coding parameters, which are with regular construction; While CAVLC is adopted to decode the quantized transform coefficients for residual blocks. Since the computation complexity of whole entropy decoding time is mainly occupied by the CAVLC decoding, the paper will focus on the CAVLC decoding procedures. In the process of CAVLC decoding, the quantized coefficients are zigzag scanned and then decoded by the five syntax elements. Figure 1 is the decoding order and definition of five syntax elements above are described as follows.

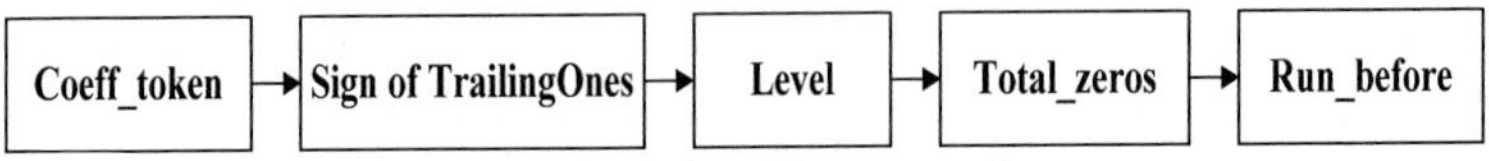

Fig. 1. Decoding order of five syntax elements.

- Coeff_token: Both the number of nonzero coefficients (Totalcoeff) and number of coefficients that absolute value is equal to one (TrailingOnes).
- Sign of TrailingOnes (T1s): Use a single sign bit, which 0 is for positive and 1 is for negative, to represent each T1s in reverse zigzag order.
- Level: The values for each nonzero coefficient except for T1s in reverse zigzag order.
- Total_zeros: The total number of zero coefficients between the DC and the last nonzero coefficient in zigzag order.
- Run_before: The numbers of zeros preceding each nonzero coefficient in reverse zigzag order.

3. Proposed Scheme

3.1. *Observe variable length tables*

As is mentioned above, the process of CAVLC decoding needs to decode five syntax elements (Coeff_token, T1s, Level, Run_before and Total_zeros). Among five elements above, decoding the syntax elements of Coeff_token, Run_before and Total_zeros need to look-up some variable length tables while the other two syntax elements of Level and Sign of T1s are decoded by the regular arithmetic operation directly without looking up some variable length tables. Through analysis the structure of variable length tables, we can find that Coeff_token has three 2D-variable length tables: $0<=NC<2$, $2<=NC<4$, $4<=NC<8$ while the Run_before and Total_zeros VLDs has one 1D- variable length tables.

Table 1. Part of 2D-UVLT for Coeff_token (2≤NC<4).

Code	Codeword	[T1,Tc]
10	0x21	[1,1]
11	0x00	[0,0]
011	0x42	[2,2]
0100	0x64	[3,4]
0101	0x63	[3,3]
00110	0x65	[3,5]
00111	0x22	[1,2]
001000	0x66	[3,6]
000111	0x02	[0,2]
...	......	......
0000000000001	0x6f	[3,5]

In table 1, the code represents for the input bit-stream of the Coeff_token syntax element (0≤NC<2). The 8-bit Codeword represents the decoded output elements. The front 3 bits of them are for total number of ones (T1) and the other tail 5 bits are for the total number of coefficients (Tc). The other elements of Coeff_token, Run_before and Total_zeros have the same code table structure as Table 1.

3.2. *Relationship between code length and numbers of 0 in code prefix*

By analyzing the codeword structure in Table 1, we find that corresponding relationships existing between numbers of 0 in code prefix and code length for Coeff_token and Total_zeros are shown as Table 2.

Table 2. The internal relationship between length of code and numbers of 0 in code prefix(2≤NC<4).

Numbers of 0 in code prefix	Code length
0	2
1	3 or 4
2	5 or 6
3	6
4	7
5	8
6	9
7	11
8	12
9	13
10	13 or14
11	14
12	13

Table 2 mean the relationship between the length of code and proposed Numbers of 0 In Code Prefix and Total_zeros. Through the relationship, we can quickly determine the length of codeword suffix from decoding input bit-stream, which can save lots of time of looking up and judging codeword suffix.

Based on this principle above, we build a hash table to express the relationship existed among code and numbers of 0 and length of code prefix and code length. The built hash tables for 2D-Coeff_token and the 1D-Total zeros are shown as Table 3.

Table 3. Part hash table of 2D-Table for Coeff_token (VLCT1, 2≤NC<4).

Code	Hash table address	Numbers of 0 in code prefix	Length of code suffix	Code length
10	0	0	2	2
11				2
011				3 Or 4
0100	1	1	1or 2	
0101				
00110				
00111	2	2	2 or 3	5 or 6
001000				
000111	3	3	2	6
…	……	……	……	……
0000000000001	12	12	0	13

3.3. *Propose tables look-up algorithm based on hash query*

Based on the analysis above, we propose a quick table look-up algorithm for CAVLC decoding based on the unit of hash table query and table look-up. The basic idea of new algorithm rests that we make use of the numbers of consecutive zero Calculated in input bit-stream to find the code length of code suffix by a hash table built, after getting the value of code suffix, we can quickly get the decoded output only looking up one CAVLC standard table through number of zero in code prefix and value of code suffix determined above.

The details of the realization process for our proposed schme can be summed up in the following steps.

Step 1: select variable length tables of Coeff_ token syntax element through the value of NC.

Step 2: read input decoding bit-stream and calculate numbers of 0 in code prefix

Step 3: find the length of code suffix by numbers of 0 in code prefix by built hash table

Step 4: determine the length of code suffix. If the length of code suffix found has two possible value , we will determine right length of code suffix after judging the first value of code suffix again.

Step 5: get code length according to the numbers of 0 in code prefix and length of code suffix above.

Step 6: look up CAVLC standar codtab to get decoded codeword according to code length determined above.

3.4. *Demonstrate method instance*

Now suppose that NC value is $0 \le NC < 2$, and the bit-stream inputted is 0000100011... Take decoding Coeff_token syntax element for example to illustrate decoding process for our supposed table look-up algorithm. The detained decoding process of an example supposed above with our proposed algorithm are shown as follows.

In step 1, select variable length table0 of Coeff_token as table entry, because NC value is $0 \le NC < 2$;

In step 2, read input decoding bit-stream and the numbers of 0 in code prefix calculated from input decoding bit-stream is 4;

In step 3, look up hash table 4 to get the length of code suffix being 1 or 2.

In step 4, as length of code suffix has two possible value, so we need to judge the first value of code suffix further and then we can directly get the length of code suffix being 2, and obtain the value of code suffix being 00.

In step 5, we can get the code length being 7 and the code being 0000100;

In step 6, as to 0000100, look up CAVLC codtab of Coeff_token, and the decoded Codeword is T1s=2 and Tc=5.

After decoding the Coeff_token, level, sign of T1s, Total_zeros and Run_before syntax elements in sequence, we can get the last decoded output. Same steps can be taken for the syntax elements of Total_zeros and Run_before.

4. Experimental Results and Analysis

In order to verify the effectiveness of our proposed method above, we take some experiments. In this section, our designed experiments mainly include experimental environment and save table memory access.

The simulation environment was conducted on an Intel 2GHz processor, 1GB memory capacity, Intel Windows XP operating system. We evaluate the table memory access savings with our proposed methods compared with TLSS[10]. We use Forman(FM), Crew(C), Paris(P), Wal(W), Football(FB), Harbour(H), Mobile(M), Soccer(S)as our test sequences in our experiment. The memory access savings comparison results in table look up can be shown in Figure 2 to Figure 5. Where the Figure 2 and Figure 3 are memory access saving comparison result in table look up with different sequences respectively in the same QP value. Figure 4 and Figure 5 are the memory access savings comparison results in table look-up with different QP for the same test sequences.

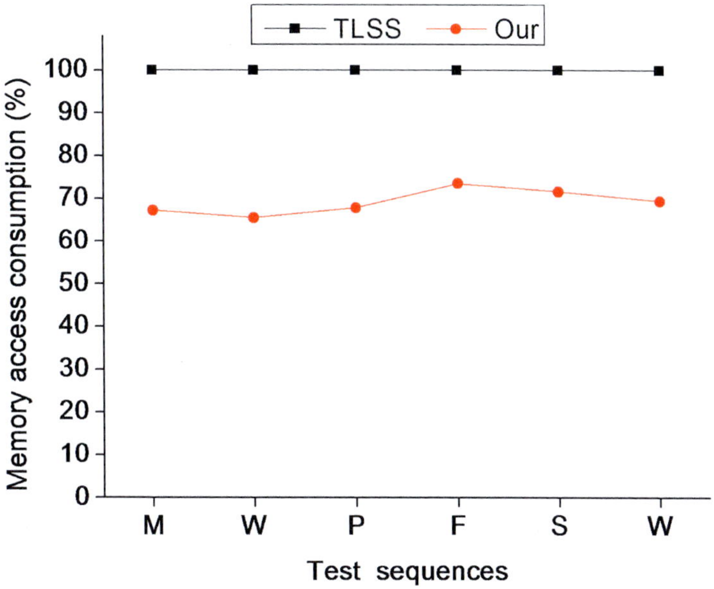

Fig. 2. Memory access consumption in table.

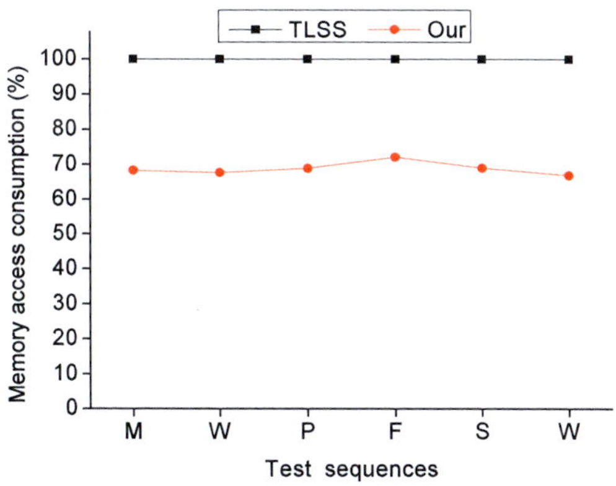

Fig. 3. Memory access consumption in table look-up for120 frames of test sequence (QP=24) ; look-up for 120 frames of test sequence (QP=32).

From Figure 2, Figure 3, Figure 4 and Figure 5, we can see that our proposed algorithms have the better table memory access saving compared with TLSS, reaching about 25% memory access saving. The reasons for them are that we use hash table query technology to instead of the sequential search code length in TLSS of CAVLC decoding respectively, which can entirely save the table memory access spent in looking up lentab() during CAVLC decoding, thus reducing the total table memory access for CAVLC decoding.

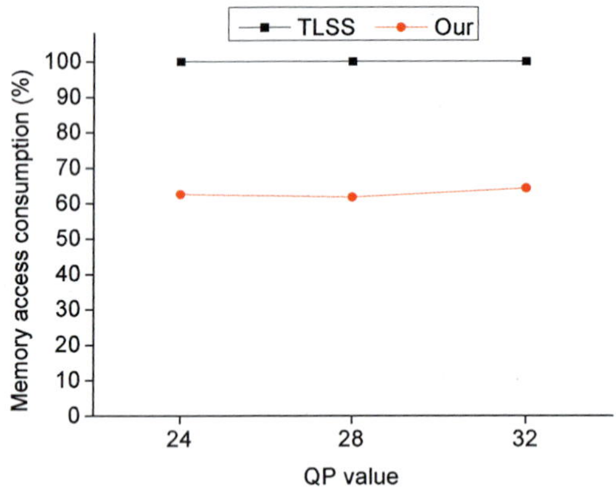

Fig. 4. Memory access consumption in table look-up for Paris(CIF).

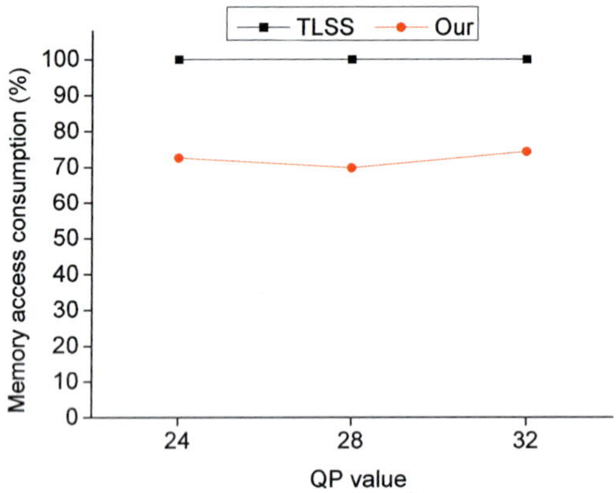

Fig. 5. Memory access consumption in table look-up for Walk(QCIF).

5. Conclusion

In this paper, we propose an efficient tables memory access scheme for CAVLC decoding in H.264 baseline profile. As our proposed table memory access method uses hash table query to quickly determine the code length and output codeword, it could lower a lot of table memory access. Simulation results in the paper show that our proposed scheme can reduce lower about 25% table memory access without degrading video quality.

References

1. Joint Video Specification (ITU-T Rec. H.264|ISO/IEC 14496-10)-Joint Committee Draft Joint Video Team (JVT) of ISO/IEC MPEG and ITU-T VCEG, Doc. JVT-G050r1.doc, Dec. 2002.
2. Lu, Da,Liu, Guofan,Zhu,Lingli , "An optimization for CAVLC code table lookup algorithm in H. 264 decoder" ,Proceedings - 2011 International Symposium on Intelligence Information Processing and Trusted Computing, IPTC 2011, p 79-82, 2011, Proceedings - 2011 International Symposium on Intelligence Information Processing and Trusted Computing, IPTC 2011
3. Junghee Lee, Chanik Park, and Soonhoi Ha "Memory Access Pattern Analysis and Stream Cache Design for Multimedia Applications", in Proc. Asia and South Pacific, Design Automation Conf (DAC), Jan. 2003, pp.21-24.
4. Y. H. Moon, G. Y. Kim, and J. H. Kim, "An Efficient Decoding of CAVLC in H.264/AVC Video Coding Standard", IEEE Trans. on Consumer Electronics, Vol.51, No3, August 2005.
5. Jun Young Lee, Jae Jin Lee, and SeongMo Park, "New Lookup Tables and Searching Algorithms for Fast H.264/AVC CAVLC Decoding" , *IEEE Trans. Circuits and Syst. Video Technol.* vol. 20, no. 7 pp. 1007-1017, July 2010.
6. Yong-Hwan Kim, Yoon-jong Yoo, Jeongho Shin, Byeongho Choi, and Joonki Paik, "Memory-Efficient H.264/AVC CAVLC for Fast Decoding", IEEE Trans On Consumer Electronics, Vol.52, No.3, August 2006.
7. Heng-Yao Lin, Ying-Hong Lu, Bin-Da Liu, and Jar-Ferr Yang, "A highly Efficient VLSI architecture for H.264/AVC CAVLC Decoder" , IEEE Transactions on multimedia, vol. 10, no. 1, January 2008.
8. Wang, Kyu-Yeul, Kim, Byung-Soo; Lee, Sang-Seol; Kim, Dong-Sun; Chung, Duck-Jin. "A novel low-cost high-throughput CAVLC decoder for H.264/AVC ", *IEICE Transactions on Information and Systems*, v E94-D, n 4, p 895-904, April 2011
9. Naofumi Uchihara, Hiroki Hayakawa and Hiroyuki Kasai. "Efficient H.264/AVC Software CAVLC Decoder based on Level Length Extraction", IEEE Transactions on Consumer Electronics, Vol. 58, No. 1, February 2012
10. K. Suhring, "JM 16.2 software," http://iphome.hhi.de/suehring/tml/.

The research investigations on the NetFPGA platform

Xing Zheng*, Juan Cao and Hai-Jun Gao
Key Laboratory of RF Circuits and Systems of Ministry of Education,
Hangzhou Dianzi University,
Hangzhou, 310018, China
**E-mail: zhengxing@hdu.edu.cn*

The Internet has experienced rapid growth in capacity, access device number and applications. However, issues such as network security, energy efficiency and network flexibility have arisen. To address these issues, a networking investigation platform, NetFPGA, is set up. This paper analyzes different research aspects of the NetFPGA and presents our findings about the NetFPGA platform.

Keywords: NetFPGA; Internet; Programmability.

1. Introduction

When the Internet enters into the new era, it faces a lot of challenges including for example, network scalability and flexibility, network and device security problem, energy efficiency issue and network reconfigurability. To address these challenges, a networking environment that can be used for investigating the Internet is urgently required. NetFPGA is one of the promising platforms that can be utilized for this aim.

NetFPGA is an open source and reprogrammable hardware platform that concentrates on designing and developing high performance networking devices and systems. In the beginning, the NetFPGA is a class tool used by the professors for teaching in the university. However, more and more researchers found that NetFPGA is a promising platform for the research purpose. Therefore, the researchers start to give up the pure software way and use NetFPGA for hardware prototyping. Currently, NetFPGA is widely used in the universities and research institutions over the world. The other reasons for adopting the NetFPGA as the network research platform include: (1) NetFPGA can offer a convenient and fast way to realize a practical networking devices such as routers and switches; (2) NetFPGA can be utilized to implement the

* Corresponding author.

optimized network algorithms; (3) NetFPGA has the feature of modular and open-source, which simplifies the task of the networking investigations; (4) NetFPGA can realize the networking devices that are similar with the commercial devices, which is helpful for their future application in industries.

The investigations on the NetFPGA includes for example: (1) the NetFPGA-based network switch and router are realized and used in vast majority of the NetFPGA researches [1]; (2) NetFPGA-based virtual network system (VNS) is proposed and virtual networks are implemented and managed by utilizing the VNS [2]; (3) Several researches have been undertaken on the network security issues [3-5]; (4) Energy efficient Green technologies are used to reduce the power consumption of the NetFPGA-based routers [6-9]; (5) NetFPGA is utilized to implement new types of routers such as rate control protocol (RCP) router [10].

In this paper, the researches on or by using the NetFPGA are overviewed, where different NetFPGA-based techniques are presented. We also provide our research investigations on the NetFPGA. The paper is organized as follows. Section 2 overviews different research aspects of the NetFPGA. Section 3 presents our research investigations on the NetFPGA. The paper is summarized in section 4.

2. The Research Development of NetFPGA

Today, the rapid growth of the Internet requires the hardware-accelerated switches and routers. Many commercial vendors use FPGA to accelerate the switching and routing speed. Therefore, the NetFPGA enabled switches and routers have been developed and used by the academic community [1]. The NetFPGA platform allows researchers and students to build practical routers and switches within several weeks, which is efficient for networking teachings and investigations. Based on the NetFPGA platform, a number of novel networking routers and switches have been developed. For example, the rate control protocol (RCP) has been implemented in the RCP router, in which the feedback from the routing path allows the sender to adjust the sending rate [10]. Due to the rapid growth on the Internet capacity, the size of the lookup tables of the routers increases significantly. Some works such as [11] proposes a hash function based parallel scheme to reduce the lookup time and enables a high speed router.

For enabling flexible networks, OpenFlow has been realized on the NetFPGA-based switches, where the flow-based switches are located in the data path and the managements are operating on the remote controller [12]. One of the most important advantages of the OpenFlow is that it allows the researchers and developers to realize new networking algorithms on the practical switches and routers and speed up the improvements on the networks. Therefore, the OpenFlow is very popular in the research and development.

VNS is another way to improve the network flexibility and reconfigurability. In VNS, the network topologies at the link layer are virtualized by mapping between VLANs. Based on this mechanism, lots of virtual networks can be emulated with a small network and any networking device such as NetFPGA can be connected to the virtual topology [2]. [2] presents a virtual system that integrates the NetFPGA and VNS. It is used by the Stanford university to teach a graduate level networking course.

When the networks become larger, network security is an important issue needs to be considered. In recent years, a number of researches have been undertaken to solve the network security problem. In [5], the NetFPGA implementation is used in the Snort, a most popular network intrusion detection system (NIDS), for accelerating Snort detection. In [3], a reconfigurable security protection system is realized on the NetFPGA and embedded soft-core technique. NetFPGA is used for packet filtering, protecting from ARP attacks, flow monitoring and hardware accelerating of the transmission. [4] implements a parallel multi-field packet classification on the NetFPGA router platform and this implementation can classify multiple packets simultaneously.

Another important topic with the development of the networks is the energy efficiency. How to estimate and reduce the power consumption of the router are also undertaken on the NetFPGA platform. [6] proposes a model for the power consumption of the NetFPGA-based router. [9] presents a fine-grained energy measurements for the NetFPGA. [7,8] offers an insight into the potential energy saving of frequency scaling for the NetFPGA-enabled routers.

3. Experimental Setup and Investigation Results

We have done some works on the energy efficient routers. The experimental testing environment is built in our labs, which consists of a packet generator, a router and an energy measurement tool kit. The NetFPGA board is configured as the router. The clock frequency of the NetFPGA board can be adjusted from high frequency to low frequency. To host the NetFPGA board, one personal computer (PC) with a 2.93GHz Intel CPU and CentOS 5.5 is used. The packet generator acts as the traffic source and sink, with four ports connected with the router. The generator is realized by configuring a second NetFPGA board.

The power consumption of the router is measured accurately through the collaboration of a PCI bus extender and a National Instruments data acquisition (NI DAQ) device. The NI DAQ is connected to the PCI bus extender. The total power consumption can be calculated by accumulating the power measured from the 5V and 3.3V pins on the PCI bus extender. For collecting the power data from the NI DAQ, LabVIEW is running on a PC and the NI DAQ is connected to this PC via a USB link.

We compare the power consumption of the RCP router and reference router under different frequencies, as shown in Fig.1. It can be seen from Fig.1 that the power consumption of the reference router is slightly lower than the RCP router, which is due to the reason that the adding of the RCP protocol into the router will introduce extra power consumption. It can also be noted that the frequency down scaling from 125MHz to 62.5MHz significantly reduce the power consumption of the router. This indicates that the frequency scaling is a promising technique to reduce the power consumption for both the RCP router and reference router.

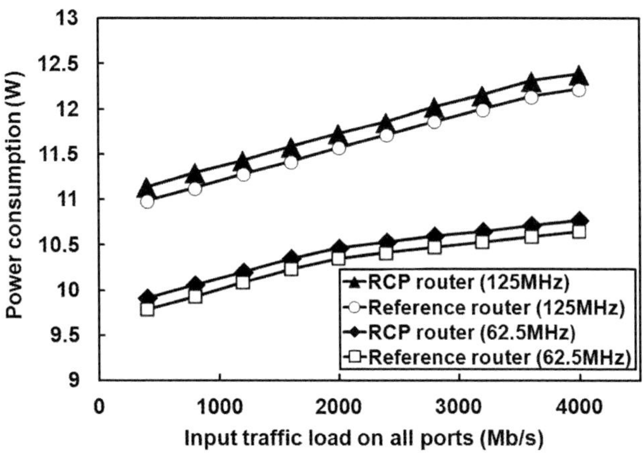

Fig. 1. Power consumption of different routers by using frequency scaling.

On the other hand, we have also done some works on the video broadcasting based on the NetFPGA router. Fig.2 shows our experimental setup for the broadcasting system. A server and a client are connected to the NetFPGA router. When the server broadcasts some data such as the video, the client can receive the real time video stream. Moreover, other traffic from/to the networks can also connect to the NetFPGA router by other Ethernet ports.

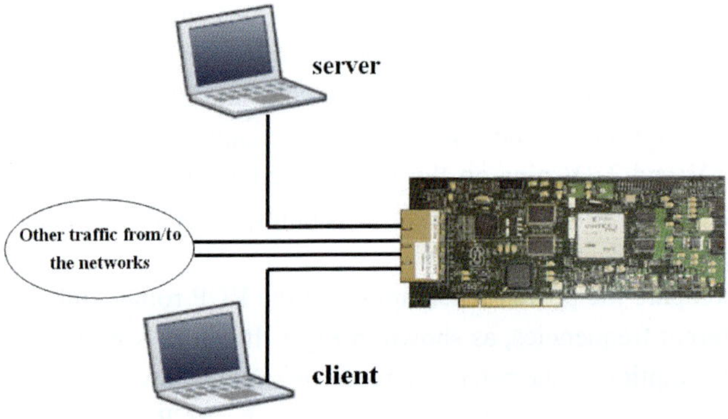

Fig. 2. Broadcasting system based on NetFPGA router.

4. Conclusions

The NetFPGA is a promising platform for investigating the network environment and devices. In recent years, the researches on the NetFPGA include for example, novel routers and switches, virtual networks based on VNS, network security and energy efficient routers. This paper has overviewed different research aspects of the NetFPGA. In addition, our investigations on the NetFPGA are also presented.

Acknowledgments

This work was supported by Zhejiang Provincial Qianjiang Talent Plan (QJD1402001), Zhejiang Provincial Natural Science Foundation of China (ZJNSF) (Grant Number: LQ13F010009), National Natural Science Foundation of China (Grant Number: 61331006), the National Natural Science Foundation of China (No.61372021), Science Foundation Ireland under the International Strategic Cooperation Award (Grant Number: SFI/12/ISCA/2496).

References

1. J. W. Lockwood, N. McKeown, G. Watson, G. Gibb, P. Hartke, J. Naous, R. Raghuraman and J. Y. Luo, NetFPGA - An open platform for gigabit-rate network switching and routing, *in Proc. IEEE Int. Conference on Microelectronic Systems Education (MSE'07)*, (San Diego, United States, 2007).
2. M. Casado, G. Watson and N. McKeown, Reconfigurable networking hardware: A classroom tool, *in Proc. the 13th Symposium on High Performance Interconnects (HOTI'05)*, (Stanford, United States, 2005).

3. K. Zhang, X. M. Ding, K. Xiong and Y.T. Zhao, Reconfigurable security protection system based on NetFPGA and embedded soft-core technology, *in Proc. Int. Conference On Computer Design And Applications (ICCDA 2010)*, (Qinhuangdao, China, 2010).
4. S. Gamage and A. Pasqual, Line rate parallel packet classification module for NetFPGA platform, *in Proc. IEEE 8th Int. Conference on Industrial and Information Systems (ICIIS 2013)*, (Peradeniya, Sri Lanka, 2013).
5. A. D. Rami, S. Khaled, A. Q. Mahmoud and O. Hadi, Framework for a NetFPGA-based Snort NIDS, *in Proc. Int. Symposium on Communication Systems, Networks & Digital Sign (CSNDSP)*, (Manchester, England, 2014).
6. A. Vishwanath, Z. Zhao, V. Sivaraman and C. Russell, An empirical model of power consumption in the NetFPGA gigabit router, *in Proc. IEEE Int.Symposium on Advanced Networks and Telecommunications Systems (ANTS)*, (Mumbai, India, 2010).
7. F. Guo, O. Ormond, L. Fialho, M. Collier and X. J. Wang, "Power consumption analysis of a NetFPGA based router," *Journal of China Universities of Posts and Telecommunications.*, vol. 19, no. 1, pp. 94-99, Jun. 2012.
8. A. Lombardo, C. Panarello, D. Reforgiato and G. Schembra, Measuring and modeling energy consumption to design a green NetFPGA giga-router, *in Proc. IEEE GLOBECOM*, (Anaheim, United States, 2012).
9. V. Sivaraman, A. Vishwanath, Z. Zhao and Craig Russell, Profiling per-packet and per-byte energy consumption in the NetFPGA gigabit router, *in Proc. IEEE INFOCOM*, (Shanghai, China, 2011).
10. N. Dukkipati, G. Gibb, N. McKeown and J. Zhu, Building a RCP (rate control protocol) test network, *in Proc. 15th Annual IEEE Symposium on High-Performance Interconnects (HOT interconnects)*, (Stanford, United States 2007).
11. G. Antichi, A. D. Pietro, D. Ficara, S. Giordano, G. Procissi and F. Vitucci, A randomized scheme for IP lookup at wire speed on NetFPGA, *in Proc. IEEE Int. Conference on Communications (ICC)*, (Cape Town, South Africa, 2010).
12. J. Naous, D. Erickson, G. A. Covington, G. Appenzeller and N. McKeown, implementing an OpenFlow switch on the NetFPGA platform, *in Proc. 4th ACM/IEEE Symposium on Architectures for Networking and Communica-tionsSystems (ANCS)*, (San Jose, United States, 2008).

Chapter 6

Mathematics and Management

An approach to process uncertain complex events

Qi-Xin Cai, Jian Liu, Qing Xu and Hui-Ling Su
Jiangsu Electric Power Company Research Institute
Nanjing, 210003, China

Hong-Sheng Zhao*
Institute of Automation Chinese Academy of Sciences
Beijing, 100190, China
E-mail: qxrzhs@hotmail.com

Complex Event Processing (CEP) often involves processing huge amounts of information in real time on the fly. In this paper, we present a new approach to deal with the uncertainty in complex events in terms of complex event patterns and provide an algorithm to demonstrate the improvement for Complex Event Processing (CEP). Two complex event detecting principles, *Strong Match Principle* and *Weak Match Principle,* adopted in different situations, were studied. Our results show that if basic events occur in random, all basic events in a rule may not completely occur in real world situations. Type I error and Type II error are also discussed when weight was introduced to support fuzzy reasoning to deal with the uncertainty in CEP. We have found that a good uncertain CEP detecting system should not only indicate which event occurred, but also when the event is going to happen with fuzzy reasoning and forecasting functions.

Keywords: Uncertain Cep; Weight Value; Strong Match Principle.

1. Introduction

In the last few years several CEP systems have been proposed [1], however, there is a same limitation of these systems is that their inability to uncertain that exists in most application [2]. There are two types of uncertainty in complex event processing (CEP), uncertainty in events and uncertainty in rules. Sometimes we cannot define all the possible situations which can interpret the occurrence of an abnormal sensor data and it is difficult or impossible to form rules precisely due to certain domain knowledge under studying. Recently the demand about the uncertainty of CEP is growing, for a large device monitoring system, the system will generate lots of real-time data flows of system devices'

* Corresponding author.

performance and status, by effectively analyzing the data flows we can implement system failure warning or scheduling, and daily devices verification.

With the advent of uncertainty of events, the situation becomes complicated; we cannot directly make decision based on uncertain information. Once a large device monitoring system [3] is shut down, cost will be great, so the research on uncertain reasoning technology becomes more important when we meet uncertainty in complex event processing and how to generate trusted complex events is becoming an urgent problem. The lack of effective mechanisms for dealing with uncertainty and imprecision of basic events may lead to detection errors for complex event in real world application.

In this paper we discuss a new approach to deal with the uncertainty in complex event processing from the viewpoint of uncertainty itself. The remainder of the paper is organized as follows. Section II we present the relevant uncertain event processing methods. In Section III, we present a new uncertain complex event processing model. A probabilistic event detection algorithm is proposed in Section IV. Then, Section V we presents an example to demonstrate our new method. Finally, in Section VI we discuss directions for further research.

2. Background

Recently some works have been carried out to study the uncertainty for CEP, the state-of-the-art work is done by Cugola.G. et al. [4] which presented a new model CEP2U (Complex Event Processing Under Uncertainty) to deal with uncertainty in CEP, CEP2U models uncertainty in events using the theory of probability, while it exploits Beyesian Network (BNs) to model uncertainty in rules.it extends the model of events to include probabilistic data into event notification, while it automatically builds a BN for each rule deployed in the system. Domain experts can extend such BNs to capture a-prior knowledge about those aspects of the environment that cannot be directly observed by source.

However, when we talk about a uncertain event, event itself has randomness, traditional concept of complex event can't satisfy this situation. For example, a traditional CEP rule r:(a,b,c)→Action means when event a, b, c are all detected from event streams in a time window, Action will be activated. Now events are all randomness, it is obvious that activate mechanism of complex event need to be modified.

3. New Approach Model

3.1. *Description of uncertain event*

Due to basic event uncertain property, complex event based on basic event also exist uncertain. We can use a tuple (*Event instance*, *uncertain degree*) to describe it. For example, (smoke, 0.75) indicates an uncertain event (smoke) occurred with probability (value is 0.75).

According to CEP mechanism, some basic event will cause a complex event, suppose BE_i $(i = 1,2 \ldots n)$ are a series of basic events and CE is a complex event, we can specify the rule as following:

$$(BE_i, P_i) \rightarrow (CE, cf) \quad i = 1,2, \ldots n \tag{1}$$

In above expressive formula, P_i is event BE_i occur probability and *cf* is confidence degree of complex event CE occurs. Formula (1) also can be expressed as following:

$$R: (BE_1, P_1) \cap (BE_2, P_2) \cap \ldots . \cap (BE_n, P_n) \rightarrow (CE, cf)$$

3.2. *Enhanced description of uncertain event*

In reality society, the relationship between conditions and conclusions are usually non-linear. However, by applying mathematical method (linear approximation) we can suppose the relationships are linear, based on it, we can know degree of influence of each event on the complex event by calculating its correlation coefficient.

A correlation coefficient is a coefficient that illustrates a quantitative measure of some type of correlation and dependence, meaning statistical relationships between two or more random variables or observed data values. Suppose we obtain *m* records about *i-th* basic event BE_i and CE from historical data. We note them as following:

$$\left((BE_i, P_{i1}), (BE_i, P_{i2}), \ldots . (BE_i, P_{im})\right)$$

And

$$\left((CE, cf_1), (CE, cf_2), \ldots . (CE, cf_m)\right)$$

From statistic knowledge, we can obtain the correlation coefficient between BE_i and CE according to following formulas.

$$r_j = \frac{\sum_{j=1}^{m} P_{ij} cf_i - \frac{\sum_{j=1}^{m} P_{ij} \cdot \sum_{j=1}^{m} cf_j}{m}}{\sqrt{\left(\sum_{j=1}^{m} P_{ij}^2 - \frac{\left(\sum_{j=1}^{m} P_{ij}\right)^2}{m}\right) \cdot \left(\sum_{j=1}^{m} cf_j^2 - \frac{\left(\sum_{j=1}^{m} cf_j\right)^2}{m}\right)}} \tag{2}$$

Now we can calculate the weight value of each basic event BE_i in rule (1) over CE, noted by w_i .

$$w_j = \frac{r_i}{\sum_{k=1}^{n} r_k}$$

Obviously, w_i reflects the relevant degree between *i-th* basic event BE_i and CE. In tradition CEP method, if there are more basic events can cause a complex event, because each event is accurate, we do not need to care the important degree of each basic event and regard them as the same during processing. However, when the basic event occurs with randomness, the important of each basic event over CE is not the same. We use a triad (*Event instance*, *uncertain degree*, *weight value*) to describe an uncertain event to replace above tuple description. And rule (1) is also changed as following:

$$(BE_i, P_{i,} w_i) \rightarrow (CE, cf) \;\; i = 1,2, \ldots n \tag{3}$$

And

$$R^*: (BE_1, P_1, w_1) \cap (BE_2, P_2, w_2) \cap \ldots . \cap (BE_n, P_n, w_n) \rightarrow (CE, cf) \tag{4}$$

According to above suppose, BE_i and CE exists linear relation, CE can be represented as BE_i linear combination, i.e.

$$cf = F(P_i, w_i) \tag{5}$$

Equation (5) indicates the uncertainty and weight of basic events will affect the confidence of complex event.

3.3. *Further explanation of weight value of basic event*

We assign a new attribute of each basic event when we process uncertain complex events. When a basic event's weight value, relative to other events' weight values, is smaller, it shows this basic event has less affect on CE and its function can be omitted. We note $\omega \ll W$.

We think it is important to consider weight values of basic events when we process uncertain events , as we know, if basic event occurs in random, all basic events in a rule may not complete occurred in real world condition, the reasons of an event has not occurred is different in nonrandom and random situations. In nonrandom situation, the event is really not occurred; but in random situation, the event may be really not occurred or event occurred but we don't catch it due to uncertainty. In such situation, method of tradition CEP can't be executed as well, we mentioned it in the end of last section.

3.4. *New complex event processing model*

First, we introduce some parameters before we begin to build new model.

P_i : uncertain degree of event BE_i occur

P_i^* : the observed uncertain degree value of *i-th* basic event.

$P_i[]$: the range of P_i,it means an interval which can ensure most of P_i are included in it. For a certain basic event BE_i $(i = 1,2 \ldots n)$, we can adopt probability and statistics methods to calculate or evaluate the value of δ_i based on event history record.

δ_i: The center value of $P_i[]$.

$fb(x, y, \delta)$: a Boolean function. When $|x - y| \leq \delta$, its value equals to *True*, otherwise, *False*.

$fb_{BE}(X, Y, \delta)$: $fb(x, y, \delta)$ for certain basic event.

Now we define new uncertain complex event processing detection (or match) rule as following:

Suppose R^* (Equation (4)) is a complex event rule, this rule will be activated (or an uncertain complex event is detected) if and only if the value of F is *True*.

$$F(R^*) = fb_{BE_1}(P_1, P_1^*, \delta_1) \cap fb_{BE_2}(P_2, P_2^*, \delta_2) \cap \ldots . \cap fb_{BE_n}(P_n, P_n^*, \delta_n) \quad (6)$$

And confidence value of complex event CE is *cf.*

We call above principle as *Strong Match Principle*.

Sometimes, perhaps in certain condition, equation (6) can not satisfied, i.e.

$\exists \varepsilon$, for certain k $(1 \leq k \leq n)$ s.t. $|P_k - P_k^*| - \delta < \varepsilon$.

In this condition, if the weight value of this basic event w_k (we define it in above section) is very small relative to weight values of other basic events, i.e. $w_k \ll W$.

Suppose R^* (Equation 4) is a complex event rule, this rule will be activated (or an uncertain complex event is detected) if and only if the value of F^* is *True*.

$$F^*(R^*) = \cap_{m=1, m \neq k}^{n} fb_{BE_k}(P_k, P_k^*, \delta_k) \quad (7)$$

We call above principle as *Weak Match Principle*.

In *Weak Match Principle,* confidence value of complex event isn't *cf,* intuitively it will be less than *cf.*

We define the method to calculate confidence value as following:

Define T_i $(i = 0,1, \ldots n)$ are series of 2-values variables.

$$T_i = \begin{cases} 1 & \text{if } |P_i - P_i^*| \leq \delta_i \\ (1 - |P_i - P_i^*|)^2 & \text{otherwise} \end{cases}$$

And

$$T = \sum_{i=0}^{n} w_i \cdot T_i$$

The confidence value of complex event can be obtained from following equation:

$$cf^* = T \cdot cf$$

For *Weak Match Principle*, we still have further questions to discuss, if a complex event rule has N basic events, *k-th* basic event is omitted due to low weight value, in fact, original rule is modified, so the rule corresponding complex event will be easily be detected.

4. Algorithm

Now we introduce algorithms focus on above match principles. *Strong Match Principle* algorithms is shown in Algorithm 1.

Algorithm 1 Strong Match Principle Detection Algorithm

Input: uncertain event stream S_0

Output: Complex Event Set CE

for all basic sets <BE,P> in uncertain event stream S_0, **do**
$\forall(A,a,\delta)$ is a basic event in rule for complex event
  if(A=BE) then a* = P
Index: = .false.;
  Calculate F(R*)
  if F(R*) then
    Index = .true. ;
    $S_0 = S_0 - \{R^*\}$
    add basic event BE to the set CE
    else
      discard {R*}
    end
    If $S_0 = \emptyset$ end
    else continue;
end

5. Conclusion

In this paper, we presented two complex event detect principles, *Strong Match Principle* and *Weak Match Principle,* they can be adopted in different situation. We presented a new concept of uncertain event, i.e. weight value of basic event. We have not found anyone ever mention this concept in uncertain complex event processing. But if basic event occurs in random, all basic events in a rule may not complete occurred in real world condition. Type I error and Type II error is also discussed due to introduce of weight value of basic event.

There are several directions for further research. A good uncertain CEP detect system should indicate not only which event happed, but also when the event is going to happen with fuzzy reasoning and forecasting functions.

Acknowledgments

This work was partially supported by the National Natural Science Foundation of China under Grant U1201251and the National Science & Technology Support Plan of China under Grant No. 2015BAF09B01.

References

1. Cugola G, Margara A.: Processing flows of information: From data stream to complex event processing. [J]. Acm Computing Surveys, 44(3),15:1-15:62 (2012). DOI 10.1145/2187671.2187677.
2. A. Artikis, O. Etzion, Z. Feldman, Event processing under uncertainty, in: Proceedings of the Sixth ACM International Conference on Distributed Event-Based Systems, DEBS 2012, Berlin, Germany, 16–20 July, 2012, pp. 32–43.
3. Dong Wang, Large data visualization technology in the static grid enterprises [J],Jiangsu Electrical Engineering, vol.33, no. 6, pp.82-84, 2014.
4. Cugola G, Margara A, Matteucci M, et al. Introducing uncertainty in complex event processing: model, implementation, and validation[J]. Computing, 2015, 97(2):103-144.

N-soliton solutions for an integrable differential-difference equation with computer symbolic computation

Xiao-Yong Wen*
Department of Mathematics, School of Applied Science,
Beijing Information Science and Technology University,
Beijing 100192, China
E-mail: xiaoyongwen@163.com

In this paper, an integrable differential-difference equation is investigated based on its Lax representation. N-fold Darboux transformation (DT) are constructed with the aid of computer symbolic computation. As applications of N-fold DT, N-soliton solutions in terms of determinants are given. Also, soliton elastic interaction behavior of those solutions are shown graphically, which might be helpful for understanding physical phenomena.

Keywords: Integrable Differential-Difference Equation; N-Soliton Solution; Elastic Interaction; Computer Symbolic Computation Maple.

1. Introduction

Nonlinear differential-difference equations (NDDEs) are taken as the models of some physical phenomena and may be used to describe such physical situations as the electrical signals in electric circuit, the feader-eater processes in biology and population dynamics so on. [1–5]. In the present paper, motivated by the work of Ref. [6], we consider the following integrable nonlinear differential-difference equation

$$u_{n,t} = \frac{1}{v_{n+1}} - \frac{1}{v_n}, \quad v_{n,t} = \frac{1}{u_n} - \frac{1}{u_{n-1}}, \tag{1}$$

where $u_n = u(n,t), v_n = v(n,t)$ are the functions of the discrete variable n and tim e variable t, and $u_{n,t} = du_n / dt, v_{n,t} = dv_n / dt$. In Ref. [6], the corresponding hierar chy associated with Eq. (1) have been proposed by a new discrete isospectral pro blem, the Hamiltonian structures have been established for that hierarchy, and t wo-soliton solutions of Eq. (1) have been given by a single DT. According to Re f. [6], the Lax pair for Eq. (1) are listed as follows:

---

* Corresponding author.

$$E\varphi_n = U_n\varphi_n = \begin{pmatrix} 1+\lambda u_n v_n & u_n \\ v_n & 1 \end{pmatrix}\varphi_n, \quad \varphi_{n,t} = V_n\varphi_n = \begin{pmatrix} \lambda/2 & 1/v_n \\ \lambda/u_{n-1} & -\lambda/2 \end{pmatrix}\varphi_n, \quad (2)$$

where $\varphi_n = (\varphi_{1,n}, \varphi_{2,n})^T$ is a basic solution of Eqs. (2) with T as the response, the shift operator E is defined by $Ef_n = Ef(n,t) = f(n+1,t) = f_{n+1}$, $n \in Z, t \in R$, λ is the spectral parameter and $\lambda_t = 0$. The compatibility condition $U_{n,t} - (EV_n)U_n + U_nV_n = 0$ between Eqs. (2) gives rise to Eq. (1). The DT method is an effective tool to solve the NDDEs, the key idea is to keep the linear eigenvalue problems invariant, it is an algebraic method, which can be used to obtain the explicit solutions in a recursive manner [7, 8]. However, to the best of our knowledge, N-soliton solutions in terms of determinants and N-fold DT for Eq. (1) have not been studied yet.

In this paper, we will make further investigation on Eq. (1) via the Lax theory and N-fold DT technique. The rest of this paper will be as follows. In Section 2, we will construct the N-fold DT of Eq. (1) according to its Lax pair. In Section 3, N-soliton solutions will be given via the resulting N-fold DT, and soliton elastic interaction will be analyzed graphically. Finally, we will address the conclusions of this paper in Section 4.

2. *N*-fold DT for Eq. (1)

In this part, we will construct the N-fold DT for Eq. (1). We introduce the following transformation

$$\widetilde{\varphi_n} = T_n\varphi_n \quad (3)$$

where $\widetilde{\varphi_n}$ is required to satisfy (2) with $\widetilde{U_n}$ and $\widetilde{V_n}$, i.e.,

$$E\widetilde{\varphi_n} = \widetilde{U_n}\widetilde{\varphi_n}, \quad \widetilde{U_n} = T_{n+1}U_nT_n^{-1}, \quad (4)$$

$$\widetilde{\varphi_{n,t}} = \widetilde{V_n}\widetilde{\varphi_n}, \quad \widetilde{V_n} = (T_{n,t} + T_nV_n)T_n^{-1}. \quad (5)$$

Hereby, we assume that T_n is of the form

$$T_n = \begin{pmatrix} 1+\sum_{j=1}^{N} a_n^{(j)}\lambda^j & \sum_{j=0}^{N-1} b_n^{(j)}\lambda^j \\ \sum_{j=1}^{N} c_n^{(j)}\lambda^j & 1+\sum_{j=1}^{N} d_n^{(j)}\lambda^j \end{pmatrix}. \quad (6)$$

Let $2N$ arbitrary parameters $\lambda_i(\lambda_i \neq \lambda_j, i \neq j, i = 1,2,..,2N)$ be the roots of the $(2N)$-th order polynomial $\det T_n$, i.e.,

$$\det T_n = a_n^{(N)} d_n^{(N)} \sum_{i=1}^{2N} (\lambda - \lambda_i). \tag{7}$$

As $\lambda = \lambda_i$, the column vectors of $\widetilde{\varphi_n}$ are linearly dependent, then we have

$$\begin{cases} \sum_{j=1}^{N} a_n^{(j)} \lambda^j + \delta_{i,n} \sum_{j=0}^{N-1} b_n^{(j)} \lambda^j = -1, \\ \sum_{j=1}^{N} c_n^{(j)} \lambda^j + \delta_{i,n} \sum_{j=1}^{N} d_n^{(j)} \lambda^j = -\delta_{i,n}, \end{cases} \tag{8}$$

where $\delta_{i,n} = \dfrac{\varphi_{2,n} - r_i \psi_{2,n}}{\varphi_{1,n} - r_i \psi_{1,n}}$ and $\varphi_n = (\varphi_{1,n}, \varphi_{2,n})^T, \psi_n = (\psi_{1,n}, \psi_{2,n})^T$ are two basic solutions of Eqs. (2), $r_i(r_i \neq r_j, i \neq j, i = 1,2,..,2N)$ are $2N$ arbitrary parameters. When the $4N$ parameters $\lambda_i, r_i(\lambda_i \neq \lambda_j, i \neq j, i = 1,2,..,2N)$ are suitably chosen so that the determinant of the coefficients for Eq. (8) is nonzero, the Darboux matrix T_n is uniquely determined by Eq. (8).

By using the above facts, we can prove the following theorem.

Theorem 1. Matrices $\widetilde{U_n}$ and $\widetilde{V_n}$ determined by (4) and (5) have the same forms as U_n and V_n respectively, where the transformations from the old potentials u_n, v_n into the new ones $\widetilde{u_n}, \widetilde{v_n}$ are given by

$$\widetilde{u_n} = u_n - b_n^{(0)} + b_{n+1}^{(0)}, \quad \widetilde{v_n} = v_n - c_n^{(1)} + c_{n+1}^{(1)}, \tag{9}$$

with $b_n^{(0)} = \dfrac{\Delta b_n^{(0)}}{\Delta_1}$, $c_n^{(1)} = \dfrac{\Delta c_n^{(1)}}{\Delta_2}$,

$$\Delta_1 = \begin{vmatrix} \lambda_1 & \lambda_1^2 & \cdots & \lambda_1^N & \delta_{1,n} & \lambda_1 \delta_{1,n} & \cdots & \lambda_1^{N-1} \delta_{1,n} \\ \lambda_2 & \lambda_2^2 & \cdots & \lambda_2^N & \delta_{2,n} & \lambda_2 \delta_{2,n} & \cdots & \lambda_2^{N-1} \delta_{2,n} \\ \cdots & \cdots & \cdots & \cdots & \cdots & \cdots & \cdots & \cdots \\ \lambda_{2N} & \lambda_{2N}^2 & \cdots & \lambda_{2N}^N & \delta_{2N,n} & \lambda_{2N} \delta_{2N,n} & \cdots & \lambda_{2N}^{N-1} \delta_{2N,n} \end{vmatrix}$$

$$\Delta_2 = \begin{vmatrix} \lambda_1 & \lambda_1^2 & \cdots & \lambda_1^N & \lambda_1 \delta_{1,n} & \lambda_1^2 \delta_{1,n} & \cdots & \lambda_1^N \delta_{1,n} \\ \lambda_2 & \lambda_2^2 & \cdots & \lambda_2^N & \lambda_2 \delta_{2,n} & \lambda_2^2 \delta_{2,n} & \cdots & \lambda_2^N \delta_{2,n} \\ \cdots & \cdots & \cdots & \cdots & \cdots & \cdots & \cdots & \cdots \\ \lambda_{2N} & \lambda_{2N}^2 & \cdots & \lambda_{2N}^N & \lambda_{2N} \delta_{2N,n} & \lambda_{2N}^2 \delta_{2N,n} & \cdots & \lambda_{2N}^N \delta_{2N,n} \end{vmatrix},$$

and $\Delta b_n^{(0)}$ is produced from Δ_1 by replacing its $(N+1)$-th column with $(-1,-1,\ldots,-1)^T$, $\Delta c_n^{(1)}$ is produced from Δ_2 by replacing its first column with $(-\delta_{1,n},-\delta_{2,n},\cdots,-\delta_{2N,n})^T$, $\Delta b_{n+1}^{(0)},\Delta c_{n+1}^{(1)}$, are derived from $\Delta b_n^{(0)},\Delta c_n^{(1)}$, respectively, by replacing their n with $n+1$.

For the proof of the form invariance for $\widetilde{U}_n,\widetilde{V}_n$ and U_n,V_n, one can refer to Refs. [9,10]. The aim of this paper is to construct the multi-soliton solutions for Eq. (1). Here, the proof is omitted for simplicity. According to Theorem 1, we see that the transformations (3) and (9) can transform the Lax pair (2) into Lax pair of the same type. Therefore, both of the Lax pairs can lead to Eq. (1). Thus, the transformations (3) and (9) are called an N-fold DT of Eq. (1). By applying the N-fold DT, multi-solitons solutions for Eq. (1) have been obtained by choosing constant seed solution. It is worth pointing out that the DT in Ref. [6] is the special case of our DT with $N=1$.

3. *N*-Soliton Solutions of Eq. (1)

In this section, we will give some explicit solutions of Eq. (1). Substituting the trivial solutions $u_n=v_n=1$ into (2), we can give two solutions of Lax pair (2) with $\lambda=\lambda_i$ as follows:

$$\varphi=\begin{pmatrix} \tau_1^{\,n}e^{\rho_1 t} \\ \tau_1^{\,n}e^{\rho_1 t}\dfrac{\sqrt{\lambda_i^4+4\lambda_i}-\lambda_i}{2} \end{pmatrix},\quad \phi=\begin{pmatrix} \tau_2^{\,n}e^{\rho_2 t} \\ \tau_2^{\,n}e^{\rho_2 t}\dfrac{-\sqrt{\lambda_i^4+4\lambda_i}-\lambda_i}{2} \end{pmatrix}, \tag{10}$$

with

$$\tau_1=\frac{\sqrt{\lambda_i^4+4\lambda_i}}{2},\tau_2=-\frac{\sqrt{\lambda_i^4+4\lambda_i}}{2},\rho_1=\frac{\lambda_i+2+\sqrt{\lambda_i^4+4\lambda_i}}{2},\rho_2=\frac{\lambda_i+2-\sqrt{\lambda_i^4+4\lambda_i}}{2}.$$

From (8) with $\lambda=\lambda_i$ ($i=1,2,..,2N$), we have

$$\delta_{i,n}=\frac{\varphi_{2,n}-r_i\psi_{2,n}}{\varphi_{1,n}-r_i\psi_{1,n}},\quad \delta_{i,n+1}=\frac{\lambda_i+\delta_{i,n}}{1+\lambda_i+\delta_{i,n}}.$$

By use of (9), we can derive new solutions. To understand Solutions (9), we plot their structure figures as shown in Figs. 1 to 4 with $N=1$ and $N=2$.

(I) when $N=1$, let $\lambda=\lambda_i$ ($i=1,2$). Solving the linear algebraic system (8) leads to $b_n^{(0)}=\dfrac{\Delta b_n^{(0)}}{\Delta_1}$, $c_n^{(1)}=\dfrac{\Delta c_n^{(1)}}{\Delta_2}$, with

$$\Delta_1=\begin{vmatrix}\lambda_1 & \delta_{1,n}\\ \lambda_2 & \delta_{2,n}\end{vmatrix},\Delta b_n^{(0)}=\begin{vmatrix}\lambda_1 & -1\\ \lambda_2 & -1\end{vmatrix},\Delta_2=\begin{vmatrix}\lambda_1 & \lambda_1\delta_{1,n}\\ \lambda_2 & \lambda_2\delta_{2,n}\end{vmatrix},\Delta c_n^{(1)}=\begin{vmatrix}-\delta_{1,n} & \lambda_1\delta_{1,n}\\ -\delta_{2,n} & \lambda_2\delta_{2,n}\end{vmatrix}.$$

By use of (9), we can obtain one and two soliton solutions of Eq.(1). Here, Solutions (9) are the results in Ref. [6] which are the special case of our results with $N = 1$, and the authors pointed out that the above results are the two-soliton solutions, in fact, when one of the spectral parameters are chosen as −4, Solutions (9) with $N = 1$ can be one-soliton and two-soliton solutions. Fig. 1(a) shows the anti-bell-shaped one-soliton structure of $\widetilde{u_n}$, Fig. 1(b) shows corresponding evolution plots for Fig. 1(a) at different time, Fig. 1(c) shows the anti-bell-shaped one-soliton structure of $\widetilde{v_n}$, Fig. 1(d) shows corresponding evolution plots for Fig. 1(c) at different time. From Fig. 1, we can find that the one-solitonic shape, amplitude and velocity have not changed during propagation. Fig. 2 show that Solutions $\widetilde{u_n},\widetilde{v_n}$ admit the two-soliton structures, Fig. 2(a) and 2(c) show the elastic interaction between two anti-bell-shaped solitons for $\widetilde{u_n}$ and $\widetilde{v_n}$, Fig. 2(b) and Fig. 2(d) show corresponding evolution plots for Fig. 2(a) and Fig. 2(c), respectively, at different time, from which we can find that the solitonic shapes and amplitudes have not changed after the interaction.

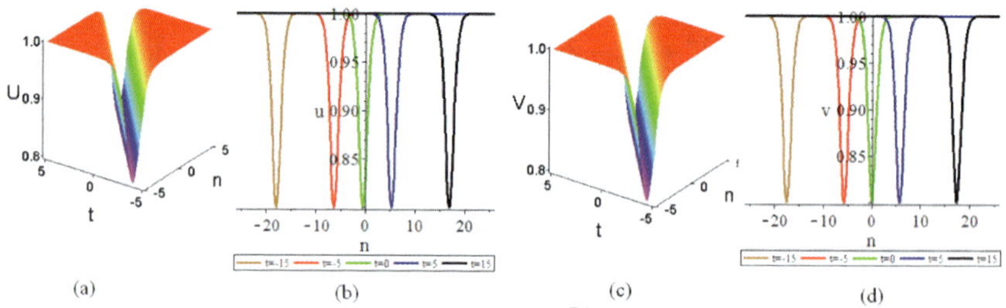

Fig. 1. (*a*) Anti-bell-shaped one-soliton solution $\widetilde{u_n}$ with the parameters $r_1=-1, r_2=1$, $\lambda_1=-4,\lambda_2=-5$; (*b*) Evolutional profile of $\widetilde{u_n}$ in (*a*) at different time; (*c*) Anti-bell-shaped one-soliton solution $\widetilde{v_n}$ with the same parameters in (*a*); (*d*) Evolutional profile of $\widetilde{v_n}$ in (*c*) at different time.

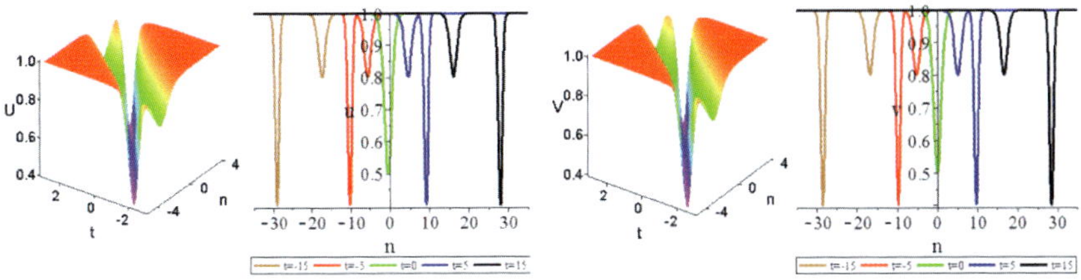

(a) (b) (c) (d)

Fig. 2. (*a*) Two-soliton solution $\widetilde{u_n}$ with the parameters $r_1 = -1, r_2 = 1, \lambda_1 = -5, \lambda_2 = -10$; (*b*) Evolutional profile of $\widetilde{u_n}$ in (*a*) at different time; (*c*) Two-soliton solution v_n with the same parameters in (*a*); (*d*) Evolutional profile of v_n in (*c*) at different time.

(II) when $N = 2$, let $\lambda = \lambda_i$ ($i = 1, 2,3,4$). Solving the linear algebraic system (8) leads to the three and four soliton solutions by use of (9) when the parameters are suitably chosen with $N = 2$. Fig. 3 show the overtaking interactions among the three anti-bell-shaped solitons with different amplitudes along the same direction of propagation for $\widetilde{u_n}, \widetilde{v_n}$, after the overtaking interaction, the final three anti-bell-shaped solitons preserve their shapes, amplitudes and velocities, so that the interactions are elastic. Fig. 4 show the elastic overtaking interaction among the four anti-bell-shaped solitons with different amplitudes along the same direction of propagation for $\widetilde{u_n}, \widetilde{v_n}$.

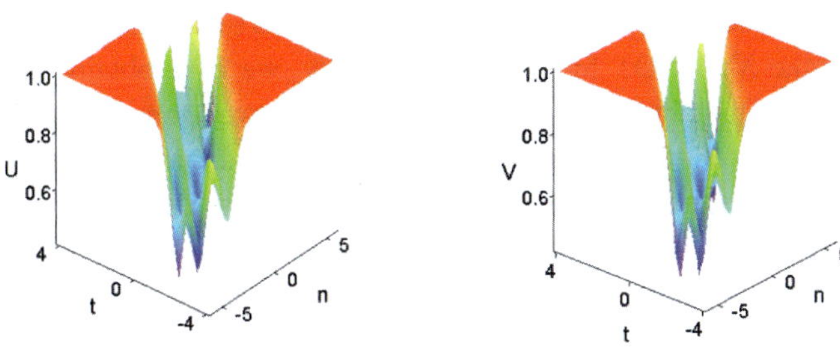

(a) (b)

Fig . 3. (*a*) Three-soliton solution $\widetilde{u_n}$ with the parameters $r_1 = -1, r_2 = 1, r_3 = -2, r_4 = 1/2$, $\lambda_1 = -4, \lambda_4 = -6, \lambda_4 = -8, \lambda_4 = -10$. (*b*) Three-soliton solution v_n with same parameters in (*a*).

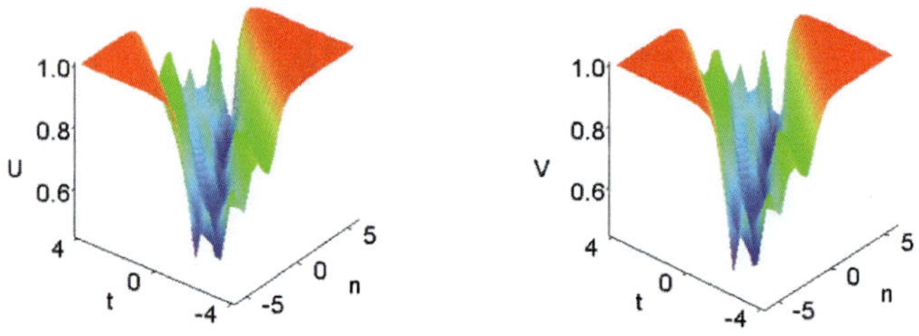

(a) (b)

Fig . 4. (*a*) Four-soliton solution $\widetilde{u_n}$ with the parameters $r_1 = -1, r_2 = 1, r_3 = -2, r_4 = 1/2$, $\lambda_1 = -5, \lambda_4 = -6, \lambda_4 = -8, \lambda_4 = -10$. (*b*) Four-soliton solution v_n with same parameters in (*a*).

With the aid of computer symbolic computation ***Maple***, it is straightforward to check that Solutions (9) satisfy Eq. (1) with N=1 and N=2. It is worth pointing out that the corresponding 2N-soliton solutions will reduce to (2N−1)-soliton solutions when one of λ is −4. N-soliton solutions of Eq. (1) are made up of the 2N-soliton and (2N−1)-soliton solutions. At last, it is particularly worth emphasizing that the DT in Ref. [6] is the special case of our DT when $N = 1$.

4. Conclusions

In this paper, Eq. (1) has been investigated via the DT method. Firstly, we have constructed the N-fold DT for Eq. (1). Secondly, we have obtained 2N-soliton solutions (9) via the obtained the N-fold DT. One-, two-, three- and four-soliton structures have been shown graphically: Figs. 1-2 exhibit the one-soliton and two-soliton structures with N = 1; Figs. 3-4 exhibit the elastic interactions among the three and four anti-bell-shaped solitons with $N = 2$; solitonic shapes, amplitudes and velocities have not changed after the interaction between/among the two, three and four solitons, which might be helpful for understanding physical phenomena.

Acknowledgments

This work is supported by Beijing Natural Science Foundation under Grant No.1153004, and China Postdoctoral Science Foundation under Grant No.2015M570161.

References

1. M.J. Ablowitz and P.A. Clarkson, Solitons, Nonlinear Evolution Equations and Inverse Scattering , (Cambridge University Press, Cambridge 1991).
2. M.J. Ablowitz and J.F. Ladik, *Nonlinear differential-difference equations*, *J. Math. Phys*. **16**, 598-603 (1975).
3. M. Toda, *Theory of nonlinear lattices*, (Springer, Berlin 1989).
4. M. Wadati, Transformation theories for nonlinear discrete systems, *Prog. Theor. Phys. Suppl.* **59**, 36-63(1977).
5. W.X. Ma and B. Fuchssteiner, Algebraic structure of discrete zero curvature equations and master symmetries of discrete evolution equations, *J. Math. Phys.* **40**, 2400-2418 (1999).
6. Y.T. Wu andX.G. Geng, A new hierarchy of integrable differential-difference equations and Darboux transformation. *J. Phys. A* **31**, L677-L684 (1998).
7. C.H. Gu, H.S. Hu and Z.X. Zhou, *Darboux transformation in soliton theory and its geometric application*,.(Shanghai Scientific and Technical Press, Shanghai 1999).
8. V.B. Matveev and M.A. Salle, *Darboux Transformations and Solitons,* (Springer, Berlin 1991).
9. X.Y. Wen, Elastic interaction and conservation laws for the nonlinear self-dual network equation in electric circuit. *J. Phys. Soc. Jpn.* **81**, 114006 (2012).
10. X.Y. Wen and X.Y. Hu, *N*-Fold Darboux transformation and solitonic interactions for a Volterra lattice system, *Adv. Diff. Equ.* **2014**, 213 (2014).

Research on knowledge management for collaborative design of aviation enterprises

Yong-Hua Wu*

College of Electronics and Information Science, Fujian Jiangxia University, Fuzhou 350108, People's Republic of China
E-mail:wuyonghua@vip.126.com
www.fjjxu.edu.cn

Bin Liu

Jiangxi Institute of Scientific and Technological Information, Nanchang 330046, People's Republic of China
E-mail: lb@jxstc.gov.cn

For aviation enterprise collaborative design process involving a large number of complicated design knowledge, knowledge leads to cumbersome screening, the use of low efficiency. The knowledge management modeling approach is proposed in aviation enterprise collaborative design. In the analysis of aviation enterprise knowledge management infrastructure on collaborative design, collaborative design management is analyzed on the basis of given aviation enterprise collaborative knowledge management model design, and an aviation enterprise collaborative design mode is given. The method solves the low utilization efficiency of collaborative design knowledge, and looks for such knowledge complex issues to meet the needs of designers.

Keywords: Knowledge Network; Aviation Enterprise; Information Sharing; Cooperative Design.

* Corresponding author.

1. Introduction

With the development of aviation enterprises in China, the design of the computer aided design and cooperative design has been introduced to the design department, and the aviation product design cycle has been shortened[1]. According to statistics, the current aviation design units of about 40% of working hours for the collection, collation, classification, retrieval, this increase the labor time, reduce the production efficiency, no doubt this is a waste. In order to shorten the design cycle and improve the competitiveness of aviation enterprises, it is urgent to manage the knowledge resources in the process of collaborative design.

Such as Deng Fumin[2] and his partners allocated the design task from the relationship between the design, and obtain the maximum of the optimal objectives of the coordination efficiency. Jing Yi[3] and his partners constructed the model of product development process planning task allocation model, close to a strategic height and development as the goal, the task allocation of collaborative product. Li Yujia[4] and others were against the task allocation problem, established a specific model to shorten the development cycle of the task planning.

The research has been carried out to study the different aspects of collaborative design, but it is not related to the knowledge management in the collaborative design of aviation enterprises. In this paper, the knowledge management model of the collaborative design of aviation enterprises is established.

2. Collaborative Design Management System Model

Collaborative design is to knowledge based design process, not only need to different areas of knowledge and expert experience, more important to designers, to the relevant knowledge to carry on the comprehensive and finishing, sent to the hands of the needs of departments and related design personnel[5], at the same time, through effective mechanisms to integrate and coordinate these multifarious knowledge to complete different design tasks[6].

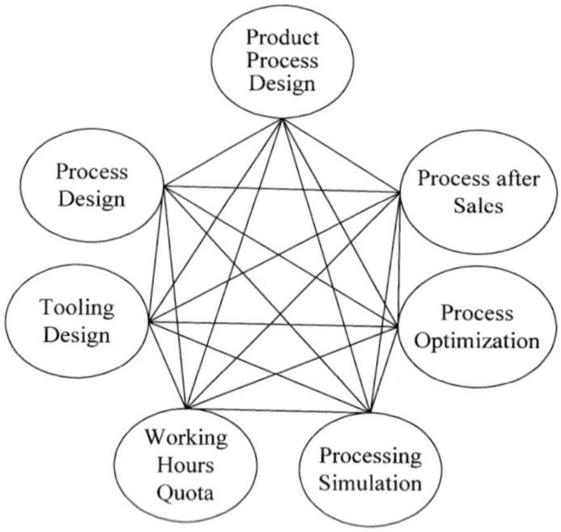

Fig. 1. Traditional point to point method.

Traditional cooperative work mode, is a network of networks, each person in the network, at the time of participation in the work, each person has to work with one or more contact, more complex work, low efficiency. In aviation manufacturing enterprises, the process of product process design includes many design phase, involving many departments and designers work together, in the traditional collaborative design, the departments more or less to contact. Figure 1 shows.

Work model of collaborative management platform, is a star network, each design personnel and collaborative design center communication, can be very clear to know the required design knowledge and tasks. The collaborative design center requires high and need based on knowledge network and platform, allowing the designers to design in the whole period of product process design and management, and don't have to consider the relationship between the design staff, makes the work more clear, to speed up the development of aviation products of collaboration and information sharing. As shown in figure 2.

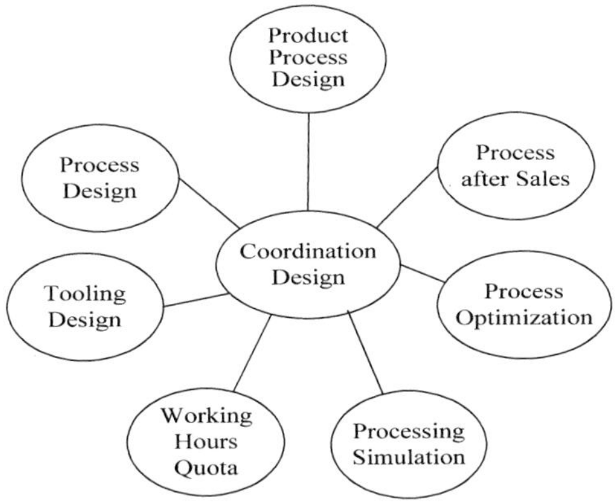

Fig. 2. Centralized shared collaborative management model.

Efficient collaborative design is to spread the knowledge of research and development process, different departments and professional disciplines, so that many departments, multi professional, multi design stage can develop and research the knowledge and management experience in the same system, provide a new design support platform.

3. Knowledge Management Model Based on Collaborative Design of Aviation Enterprise

In large data environment, collaborative design can be through the knowledge management to link in different places of the enterprise or the organization of knowledge, so the enterprises and organizations can rapidly acquire needed knowledge resources, quickly adapt to market changes, in the shortest time realize the needs of users of the service, to achieve win-win cooperation, enhances the overall competitiveness among different enterprises[7].

With the existing knowledge and knowledge management, this paper puts forward the function model of the knowledge management system of aviation enterprise collaborative design, which mainly consists of 3 levels: aviation enterprise knowledge resource layer, tool layer, aviation enterprise user service layer, as shown in figure 3.

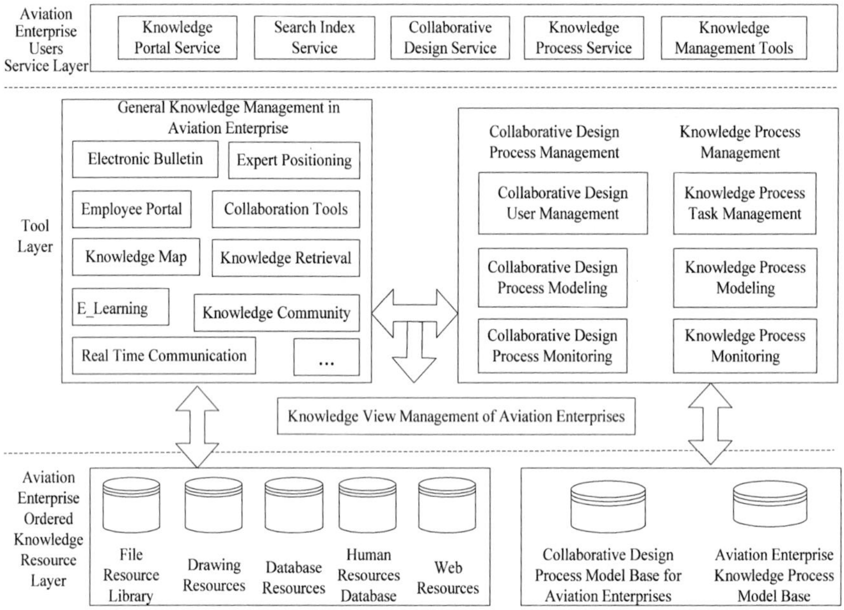

Fig. 3. Knowledge management model of the collaborative design of aviation enterprises.

The user service layer of the aviation enterprise is mainly for the design work of the design staff, including knowledge portal service, search index service and so on. Collaborative design service is an integrated process management of aviation enterprises based on knowledge service, knowledge management, knowledge application and knowledge sharing.

The aviation enterprise tool layer is the middle layer of the design system. In the general knowledge management panel, there are enterprises' electronic bulletin, employee portal, collaborative tools, knowledge retrieval, etc.

The aviation enterprise knowledge resource layer is a support layer for collaborative design knowledge management. Provide and preserve a variety of resources, such as documents, paper, database, etc., to provide a large number of resource services for aviation enterprise knowledge view management.

4. Collaborative Design Work Mode of Aviation Enterprise

4.1. *Design & design collaboration of aviation enterprise*

In the stage of the aviation enterprise collaborative design, Design & Design Collaboration between enterprises is very important, it runs through in departments and disciplines and between different professional entire collaborative design process. In the knowledge and knowledge resources for the development of the whole enterprise collaborative, makes the product to achieve the best.

(1) Overall design and collaborative design of the system. Overall design is determine the products of the generation process, and the frame, and of the system planning, between the general and the different component design, system, overall and between the components, component and component in the design, assembly and components between deposit in collaborative design optimization, and so on.

(2) Collaboration between disciplines and majors. Aviation product is a multi discipline and multi professional collaboration, which is needed to study the collaborative modeling and data management problems of multiple disciplines, including the application of data sharing in each discipline.

(3) The relationship of multi subject model. It mainly manifests in the geometry, the performance and the information relations. For aviation enterprises, the overall design is a meaningful geometric figure, is the coordination between the whole and the parts. Performance relationship is a key parameter that is related to a product's quality. Information relation is a relation of information among various subjects. The synergy between product performance optimization will affect the whole performance of the product.

4.2. *Design & process collaboration of aviation enterprise*

With the integrated collaborative design of aviation enterprise, the design & process cooperation has produced in the product design process. It mainly include:

(1) Collaborative modeling of process & tooling & design for aviation enterprises. Include CAD graphics, 3D models and other geometric models that are required by the enterprise, and the relationships between these models are identified by the collaborative modeling.

(2) The design of the collaborative process of aviation enterprise. Including the whole process of the planning, process change management and the adjustment and processing of industrial feedback information.

(3) Collaborative design of aircraft products for aviation products. In aviation products, the design of industrial equipment is in a special position in the product design. It is important to cooperate with the whole product cycle. For example, in the overall framework of the aircraft design and aircraft sheet metal design, the use of products and the relationship between the assembly, to build a module sharing and information integration.

5. Conclusions

Collaborative design is a design process based on knowledge, which is used by different disciplines and different design departments to exchange knowledge and knowledge in different areas, which leads to a large amount of knowledge acquisition and low utilization efficiency.

According to the idea of collaborative design, the paper establishes a model of collaborative design knowledge management for aviation enterprises. This method solves the problem of low efficiency of knowledge utilization, complex knowledge search and other problems, which can meet the needs of designers and reduce the work of repetitive search, which makes the design more reliable and accurate.

References

1. Qian Yadong, Gu Xinjian, Ma Jun et al. Research on knowledge management of collaborative design [J]. Journal of Zhejiang University: Engineering Edition, 2007,47 (2): 304-309.
2. Deng Fumin, Liang Xuedong, Xiao Tingdan. Collaborative designers' efficiency based on adaptive dependency relationship[J]. Systems engineering theory & practice, 2011, 31 (11): 2122-2130.
3. Jing Yi, Wang Xu, Li Wenchuan. Task Allocation Optimization for Supplier Involvement in Cooperative Product Development[J]. Mechanical Engineering of China, 2011,22(21):2566-2571.
4. Li Yujia, Hu Zongwu, Jin Ye. Research on Task Distribution in Concurrent Product Development[J]. Mechanical Engineering of China, 2002, 13(7):580-583.
5. Huang Hongzhong, Liu Wei, Li Li. Process Modeling of Product Collaborative Design [J]. Computer Integrated Manufacturing Systems, 2003, 9(11):955-959.
6. Su Caimao, Ke Yinglin. Task planning and decoupling strategy for collaborative design[J]. Computer Integrated Manufacturing Systems, 2006, 12(1): 21-26.
7. Ramsey C E, Rickson R E. Environment Knowledge and Attitude [J]. The Journal of the Environment Education Studies, 1976, 23(2): 473-477.

Author Index